AF602065

Integrated Environmental Management

Integrated Environmental Management

Editor
Professor (Dr.) Arvind Kumar
FLS (London), FASc (Swiss), FISEC, FSESc, FAZ, FZA (Gold Medalist)
Pro-Vice Chancellor
S.K.M. University, Dumka – 814 101 (Jharkhand)

Daya Publishing House®
A Division of
Astral International Pvt. Ltd.
New Delhi – 110 002

Reprinted, 2021

ISBN 9789351240822 (Int. Edn.)

Publisher's Note:

Every possible effort has been made to ensure that the information contained in this book is accurate at the time of going to press, and the publisher and author cannot accept responsibility for any errors or omissions, however caused. No responsibility for loss or damage occasioned to any person acting, or refraining from action, as a result of the material in this publication can be accepted by the editor, the publisher or the author. The Publisher is not associated with any product or vendor mentioned in the book. The contents of this work are intended to further general scientific research, understanding and discussion only. Readers should consult with a specialist where appropriate.

Every effort has been made to trace the owners of copyright material used in this book, if any. The author and the publisher will be grateful for any omission brought to their notice for acknowledgement in the future editions of the book.

Published by : **Daya Publishing House®**
A Division of
Astral International Pvt. Ltd.
– ISO 9001:2015 Certified Company –
4736/23, Ansari Road, Darya Ganj
New Delhi-110 002
Ph. 011-43549197, 23278134
e-mail: info@astralint.com
Website: www.astralint.com

Preface

Environmental and Integrated Management is not, as the phrase could suggest, the management of the *environment* as such, but rather the management of interaction by the modern human societies with, and impact upon the environment. The three main issues that affect managers are those involving politics (networking), programs (projects), and resources (*i.e.* money, facilities, etc). The need for environmental management can be viewed from a variety of perspectives. A more common philosophy and impetus behind environmental management is the concept of carrying capacity. Simply put, carrying capacity refers to the maximum number of organisms a particular resource can sustain. The concept of carrying capacity, whilst understood by many cultures over history, has its roots in Malthusian theory. Environmental management is, therefore, not the conservation of the environment solely for the environment's sake, but rather the conservation of the environment for humankind's sake.

Environmental management involves the management of all components of the bio-physical environment, both living (biotic) and non-living (abiotic). This is due to the interconnected and network of relationships amongst all living species and their habitats. The environment also involves the relationships of the human environment, such as the social, cultural and economic environment with the bio-physical environment.

Therefore, it is high time for Environmental biologists of India to develop certain pertinent strategies for integrated management so that future generation should get pollution-free environment. This book deals with the scope, feasibility and applicability of environmental management. However, in spite of great importance of environmental management, there is a paucity of literature in India. To fill this gap, I have ventured to compile some innovative recent research articles of eminent biologists in a form of the present book.

My special thanks and appreciation go to the scientists whose contributions have enriched this volume. I wish to express my sincere gratitude to Dr. Victor Tigga, Hon'ble Vice Chancellor, S. K. M. University, Dumka who has been a source of constant inspiration. I owe my special thanks to Professor

M. C. Dash, Hon'ble former Vice Chancellor of Sambalpur University, Professor N. C. Datta of Calcutta University, Professor K. C. Pandey of Lucknow University, Professor S. K. Konar of Kalyani University, Professor D. K. Belsare of Bhopal University, Professor Professor U. C. Goswami of Gauhati University, Professor P. C. Mishra of Sambalpur University, Professor P. Natarajan of Kerala University, Professor P. S. Murthy of Bangalore University, Professor Ajit Varma of JNU, Professor A. L. Bhatia of Jaipur University, Professor A. K. Mittal of BHU, Professor S. P. Hosmani of Mysore University, Professor K. Kapoor of Udaipur University, Professor K. B. Reddy of Nagarjuna University, Professor M. Vikram Reddy of Pondicherry University, Professor B. K. Tiwari of NEHU, Professor G. Tripathi of Jodhpur University, Professor R. Ramalingam of Annamalai University, Professor Sharif U. Ahmad of Nagaland University, Professor G. K. Kulkarni of Aurangabad University, Professor S. U. Mehram of Nagpur University, Professor S. K. Battish of PAU, Professor B. M. Sharma of Manipur University, Professor B. D. Joshi of Hardwar University, Professor G. C. Pandey of Faizabad University, Professor K. C. Sharma of Ajmer University, Professor M. Raziuddin of Hazaribag University, Professor U. S. Bagde of Mumbai University, Professor Gurdeep Singh of I. S. M., Dhanbad, Dr. P. K. Goel of Karad, Professor S. P. Roy and Shri Tribhuwan Poddar of Bhagalpur University, Bhagalpur.

I also express my deep sense of gratitude to my parents whose blessings have always prompted me to pursue academic activities deeply. I am also thankful to my sweet wife, Kumari Bimla and my two lovely sons, *Kumar Pallav Shivshankaran* and *Kumar Prasun Ramakrishnan* whose natural smiles extended to me relief all through this tiresome endeavour.

Last but not the least, I am also thankful to Mr. Anil Mittal, Proprietor, Daya Publishing House, New Delhi for taking keen interest in bringing out of this book. Finally, I will always remain a debtor to all my well-wishers for their blessings, without which this book would not have come into existence.

Arvind Kumar

Contents

Chapter 1

Ecology and Avian Biodiversity of the Wetland (Goga Beel) of North Bihar, India

Arvind Kumar and Ajay Kumar

Environmental Biology Research Unit, P.G. Department of Zoology, S.K.M. University, Dumka – 814 101, Jharkhand (India)

ABSTRACT

The systematic of avian fauna colonizing the wetland; Goga Beel, Katihar (Bihar) has been studied along with physico-chemical and biological parameters. 36 species of birds belonging to nine orders have been identified and their diagnostic characters with notes on their field status are enumerated. Macrophytic infestation, abundance of molluscan and Ichthyo-fauna were found as the main causes of rich avain biodiversity. The physico-chemical parameters shows that the eutrophication process has set in the wetland (Goga Beel). Apart from these, a checklist of aquatic macrophytes and fishes was also prepared during investigations.

***Keywords**: Ecology, Biodiversity, Goga beel, Avain fauna, Birds, Physico-chemical/Biological parameters.*

Introduction

Wetlands (beels/jheels) are links between terrestrial and aquatic ecosystems. According to Beadle and lind (1960), sufficiently shallow, slow moving water which permits the establishment of characteristic vegetation are wetlands. Dehardri and Triparti (1976) defined it as the water logged, shallow water areas with a loose peaty bottom, rich in decaying organic matter, retaining water either periodically or shrinking or drying completely during summer months. And according to Datta Munshi ande Hughes (1992) wetlands are shallow water areas with profuse growth of macrophytic vegetation.

In North Bihar many wetlands/beels exist which are larger and smaller tributaries or cut off parts of the complex river system of Kosi, Bagmati, Gandak and Mahananda. Considerable investigation has been done on the hydrobiology of various wetlands/beels (Das and Srivastava, 1956; Michael, 1969; Munawar, 1970; Bhatnagar and Sharma, 1978; Rai and Datta Munshi, 1979; Prakash *et al.*, 1983; Singh, 1986; and Choudhary and Choudhary, 2007). But particularly no information is available on the Eco-biodiversity of avain fauna associated with wetlands/beels except the work of Hancock (1984), Vijayan (1986) and Sahu and Dutta (2005) who studies some aspects on conservation of the avain fauna of wetlands. In the present work an attempt has been made to investigate the ecotaxo-biology of birds inhabition of the Goga Beel, Katihar (Bihar).

Study Area

The Goga beel (25°19′ N and 87°45′ E) is a close type beel located in Ambabad block area, 6 km east of Manihari and about 19 km south-east to district head quarter of Katihar. It is 'X' shaped, perennail beel surrounded by human habitation and agricultural lands (Figure 1.1). Its average depth is 2.3–3.7m and highly infested by macrophytes. The anthropogeneic pressure in the beel area is very

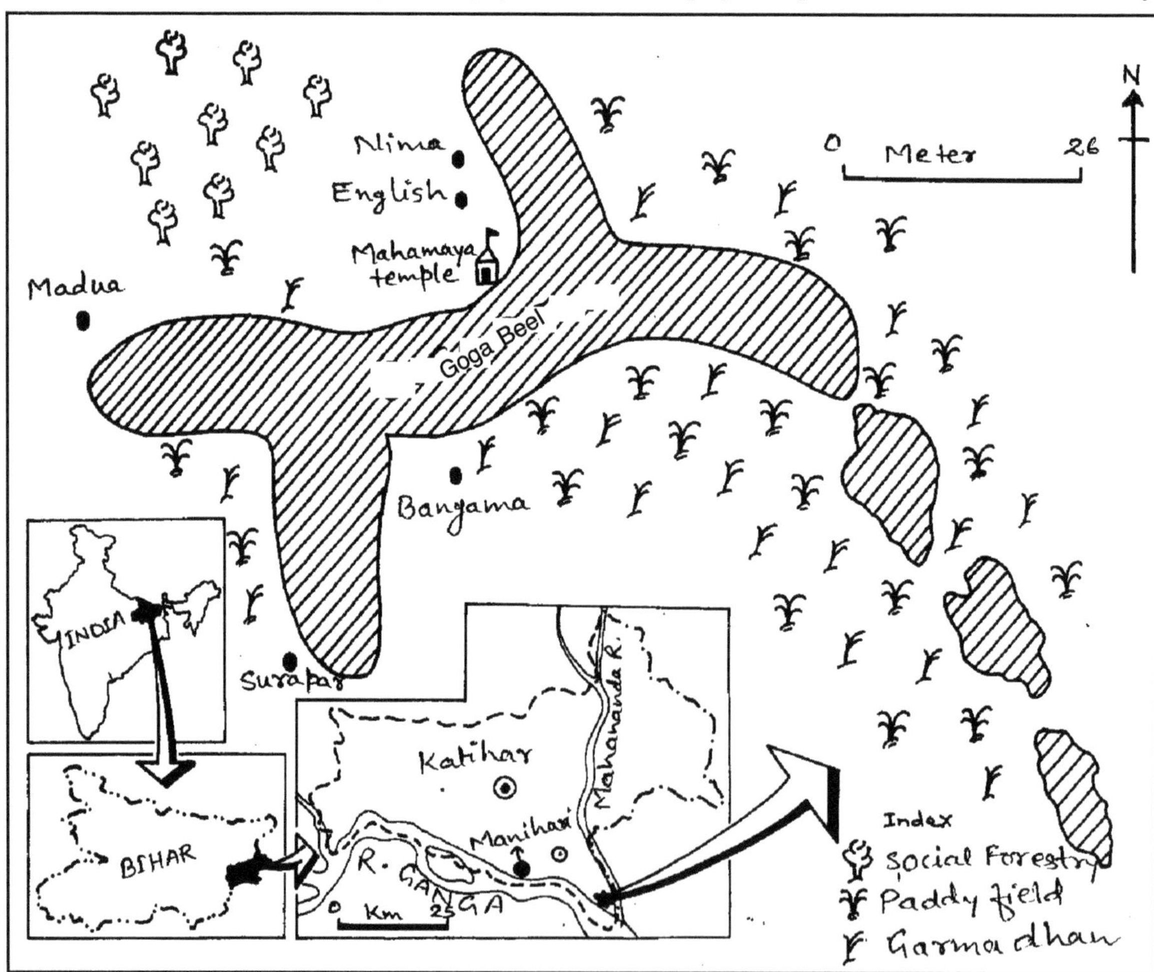

Figure 1.1: Location Map of Goga Beel, Katihar (Bihar)

high. Local inhabitants practices navigation, irrigation and fishing throughout the year in bee1 area. Goga bee1 gets mostly all its water from rain between July and September from the catchment area but during flood month if also gets water from Mabananda and Ganga river. Water depth varies from 26 cm to 4.8 cm in different seasons of the year. Soil is sandy loam, rich with humus and very fertile and suitable for luxuriant growth of macrophytic vegetations in the beel.

Material and Methods

Fortnightly census of birds was conducted for estimation of the avain population of the beel during the period November, 2001 to October 2002. The methodology adopted by Vijayan (1986) of Keoladeo Ghana National Park, Bharatpur was followed with certain modification. Identification of birds was done in the field with the help of Ali (1945, 1949), Ali and Ripley (1983), Wood Cock (1984) and Hancock (1984). Sampling of biota and analysis of physico-chemical parameters were performed monthly. Physico-chemical analysis was done using the standard methods of APHA (1989) and Nath and Sammant (1999).

Results

The present, investigation on the eco-taxonomy of aves inhabiting the Goga beel, Katihar suggests that the physico-chemical parameters of their habitat are the major regulating force of their population density. It was observed that the seasonal changes in the population of avain fauna of the beel are caused by a complex interaction of some extrinsic and intrinsic factors.

The species of the nine order of birds have been found associated with this beel out of them Ciconiiformes, Pelecaniformes and Passeriformes contribute substantially the avian fauna of the beel throughout the year. The Anseriformes and Charadriiformes are migratory orders who utilize the beel as foraging medium during the winter months. However, the local migrants are found scattered throughout the different avian orders, moving to and fro utilizing the best resources available to them.

The systematic account of the most dominant and abundant birds of the beel with their diagnostic features are presented below:

Order–Passeriformes

Claws present and well developed for perching; primarily terrestrial but some species are partially dependent on beel (wetland) for their food.

Family–Hirundinidae

***Hirundo rustica*, Scopali (Eastern Swallow: Ababeel)**

Size about a sparrow (±18 cm) with long forked tail. The forehead, chin and throat chestnut and blue black breast band. Above glistening metallic blue black and has pale pinkish white below. They are highly gregarious in winter and found almost skimming the surface of water. Roost in partially submerged reed. Feeds on terrestrial and aquatic insects and breed between May to June to July and nests on branches of trees. Nest is made of soil. Alarm call high pitched 'swee' or contact call 'tvee-veet' rapid, constant twitter (Figure 1.2).

Figure 1.2: ***Hirundo rustica***

***Hirundo fluvicola*, Blyth (Cliff Swallow: Ababeel)**

Small swallow of size about 10 cm with slightly short and forked tail. Head and crown dull chestnut, iridescent steel blue wings and back. Rump pale brown .Throat, breast and sides of the neck streaked with blakish and rest of under parts whitish. Resident, often found in the vicinity of water. Feeds on Gnats, Midges and other dipteran insects. Breeds in two successive brood in a year *i.e.*, December to April and July to October and nest is made of mud (Figure 1.3).

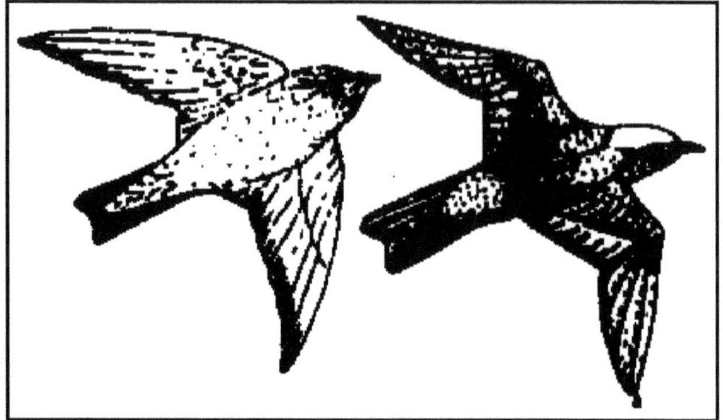

Figure 1.3: ***Hirundo fluvicola***

Order–Charadriiformes

Family–Charadriidae

Sub-Family–Charadriinae

Bill short to medium long, never decurved or recurved; a swollen area at the tip of upper mandible, contracted in middle portion.

***Pulvialis apricaria* (Golden Plover: Tithi)**

Winter visitor, gregarious, tarsus covered with hexagonal scale, outer and middle toes connected by a small web at the base, hind toe absent, tail white, back spotted and wing long and pointed. Feeds on tiny molluscs, crustaceans and worms and also take berries and seeds of marsh plants. Breeding extralimital.

Sub-Family–Scolopacinae

Bill short to very long, straight, decurved or recurved.

***Gallinago gallinago*, (Common Snipe: Chacha)**

Common and abundant winter visitor with straight slender bill, eye is high in the side of head so as to give maximum visibility when feeding. Feeds chiefly on worms, larvae and tiny molluscs and breeding extralimital.

Sub-Family–Rostratulae

***Rostratula bengalensis*, Linn. (Painted Snipe: Rajchaha)**

It is very similar to common snipe in size (about 25–28 cm) and shape but slightly longer legs. A true relative of rail with long slender bill slightly down curved at the tip. A conspicuous white line from side of breast joins yellowish scapular V-line. Upper plumage olivaceous-brown faintly barred with black and distinct dark green bars along scapulars and inner wing coverts. Female is stouter and larger than male and has the head and breast a rich maroon chestnut with distinct white spectacles and creamy mesial line on the top of the crown. Males are small, dull with eye distinctly framed by broad whitish spectacle mark extending behind the eye. Its call is deep mellow and call of

Figure 1.4: ***Rostratula bengalensis***

female consists of a soft hollow sounding 'hoop-houp-houp'. Resident, singly or small parties are seen active at down of dusk than in the heat of the day. Feeds on aquatic insects, molluscs, crustaceans, and vegetable matters. Breeds between July to September and nests on grass of on bund near water (Figure 1.4).

Metopidius indicus, **Latham (Bronze Winged Jacana: Karauwa)**

Common resident bird found walking on weed and aqatic plants on the surface of the water of beel. A blakish rail like bird of about 28–31 cm size with long toes and claws. A lappet at the base of bill resting against forehead; head, neck and breast glossy black, back and wing bronze, a broad white stripe from eye to nape; vent and tail chestnut. Feeds on aquatic insects, molluscs, seeds and roots of hydrophytes. Females are polyandrous and breeds in between June to September and nests on Singhara or water hyacinth leaves. Call harsh grunt, also a wheey piping 'seek-eek-eek' (Figure 1.5).

Figure 1.5: ***Metopidius indicus***

Order–Gruifonnes (Cranes, Rail, Coot)

Family–Rallidae (Moorhen, Coot)

Sulking marsh hunting birds with stubby tail, rounded wing, longish bare legs and free toes.

Porphyrio porphyrio, **Latham (Purple Moorhen: Karan)**

A purplish blue rail with long red leg and enormous toes and the size of bird is about 43 cm. Bill and frontal shield and vent is white. The bare red forehead running back from the short heavy red bill. There is a distinct white patch present on the lower side of the stumpy tail. Its call is a series of clucks and an alarm 'kuk' and a contact call 'chuk-chuk'. Local migrant and found in marshy reed-beds or wades over floating hydrophytes. Feeds mostly on shoots of paddy or marsh plants. Aquatic insects and snail are also eaten and breeds between June to September and nests matted aquatic plants within flooded reed beds (Figure 1.6).

Figure 1.6: ***Porphyrio porphyrio***

Gallicrex cinerea, **Gmelin (Water Cock: Kora)**

Resident, Rail-like swamp bird with long leg and a small triangular yellowish horny shield on forehead. Male in summer season very black and large, fleshy red horn protruding above crown. Female and winter male are brownish, paler below with irregular bars. In flight, appears slimmer than *porphyria porphyria.* Its call is hollow metallic booming 'utumb-utumb-utumb' or 'kok-kok' and rapidly repeated. Crepuscular feeder and flight feeble. Feeds on seeds and shoots of green crops and also take water insect and their larvae, molluscs, worms etc. Breeds in monsoon month. Nests on tangled reed beds in water bodies (Figure 1.7).

Figure 1.7: *Gallicrex cinerea*

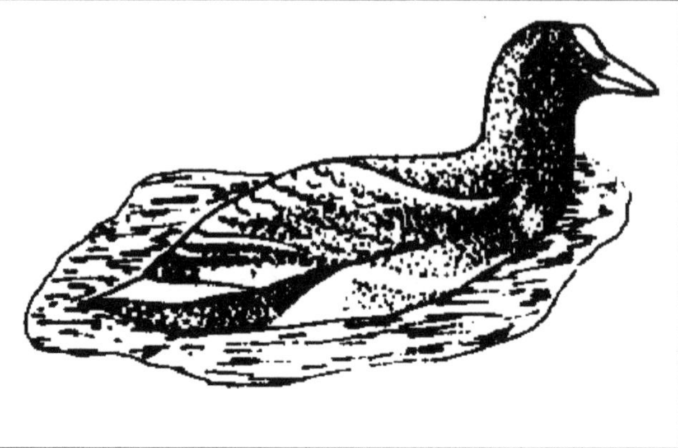

Figure 1.8: *Fulica atra*

***Fulica atra*, Linn. (Coot: Sarair)**

A plump, slaty black bird of size about 40 cm with white frontal shield, bill ivory white painted. Toes are lobed or scalloped, flat, rounded horny shield covering forehead. Practically tailless gregarious birds, diving expertly, flight feeble and swims also. Chiefly vegetarian but also feed on water insects, worms and molluscs and breeds between June to September and nests on reeds, slightly above water. Its call is clear and loud trumpet like cry, also 'Kluk' and series of soft notes. (Figure 1.8)

Order–Coraciformes

Only members of family Alcedinidae are partially dependent on water of beel and are characterised by short stumpy tail and long straight dagger-shaped (Kingfisher) bill.

Family–Alcedinidae

***Ceryle rudis* (Pied Kingfisher: Dhobinia)**

Common resident bird of size in between Myna and a Pigeon (about 31cm). Plumage speckled, black and white and bill dagger-shaped. Black nuchal crest finely streaked with white, supercilium and collar on hindneck white. Double black gorget across the breast in male. In female breast garget is broken. Its call is a repeated 'Chick-kuk' and also 'Chirrwk-chirruk'. A specialist at fishing in air, besides taking small fishes and tadpoles it can also feeds on water insects and nests in horizontal tunnel dug into an earth bank (Figure 1.9).

Figure 1.9: *Ceryle rudis*

***Ceryle lugubris* (Large Pied Kingfisher)**

Common resident bird of about 41 cm in size. It is a large crested black and white kingfisher with cross-barred back. Crest erectile, wings and tail blackish grey, boldly spotted and barred with white. Abroad with half collar on the neck and under parts whitish. There is a broad pectoral band of black spots mixed with brown.

Its call is like a sharp click or klick. It is a specialists at fishing in air and feeds on small fishes, tadepoles and water insects. Makes their nests in horizontal tunnel dug into an earth bank (Figure 1.10).

Figure 1.10: ***Ceryle lugubris***

Halcyon smyrnensis **Linn. (White Breasted Kingfisher: Kilkila or Kourilla)**

Common resident kingfisher of about 28 cm in size. A kingfisher with brilliant turquoise-blue upper parts including rump and tail. It has head and neck chocolate brown colour. Under parts, chin, throat and centre of breast white, rest chocolate brown. Beak and legs red. Always keep near water, hunt schooling fish in transparent water. Its call is a loud 'klik', loud long 'kil-kil'. Generally observed on small trees near beel or wandering over water surface for fish catching (Figure 1.11).

Pelargopsis capensis **(Brown Headed Stroked Billed Kingfisher: Badami Kourilla)**

A large resident kingfisher of about 35 cm size with reddish dagger like bill. Head dark grayish brown, under parts yellow. Yellowish colour on hind neck, rest of upper parts pale greenish blue. Its call is a loud 'krrr-ley-you-koo' and also 'ke-ke-ke-ke-ke'. Generally observed on large trees near beel or wandering over water for fish catching (Figure 1.12).

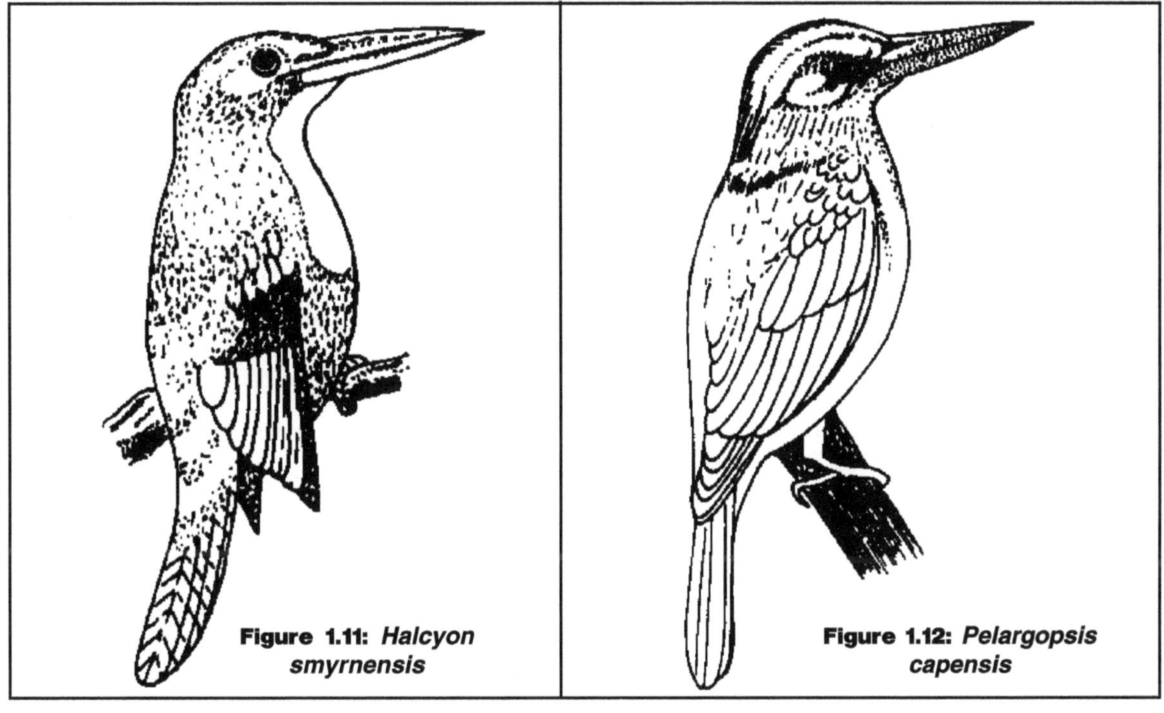

Figure 1.11: ***Halcyon smyrnensis***

Figure 1.12: ***Pelargopsis capensis***

Order–*Ciconiiformes*

Family–*Ardeidae* (Herons, Egrets)

Wader with long leg, long flexible neck, long straight dagger like bill, tarsi very long, the middle and outer toes united by a small web at the base, claw of middle toe pectin ate and hind toe well developed.

***Ardea cinerea*, L. (Grey Heron: Kubud/Khaira)**

Resident tall grey bird of about 75 cm size. Well marked with black and white plumage and a long sharp dagger like bill. Crown white, occipital crest black, a black dotted line down the middle of the fore neck. Neck is white tinged with greyish iliac, flight feather bluish black. A bare patch of greenish skin in front of eye, iris golden yellow, leg greenish brown. It has a long black dropping occipital crest. This heron has special powdery down on the side of its chest. Its call is frank and a verity of harsh croaks like 'Kraank' heard in flight. It is usually solitary bird standing motionless in knee deep water, head sunk between the shoulder. Feeds mostly on fishes and frogs. Nests on trees near water and breeds between July to October (Figure 1.13).

Figure 1.13: *Ardea cinerea*

***Ardea purpurea*, Meyex (Purple Heron: Khaira)**

Common resident bird of about 70 cm in size with long neck and legs. It is similar in general appearance and habit to *Ardea cinerea* but it is purplish in colour. Upper part purple blue, blackish on wing and tail. Crown and crest slaty black. Under parts, chin and throat white, upper breast has long droping buffy white plumes. Female has undeveloped crest and pectoral plumes. This bird actually looks bright purple in direct sun light. Its call is frog like crak; loud 'Kaa-ka-row-ka-row' when nesting. Similar to Grey Heron in food habit and breed between June to September to October. It nests in reed in wetlands (Figure 1.14).

Figure 1.14: *Ardea purpurea*

***Ardeola grayli*, Skyes (Paddy Bird: Andha Bagula)**

Common resident bird found on mud flats and wet fields, live near reed bed of wetland near grooves of Babul and Sissam. It is about the size of a village hen and earthy brown in non-breeding plumage, in breeding plumage the back is covered with darty maroon hair-like plumes and white occipital crest developed. Feeds on frogs, fishes, crustaceans and aquatic insects. Very active bird feeding by day and night. They are solitary hunters. Roost in community during winter. Breeds between May to September and nests on trees not necessarily near water (Figure 1.15).

Nycticorax nycticorax **(Ninght Heron: Kwak or Wak)**

Common resident and seasonal migrant bird of about 60 cm in size. It is a small heavy heron with greyish brown colour with white on the face and under surface, and greenish black crown and back are quite distinctive. During young stage this bird resemble adult *Ardeola grayli* but are darker and lack white wings. It is largely nocturanal in habits and sleeping by day in dense vegetation near beel. Nests in reeds or tall trees. Its call is a harsh raucous wock (whistler). Sometimes emits throaty quawk at several second intervals, also a rapid 'wuk-wuk-wuk' (Figure 1.16).

Figure 1.15: ***Ardeola grayli***

***Bubulcus ibis*, Boddaert (Cattle Egret: Gai Bagula)**

Common resident bird of about 50 cm. in size and commonly seen scattered flocks near grazing cattle. It is similar in general appearance with *Egretta garzetta.* It is a slender white bird with long neck and legs. It non-breeding stage it has pure white plumage and has yellow colour of bill but in breeding stage plumage feathers at its head, neck and back becomes hairy, assume orange buff colour. Its call is croak. It is comparatively less dependent on water and feeds mostly on tettestrial insects. Breeds between June to August and nests on trees near or away form water (Figure 1.17).

Figure 1.16: ***Nycticorax nycticorax***

***Egretta alba*, Linn. (Large Egret: Bada Bugula)**

Large egret is about the size of purple Heron (about 90 cm in length). It is a large snow white egret. Legs bare black, bill black and yellow or yellow. In breeding season this egret develop a cluster of flimsy ornamental dorsal plumes so called 'aigrettes' falling over beyond the tail. Its bill is black in summer and yellowish in winter. Large egret can be distinguished easily from intermediate egret by larger size and longer head and neck. It is a common resident bird of beel area (Figure 1.18).

***Egretta intermedia*, Wagler (Intermediate Egret: Manjholla Bagula)**

Median egret is slightly smaller (about 80 cm in length). A handsome snow white egret with thinner neck and shorter bill. In breeding season bill looks black but in winter bill become yellow tipped with black colour. Leg joints faintly greenish yellow. The voice is long drown out gargle like or deep throaty call g-a-a-a-rh. It is also common resident bird of beel area but exhibits some extent of local migration also (Figure 1.19).

Figure 1.17: ***Bubulcus ibis***

Figure 1.18: ***Egretta alba*** **Figure 1.19:** ***Egretta intermedia***

***Egretta garzetta*, Linn. (Little Egret: Chota Bagula)**

Little egret is slightly smaller and the size of a domestic hen and its length varies in between 50–63 cm. It has black bill, long and long neck. It is white bird with general plumage pure white. In breeding season, the feather of the breast get overgrown and assume, lanceolate shape and there is a crest of two long attached feathers. It resembles to Bubulcus ibis but black bill and yellow feet differentiate it from other egrets. Dropping crest in breeding season and the wonderful nuptial plumes from which aigrettes are derived are remarkable features. Its croaking call are of wide variety like 'Kaak-kaak'. It is a common resident bird of beel area and most commonly found in paddy fields and irriagated fields (Figure 1.20).

Figure 1.20: ***Egretta garzetta***

Family–Ciconiidae (Storks)

Chiefly terrestrial and marsh hunting birds. Wings long and broad; tail short, legs very long; toes of moderate length, webbed at the base, middle toe not pectinate as in Heron. Unlike Herons they fly with neck outstretched like Cranes, Ibises and Spoonbills.

***Anastomus oscitans* (Open Billed Stork: Ghonghail)**

Common resident storks of about 68 cm in length. Their seasonal movement depends on water level and aquatic prey species. The peculiar reddish black bill and narrow open gap in between mandible is the diagnostic feature. It is highly gregarious, powerful flier and feed chiefly on molluscs, therefore, commonly called as Ghonghail. It is whitish or pale gray stork with black wings and short square black tail. Its call is deep moon. Breeds between July to September. Nests on partly submerged hydrophytes (Figure 1.21).

Figure 1.21: *Anastomus oscitans*

Family–Threskiomithidae

***Platalea leucoroidea,* T. and Schlegel (Spoon bill: Khurpia Dabil)**

A tall white long legged and long necked, marsh bird with black legs. In this bird a distinct crest develops in breeding season. A cinnamon-yellow patch at the base of fore neck is quite distinctive. It is a snow-white bird with large, flat, black and yellow speculate bill adapted for feeding in shallow water. Food, breeding season and nesting is similar to white Ibis except that it can also take some vegetable matters. (Figure 1.22)

Figure 1.22: *Platalea leucoroidea*

Order–*Anseriformes*

Family–Anatidae (Duck, Geese)

Popular group of water bird from sporting and food resource point of view. Legs short, feet webbed, adapted for swimming. Bill broad, flat, rounded at the tip and with a comb like fringe for straining out food particles from the water. Wings narrow and pointed, adapted for swift and long ranging flight. Tail short. Many of them are migratory .

***Anas poecilorhyncha,* Oates (Spot bill of Grey Duck: Kapila)**

Common resident and as large (about 40 cm tall) as domestic duck. Distinctive large duck in which male and female have greyish brown colour and have speculum narrowly boarded with white. Bright red spots at the base of forehead present in male. Distinctive black and yellow bill tipped black is a good field marks. Legs coral red. Its call is just similar to 'quack-quack'. Chiefly vegetarian feeding on shoots of hydrophytes, seeds and paddy grains with the addition of molluscs, aquatic insects and worms. Breeding throughout the year and nests concealed under herbage on the edge of beel/ wetland (Figure 1.23).

Figure 1.23: *Anas poecilorhyncha*

***Dendrocygna javanica*, Horsfield (Lesser Whisling Teal: Silli)**

Small common resident duck of about 43 cm in length and have uniform chestnut colour. The wings are round, crown dark brown, neck and sides of the head light pinkish brown and uniformly chestnut upper tail-coverts. Neck often outstretched when alarmed. Its call is as a wheezy 'seasick-seasick'. Mostly nocturnal feeder and perches freely on surrounding trees of beel. Exclusively vegetarian but also take small fishes, frogs, snails, worms, etc. Breeds between June to October and nests in the holes of old trees, near water (Figure 1.24).

Figure 1.24: ***Dendrocygna javanica***

***Anas creca*, L. (Common Teal: Kerra)**

Commonest and most abundant winter visitor duck, twisting and twinning in flight and springing off the water characteristic dash when alarmed. It is smallest (size about 38 cm long) duck with brisk wader like flight. Male ashy grey with chestnut head and a distinct metallic green band extending eye to nape, bordered above and below by whitish lines. *Anas creca* distinguished from other by green and black speculum. It has multicoloured wing speculum-black, green and buff particularly conspicuous in flight. Female usually mottled dark brown with pale belly and black and green speculum. Its call is noisy duck emit 'krit-krit'. Entirely vegetarian and feeds generally on shoots, tubers, and seeds of hydrophytes and grains of cultivated rice. Breeding extralimital (Figure 1.25).

Figure 1.25: ***Anas creca***

***Anas acutta*, Linn. (Pintail: Dighaunch)**

Commonest winter visitor with long tail being distinctive at rest when flying. It is an asristocratic duck of about 55 cm long in size with striking colour. It is slender in build, with long neek, long paralled side bill and long pointed tail. Upper plumage pencil grey. Head chocolate brown, with a white band on either side running down into the white neck and under parts. The upper plumage and flanks vermiculated with black. Long, pointed pin like feathers projecting well beyond the tail distinctive in both sexes. Speculum metallic bronze green. Female mottled brown and buff with pointed pinless tail. A distinct whistle call from male and short quacks from female and are silent during flight. Alert, worry and one of the most elegant ducks. It is a good walker holding the neck erect; wings make a distinctive hissing noise on flight. Feeds on grass, corn, shoots and seeds of hydrophytes and wild and cultivated rice. Breeding extralimital (Figure 1.26).

Figure 1.26: ***Anas acutta***

***Anas quarquedula*, Linn. (Garganey: Khaira)**

It is earliest to arrive in winter and last to leave. It is a small handsome gregarious, non-diving duck of about 40 cm long body size. It has broad white eye stripe, the pale blue shoulder to the wing and the elongated scapulars in the male. Both male and female have a green speculum edged with white bars, but in female the speculum largely hidden. Male shows conspicuons eye stripe which is weak in female. Both sexes show blue-grey forewing especially noticeable in flight. Female quack quietly, some time harsh cackle, 'Krrrr' and male produce rattly call. Mainly vegetarian but also take some water insects, larvae, worms, molluscs, etc. Breeding extralimital (Figure 1.27).

Figure 1.27: ***Anas quarquedula***

***Anas anser*, Swinhoe (Greyleg Goose: Kaj)**

Winter visitor migratory bird of about 76–90 cm long in size. It is a large, gray-brown goose with flesh pink bill and legs, in flight note white upper tail coverts and vents. It is an ashy and wary bird on its wintering ground. Its call is a fine honk on the wing 'aahng-ung-ung' during flight gurgles 'gay-gay-gay' while feeding. Gregarious and nocturnal feeder and spend the day time squatting belly to ground or resting on one leg on mud or floating on water. Exclusively vegetarian. Breeding extralimital *i.e.*, outside the Indian territory (Figure 1.28).

Figure 1.28: ***Anas anser***

***Anas strepera*, Linn (Gadwall: Malkai)**

Winter visitor bird of about 50 cm long in size. It is a large duck which may be recognized at once by white speculum divided by black bar. Distinct chestnut patch on wing coverts. Duck has the tail set in patch of velvety black,with breast will marked with brown and white crescents, under tail-coverts black. The black and white speculum of the female is unique. Male in eclipse like female but grayer. It produce chunkling croak 'quak-quak-quack' call. Feeds mostly on seeds, shoots, and tubers of marsh plant, aquatic weed and grains of wild and cultivated rice, occasionally insects, worms, molluses etc. Breeding extralimital (Figure 1.29).

Figure 1.29: ***Anas strepera***

***Netta rufina* (Red Crested Pochard: Lalsar)**

It is a handsome diving duck of about 55 cm long in size. Drake has orange chestnut head with darker brown colour on the cheeks and lower-parts. Crest moplike erectile, bushy and orange coloured. Bill bright red, iris crimson. The breast orange red. Upper plumage light brown with white patches on shoulders and white wing-mirror. Under-part black and flanks white. Female dull sooty brown above, largely whitish below, with dark brown crown and nape sharply separated from whitish face and fore neck and has whitish speculum. It produce soft call 'guk-guk-guk'. Winter visitor and leave the beel earlier.

Largely vegetarian and feeds on shoots, buds, rhizomes and seeds of aquatic weeds and grasses. It can also take aquatic insects, molluscs, tadpoles and small fishes. Breeding extralimital (Figure 1.30).

Figure 1.30: ***Netta rufina***

Figure 1.31: ***Podiceps rufficolis***

Order–*Podicepediformes*

Short-winged almost tailless aquatic birds with the legs placed far back and the front toes lobed laterally with a fringe of skin; hide toe small and vertically lobed, tarsi scutellated in tront.

Family–*Podicepedidae*

***Podiceps rufficolis*, Sal (Little Grebe: Pandubi)**

Common resident small (size of about 25 cm long) bird. It has buoyant rounded body with shout downy feathers. In breeding plumage top of head and hind neck darker brown, blackish round bill and chin. It has yellow on gape. The sides of head, neck and throat chestnut. Upper parts dark brown, under part including abdomen silky white. Its reddish neck changes to pale rufous in non-breeding plumage. It produce a distinct high pitched titering giggle or repeated 'klik' and a trill. Swimmer and constant diver. Feeds mostly under water by diving but also seizes food on surface escaping from under water floating vegetation. Feeds mostly on aquatic insects, *Crustacea,* tadpoles, snails and tiny fishes. Breeds between April to October in beel area. Nests are on partly submerged floating plants (Figure 1.31).

Order–*Pelecaniformes*

Family–Pelecanidae (Pelicans, Darter)

Short stout legs and fully webbed feet, tarsus reticulate in front, bill long, heavy, upper mandible flattened and hooked at the tip, presence of gullar pouch, wings large and broad, help in swimming and strong flight, tail short, cannot dive.

Figure 1.32: ***Pelecanus philipensis***

***Pelecanus philipensis*, Gmelin (Spotted billed/Grey Pelican: Bherwa)**

The huge (size about 152 cm long) squat bird with enormous beak and elastic pouch. The colour of this bird is mainly greenish white with distinct brown nuchal crest. It has stout heavy legs, large webbed feet and large flattened bill, hang down throughout its length by an elastic bag of dull purplish skin. Large blue black spots along edge of upper mandible, blackish wing quills and greyish brown tail diagnostic. Voracious fish eater and nests on tall leafy trees between October to March (Figure 1.32).

Family–Phalacrocoracidae

Bill long, slender, laterally compressed and hooked at the tip, wings of moderate length, tail long and stiff, legs pelecaniform, claws much curved, plumage black, expert divers, swimming under water with the use of wing unlike pelican.

Figure 1.33: ***Phalacrocorax niger***

***Phalacrocorax niger*, Vieillot (Little Carmorant: Pan-Kowa)**

Common winter visitor bird of size about 48 cm long. A shiny bronze-black water bird with upright carriage, has scattered white feathers growing on the head and a few hairy plumes on the sides of neck in breeding plumage which disappears and throat becomes white in non-breeding season. Bill brown, livid purple. Eyelid and gular pouch and legs blackish. Usually seen perched on trees. Feeds mostly on fishes and breeds between July to September. Nests on trees standing in or near water (Figure 1.33).

Order–Falconiformes

Family–Acipitridae

Bill short with upper mandible longer than lower, curved and strongly hooked at the tip, basal portion covered with a cere, in which nostril are situated; powerful hooked claws, hallux always present. They are not aquatic bird but soon are dependent on aquatic organisms for food.

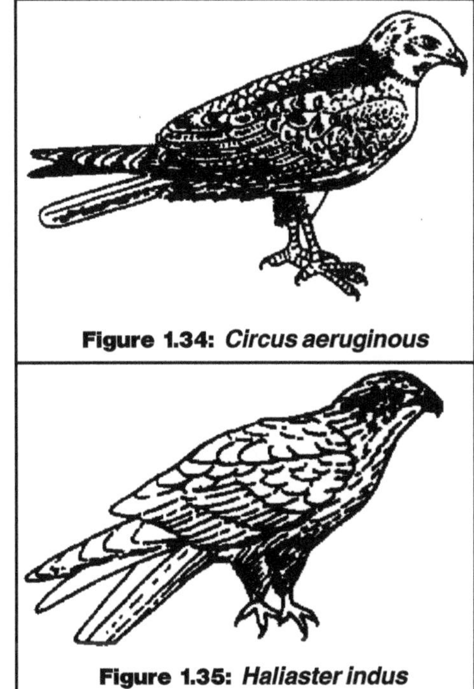

Figure 1.34: ***Circus aeruginous***

Figure 1.35: ***Haliaster indus***

***Circus aeruginous*, Linn. (Marsh Harrier: Safed Sira)**

Winter visitor and marsh inhabitants. It is one of the heavily (size about 53–58 cm long) built bird. The general body colour is chocolate brown but adult male has pale grey tail. It has pale rufous head, neck and breast. Pale forehead and shoulders, of female are diagnostic. Juvenile bird entirely chocolate brown except for pale throat and crown. Male is smaller than the female. It produce harsh note 'Kee-kee' and alarm call 'kek-kek'. It is highly territorial during the breeding season. Feeds on frogs, fishes, field mice and voles, wounded bird and large insects. Breeds between April to June. Nests on the ground near a marsh (Figure 1.34).

***Haliaster indus*, B. (Brahrniny Kite: Dhobia Cheel)**

Resident bird of size about 48 cm long, seen solitary or in pairs. Bright chestnut coloured bird of prey with black wing tips and white head and breast. The upper parts rusty red with white head neck, and breast down to abdomen. Adults have yellow feet, tarsi and cere. Young birds are chocolate-brown and having rounded tail. It produce a loud, wheezy squeal. Prefers to scoop up its food from the surface of water rather than land. Feeds on died fishes, frogs, terrestrial insect, rarely carion. Breeds in cold weather. Nests on large trees (Figure 1.35).

Physico-chemical Parameters

The physico-chemical characteristics reflect the water quality and the productivity of the aquatic system including beel/wetland and also determine their biota. The average range of different physico-chemical parameters have been presented in Table 1.1. Atmospheric temperature ranged in between 19.5°C to 32.9°C with the minimum and maximum value recorded in November and July respectively. Water temperature also ranged between 16.7°C to 29.8°C with the minimum and maximum values in November and July respectively. Low depth coupled with high intensity of sunlight during the major part of the year provide high input of heat budget for initiating rapid decomposition of organic matters in substrate releasing nutrients into the beel accelerating the process of eutrophication. pH varies from 6.7 to 7.9. The range of pH recorded exhibit symptom of highly productive beel (Moyle, 1946). Maximum conductivity (462 µmho) was recorded during the monsoon. The higher conductivity exhibit higher concentration of ions in water in monsoon months. Dissolved oxygen values was recorded between 4.3 mg/L to 8.4 mg/L. The higher level of dissolved oxygen throughout the year are indicative of eutrophication effects (Arumugan and Furtad, 1979). The maximum record of DO_2 in summer season are due to increased photosynthetic activities (Bhatnagar and Sharma, 1978). The value of free Carbon dioxide (FCO_2) various from nil to 12.2 mg/L. The main source of FCO_2 in certain months seem to be due to decomposition of organic matters and respiration of hydrophytes and animals in the beel. Carbonate alkalinity ranged from nil to 11.9 mg/L. FCO_2 and CO_3 cannot coexist together since they neutralise each other to form bicarbonate. The presence of CO_3–alkalinity is an indication of high rate of carbon assimilation. Bicarbonate alkalinity varies from 39.8 mg/L to 74.5 mg/L. It was maximum in summer and monsoon month. It fluctuates more or less inversely with the carbonate alkalinity. The total hardness range indicates that water of the beel is hard throughout the year and favourable for the growth of Gastropods. Range of calcium hardness is suitable for the formation of shell of molluscan population. This calcium is carried and transported to the Mahananda by the river of Nepal, which passes through the calcium hillock of limestone in Nepal (Dutta Munshi and Dutta Munshi, 2000). Chloride content of water ranged between 31.7 mg/L to 56.0 mg/L. The higher chloride content of water is an index of water pollution of animal origin (Thresh *et al.*, 1994). Silica contents were high and recorded throughout the year. Silica is an essential nutrient for diatoms and for some plankton (Cole, 1979).

Table 1.1: Physico-chemical Properties of Water of Goga Beel, Katihar

Sl.No.	*Parameters*	*Minimum*	*Maximum*
1.	Atmospheric temperature (°C)	19.5	32.9
2.	Water temperature (°C)	16.7	29.8
3.	pH	6.7	7.9
4.	Conductivity (µmho)	278	462
5.	DO_2 (mg/l)	4.3	8.4
6.	FCO_2 (mg/l)	Nil	12.2
7.	Carbonate alkalinity (mg/l)	Nil	11.9
8.	Biocarbonate alkalinity (mg/l)	39.8	74.5
9.	Total hardness (mg/l)	69.4	101.4
10.	Ca hardness (mg/l)	36.9	58.8
11.	Chloride (mg/l)	31.7	56.0
12.	Silicate (mg/l)	32.0	44.8

Thus, the physico-chemical parameters considered in the present work, operating in beel establish their eutrophic character and initiate eutrophication process. Eutrophication is slow progressive process leading to terrestrialization of the beel affecting the avian fauna considerably in course of time. Dugan, (1994) state that over 50 per cent of the wetlands of the world have been lost due to conversion into agricultural field due to sum total effect of eutrophication.

Biological Parameters

In addition to physico-chemical factors the biological parameters are the major regulating forces present in the Goga beel and were found to affects the avian biodiversity. Vegetations present in the wetland forms important element of aquatic environment and providing food and shelter for avian fauna along with various other forms of wildlife which are the natural food of birds (Winter-Bourn and Lewis, 1975). A complex flora allows for a greater diversity of faunal components and more complex food webs (Boyd, 1971).

Luxuriant growth of macrophytes are found in Goga beel are grouped into three heading: Floating, Submerged and Emergent forms (Table 1.2). The floating vegetations constitute a biotope with so many ecological niches, that almost all the aquatic faunal groups are represented in this vegetation type; which may be considered the most species rich biotope of the aquatic ecosystem. The submerged plants form the habitat for the periphytic organisms, mining animals and periodically used by various free swimming species of insects and fishes (Piecynska and Ozimek, 1976). The emergent forms of vegetation of the Goga beel afford shelter to the adult stages of insects, molluscs and birds. General vegetation is poorly developed in the Goga beel area. The most common are: *Delbergia sisso* (Sissam), *Acacia nilotica* (Babul), *Mangifera indica* (Mango), *Shorea robusta* (Sal), *Cocos nuciftra* (Coconut), *Phoenix dactylifera* (Khajoor), *Tamarindus indica* (Imli), *Artocarpus altilis* (Jackfruit), *Syzygium cumini* (Jamun), *Dendrocalamus strictus* (Bamboo), *Musa paraisiaca* (Banana), *Borassus flabellifer* (Tar). These trees are planted in beel area by the local inhabitants and they are able to provide excellent site for perching, roosting, nesting, brooding, migratory stop-over and for other life activities of the birds (Wood Cock, 1984).

Table 1.2: Common Macrophytes of Goga Beel, Katihar

Floating Forms	*Submerged Forms*	*Emergent Forms*
Eichornia crassipes Solms	*Hydrilla verticillta* Royle	*Oryza sativa* Linn.
Nymphaea strellata Wild	*Najas minor* All	*Ipomea aquatica* Forsk
Nymphaea nouchali Burm	*Najasgraminea* Dal	*Ipomea chrupoides* Ker-Gawl
Nymhpoid cristatum Roxio	*Potamogeton cirspus* Linn.	*Ipomea fistulosa*
Nymphoid indicum Linn.	*Potamogeton nodosus* Poir	*Scriptus supinus* Linn.
Hydrorhiza aristata Refz	*Vallisnaria spiralis* Linn.	
Nelumbo nucifera Giertn.	*Utricularia stellaris* Linn.	
Trapa bispinosa Roxb.		
Lemna minor Linn.		
Pistia stratiotes Linn.		

Wetlands are considered as the most productive ecosystem and rich in faunal diversity also. Various groups of faunal forms *viz.*, molluscs, arthropods, fishes are abundant in Goga beel. The most abundant and dominent molluscan fauna found in Goga beel are: *Viviparous bengalensis, Pila globosa, Indoplanorbis exustus, Malanoides lineatus, Lymnae rufescens, Unio marginalis* and *Perreysia* sp. During present investigation, it was found that Gastropods predominate the molluscan population with a restricted species of pelecypods. Arthropods are represented by prawns, crabs and insects. The insects population of Goga beel is dominating and found distributed in five orders (*Coleoptera, Hemiptera, Odonata, Ephemeroptera* and *Diptera*). The Goga beel was found very rich in Ichthyo fauna comprising air-breathing fishes, major carps and minor carps (Table 1.3). Conclusively we state that the physico-chemical characteristics reflect the water quality and the productivity of the aquatic system and also determine their biota. The biotic components of Goga beel are diverse and varied. Actually, certain physico-chemical parameters, domination of macrophytic vegetation and molluscan fauna along with insects and fishes in Goga beel and social forestry with numerous bamboo orchids are the main fundamental factors, regulating the population richness and species diversity of avian fauna in Goga beel.

Table 1.3: Common Fishes of Goga Beel, Katihar

Air-breathing Fishes	*Carp*	*Other Fishes*
Notopterus chitala (Ham.)	*Labeo rohita* (Ham.)	Chanda nama (Ham.)
Notopterus notopterus (Pallas)	*Labeo gonius* (Ham.)	*Chanda ranga* (Ham.)
Heteropneustes fossilis (Block)	*Labeo bata* (Ham.)	*Tetradon cutcutia* (Ham.)
Monopterus cuchia (Ham.)	*Labeo calbasu* (Ham.)	*Xenetodon cancila* (Ham.)
Clarias batrachus (Linn.)	*Cirrhinus mrigala* (Ham.)	*Gudusia chapra* (Ham.)
Colisa fasciatus (Bloch and Schn.)	*Cirrhinus reba* (Ham.)	*Setipinna phasa* (Ham.)
Anabas testudineus (Bloch)	*Catla catla* (Ham.)	*Wallago attu* (Schn.)
Channa punctatus (Bloch)	*Puntius conchonius* (Ham.)	*Mystus cavasius* (Ham.)
Channa gachua (Ham.)	*Puntius sarana* (Ham.)	*Mystus vittatus* (Bloch)
Channa stratisu (Bloch)	*Puntius ticto* (Ham.)	
Channa marulius (Ham.)		
Mastacembelus sp.		
Macrognathus sp.		

References

Ali, Salim, 1945. *The Bird of Kutch*. Oxford University Press, Bombay.

Ali, Salim and Ripley, S.D., 1983. *Handbook of the Birds of India and Pakistan,* complied edition of 10 Volume. Oxford University Press, New Delhi.

APHA, 1989. *Standard Methods for Examination of Water and Wastewater*. American Public Health Association, Washington, D.C.

Arumugan, P.T. and Furtad, J.I., 1979. Eutrophication of Malaysian reservoir effects of agro industrial effluents. In: *Presented at the 5th International Symposium of Tropical Ecology ISTR*, Kualaumpur.

Beadle, L.C. and Lind, E.M., 1960. Research on the swamps of Uganda. *Uganda Journal*, 24: 84–87.

Bhatnagar, G.P. and Sharma G.P., 1978. Physico-chemical feature of sewage polluted lower lake, Bhopal. In: *Proc. Int. Symp. on Environ. Agents and their Biological Effect*. In: *Heredity Suppl.*, Vol. II, 1979.

Boyd, C.B., 1971. The limnological role of aquatic macrophites and their relationship to reservoir management. *Reser. Fish. Limn.*, Spl. Pub., 8: 153–166.

Choudhary, D.N. and Choudhary, N., 2007. Mortality of water birds due to chemical poisoning to Kawar lake bird sanctuary, Bihar. In: *Presented in International Symposium on Recent Advances in Contemporary Biology, Environmental Issues and Sustainable Development* at S.K.M. University, Dumka, Jharkhand, 29–30 September.

Das, S.M. and Srivastava, V.K., 1956. Some new observation on plankton from freshwater ponds and tanks of Lucknow. *Sci. and Cult.*, 21(18): 466–467.

Datta Munshi, J.S. and Huges, G.M., 1992. *Air-breathing Fishes of India: Their Structure, Function and Life History*. Oxford and IBH Publishing Co.

Datta Munshi, J.S. and Datta Munshi, J., 2000. The sustainability of hydrological cycle of wetlands of Kosi river basin of North Bihar, India. In: *Waste Recycling and Resource Management in the Developing World*, (Eds.) B.B. Jana, R.D. Banerjee, B. Guterstam and J. Heeb. India and International Eco. Soc., Switzerland, p. 665–674.

Dehadrai, P.V. and Tripathi, S.D., 1976. Environment and ecology of freshwater air breathing teleosts. In: *Respiration of Amphibious Vertebrates*, (Ed.) G. M. Huges. Academic Press, London, New York, p. 39–72.

Hancock, J., 1984. *The Birds of the Wetlands, Delhi*. Oxford University Press, Bombay, pp. 1–176.

Kumar, Ajay, 1999. Pen culture prospects in reservoirs and beels: Success and constraints with particular reference to Akaipur Beel of 24 Parganas, West Bengal. *J. Freshwater Biol.*, 11(1–2): 47–52.

Kumar, Ajay and Singh, S., 2001. Prospects of pen culture in North Bihar with special reference to Akaipur beel (24-Parganas) West Bengal. In: *Presented in Nat. Sem on Fish Health and Management*, March 25–27, Department of Zoology, L.N.M.U. Darbhanga, Bihar.

Michael, R.G., 1969. Seasonal trends in physico-chemical factors and plankton of freshwater fish pond and their role in fish culture. *Hydrobiologia*, 33(1): 144–160.

Moyle, J.B., 1946. Some indices of lake productivity. *Trans. Amer. Fish. Soc.*, 76: 332–334.

Munawar, M., 1970. Limnological studies of freshwater pond of Hydreabad, India. I. The biotope. *Hydrobiologia*, 35(1): 127–192.

Pieczynska, E. and Ozimek, T., 1976. Ecological significance of lake macrophites. *Inst. J. Environ, Sci.*, 2: 115–128.

Prakash, S., Joshi, H.C. and Karmachandani, S.J., 1983. On the trophic condition of Govindgarh lake in Madhya Pradesh. *J. Scientific Research*, 23: 369–378.

Rai, D.N. and Datta Munsi, J.S.D., 1979. The influence of thick floating vegetation (Water hyacinth: *Eichornia crassipes*) on the physico-chemical environment of freshwater wetland. *Hydrobiologia*, 62(1): 65–67.

Sahu, H.K. and Dutta, P.K., 2005. Status of aquatic bird in Mayurbhanj district, Orissa, India. *Indian J. Environ. & Ecoplan.*, 10(3): 833–888

Singh, S.R., 1986. On the trophic characteristic of Dah–Lake (Ballia). *Biol. Bull. of India*, 8(2): 125–135.

Thresh, J.C., Suckling and Beale, 1944. *The Examination of Water and Water Supplies*. Ed. Taylor, E.W. (1949).

Vijayan, V.S., 1986. On conserving the bird fauna of Indian wetland. *Proc. Ind. Acad. Sci. (Animal Sci./ Plant Sci.) Suppl.,* Nov. p. 91–101.

Winterboum, M.J. and Lewis, M.H., 1975. Littoral fauna. In: *New Zealand Lakes,* (Eds.) V.H. Jolly and J.M.A. Brown. Auckland University Press, Oxford University Press, p. 271–280.

Wood Cock, M., 1984. *Collins Hand Guide to the Birds of the Indian Sub-continent*. South China Printing Co., Hong Kong, pp. 1–176

Chapter 2

Particulate Transport of Trace Metals in a Riverine Ecosystem–Karamana River, South Kerala, India

P.R. Jayaraman[1], *C.G. Radhika*[2], *T. Ganga Devi*[3] *and T. Vasudevan Nayar*[4]

[1]*Department of Botany, Government College for Women, Thiruvananthapuram, India*
[2]*Department of Botany, University College, Thiruvananthapuram, India*
[3]*Principal (Retired), Government College for Women, Thiruvananthapuram, India*
[4]*Scientific Officer (Retired), Division of Marine Chemistry, Department of Aquatic Biology and Fisheries, University of Kerala*

ABSTRACT

Suspended Particulate Matter (SPM) and trace metal fluxes in Karamana River, the major source of drinking water for Thiruvananthapuram, the capital city of Kerala, South India, were monitored for a period of one year from February 1998 to January 1999. Six trace metals, *viz.*, copper, lead, cadmium, zinc, manganese and iron were estimated in the particulate matter using Atomic Absorption Spectroscopy. The total particulate flux (15.68 mg L^{-1}) in the river was very low as compared to Indian and world averages. Significant spatial and temporal variations were evident in the total flux of SPM and associated trace metals to varying extents. Co-transportation of trace elements in different combinations during different seasons was noticed. SPM, associated transport of cadmium (155.96 μg g^{-1}), lead (464.01 μg g^{-1}) and iron. (56,633 μg g^{-1}) remained high, whereas that of copper (212.91 μg g^{-1}), zinc (623.58 μg g^{-1}) and manganese (957.13 μg g^{-1}) was found low in comparison with standard values as suggested by Aston and Chester (1976). Particulate flux of all the six metals was found higher than the standard values for average shale, upper continental crust, average mud and average values for Indian rivers, while a slightly lower value was noticed in the case of manganese when compared to world river SPM and average river particulate. Particulate load of cadmium was very high as compared to the tentative threshold values, while the values of particulate cadmium, lead and iron were found low compared to metal content in world river borne detritus. Leaching of heavy metals into the lotic ecosystem resulting from plantations along with other anthropogenic activities have been highlighted. Toxic metals cadmium and lead were detected in obvious quantities indicating the association of these elements in the river water. In general, the levels of these trace metals have not reached any alarming level yet, while cadmium and lead pose great concern.

Keywords: *Fluvial systems, Suspended particulate matter, Biogeochemical cycles, Trace metals.*

Introduction

Suspended sediments play a significant role in the biological and biogeochemical cycling of trace elements in fluvial systems (Horowitz, 1996). Riverine flora ranging from microscopic phytoplankton to angiosperms depend largely on river borne particulates for their nutrients. The allochthonous materials in the river are rapidly adsorbed and transported by particulates (Walling and Webb, 1985) and therefore Suspended Particulate Matter (SPM) reflects the effect of pollution/deterioration in a given area. Hence, spatial and temporal variations of trace metals in Suspended Particulate Matter (SPM) in fluvial environments have been recognized as an area, which demands serious attention.

Suspended particulate matter in aquatic systems according to Burton (1976) can be considered as suspended matter retained by a 0.45 µm filter. A large proportion of the fluviatile transport of matter occurs in the form of suspended material with mean grain sizes of 2–63 µm (silt grain size) as opined by Forstner and Wittman (1983). Particulate matter is an extremely important, but poorly studied substrate for the transportation of trace metals in fluvial systems and particulate transport represents a major pathway in the biogeochemical cycling of trace contaminants (Allen, 1979). Sediments and suspended matter in aquatic systems absorb and concentrate comtaminants such as trace metals. Rivers are the important transporting agents for continental material (both natural as well as anthropogenic) to the oceans that happen mostly in particulate phase amounting to 13 billion tons (Milliman and Meade, 1983). Studies on the chemical composition of particulates have been found to contribute much to the field of geochemical prospecting in riover beds (Hawkes and Webb, 1962).

Although a number of studies have been carried out to assess the role of particulate matter in riverine transport in India (Subramanian, 1979; Borole *et al.*, 1982; Subramanian *et al.*, 1985; Subramanian *et al.*, 1987; Biksham and Subramanian 1988; Alagarsamy and Zhang, 2005; Jain and Sharma, 2006) no serious attempt in this regard has yet been made on rivers in Kerala but for the studies by Paul and Pillai (1983) and Shibu *et al.* (1990) on the lower reaches of Periyar and Muvattupuzha rivers. A systematic study on the flux and dynamics of trace metals in lotic water bodies is important to understand their transport and role in biogeochemical processes as well as in detecting their sources. Hence, the present study was undertaken to evaluate the occurrence and abundance as well as the spatial and temporal variations of selected trace metals in the waters of Karamana River, which is the main source of drinking water for Thiruvananthapuram, the capital city of Kerala.

Karamama River originates from Chemmungi Mottai, a peak in Western Ghats and after flowing for about 68 km through the Peppara Wildlife Sanctuary, plantations, rural and urban areas, it debauches into the Lakshadweep Sea at Panathura, through a barmouth which remains closed during summer. The river along with its tributaries lies between 8°21′ and 8°42′N and 76°52′ and 77° 15′E, within the administrative boundaries of Thiruvananthapuram District, South Kerala. 10 stations were identified along the river for the collection of samples (Figure 2.1). Station 1 was located down to Peppara dam in the upper reaches of the river under the influence of land runoff from thick forests and

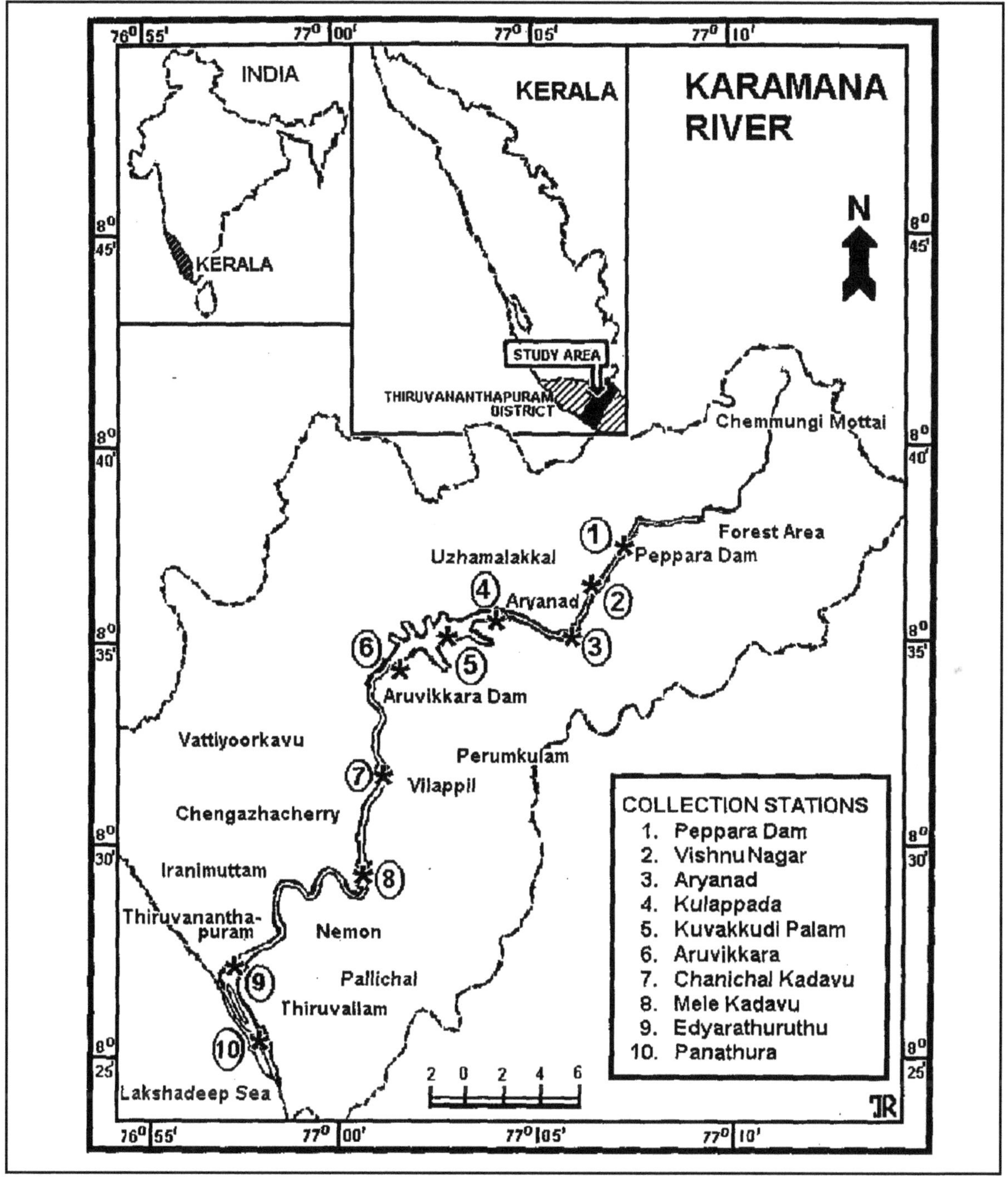

Figure 2.1: Map of Karamana River Showing Sampling Sites

discharge from the dam. Station 2 was identified at Vishnu Nagar, a housing colony that receives discharges from plantations and agricultural lands and wastes from nearby hutments. Station 3 was fixed at Aryanad Township influenced by domestic effluents. Station 4 was at Kulappada, where a ferry is being operated across the deep waters of the river. Station 5 was set at Kuvakkudi Palam (bridge) at the upper reaches of Auuvikkara dam characterized by turbid and stagnant water throughout the year. Station 6 was established down to Aruvikkara dam floodgate, where the river flows very fast through rocky terrain. Station 7 was placed at Chanichal Kadavu, a bathing ghat with more or less stagnant water that is under the influence of sand-mining. Station 8 was at Mele Kadavu, another bathing Ghat, where the river is shallow and slow flowing and under the influence of a granite quarry nearby. Station 9 was located at Edyarathuruthu where the Parvathyputhenar loaded with effluents from Thiruvananthapuram city joins the river. Besides, the station has estuarine features and is under the influence of retting of coconut husk and sand-mining. Station 10 was at Panathura, where the barmouth remains closed during summer and sea water enters the river through tidal effect during rainy season.

Materials and Methods

Water samples were collected in acid-cleaned, non-reactive plastic containers once a month from the 10 stations along the Karamana River for a period of one year from February 1998 to January 1999. Known volumes of water samples were filtered to collect the particulate matter on acid-treated, dried and pre-weighed 0.45 micron Millipore filter papers using an all-glass Millipore filtration unit. Filter papers were then oven dried and weighed to find out the particulate load. The particulate matter was acid digested following standard procedures (APHA, 1992) and subjected to Atomic absorption spectrophotometric analysis for the estimation of six trace metals *viz.*, copper, cadmium, lead, zinc, manganese and iron and values were expressed in $\mu g\ g^{-1}$ (A.A.S. Perkin Elmer Model 2380). Monthly data were pooled and expressed on seasonal basis as pre-monsoon (February to May), monsoon (June to September) and post-monsoon (October to January). Annual average transport of SPM and metals was computed based on the annual discharge of water by the Karamana River, *viz.*, 1324 million m^3 (Kerala State Land Use Board, 1995). The results were compared with standard values for average shale (Turekian and Wedepohl, 1961), the tentative threshold values (Aston and Thornton, 1977), geochemistry of average river borne detritus (Aston and Chester, 1976), average river particulate (Martin and Meybeck, 1979), average mud and upper continental crust (Taylor and McLennan, 1985) world river SPM (Martin and Windom, 1991) and also with known reports on particulate metal content in rivers in India and abroad. Enrichment Factors (EF) were computed for each metal in the suspended particulate matter at Stations 2 to 10 that are apparently exposed to anthropogenic inputs by weighing the respective concentrations against background values at Station 1 located in the comparatively pristine forest environment. The results were subjected to regression analysis and ANOVA test to evaluate the relationships among the various factors involved.

Results and Discussion

The data on the season-wise distribution of SPM and associated heavy metal concentrations in the Karamana River are presented in Figures 2.2 to 2.8. Table 2.1 compares the SPM content and annual flux of SPM in the Karamana River with that of other rivers. Table 2.2 summarizes geochemistry of SPM from the Karamana River compared with other rivers of India and abroad, other water bodies and standard values for rocks and sediments. Metal concentration in SPM of the Karamana River in comparison with that of the tentative threshold values as suggested by Aston and Thornton (1977) is presented in Table 2.3. Enrichment Factors (EF) computed for each metal in the SPM at Stations 2 to 10 are depicted in Table 2.4.

Table 2.1: SPM Content and Annual Flux of SPM in the Karamana River Water Compared to Indian and World Rivers

Name of River	*SPM (mg L^{-1})*	*Annual Flux (109 kg yr^{-1})*
Ganges	1250.0[r]	460.10[su2]
Brahmaputra	1370.0[r]	711.20[su2]
Godavari	122.0[r]	16.20[su2]
Krishna	130.0[r]	8.50[su2]
Narmada	10.0[r] (1154[b])	6.20[su2]
Tapti	13.0[r] (445[b])	2.70[su2]
Cauvery	48.0[r]	0.71[su2]
Netravati	54.0[sh]	1.40[kib]
Gurpur	52.0[sh]	0.10[kib]
Mahanadi	93.8-596[ry]	–
Krishna	600[sul]	–
Godavari	2000[sul]	–
Mulki-Pavanje	38[k]	–
Danube	20[g]	–
Global flux	500[h]	12000–13000[mm]
US rivers	400[t]	–
Karamana (present study)	15.68	0.02

h: Holeman (1968), t: Turekian (1969), r: Rao (1975), sul: Subramanian (1979), b: Borole *et al.* (1982), mm: Milliman and Meade (1983), ry: Ray *et al.* (1984), su2: Subramanian *et al.* (1985), kib: Karnataka Irrigation Department (1986), k: Karbassi (1989), sh: Shankar and Manjunatha (1997), g: Guieu *et al.*(1998).

Table 2.2: Geochemistry of Particulates from the Karamana River Compared with Other Indian and Foreign Rivers, Other Water-bodies, Rocks and Sediments

Sample Description	*Elements (μg g^{-1})*					*Fe (%)*
	Cu	*Cd*	*Pb*	*Zn*	*Mn*	*Fe*
World river SPM[mw]	100	1.20	35	250	1050	4.80
Average river particulate[mm]	100	1.00	150	350	1050	4.80
World river–borne detritus[ac]	2500	18.00	280	950	4300	2.20
Average shale[tw]	45	0.30	20	95	850	4.67
Upper continental crut[tm]	25	0.098	20	71	600	3.50
Average mud[tm]	50	–	20	85	850	5.10
Indian average[su]	28	–	–	16	605	2.90
Ganga river SPM[sa]	75	–	–	–	1000	7.97

Contd...

Table 2.2–Contd...

Sample Description	*Elements (µg g⁻¹)*			*Fe (%)*		
	Cu	*Cd*	*Pb*	*Zn*	*Mn*	*Fe*
Ganga river SPM[az]	252	–	37	643	3450	9.00
Brahmaputra river SPM[az]	108	–	–	916	522	10.90
Ganga and Brahmaputra rivers SPM[su]	19	–	–	46	522	2.53
South West coast rivers of India SPM[mj]	71	–	–	90	724	6.50
Krishna[az]	220	–	45	270	2540	13.20
Cauvery[az]	60	–	40	500	1300	6.20
Netravathi river SPM[sh]	75	–	43	95	884	6.40
Gurpur river SPM[sh]	70	–	57	86	546	6.59
Tapti river SPM[b]	139	–	–	157	1075	7.10
Tapti river SPM[az]	136	–	–	157	1304	7.90
Mulki–Pavanje river SPM[k]	88	–	55	324	884	7.50
Narmada river SPM[b]	133	–	–	143	1182	7.40
Narmada river SPM[az]	128	–	–	143	1125	7.60
Periyar and Muvattupuzha rivers and Cochin estuary SPM[sb]	20–300	8–60	–	60–1550	–	–
Karapad creek, Tuticorin (polluted) SPM[ch]	0.9–34.5	0.4–4.5	5.89–28.1	21.3–110	–	0.93–3.56
Major American rivers SPM[c]	–	–	–	–	1500	4.41
Mississipi[az]	38.5	–	39	193	1260	4.43
Amazon[az]	266	–	–	–	1033	5.55
Yellow river SPM[l]	33	–	35	75	800	3.20
Yangtze river SPM[l]	71	–	87	115	960	5.40
Huanghe river SPM[az]	26.7	–	16.5	69.8	767	3.72
Danube river SPM[g]	115	1.1	84	248	1704	3.60
Patuxent river (USA) SPM[f]	17–40	0.2–7.3	–	–	123–230	
Lena river (Russia) SPMm[g]	28	–	23	143	–	
Wisconsin rivers (USA) SPM[sop]	22.9	0.7	36.2	–	128	–
St. Lawrence[az]	40.2	–	36.2	342.7	1311	5.07
Orinoco[az]	60.8	–	–	76.5	588	7.40
Zaire[az]	100	–	220	300	1400	7.10
Taylor Creek, southern Nigeria[oo]	–	5.9	19	83	90	0.12
Karamana River SPM (present study)	212.91	155.96	464.01	623.58	957.13	5.66

tw: Turekian and Wedepohl (1961), ac: Aston and Chester (1976), mm: Martin and Meybeck (1979), sa: Sarin *et al.* (1979), b: Borole (1980), l: Li *et al.* (1984), su: Subramanian *et al.* (1985), tm: Taylor and McLennan (1985). k: Karbassi (1989), sb: Shibu *et al.* (1990), mw: Martin and Windom (1991), mg: Martin *et al.* (1993), c: Canfield (1997), sh: Shankar and Manjunatha (1997), g: Guieu *et al.* (1998), mj: Manjunatha *et al.* (1998), sop: Shafer *et al.* (1999), f: Riedel *et al.* (2000), ch: Chandrasekhar (2001), az: Alagarsamy and Zhang (2005), oo: Okafor and Opuene (2006).

Table 2.3: Metal Concentration in SPM of the Karamana River in Comparison with that of the Tentative Threshold Values (The concentration level in stream sediments at which it is likely that associated waters may, on occasion, exceed the highest desirable levels (HDL)-as suggested by Aston and Thornton, 1977)

Elements	*HDL (μg L^{-1})*	*Tentative Threshold Value*	*Metal in SPM of the Karamana River*
Cu	50	1000 μg g^{-1}	212.91 μg g^{-1}
Cd	10	10 μg g^{-1}	155.96 μg g^{-1}
Pb	50	500 μg g^{-1}	464.01 μg g^{-1}
Zn	5000	2000 μg g^{-1}	623.58 μg g^{-1}
Mn	500	1000 μg g^{-1}	957.13 μg g^{-1}
Fe	100	6 per cent	5.66 per cent

Table 2.4: Enrichment Factors (EF) Computed for Trace Metals in SPM at Stations 2 to 10 Against the Background Values at Station 1

Trace Metals	*Stations*								
	2	*3*	*4*	*5*	*6*	*7*	*8*	*9*	*10*
Cu	0.76	0.66	1.10	0.73	0.85	0.62	0.50	0.64	0.58
Cd	1.20	0.93	0.95	0.76	0.91	0.95	1.05	0.93	0.96
Pb	1.12	0.95	1.82	0.55	1.35	0.35	0.71	0.41	1.42
Zn	0.97	0.79	0.83	0.92	0.92	0.81	0.73	0.68	0.47
Mn	1.89	1.59	1.47	1.92	2.14	2.00	1.58	1.42	0.75
Fe	1.14	1.21	0.92	0.82	0.79	0.66	0.71	0.57	0.60

Transport of Suspended Particulate Matter (SPM)

Seasonal variations in SPM flux in the river is depicted in Figure 2.2. The maximum monthly SPM load was observed during May (24.91 mg L^{-1}), may be due to the impact of the first summer showers and the minimum was during February and March (8.50 mg L^{-1} and 9.95 mg L^{-1} respectively) characterized by slow flow rate of the river in summer. The maximum yearly SPM flux was registered at Stations 9 (20.87 mg L^{-1}), 7 (20.84 mg L^{-1}) and 5 (19.63 mg L^{-1}) owing to plankton growth within the dam at Station 5 and the combined effect of plankton growth and sand mining at Stations 7 and 9. The impact of tidal fluxes might also have been a contributing factor at Station 9. The minimum annual SPM load exhibited at Stations 1 and 2 (10.43 mg L^{-1}) may be attributed to the settling of SPM in the reservoir beyond Station 1.

Based on the annual discharge of water, the Karamana River was found to transport an annual average load of 15.68 mg L^{-1} SPM that was lower by a factor of 32 than the global average and by a factor of 25.5 than the average for US rivers. The SPM load was also much lower compared to Indian rivers the Ganges, Brahmaputra, Godavari, Krishna, Cauvery, Netravati, Gurpur, Mahanadi and Mulki-Pavanje but slightly higher than that of rivers Narmada and Tapti (Table 2.1). It could evidently be a positive index of the drinking quality of water and at the same time it may also indicate the poor status of the river as a potential transporter of nutrients and toxic elements. The river was found to contribute an yearly load of 20,786.8 metric tons of SPM to the Lakshadweep Sea that was much lower

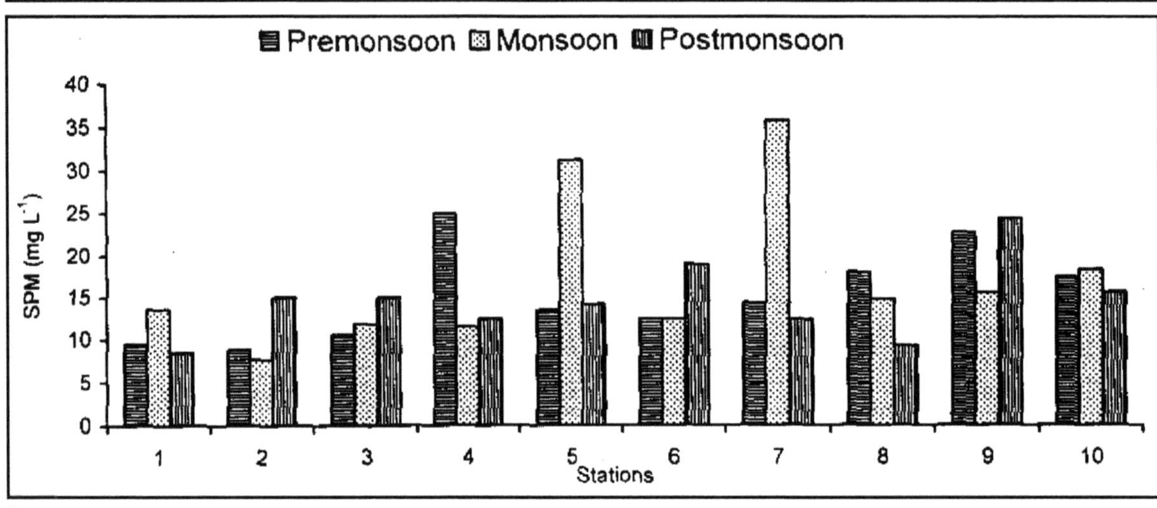

Figure 2.2: Seasonal Distribution of SPM in Water

than the annual flux of SPM by other Indian rivers. SPM transport in the river did not exhibit any significant season-wise and station-wise variations, but month-wise variations were significant (ANOVA: $df = 11, F = 2.378, P \leq 0.012$) indicating that the SPM flux in the river may be independent of spatial and temporal environmental attributes other than month-wise variations. Short residence time due to fast flow rate may account for the absence of significant station-wise variations. Total SPM load showed weak, yet significant negative correlations with particulate metals iron ($r = -0.322, y = 21.184 - 0.0001x, p \leq 0.0003, n = 120$), manganese ($r = -0.332, Y = 20.091 - 0.005x, P \leq 0.0002, n = 120$), zinc ($r = -0.393, y = 24.424 - 0.140x, P \leq 0.00001, n = 120$), cadmium ($r = -0.326, Y = 18.432 - 0.018x, P \leq 0.0003, n = 120$) and lead ($r = -0.219, Y = 17.274 - 0.0003x, P \leq 0.02, n = 120$). The apparent negative relationships may be due to the normal component of small sized metal-rich particulates occasionally getting mixed up with large sized metal poor sediment fractions by resuspension and advection resulting in higher rates of particulate flux. The inverse relationship between particle size and metal content is well documented (Forstner and Wittmann, 1979; Salomons and Forstner, 1984; Horowitz, 1988; Combest, 1991; Verma *et al.*, 1993: Bertin and Bourg, 1995; Mohan and Hosetti, 1997). No correlation was evident in the case of copper suggesting an organic association of particulate copper as reported elsewhere (Hasle and Abdullah, 1981; Shibu *et al.*, 1990; Pardo *et al.*, 1990). Significant correlations were not evident in the relationship of SPM with hydrographical parameters (Jayaraman *et al.*, 2003) but for the weak, yet significant positive correlations with nitrate-nitrogen ($r = 0.305, Y = 11.601 + 22.460x, P \leq 0.0007, n = 120$) and silicate-silicon ($r = 0.244, Y = 9.620 + 1.706x, p \leq 0.007, n = 120$) and it may be due to the dynamic nature of the lotic system. The weak correlation with nitrate–nitrogen and silicate–silicon may be attributed to terrigenous land run-off preceding particulate flux.

Particulate Transport of Trace Metals

Colloidal surfaces are greater in fluvial particulates as compared to bottom sediments (Hart, 1982) and hence particulates may have a greater role in physico-chemical sorption reactions. Sedimentation of enriched particulate matter is a potentially important mechanism by which sediments may concentrate trace metals. It is well established that transport of elements in rivers, except that of mercury, is through solid phase rather than in dissolved form (Blachford and Ongley, 1984). The ability to adsorb elements onto SPM depends on surface charges and it was investigated by several workers (Gibbs, 1973; Aston and Chester, 1973; Bowares and Huang, 1987).

Copper

Most copper minerals are of low solubility as it is strongly immobilized in the crystal lattices of mineral micelles (Burton, 1976) and hence it is transported dominantly or exclusively by SPM. Only 5.1 per cent of total copper flux happens in dissolved form as reported for Amazon and Yukon (Gibbs, 1973) and only 30 per cent is leachable into dissolved phase (Duursma, 1976).

Seasonal distribution of particulate copper in the river is presented in Figure 2.3. Analysis of the monthly data at various stations showing the highest value of SPM associated copper during July (479.29 µg g^{-1}) may be due to the influence of monsoon rains, whereas the lowest was during December (59.41 µg g^{-1}) owing to the reduced rate of water discharge. The maximum copper content at Stations 1, 2, 3, 7, 8 and 9 was experienced during monsoon, while at Stations 4, 5, 6 and 10 the maximum was during pre-monsoon. The pre-monsoon and monsoon seasonal maxima may suggest the dominant role of particulate adsorption and terrigenous inputs respectively. The maximum annual flux of particulate copper was at Station 4 (315.25 µg g^{-1}), whereas Station 8 (141.73 µg g^{-1}) recorded the least. A slightly higher rate was observed at Station 4 followed by Stations 1, 6 and 2 which may suggest the role of pesticides and fungicides used in rubber plantations.

The annual average rate of particulate copper transport was found to be 212.91µg g^{-1} with an annual load of 4.43 metric tons. The annual average rate was higher by a factor of two than that of world river SPM and average river particulate, twelve times lower than that of world river-borne detritus, five times higher than that of baseline levels in average shale, more than eight times higher than that of upper continental crust and four times higher than that of average mud. It was also much higher in comparison with Indian average and of most Indian rivers and other aquatic systems (Table 2.2). Copper content in SPM was five times lower than that of the tentative threshold value of 1000 µg g^{-1} at which the related water may exceed the highest desirable level of 50 µg L^{-1} (Table 2.3). The seasonal and annual average station-wise distribution was without any pattern or trend that ruled out the possibility of any point source for particulate copper. The Enrichment Factor (EF) computed for particulate copper on an annual cycle remained less than unity and did not indicate any appreciable degree of enrichment and this may be due to removal under the impact of terrigenous leachates (Table

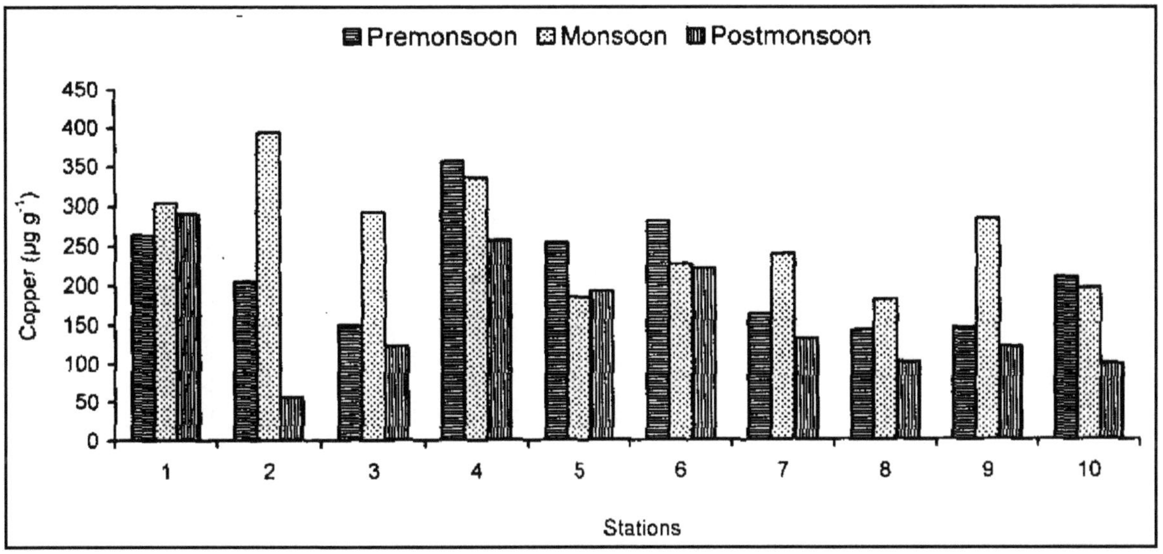

Figure 2.3: Seasonal Distribution of Particulate Copper

2.4). Hence, it is reasonable to state that the water of the Karamana River did not pose any threatening level of pollution with respect to particulate copper.

The distribution of SPM-associated copper showed statistically significant temporal variations (ANOVA: month-wise –df= 11, F = 8.560, $p \leq$ 2E-10; season-wise df = 2, F = 7.814, $p \leq$ 0.004), while significant station-wise variations were evident in season-wise distribution alone (ANOVA: df = 9, F = 2.555, $P \leq 0.04$) reflecting the extent of the influence of spatial and temporal environmental variables. Particulate copper showed low, but significant positive correlations with other metals, *viz.*, zinc (r = 0.439, $y = 55.510 + 0.253x$, $p \leq$ 1E-06, n = 120), cadmium (r = 0.420, $Y = 156.031 + 0.366x$, $P \leq$ 1.76E-06, n = 120) and lead (r = 0.181, $Y = 192.261 + 0.045x$, $p \leq 05$, n = 120), but such affinities were not evident in the case of iron and manganese. Copper in SPM of the river may at least partially be associated with zinc, cadmium and lead suggesting a mineralogical and or anthropogenic origin. A sympathetic association of copper with organic matter in the sediments of the Karamana River was substantiated by its positive correlation with SOC, carbohydrates, proteins and lipids (Jayaraman, 2003) suggesting an organic association of particulate copper as reported elsewhere (Hasle and Abdullah, 1981; Shibu *et al.*, 1990; Pardo *et al.*, 1990). Hence, it is reasonable to hypothesize that copper content in the SPM of the Karamana River may be partly anthropogenic (as copper fungicides applied in the plantations and farmlands in the catchment area) and partly mineralogical in origin. Regression analysis revealed that hydrological parameters did not exert any significant influence on SPM-associated copper transport in the river.

Cadmium

Seasonal pattern of particulate flux of cadmium is depicted in Figure 2.4. Month-wise data revealed the maximum particulate load of cadmium during March (548.03 µg g^{-1}) and the minimum during December (0.12 µg g^{-1}). Season-wise analysis showed a uniform pre-monsoon > monsoon > post-monsoon pattern at all the 10 stations that clearly indicated a pre-monsoon enrichment of cadmium by adsorption onto SPM and a subsequent desorption into dissolved phase under the impact of terrigenous bioorganic leaching agents as opined by Gardiner (1974) and Duursma (1976) during monsoon and post-monsoon.

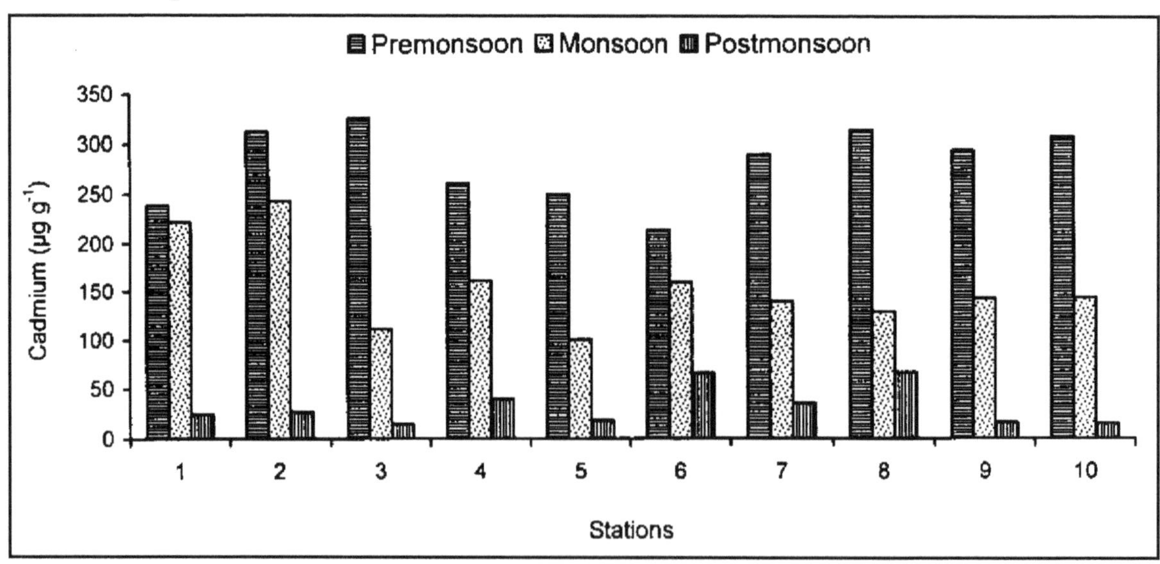

Figure 2.4: Seasonal Distribution of Particulate Cadmium

An annual average load of 155.96 µg g^{-1} with an yearly load of 3.24 metric tons of particulate flux of cadmium was observed in the present study. The annual particulate cadmium load estimated was 130 times higher than that of world river SPM, 156 times greater than average value for river particulate; approximately 9 times greater than the average value for river borne detritus, 520 times greater than the baseline level for average shale and 1591 times higher than that for upper continental crust (Table 2.2). The values were also much higher in comparison with most of the reported cases. Cadmium content in SPM was more than 15 times as high as the tentative threshold value of 10µg g^{-1} at which the related water may exceed the highest desirable level of 10 µg L^{-1} (Table 2.3). All these findings are indicative of high level of particulate cadmium in the river. Enrichment Factor (EF) computed for particulate cadmium revealed slight enrichment at a few stations during pre-monsoon and post-monsoon, however on an annual basis there was no discernable incidence of enrichment, which is likely to be under the influence of terrigenous leachates (Table 2.4).

The distribution of cadmium in the riverine particulates showed significant month-wise ($df = 11$, $F = 90.551$, $p \leq$ E-46) and season-wise (ANOVA: $df = 2$, $F = 107.111$, $p \leq 1E\text{-}10$) variations, but station-wise variations were not significant during all the three seasons that indicated non-point particulate cadmium input into the lotic ecosystem. Cadmium content in SPM correlated significantly, but with small coefficients to other metal elements such as iron ($r = 0.193$, $y = 94.673 + 0.001x$, $p \leq 0.04$, $n = 120$), manganese ($r = 0.273$, $Y = 88.812 + 0.070x$, $p \leq 0.003$, $n = 120$), zinc ($r = 0.299$, $y = 36.462 + 0.193x$, $p \leq 0.001$, $n = 120$), lead ($r = 0.251$, $Y = 122.430 + 0.072x$, $p \leq 0.006$, $n = 120$) and copper ($r = 0.420$, $y = 52.824 + 0.483x$, $p \leq 1.7E\text{-}06$, $n = 120$) indicating very low level of affinity. It also showed significant yet weak correlation with BOD ($r = 0.322$, $Y = 90.489 + 2.173x$, $p \leq 0.0003$, $n = 120$), which may be an indirect indication of organic association. It exhibited small coefficients of negative correlation with nitrate-nitrogen ($r = -0.309$, $y = 232.873 - 422.017x$, $p \leq 0.0006$, $n = 120$) and silicate-silicon ($r = -0.369$, $y = 323.68 - 46.068x$, $p \leq 4E\text{-}05$, $n = 120$), may be due to a negative relationship with land runoff from agricultural areas well reflecting the seasonal behaviour of particulate cadmium. The absence of strong and clear correlations experienced in the present study may indicate the possibility of several types of geochemical support phases and or the chance for mixed associations that are in close agreement with that of the findings by Bertin and Bourg (1995) in Lot River system in France. Significant relationships not evident in the case of other hydrographical parameters may be due to the short residence time and fast flow rate.

Although the particulate level of cadmium was very high, the present data clearly indicated that the amount of cadmium available for biological cycling in the Karamana River has not reached any alarming level due to the very low particulate matter content in the river. But it deserves serious attention as a potential contaminant in the years to come.

Lead

The seasonal distribution profile of particulate lead in the Karamana River is presented in Figure 2.5. Data on SPM-associated lead revealed the maximum flux during March (1142.48 µg g^{-1}) followed by January (762.92 µg g^{-1}) that may indicate pre-monsoon accumulation due to phytoplankton growth. Minimum month-wise values observed during September (82.52 µg g^{-1}) and October (136.61 µg g^{-1}) may be under the impact of flood borne organic leaching agents.

The annual average rate of particulate lead transport was found to be 464.01 µg g^{-1} with an yearly flux of 9.64 metric tons that was about 13.25 times greater than that of world river SPM, 3.09 times higher than that of average river particulate, 1.66 times greater than the world average lead content for riverine detritus and 23 times greater than baseline levels in average shale, upper continental crust

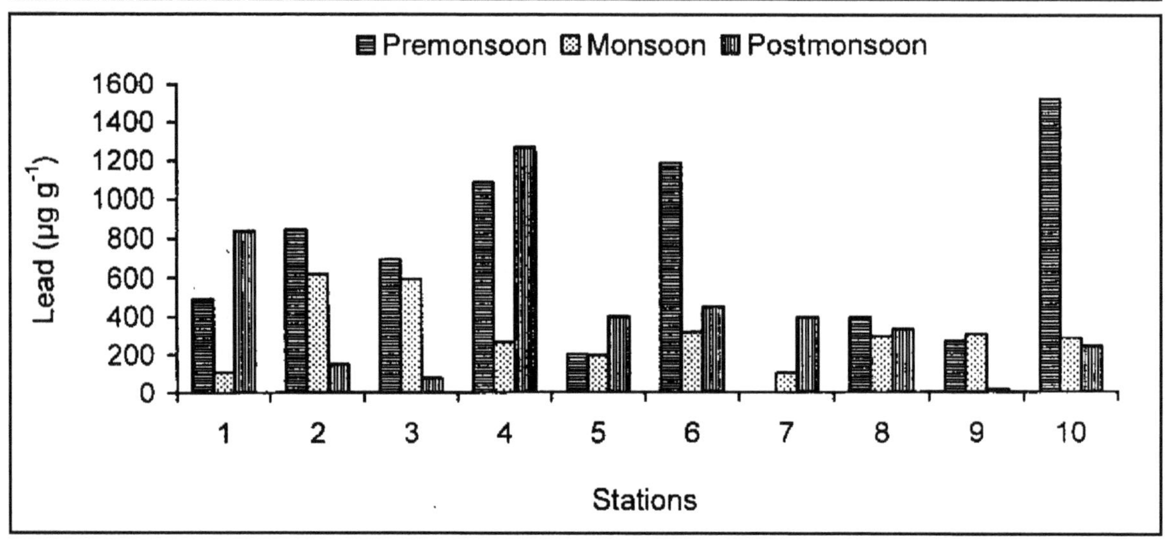

Figure 2.5: Seasonal Distribution of Particulate Lead

and average mud. It was also much higher in comparison with most of Indian and world rivers (Table 2.2). The particulate load of lead was slightly higher than that of the tentative threshold value (500 µg g^{-1}) corresponding to the highest desirable level of 50 µg L^{-1} (Table 2.3). Enrichment Factor (EF) computed for SPM-associated lead revealed mild enrichment at various stations especially during pre-monsoon and monsoon indicating the input of terrigenous particulates (Table 2.4).

Significant month-wise, season-wise and station-wise variations were not evident in the particulate transport of lead in the Karamana River suggesting non-point source, possibly air borne particulates from automobile exhausts from Thiruvananthapuram metropolitan area that is brought in by surface runoff as recorded elsewhere (Knoppers *et al.*, 1990; Bertin and Bourg, 1995). Significant, yet weak positive correlations were evident in the relationship of lead with iron ($r = 0.217$, $Y = 223.765 + 0.004x$, $p \leq 0.02$, $n = 120$), copper ($r = 0.0.181$, $Y = 309.036 + 0.0.728x$, $p \leq 0.05$, $n = 120$) and cadmium ($r = 0.251$, $Y = 337.606 + 0.0.876$, $p \leq 0.006$, $n = 120$) indicating weak mineralogical affinity with these metals. The results were also indicative of involvement of several forms of geochemical support phases and/or mixed associations (Bertin and Bourg, 1995). Hydrographical parameters except nitrate nitrogen ($r = -0.238$, $Y = 671.38–1134.53x$, $p \leq 0.01$, $n = 120$) were found to be less influential on the disposition of sedimentary lead owing to the dynamic nature of the river. Hence, it seems reasonable to suggest an anthropogenic input of SPM-associated lead into the riverine environment rather than a mineralogical origin. Transport of particulate lead from automobile exhaust through surface runoff might be a significant factor in this respect.

Although particulate lead concentration was higher than the standard levels for SPM, as the total flux of particulate matter is much less in the river water it can be concluded that lead concentration has not reached any alarming level in the river yet. However, the possibility of reaching higher levels due to anthropogenic influences in the near future can never be overlooked.

Zinc

Seasonal distribution profile of SPM associated Zinc at Stations 1 to 10 is presented in Figure 2.6. Month-wise data showed a maximum load of particulate zinc during January (1050. 60 µg g^{-1}) and the

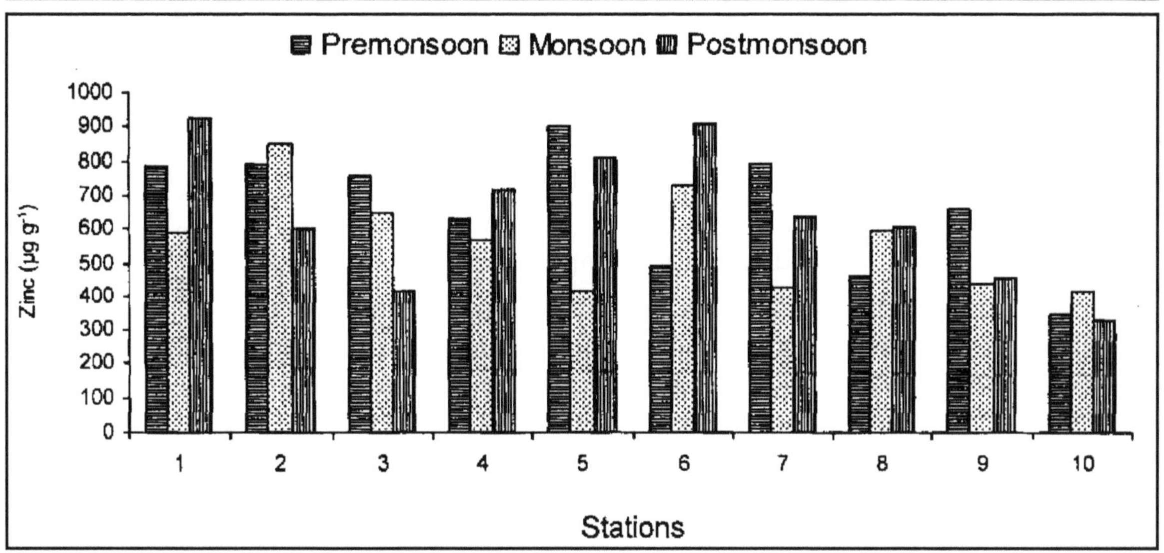

Figure 2.6: Seasonal Distribution of Particulate Zinc

minimum during October (378.43 µg g^{-1}). Particulate flux of zinc was generally low at Station 10 with a Station 9 > Station 10 pattern owing to salinity dependent desorption as suggested by Evans and Cutshall (1973).

Particulate transport of zinc in the river was found to be at a rate of 623.58 µg g^{-1} with an annual load of 12.96 metric tons. The average rate was about 2.5 times greater than that of world river SPM, 4.16 times higher than that of average river particulate, 2/3rd of the world river-borne detritus, 6.6 times greater than zinc content in average shale, 8.8 times higher than that of upper continental crust, 7.3 times greater than that of average mud, while it remained 39 times higher than the Indian average for riverine SPM. It was generally higher in comparison with most of the Indian and world rivers, but within the range of values recorded for Periyar and Muvattupuzha Rivers and Cochin estuary (Table 2.2). Besides, it was about 3.2 times lower than the tentative threshold value of 2000 µg g^{-1} at which the related water may exceed the highest desirable level of 5000 µg L^{-1} (Table 2.3). Enrichment Factor (EF) computed for SPM-associated zinc did not indicate any appreciable extent of enrichment in particulate phase as the values mainly remained below unity which is suggestive of mobilization into dissolved phase under the influence of river-bome leachates (Table 2.4).

SPM-associated zinc in the riverine system showed significant month-wise (ANOVA: $df = 11, F = 4.167, p \leq$ IE-05) and station-wise (ANOVA: $df = 9, F = 2.235, p \leq 0.03$) variations in its distribution, but seasonal variations were not significant reflecting the extent of the influence of spatial and temporal environmental variables. Regression analysis showed that particulate zinc flux in the river had significant positive correlations with iron ($r = 0.236$, $Y == 510.497 + 00.002x$, $p \leq 0.0005$, $n = 120$), manganese ($r = 0.360$, $Y = 489.867 + 0.139x$, $p \leq 0.0001$, $n = 120$), copper ($r = 0.439$, $Y = 460.985 + 0.764x$, $p \leq$ 5.25E-07, $n = 120$) and cadmium ($r = 0.289$, $Y = 555.457 + 0.438x$, $p \leq 0.001$, $n = 120$), however the coefficients were low. The relationship with iron and manganese is consistent with earlier observations on the scavenging properties of iron and manganese oxides in metal ion accumulation. The relationship with copper and cadmium may suggest mineralogical association. It showed negative correlation with low coefficients to hydrographical parameters, *viz.*, COD ($r = -0.213$, $Y = 652.703 - 0.445x$, $p \leq$

0.02, $n = 120$), nitrite-nitrogen ($r = -0.260$, $Y = 675.224 - 1990.623x$, $p \leq 0.004$, $n = 120$), nitrate-nitrogen ($r = -0.268$, $Y = 725.066 - 555.202x$, $p \leq 0.003$, $n = 120$) and hardness ($r = -0.209$, $Y = 644.832 - 0.267x$, $p \leq 0.02$, $n = 120$) indicating their significant roles in particulate zinc transport. Other water parameters were without any significant relationships. Results of correlation studies revealed a dominant mineralogical association of particulate zinc.

Although the SPM associated zinc content in the river was generally higher than that of most of the reported cases, considering the low rate of total particulate flux in the river it can reasonably be concluded that there is no immediate threat of contamination by particulate zinc.

Manganese

Seasonal variation in SPM associated manganese in the river is depicted in Figure 2.7. Monthly data on particulate manganese content at various stations showed the highest value during April (1392.74 µg g^{-1}) and the lowest during December (374.62 µg g^{-1}).

Particulate manganese in the Karamana River was found to be mobilized at a rate of 957.13 µg g^{-1} with an annual transport of 19.89 metric tons. The observed mean value of SPM associated manganese was slightly lower than that of world river SPM and average river particulate, about 4.5 times lower than the average for world river borne detritus and higher than that of average shale, upper continental crust, average mud and average for Indian rivers. The particulate manganese load was higher than or within the range of the reported values for many of the Indian and world rivers. However, the values remained lower than that for major American rivers and Danube River (Table 2.2). Manganese content in SPM was slightly lower than the tentative threshold value of 1000 µg g^{-1} at which the related water may exceed the highest desirable level of 500 µg L^{-1} (Table 2.3). Enrichment Factor (EF) computed for manganese revealing incidence of enrichment at various stations may be due to adsorption on hydroxy coatings on sediment grains (Gibbs, 1977; Wilbur and Hunter, 1979; Tessier *et al.*, 1982; Brook and Moore, 1988) in the conducive environment during pre-monsoon. The slight enrichment recorded during monsoon and post-monsoon may be under the impact of floodwater.

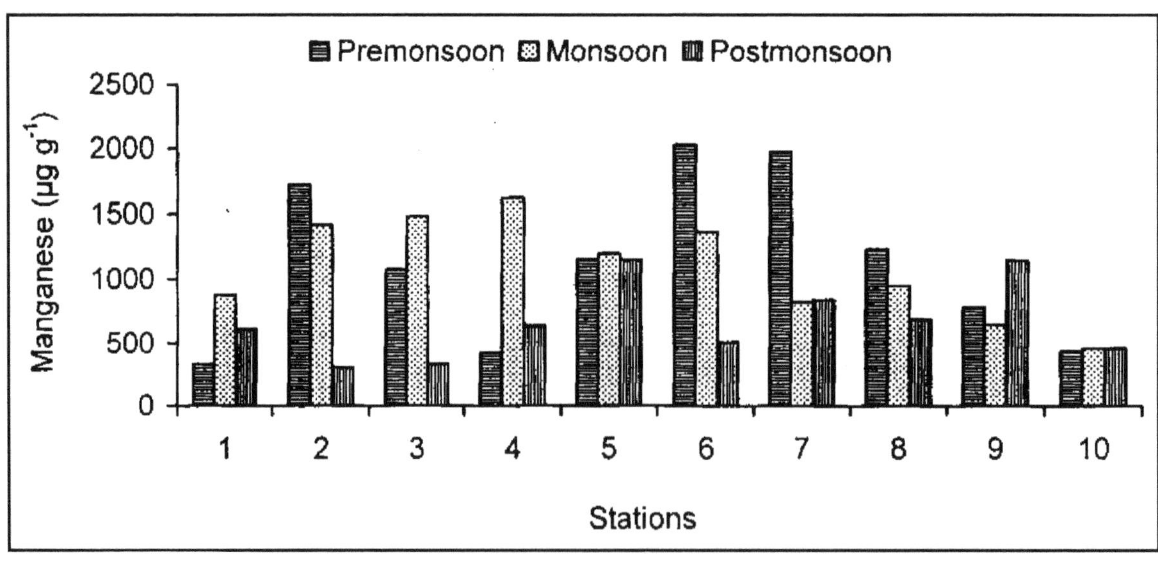

Figure 2.7: Seasonal Distribution of Particulate Manganese

In the present study, particulate manganese showed statistically significant month-wise variations (ANOVA: $df = 9$, $F = 2.427$, $P \leq 0.01$) but season-wise as well as station-wise variations were not significant reflecting the limited influence of spatial and temporal environmental variables. Particulate manganese showed significant positive correlations with low coefficients to various factors such as iron ($r = 0.262$, $Y = 634.171 + 0.006x$, $p \leq 0.004$, $n = 120$), zinc ($r = 0.360$, $Y = 377.782 + 0.929x$, $p \leq 5.3E\text{-}05$, $n = 120$), cadmium ($r = 0.273$, $Y = 791.52 + 0.064x$, $p \leq 0.0003$, $n = 120$), pH ($r = 0.189$, $y = -2357.86 + 482.39x$, $p \leq 0.04$, $n = 120$) and silicate-silicon ($r = -0.199$, $Y = 1317.95 - 101.114x$, $p \leq 0.03$, $n = 120$). The apparent weak correlations with iron, zinc and cadmium suggested the possibility of several types of geochemical support and or mixed associations (Bertin and Bourg, 1995). The weak relationships with pH and silicate-silicon showed at least a partial involvement of these hydrographical parameters in the mobilization of particulate manganese. The other metals and hydrographical parameters did not show any significant relationships. Results of regression analysis suggested a dominantly mineralogical association of SPM associated manganese.

Considering the very low level of SPM flux in the river, it could reasonably be concluded that for the time being there is no immediate threat of contamination by particulate manganese in the Karamana River.

Iron

Seasonal variations in SPM associated iron flux in the Karamana River is depicted in Figure 2.8. Month-wise data revealed maximum particulate iron content during March (111,136 µg g^{-1}) and minimum during February (36,375 µg g^{-1}). On an annual basis, maximum SPM-associated iron was registered at Station 3 (81,566 µg g^{-1}) followed by Stations 2 (76,945 µg g^{-1}) and 1 (67,223 µg g^{-1}), whereas the minimum was recorded at Station 9 (36,180 µg g^{-1}) and at Station 10 (40,067 µg g^{-1}) showing a decreasing trend towards downstream reflecting the increasing influence of organic leaching agents (Duursma, 1976) and salinity dependent desorption at the last two stations (Evans and Cutshall, 1973).

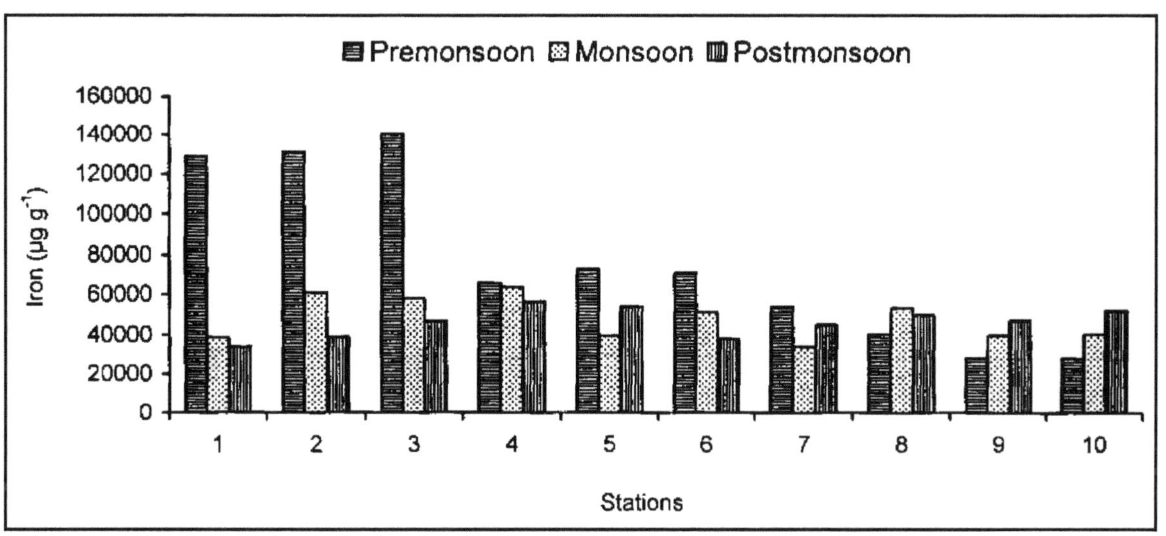

Figure 2.8: Seasonal Distribution of Particulate Iron

Particulate iron flux in the river was found to be at a rate of 56,633 $\mu g\ g^{-1}$ (5.66 per cent) with an annual load of 1177.22 metric tons. The average rate of particulate iron flux was found higher in comparison with world river SPM, world river-borne detritus, average river particulate, upper continental crust, average mud, average shale, and the average for Indian rivers. The values generally remained higher in comparison with most of the rivers abroad, but lower than those of the rivers in the south west coast of India and River Ganga (Table 2.2). Particulate iron content in the Karamana River was slightly lower when compared to the tentative threshold value (6 per cent) at which the related water may exceed the highest desirable level of 100$\mu g\ L^{-1}$ (Table2. 3). Enrichment Factor (EF) computed for SPM-associated iron revealed slight enrichment at various stations during monsoon and post-monsoon, however on an annual basis there was no discernable incidence of enrichment, except at Stations 2 and 3 (Table 2.4). Terrigenous organic leachates might have played a significant role in this respect.

Statistically significant station-wise (ANOVA: $df = 9, F = 2.519, p \leq 0.01$), month-wise (ANOVA: $df = 11, F = 4.563, p \leq 1E\text{-}05$) and season-wise variations (ANOVA: $df = 2, F = 4.251, p \leq 0.03$) were evident in the distribution of SPM associated iron in the riverine environment reflecting the influence of temporal variables supplemented by local disturbances such as sand mining and illegal prospecting for gemstones. Particulate iron showed weak yet significant positive correlations with lead ($r = 0.217$, $Y = 51502.51 + 11.128x, p \leq 0.02, n = 120$), cadmium ($r = 0.193, y = 51291.42 + 34.511x, p \leq 0.03, n = 120$), zinc ($r = 0.236$, $Y = 39242.27 + 27.941x$, $p \leq 0.01$, $n = 120$) and manganese ($r = 0.262$, $Y = 45180.74 + 11.999x, p \leq 0.004, n = 120$) indicating weak mineralogical affinity with these metals or the involvement of several forms of geochemical support phases and or mixed associations as suggested by Bertin and Bourg (1995). Among the various hydrographical parameters studied, significant relationships were evident only in the case of nitrate-nitrogen ($r = -0.207$, $y = 61524.5 - 18734.7x$, $p \leq 0.03$, $n = 120$). Hydrographical parameters may have only a limited influence on the disposition of SPM associated iron owing to the dynamic nature of the river. Results of correlation studies suggested a dominantly mineralogical association of particulate iron.

Although the SPM load of iron was higher than world standards considering the very low level of SPM flux in the river, it can very well be pointed out that there is no immediate threat of contamination by particulate iron in the Karamana River.

Conclusions

The present study provides a relatively comprehensive set of measurements of the concentrations of selected particulate trace metals in the Karamana River. The flux of total SPM was very low in the river as compared to reference standards and world averages. The spatial and temporal variations noticed in the composition of SPM maybe due to the influence of seasonal changes and local conditions. Co-transportation of trace elements in different combinations during different seasons was noticed. Particulate flux of all the six metals was found higher than that of standard values for average shale and Indian rivers, whereas a slightly lower value was noticed in the case of manganese in comparison with world river SPM and average river particulate, while the particulate flux of copper, zinc and manganese remained lower in comparison with world river-borne detritus. The particulate load of manganese was slightly higher and that of cadmium was much higher as compared to the tentative threshold values. Possibility of leaching of heavy metals into the lotic ecosystem originating from the plantations along with other anthropogenic activities can not be overlooked. Toxic metals, *viz.*, cadmium and lead detected in significant quantities pose great concern as they cause deleterious effects on biological systems.

Acknowledgement

The authors are thankful to the University of Kerala and Principal, University College, Thiruvananthapuram, Kerala for the facilities provided.

References

Alagarsamy, R. and Zhang, J., 2005. Comparative studies on trace metal geochemistry in Indian and Chinese rivers. *Current Sci.*, 89: 299–309.

Allen, R.J., 1979. Sediment related fluvial transmission of contaminants: Some advances in 1979. Scientific Series No. 107. *Environment Canada*, Ottawa, pp. 24.

APHA (American Public Health Association), 1992. *Standard Methods for the Examination of Water and Wastewater*, 18th edn. APHA, AWWA, WPCF, pp. 1134.

Aston, S.R. and Chester, R., 1976. Estuarine sedimentary process. In: *Estuarine Chemistry*, (Eds.) Burton, J.D. and Liss, P.S. Academic Press Inc. (London) Ltd., pp. 37–91.

Aston, S.R. and Thornton, I., 1977. Regional geochemical data in relation to seasonal variations in water quality. *Sci. Total Environ.*, 7: 247–260.

Bertin, C. and Bourg, A.C.M., 1995. Trends in the heavy metal content (Cd, Pb, Zn) in the drainage basin of smelting activities. *Wat. Res.*, 29: 1729–1736.

Biksham, G. and Subramanian, V., 1988. Elemental composition of Godavari sediments (Central and southern Indian subcontinent). *Chem. Geo.*, 70: 275–286.

Blachford, D.P. and Oangley, E.D., 1984. Biogeochemical pathways of phosphorus, heavy metals and organochlorine residues in the Bow and Oldman rivers, Alberta. Inland Water Directorate (Canada) Scientific Series No. 138, pp. 1980–81:

Borole, D.V., 1980. Radiometric and trace elemental investigations on Indian estuaries and adjacent seas. *Ph.D. Thesis* (Unpubl.), Gujarat University, pp. 155.

Borole, D.V., Sarin, M.M. and Somayajulu, B.L.K., 1982. Composition of Narmada and Tapti estuarine particles and adjacent Arabian Sea sediments. *Indian J. Mar. Sci.*, 11: 51–62.

Bowares, A.R. and Huang, C.P., 1987. Role of Fe (III) in metal complex absorption by hydrous solids. *Wat. Res.*, 21: 757–764.

Brook, E.J. and Moore, J.N., 1988. Particle size and chemical control of As, Cd, Cu, Fe, Mn, Ni, Pb and Zn in bed sediments from the Clark Fork River, Montana (U.S.A.). *Sci. Total Environ.*, 76: 247–266.

Burton, J.D., 1976. Basic properties and processes in estuarine chemistry. In: *Estuarine Chemistry*, (Eds.) Burton, J.D. and Liss, P.S. Academic Press Inc. (London) Ltd., pp. 1–36.

Canfield, D.E., 1997. The geochemistry of river particulates from the continental USA: Major elements. *Geochem. et Cosmochim. Acta*, 61: 3349–3365.

Chandrasekhar, N., 2001. Trace elements in the suspended sediments of salt marsh area of Karapad creek, Tuticorin. *Indian J. Environ. and Ecoplan.*, 5: 81–86.

Combest, K.B., 1991. Trace metals in sediment: Spatial trends and sorption processes. *Wat. Res. Bul.*, 27: 19–28.

Duursma, E.K., 1976. Radioactive tracers in estuarine chemical studies. In: *Estuarine Chemistry*, (Eds.) Burton, J.D. and Liss, P.S. Academic Press Inc. (London) Ltd., pp. 159–181.

Evans, D.W. and Cutshall, N.H., 1973. Effects of ocean water on the soluble-suspended distributions of Columbia River radionucleids. In: *Radioactive Contamination of the Marine Environment*. International Atomic Energy Agency, Vienna, pp. 125–140.

Forstner, U. and Wittman, G.T.W., 1983. *Metal Pollution in the Aquatic Environment*, 2nd Revised Edn. Springer-Verlag, Haeidelberg, pp. 486.

Gardiner, J., 1974. The chemistry of cadmium in natural water–I. A study of cadmium complex function using the cadmium specific-ion electrode. *Wat. Res.*, 8: 23–30.

Gibbs, R.J., 1973. Mechanisms of trace metal transport in rivers. *Science*, 180: 71–73.

Gibbs, R.J., 1977. Transport phases of transition metals in the Amazon and Yukon Rivers. *Geol. Soc. Am. Bull.*, 88: 829–843.

Guieu, C., Martin, J.M., Tankere, S.P.C., Mousty, F., Trincherini, P., Bazot, M. and Dai, M.H., 1998. The trace metal geochemistry in the Danube river and western Black sea. *Estuar. Coast. Shelf Sci.*, 47: 471–485.

Hart, B.T., 1982. Uptake of trace metals by sediments and suspended particulates: A review. *Hydrobiologia*, 91: 299–313.

Hasle, J.R. and Abdullah, M.I., 1981. Analytical fraction of dissolved copper, lead and cadmium in coastal seawater. *Mar. Chem.*, 10: 487–503.

Hawkes, H.E. and Webb, J.S. 1962. *Geochemistry in Mineral Exploration*. Harper and Row, New York, pp. 415.

Holeman, J.N., 1968. The sediment yield of major rivers of the world. *Wat. Resour. Res.*, 4: 737–747.

Horovitz, A., 1988. A review of physical and chemical partitioning of inorganic constituents in sediments. In: *The Role of Sediments in the Chemistry of Aquatic Systems. Proc. Sediment Chemistry Workshop*, (Eds.) Horowitz, A. J. and W.L. Bradford. US Geological Survey, pp. 969.

Horowitz, A.J., 1996. Spatial and temporal variations in suspended sediment and associated trace elements-requirement for sampling, data interpretation and determination of annual mass transport. In: *Suspended Particulate Matter in Rivers and Estuaries, Proceedings of an International Symposium*, (Eds.) Kausch, H. and Michaeils, W. Stuttgart, FRG. Schweizerbart' Sche Verlagsbuchhandlung. 47: 515–536.

Jain, C.K. and Sharma, M.K., 2006. Heavy metal transport in Hindon river basin, India. *Environ. Monit. Assmt.*, 112: 255–270.

Jayaraman, P.R., 2003. Impact of environmental factors on the macroflora of a lotic ecosystem–Karam ana River: A case study. *Ph.D. Thesis*, Department of Aquatic Biology and Fisheries, Kerala University, Thiruvananthapuram, pp. 405.

Jayaraman, P.R., Gangadevi, T. and Nayar, T.V., 2003. Water quality studies on Karamana River, Thiruvananthapuram District, South Kerala, India. *Poll. Res.*, 22: 89–100.

Karbassi, A.R., 1989. Geochemical and magnetic studies of marine, estuarine and riverine sediments near Mulki (Karnataka), India. *Ph.D. Thesis*, (Unpubl.), Mangalore University, Mangalore, pp. 196.

Karnataka Irrigation Department, 1986. River discharge data. Unpubl. No. 5. Gayuging Sub-division, Shimoga, India.

Kerala State Land Use Board, Thiruvananthapuram, 1995. *Land Resources of Kerala*, pp. 150.

Knoppers, B.A., Lacerda, L.D. and Pachineelam, S.R., 1990. Nutrients, heavy metals and organic micropollutants in an eutrophic Brazilian lagoon. *Mar. Poll. Bull.*, 21: 381–384.

Li, Y.H., Teraoka, H., Yang, T.S and Chen, J.S., 1984. The elemental composition of suspended particles from the Yellow and Yangtze rivers. *Geochim. Cosmochim. Acta*, 48: 1561–1564.

Manjunatha, B.R., Yeats, P.A., Smith, J.N., Shankar, R, Narayana, A.C. and Prakash, T.N., 1998. Acumulation of heavy metals in sediments of marine environments along the southwest coast of India. *Extended Synopses, International Symposium on Marine Pollution, Monaco*, International Atomic Energy Agency, pp. 93–94.

Martin, J.M. and Meybeck, M., 1979. Elemental mass balance of materials carried by major world rivers. *Mar. Chem.*, 7: 173–206.

Martin, J.M., Guan, D.M., Elbaz-Poulichet, F., Thomas, A.J. and Gordeev, V.V., 1993. Preliminary assessment of the distribution of some trace elements (As, Cd, Cu, Fe, Ni, Pb, and Zn) in a pristine aquatic environment: The Lina River Estuary (Russia). *Mar. Chem.*, 43: 185–199.

Martin, J.M. and Windom, H.L., 1991. Present and future roles of ocean margins in regulating marine biogeochemical cycles of trace elements. In: *Ocean Margin Processes in Global Change*, (Eds.) Mantoura, R.F.C., Martin, J.M. and Wollost, R. Wiley, New York, pp. 45–67.

Milliman, J.D. and Meade, R.H., 1983. Worldwide delivery of river sediments to the oceans. *J. Geology*, 19: 1–22.

Mohan, B.S. and Hosetti, B.B., 1997. Potential phytotoxicity of lead and cadmium to *Lemna minor* grown in sewage stabilization ponds. *Environ. Poll.*, 98(2): 233–238.

Okafor, E.C. and Opuene, K., 2006. Correlations, partitioning and bioaccumulation of trace metals between different segments of Taylor Creek, southern Nigeria. *Int. J. Env. Sci. and Tech.*, 3: 381–389.

Pardo, R., Barrado, E., Perez, L. and. Vega, M., 1990. Determination and speciation of heavy metals in the sediments of the Pisuerga River. *Water Res.*, 24: 373–379.

Paul, A.C. and Pillai, K.C., 1983. Trace metals in a tropical river environment Distribution. *Water, Air, Soil Poll.*, 19: 63–73.

Rao, K.L., 1975. *India's Water Wealth*. Oxford University Press, Delhi, pp. 262.

Ray, S.B., Mohanti, M. and Somayajulu, B.L.K., 1984. Suspended matter, major cations and dissolved silicon in the estuarine waters of the Mahanadi River. *India J. Hydrol.*, 69: 183–196.

Riedel, G.F., Williams, S.A., Riedel, S.G., Gilmour, C.C. and Sanders, J.G., 2000. Temporal and spatial patterns of trace elements in the Pantuxent River: A whole watershed approach. *Estuaries*, 23: 521–535.

Salomons, W. and Forstner, U., 1984. *Metals in the Hydrocycle*. Springer-Verlag, New York, pp. 349.

Sarin, M.M., Borole, D.V. and Krishnaswamy, S., 1979. Geochemistry of sediments from the Bay of Bengal. *Proc. Ind. Acad. Sci.*, 88: 131–154.

Shafer, M.M., Overdier, J.T., Phillips, H., Webb, D., Sullivan, J.R. and Armstrong, D.E., 1999. Trace metal levels and partitioning in Wisconsin rivers. *Water, Air and Soil Poll.*, 110: 273–311.

Shankar, R. and Manjunatha, B.R., 1997. Geochemistry and isotopic studies of riverine, estuarine and marine environments of the western India: A review. In: *Proc. First Regional GEOSAS Workshop on Quaternary Geology of South Asia*, Anna University, Madras, pp. 195–220.

Shibu, M.P., Balachand, A.N. and Nambisan, P.N.K., 1990. Trace metal speciation in a Tropical estuary-significance of environmental factors. *Sci. Total Environ.*, 97/98: 267–282.

Subramanian, V., 1979. Chemical and suspended sediment characteristics of rivers of India. *J. Hydrol.*, 44: 37–35.

Subramanian, V., Biksham, G. and Ramesh, R., 1987. Environmental geology of peninsular river basins of India. *J. Geol. Soc. Ind.*, 30: 393–401.

Subramanian, V., Van' T Dack, L. and Van Grieken, R., 1985. Chemical composition of river sediments from the Indian sub-continent. *Chem. Geol.*, 48: 271–279.

Taylor, S.R. and McLennan, S.M., 1985. *The Continental Crust: Its Composition and Evolution*. Blackwells, pp. 312.

Tessier, A., Campbell, P.G.C. and Bisson, M., 1982. Particulate trace metal speciation in stream sediments and relationships with grain size: Implications for geochemical exploration. *J. Geochem. Explor.*, 16: 77–104.

Turekian, K.K. and Wedepohl, K.H., 1961. Distribution of the elements in some major units of earth's crust. *Bull. Geol. Sve. Amer.*, 72: 175–191.

Turekian, K.K., 1969. The oceans, streams and atmosphere. In: *Handbook of Geochemistry*, Vol. 1, (Ed.) K.H. Wedepohl. Springer-Verlag, Berlin, pp. 297–323.

Verma, D.D., Rao, K.S.R., Rao, A.T. and Desari, M.R., 1993. Trace element geochemistry of clay fractions and bulk sediments from Vamsadhsra River basin, east coast of India. *Indian J. Mar. Sci.*, 22: 247–251.

Walling, D.E. and Webb, B.W., 1985. Estimating the discharges of contaminants to costal waters by rivers: Some cautionary comments. *Mar. Poll. Bull.*, 16: 488–492.

Wilbur, W.G. and Hunter, J.V., 1979. Impact of urbanization on the distribution of heavy metals in bottom sediments of Saddle River. *Water Resour. Bull.*, 15: 790–800.

Chapter 3

Limnological Studies on Yellamallappa Chetty Lake, Bangalore, Karnataka

C. Maya[1], *G. Raviprasad*[1], *H. Krishnaram*[2] *and M. Lakshman*[3]

[1]*Department of Botany,* [2]*Department of Zoology, Bangalore University, Bangalore – 560 056*
[3]*BMS College, Bangalore*

ABSTRACT

Water quality plays an important role in the survival and distribution of aquatic organisms. It is dependent on physico-chemical and microbiological parameters. *Yellamallappa Chetty Lake* (YCL) has historically clear water and used to supply for the drinking purpose. Local residents explain that water clarity has decreased appreciably in recent years. The study was initiated in response to concern over deteriorating water quality, probability of increased nutrient loading from non point sources and municipal effluents. Water quality data were collected at 8 sampling sites from February 2004 to June 2004. Noticeable changes in many water quality parameters were observed. High nutrient concentrations, the increasing incidence of nuisance algal blooms, and trophic state index values indicate that it can be currently classified as *Eutrophic*.

Keywords: *Domestic sewage, Nutrients, Biological Oxygen Demand (BOD), Chemical Oxygen Demand (COD), Eutrophication.*

Introduction

Water is considered as one of the most basic and important element for human civilization. Our dependence on fresh water resources has accelerated in the last century due to rapid world population growth and economic development (Postel, 1992). As a result, fresh water resources have deteriorated both in quantity and quality in many areas of the world. Lakes providing the majority of easily accessible fresh water resources have been most influenced, suffering environmental problems such as man-made eutrophication, chemical contamination, etc., consequently causing damage to ecosystem and human health. The presence of non pathogenic organisms is not of major concern, but intestinal contaminants of faecal origin are important. These pathogens are responsible for intestinal infections such as *Bacillary dysentery, Typhoid fever, Cholera* and *Paratyphoid fever*, etc. The purpose of this study was to collect sufficient baseline water quality data to define current limnological conditions and to provide a basis for future water quality protection and monitoring.

Study Area

Yellamallappa Chetty Lake (YCL) is located at about 20 kms away from, on the eastern side of the Bangalore City, on the National High way No. 4, of old Madras Road. The inflow received from the sewage water, industrial waste water, and the surface runoff generated from the Hebbal valley located on the northern side of the Bangalore city. The other inlets shown are seasonal and flow can be seen only during heavy monsoon season.

The Physical features of Yellamallappa Chetty Lake

Location:

Longitude: 77°44′ 00″

Latitude: 13°2′ 00″

Elevation above mean sea level: 900 meters

Area of lake in ha: 46.00

Watershed area: 178.71 square miles

Maximum depth: 12 feet

Capacity of the lake: 110.8 Mcft or 9.34 cusecs

Material and Methods

During this period of investigation, in order to cover the whole topography of the lake, eight sampling sites were selected and monthly water samples were collected according to the standard methods. The physico-chemical parameters of the water samples analysed were, *viz.*, Temperature, Dissolved oxygen, pH, Turbidity, Total Dissolved Salts, Conductivity, Chloride, Total Alkalinity, Total Hardness, Sulphate, BOD, COD, Total Nitrogen, Total Phosphorus, Sodium, Potassium with the aid of APHA (1989), Trivedy and Goel (1986) and Mathur (1991) and biological parameters such as *T. coli* and *F. coli* were analysed by Membrane Filtration (MF) method. These factors are discussed here in detail.

Results and Discussion

Limnological survey of *Yellamallappa Chetty Lake* was undertaken with a view to investigate various changes in its hydrobiological features. The overall means and ranges for all water quality parameters measured during this study are presented in Table 3.1 In the present investigation, the maximum temperature 26.7°C were recorded in Site–1 and minimum temperature 25°C were recorded in the Sites–5, 6 and 8 (Table 3.1). Temperature is a vital parameter for growth of organisms and also plays an important role both in the physico-chemical and physiological behavior of the aquatic ecosystem.

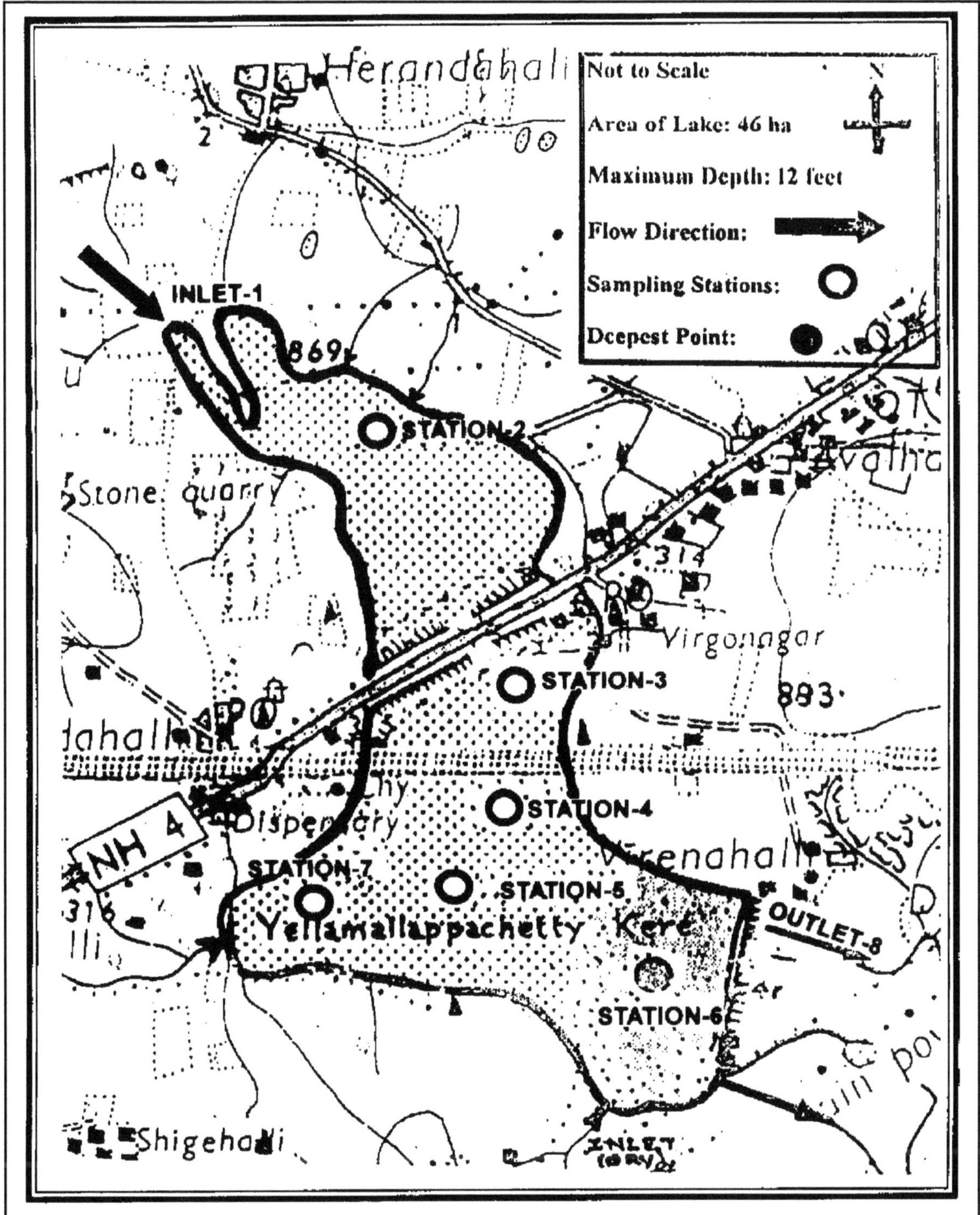

Figure 3.1: Details of Sampling Stations Along with Inlet Flow, Outflows and Stations within the Yellamallappa Chetty Lake

Table 3.1: Study Period Arithmetic Mean Values of Water Quality Parameters Collected from Yellamallappa Chetty Lake (February 2004 to June 2004)

Sl.No.	Parameters	Units	Sampling Stations							
			S1	S2	53	S4	S5	S6	S7	S8
1.	Temperature	0° C	26.7	25.8	26.0	25.7	25.0	25.0	26.1	25.0
2.	D.O.	mg/l	5.4	8.5	8.0	7.8	7.3	7.1	7.1	4.9
3.	pH	-	7.7	8.3	8.6	8.5	8.4	8.5	8.6	8.5
4.	Turbidity	NTU	45.2	34.2	36.8	40.6	36.1	37.4	36.4	31.4
5.	TDS	mg/l	1008.4	924.6	839.3	837.1	828.9	830.8	842.9	831.9
6.	Conductivity	μmohs/cms	1737.5	1600.0	1483.3	1462.5	1450.0	1425.0	1437.5	1431.3
7.	Chloride	mg/l	301.4	280.3	264.5	267.5	270.4	270.2	272.3	277.1
8.	T. alkalinity	mg/l	368.0	349.0	333.0	332.0	334.0	336.0	343.0	340.0
9.	Total Hardness	mg/l	321.0	304.0	299.0	291.0	299.0	306.0	304.0	297.0
10.	Sulphate	mg/l	20.5	19.1	16.3	17.5	17.0	16.0	16.5	15.0
11.	B.O.D.	mg/l	3.5	5.1	3.5	3.9	3.9	4.0	5.0	3.4
12.	C.O.D.	mg/l	74.6	81.2	83.7	73.7	79.5	76.6	87.9	66.5
13.	T. Nitrogen	mg/l	12.4	11.9	9.4	8.9	8.6	9.1	9.0	8.4
14.	T. Phosphorous	mg/l	0.8	0.8	0.7	0.6	0.6	0.5	0.6	0.4
15.	Sodium	mg/l	190.0	205.0	202.0	203.0	192.0	196.0	198.0	182.0
16.	Potasium	mg/l	32.0	33.0	34.0	34.0	32.0	31.0	33.0	30.0
17.	T. Coli	No/100 ml	1600.0	586.0	266.0	332.0	271.0	191.0	345.0	958.0
18.	F. Coli	No/100 ml	1250.0	222.0	167.0	187.0	127.0	101.0	125.0	363.0

The maximum Dissolved oxygen 8.5 mg/l were recorded in site–2 and minimum 4.9 mg/l was recorded in site–8 (Table 3.1). Amount of oxygen in lake is its purity indicator, also acts as an indicator of trophic status and magnitude of eutrophication in fresh water ecosystem.

pH readings were recorded at all sites throughout the study, readings varied from 7.7 to 8.6, maximum pH 8.6 were recorded in site–3 and minimum 7.7 in site–1. pH of water has been regarded as the most important factor in maintaining the bicarbonate and carbonate system, regulating the nutrients uptake of macrophytes and phytoplankton.

The maximum Turbidity 45.2 NTU; Conductivity 1737.5 μmohs/cm; Chlorides 301.4 mg/l; Total alkalinity 368 mg/l; Total hardness 321 mg/l; Sulphate 20.5 mg/l; were found in site–1, because of the entry of domestic and municipal sewages, industrial effluents through the inlets.

The maximum BOD 5.1 mg/l were recorded in site–2 and minimum 3.4 mg/l in site 8, the BOD of the water are high and thus, indicating that untreated domestic sewage are being dumped into the lake, resulting in accumulation of large amounts of organic matter in the lake.

The maximum COD 87.9 mg/l were recorded in site–7 and minimum 66.5 mg/l were recorded in site–8, it indicates the higher microbial activities and presence of oxidizable organic matter.

The maximum Total Nitrogen 12.4 mg/l was recorded in site–1 and minimum 8.4 mg/l was recorded in site–8. Nitrogen is required by all organisms for the basic processes of life *viz.*, for the synthesis of proteins, for growth and reproduction. Nitrogen is very common and found in many forms in the environment. Concentration of nutrients sometimes has a seasonal pattern in lakes, often highest during storm events soon after fertilizers are applied upstream (Mueller and Helsel, 1999).

The maximum Total Phosphorus 0.8 mg/l was recorded in sites–1, 2 and minimum 0.4 mg/l were recorded in sites–8. Phosphorus is a natural element found in rocks, soils and in organic materials, exists in water either in a particulate phase or a dissolved phase. Particulate matter includes living and dead plankton.

The maximum sodium 205 mg/l were recorded in site–2 and minimum 182 mg/l in site–8, and the maximum potassium 34 mg/l were recorded in sites–3, 4 and minimum potassium 30 mg/1 were recorded in site–8. Sodium is present in number of minerals, the major one being Rock salt (sodium chloride). In surface water, sodium concentration is usually non-toxic, excess intake of sodium chloride causes vomiting.

The water containing high sodium content is not suitable for agriculture. The concentration of potassium is usually lower in natural waters than sodium. The chemical properties of both sodium and potassium are similar as both are alkaline group metals (Ramachandra *et. al.*, 2006).

The results for the presumptive coliform count (Most Probable Number) are shown in Table 3.1. The lakes are replenished through rain, which comes in contact with particles of suspended dust and smog that immediately contaminate the moisture as it falls to the earth and so their was an increase in flush of coliforms during rain and hence exhibited maximum number of *T. coli* (1600/100 ml) and *F. coli* (1250/100 ml) were obtained in inlet *i.e.*, station–1. Total coliform bacteria test is a primary indicator of 'Potability', suitable for consumption of drinking water.

Thus, the heavy concentrations that have been reported from the *Yellamallappa Chetty Lake* water might have been contributed by domestic sewage being drained into the lake waters. The lake water quality survey conducted at the different sites of the lake during the present investigation indicates its polluted state, which is obviously because of the domestic sewage brought into the lake waters.

References

APHA, 1989. *Standard Methods for Examination of Water and Wastewater*, 17th Edition. APHA, AWWA, WPCA.

Kaushik, S. and Saxena, D.N., 1999. Physico-chemical limnology of certain water bodies of Central India. In: *Freshwater Ecosystems of India*, (Ed.) K. Vijayakumar. Daya Publishing House, New Delhi, p. 1–58.

Mueller, David, K. and Helsel, Dennis R., 1999. Nutrients in the nations waters–Too much of a good thing?", U.S. Geological Survey Circular 1136. National Water–Quality assessment program. http://water.usgs.gov/nawqa/circ–1136.html.

Postel, S., 1992. *Last Oasis: Facing Water Scarcity*. W.W. Norton and Company, Inc.

Ramachandra, T.V., Rishiram, R., and Karthik, 2006. Zooplankton as bioindicators: Hydrobiological investigations in Bangalore lakes. Technical Report: 115, Centre for Ecological Sciences, Indian Institute of Science, Bangalore, pp. 53, 56–57, 60.

Trivedy, R.K. and Goel, P.K., 1986. *Chemical and Biological Methods for Water Pollution Studies*. Environmental Publications, Karad, India.

Chapter 4

Some Studies on the Water Quality Parameters of Shriramnagar (Garividi) Vizianagaram District

A.V.L.N.S.H. Hariharan

Department of Engineering Chemistry, College of Engineering, GITAM, Visakhapatnam – 530 045

ABSTRACT

Physico-chemical analysis of bore well water samples was carried out from six sampling stations of Shriramnagar (Garividi),Vizianagaram district during the month of August 2006. The analysis of different parameters namely- pH, turbidity, color, total alkalinity, total hardness, chloride, sulphate, nitrate, TDS, DO, BOD were carried out as per standard methods. The parametric ratios and variation of temperature with D.O. and nitrate were also studied. Water quality index has been calculated for the samples collected and the corresponding values obtained from the study indicate that the water is safe for human consumption.

Keywords: *Water quality index, Shriramnagar, Parameters, Garividi, Pollution.*

Introduction

With beginning of life on earth, there was no pollution. Rapid urbanization and industrialization which are the advance tools and tips of the modem civilization have culminated into water, air and land pollution problems. Water pollution is a phenomenon which is characterized by deterioration of the quality as a result of various human activities (Kudesia, 1985; Todd, 1995), The poor quality of drinking water in our country is more due to contamination than due to the inferiority of the source

(Gibbons, 1984). Therefore, a continuous periodical monitoring of water quality is necessary so that appropriate steps may be taken for water resource management practices (Rajvaidya, 1998). In this study an attempt is made to carryout a systematic investigation of the physico-chemical parameters of samples collected from Shriramnagar (Garividi), Vizianagaram District in order to assess its suitability for drinking purposes.

Materials and Methods

Water samples collected from six sampling stations selected for the analysis were given bellow: S_1: Chalapathinagar (Bore Water), S_2: Facor colony (Bore Water), S_3: Industrial colony(Bore Water), S_4: G.P. road (Bore Water), S_5: MIG colony (Bore water), S_6: Sainagar (Bore Well). Samples for analysis were collected in sterilized bottles using the standard procedure for grab (or) catch samples in accordance with standard methods of APHA (1995). The analysis of various physico-chemical parameters namely pH temperature, total hardness, alkalinity, calcium hardness, magnesium hardness, chloride, sulphate, nitrate, DO, BOD, COD, TDS etc., were carried out as per the methods described in APHA (1995). All the chemicals and reagents used were of analytical grade. DD water was used for the preparation of solutions.

Results and Discussion

The results of various physico-chemical parameters are summarized in Tables 4.1 to 4.4. Samplings were also done at an interval of three hours commencing from 08.00 hrs of 19th August 2006 to 08.00 hrs of the next day. The measurement of water temperature, DO, carbonate and bicarbonate alkalinity were done at the sites immediately after collecting samples. However analysis of other parameters were carried out at laboratory soon after the collection.

Temperature

Temperature of water is basically important because it effects biochemical reactions in aquatic organisms. The highest temperature being at 14.00 hrs and the lowest at 05.00 hrs. The average temperature of the present study ranged from 27.10–29.39°C. It is known that the hydrogen ion concentration affects the taste of water. The pH values of the present investigation were within the ICMR standards (7.6–8.12). Electrical conductivity is a total parameter for determining the water quality for drinking and agricultural purposes. Many dissolved substances may produce aesthetically displeasing colour, taste and odour. The average values obtained are in the range 0.7 to 1.8 µmhos (Table 4.1).

Total Dissolve Solids

TDS values beyond the prescribed limit impart taste to water and reduce its palatability. The TDS values in the present study ranged within 473.5 to 935.2 mg/lt. The high TDS values (above 600 mg/lit) may be due to proximity of sea coast.

Dissolved Oxygen (DO)

Presence of DO in water may be due to direct diffusion from air and photosynthetic activity of autotrophs (Shanti *et al.*, 2002). Oxygen can be rapidly removed from the waters by discharge of oxygen demanding wastes. Higher values of DO may cause corrosion of iron and steel. The DO values obtained in the present study are within ICMR standards.

Table 4.1: Physico-chemical Parameters of Water Samples of Sriramanagar (Garividi) Colony on 19-08-2005

Parameter	S_1	S_2	S_3	S_4	S_5	S_6
Temperature °C	27.05	26.55	27.10	27.02	27.12	28.30
pH	7.83	7.75	7.68	7.82	8.03	7.87
Electrical Conductivity	0.5	0.4	1.2	1.7	1.9	0.5
TDS	309	300	585	690	632	360
TSS	26.3	45.7	57.9	34.7	44.6	28.5
Hardness	279.6	220.4	584.8	471.6	380.8	190.6
Calcium	119.2	35.4	87.2	78.8	55.3	39.04
Magnesium	28.3	27.4	68.5	66.6	51.0	15.7
Chloride	118.2	103.5	209.2	142.7	126.5	89.3
DO	4.21	5.20	5.02	4.54	4.34	5.92
Sulphate	128.5	141.6	132.8	156.7	135.3	72.8

All the parameters expressed in mg/lit. except pH and EC (mhos)

* All the values are the average of 3 determinations.

Alkalinity

Most of the alkalinity in natural water is formed due to dissolution of carbon dioxide in water. Large amounts of alkalinity imparts a bitter taste to water. Highly alkaline waters are usually unpalatable. The total alkalinity of the water samples was found in the Range 167.0 to 236.3 mg/lt.

Hardness

The hardness of water reflects the nature of the geological information with which it has been in contact. Water with Hardness above 200 mg/lit may cause scale deposition in the distribution system and results in excessive soap consumption and subsequent scum formation. Soft water with hardness of less than 100 mg/lit may have lower buffer capacity and more corrosive for water pipes (WHO, 1984). The hardness values of the present study were found to range between 238.02 to 434.7 mg/lit.

Calcium is one among the essential metals which play an important role in biological system. Magnesium though an essential and beneficial metal is toxic at higher concentrations. Magnesium hardness particularly associated with sulphate ion has laxative effect on persons unaccustomed to it (Khurshids, 1998). Principal sources of magnesium in natural waters are rocks. In the present study calcium and magnesium contents are found in the range of 57.72–108.48 and 22.08–36.82 mg/lt respectively.

Chloride

High concentration of chloride in water is considered to be an indication of pollution by sewage waste of animal origin. (Goel, 1996). Industries are also important sources of chloride in water. Chloride values obtained in the study are found in the range 48.2 to 145.4 mg/lit.

BOD and COD

BOD and COD are the parameters used to asses the pollution of surface water and groundwaters. Both these of the parameters (BOD and COD) values obtained in the present study are within permissible levels (Table 4.1).

Sulphate

Sulphate ion if present in excess amount produce cathartic effect upon human beings (Dhembare, 1998). The sulphate ion concentration in the present investigation varied from 72.0–122.5 mg/lt.

Nitrate

Water bodies polluted by organic matter exhibit higher values of nitrate (Patnaik, 2003). The maximum value for the observed at 5.00 hrs while minimum at 20.00 hrs which is inversely related to surface water temperature (Table 4.2). In the present study water samples from the stations (S_1 to S_6) showed low concentrations of nitrate (0.66 to 1.38 mg/lt) well below permissible levels as per the standards.

Table 4.2: Mean Values of Physico-chemical Parameter of Water Samples Collected in August 2005

Sl.No.	Parameter	8.00 hrs	11.00	14.00	17.00	20.00	23.00	2.00	5.00	8.00 hrs
1.	Temperature °C	27.84	28.50	29.38	29.35	28.52	28.05	27.85	27.24	27.83
2.	D.O	4.21	4.56	4.84	4.65	4.52	4.40	4.23	4.13	4.21
3.	Nitrate	1.08	0.96	0.85	0.62	0.54	0.72	1.03	2.07	1.08

Water Quality Index (WQI)

It is defined as a rating reflecting the composite influence of different water quality parameters on the overall quality of water.

Calculation of WQI

Let there be n water quality parameters and quality rating (q_n) corresponding to nth parameter is a number reflecting relative value of this parameter in the polluted water with respect to its standard permissible value. q_n values are given by the relationship.

$q_n = 100\ Vn/Vs$ except for pH and DO.

v_s = Standard value

v_n = Observed value

$q_{pH} = 100\ (v_{pH}–7.0)/(8.5–1.0)$ and

$q_{DO} = 100\ (V_{DO}–14.6)/(5.0–14.6)$.

Calculation of Unit Weight

The Unit weight (w_n) to various water Quality parameters are inversely proportional to the recommended standards for the corresponding parameters.

$$W_n = k/s_n.$$

where,

W_n: Unit weight for nth parameter

S_o: Standard permissible value for *n*th parameter

k: proportionality constant.

The unit weight (w_n) values in the present study are taken from Krishna *et al.*, 1995

Table 4.3: Parametric Ratios of Water Samples–VUDA (Garividi) Sriramanagar

Parameter	*Min*	*Max*	*Average*	*Parametric Ratio*	*Values Found*
Temperature	26.23	28.3	27.265	pH/TDS	0.05438
pH	7.58	7.83	7.705	pH/Chloride	0.051625
E.C	0.4	1.9	1.15	pHi Alkalinity	0.05206
TDS	309	702	505.5	pH/calcium	0.09968
TSS	26.3	57.9	42.1	pH/DO	1.5257
Hardness	190.6	584.8	340.4	pH/COD	1.5724
Alkalinity	103.4	192.6	148.0	EC/calcium	0.01488
Calcium	35.4	119.2	77.3	EC/Chloride	0.007705
Magnesium	15.7	76.0	45.85	EC/Sulphate	0.72108
Chloride	89.3	209.2	149.35	Chloride/TDS	0.29525
Sulphate	72.8	141.6	107.2	Hardness/TDS	2.28074
Nitrate	1.05	2.53	1.79	–	–
DO	4.2	5.9	5.05	–	–
COD	3.2	6.0	4.9	–	–

Table 4.4: Water Quality Index of Bore Well Waters of (Garividi) Sriramanagar

Parameter	*ICMR Standard*	*Unit Weight (wn)*	S_1	S_2	S_3	S_4	S_5	S_6
pH	7.74	0.07164	7.330	7.6545	7.358	7.034	7.238	7.516
TDS	500	0.00100	0.187	0.08934	0.0947	0.1254	0.1505	0.11746
TH	200	0.00167	0.354	0.3629	0.2448	0.2678	0.2475	0.1987
DO	5	0.10030	4.622	7.703	1.6	1.8	6.4192	1.4
BOD	5	0.10030	1.6048	3.8114	3.6108	10.43	1.6	1.6048
Chloride	250	0.00200	0.1163	0.09584	0.0548	0.762	0.0386	0.0626
Total Alkalinity	120	0.00417	0.580	0.8211	0.787	0.608	0.647	0.7535
NO_3	45	0.01111						
COD	20	0.02507						
Sulphate	200	0.007418						
Iron	0.3	1.6666	2.222	0.555	1.111	2.222	0.555	2.777
W_n	–	1.95879	17.0161	21.09308	16.4707	24.3742	18.5054	15.80846
q_n	–	–						
WQI	–	–	33.33096	41.31691	32.26264	47.744	36.2482	30.9654

WQI is calculated by the following equation.

$$WQI = \sum_{n=1}^{n} q_n w_n / \sum_{n=1}^{n} w_n$$

The suitability of WQI values for human consumption according to Mishra and Patel, 2001 are rated as follows. 0–25, Excellent; 26–50, Good; 51–75, Bad; 76–100, Very Bad; 100 and above, Unfit.

The WQI values of the present investigation from different sampling stations are given in Table 4.4.

Conclusions

Some of the samples have total dissolved solids, hardness and calcium values exceeding the permissible limits as prescribed by Indian standards. However the WQI values (30.96–47.7) calculated for the different samples indicate that the water is safe for human consumption.

References

APHA, 1995. *Standard Methods for the Examination of Water and Wastewater*, 19th edn. AWWA, WPCP, New York, USA.

Brown, R.M., Mccleiland, N.J., Deiniger, R.A. and O'Connor, M.F., 1972. A water quality index-crossing the physical barrier. In: *Proc. Intl. Conf. on Water Poll. Res.*, Jerusalem, (Ed.) Jenkis, S.H. 6: 787–797.

Dhembare, A.J., Pondhe, G.M. and Singh, C.R., 1998. Groundwater characteristics and their significance with special reference to public health in pravara area, Maharastra. *Poll. Res.*, 17(1): 87–90.

Gibbons, J.H., 1984. *Protecting the Nations Groundwater from Contamination*. Congress of Limited States, Office of the Tech. Assessment, Washington, DC.

Goel, P.K. and Sarma, K.P., 1996. *Environmental Guidelines and Standards in India*. Techno Sci., Pub.

Khursid, S., Zaheeruddin and Basheer, A., 1998. *Ind. J. Env. Prot.*, 18(4): 246–249.

Krishnan, J.S.R., Rambabu, K. and Rambabu, C., 1995. Studies on water quality parameters of bore waters of Reddigudum Mandal. *Ind. J. Env. Prot.*, 16(2): 91–98.

Kudesia, V.P., 1980. *Water Pollution*. Pragathi Prakashan, Meerut, India, pp. 1–12.

Mishra, P.C. and Patel, R.K., 2001. Quality of Drinking water in Rourkela, outside the steel township. *J. Env. Poll.*, 8(2): 165–169.

Patnaik, K.C. and Nayak, L., 2003. Study of diurnal variation of some physico-chemical properties of Gopalpur coastal waters of Orissa during the monsoon period Nature. *Env. and Poll. Tech.*, 2(4): 437–440.

Shanthi, K., Ramasamy, P. and Lashmanperumalsamy, 2002. Hydrological study of Singanallur lake at Coimbatore, *Nature Environment and Pollution Technology*, 1(2): 97–101.

Todd, D.K., 1995. *Groundwater Hydrology*, 2nd Edition. John Wiley and Sons Inc., Singapore.

Chapter 5

Vertical Distribution of Forms of Potassium in Different Soil Profiles in Tunga Command Areas of Karnataka

K.T. Gurumurthy, B.L. Shivaprakash, H.C. Prakasha, N.S. Mvarkar and C.J. Sridhar

Department of Soil Science and Agricultural Chemistry, College of Agriculture Navile, Shimoga – 577 204, Karnataka, India

ABSTRACT

Soil samples from four representative soils profiles in rice growing areas of Tunga command of Karnataka were studied for the vertical distribution of different forms of potassium and for their relationship with some soil properties. Water soluble K was positively correlated with all the properties studied except EC, but was non-significant. Exchangeable-K was positive and significant relationship between non exchangeable K, clay, pH, and OC EC. A significant positive correlation was recorded between non exchangeable K and clay, pH and CEC. Among the soil properties studied, a significant correlation was noticed for latice K with total K. Non-exchangeable K was positively correlated with all the forms of K studied indicates the existence of dynamic equilibrium between these forms of K.

Keywords: *Distribution pattern, Forms of soil K, Soils profile, Correlated.*

Introduction

A knowledge of different forms of potassium in soil together with their distribution in the zone of root penetration is of much relevance in assessing the long-term availability of the nutrients to crops and in formulating sound fertilizer recommendations. The pattern of the potassium distribution in soil–profile is dependent upon the homogeneity of parent material with regard to type and abundance of potassium bearing minerals and the relations like fixation and release. The various forms of potassium in soil exist in equilibrium with one another and depletion of one form is replenished by other forms (Chandel *et al.*, 1976). The present study was therefore, undertaken to find out the vertical distribution of different forms of potassium and their relationships with important soil parameters.

Materials and Methods

Five representative soil profiles in rice growing areas of Tunga command where two profiles from Shimoga (Profiles I and II), one each from Honnali (Profile III) and Hirekerur (Profile IV) taluks of Karnataka were collected depth wise. The soil samples were analyzed fore their particle size distribution by hydrometer method, pH, EC and CEC by standard laboratory methods (Jackson 1973). The different forms of potassium were determined using the procedures as outlined by Pratt (1982). Simple correlation's regression analyses were calculated between different forms of K and soil properties by adopting statistical procedure.

Results and Discussion

The soil characteristics estimated in the present study are given in Table 5.1. In all the profiles, the texture varied from sandy clay loam to sandy loam. There was no definite trend in the distribution of clay with depth except profile IV from Hirekerur taluk. The profiles from Shimoga and Honnali taluks were moderate to slightly acidic, while the profile from Hirekerur taluk was alkaline in soil reaction. The salt content ranged from 0.09 to 0.52 ds m^{-1}, which did not show any relation with depth. The CEC was in accordance with the distribution of clay content.

Table 5.1: Physico-chemical Properties of the Soils

Name of the Profile/Depth (cm)	pH	EC (ds m^{-1})	CEC [cmol (p+) kg^{-1}]	Coarse Sand (%)	Fine Sand (%)	Silt (%)	Clay (%)	Textural Class
				Profile I				
00–15	4.85	0.11	9.30	37.20	33.97	09.97	19.22	Sandy loam
15–30	6.83	0.14	12.40	33.16	39.62	12.81	14.41	Loamy sand
30–60	7.09	()'11	13.00	35.46	37.32	08.01	19.21	Sandy loam
60–100	6.70	0.12	15.10	46.04	28.34	09.61	16.01	Sandy loam
				Profile II				
00–15	6.48	0.52	23.50	38.10	33.08	12.81	16.01	Loamy sand
15–30	6.58	0.30	15.30	37.90	30.08	09.62	22.40	Sandy clay loam
30–60	6.63	0.20	13.00	37.48	41.70	09.61	11.21	Loamy sand
60–100	6.65	0.16	17.20	32.30	45.67	09.60	12.43	Loamy sand
				Profile III				
00–15	5.50	0.10	11.90	37.20	38.46	05.12	19.22	Sandy loam
15–30	7.30	0.09	12.50	37.50	33.04	04.48	24.98	Sandy clay loam
30–60	6.37	0.09	13.10	45.38	26.60	05.76	26.26	Sandy clay loam
60–100	6.32	0.15	10.20	56.62	17.70	06.09	19.53	Sandy loam

Contd...

Table 5.1–Contd...

Name of the Profile/Depth (cm)	pH	EC (ds m^{-1})	CEC [cmol (p+) kg^{-1}]	Coarse Sand (%)	Fine Sand (%)	Silt (%)	Clay (%)	Textural Class
				Profile IV				
00–15	7.33	0.30	37.10	11.98	44.79	16.01	27.22	Sandy clay loam
15–30	7.66	0.16	34.30	12.68	45.55	12.95	28.82	Sandy clay loam
30–60	7.41	0.15	33.90	12.20	47.77	11.11	28.92	Loam
60–100	7.27	0.15	27.20	15.98	44.95	06.09	32.98	Sandy clay loam

Forms of Soil Potassium

Water Soluble K

The water soluble K content in surface soil was higher than the lower depths in almost all the soils. Except Profile IV. The mean values of water soluble K in profile was ranged from 3.92 to 10.86 mg kg^{-1} in Profile III and Profile I respectively (Table 5.2). Irregular distribution of water soluble potassium was observed in Profile IV. However, the distribution of potassium within the soil body is said to be controlled by mineralogical makeup of the soil (Chandel *et al.*, 1976).

Correlation study (Table 5.3) indicated that, positive correlation of water soluble potassium with other forms. The water soluble K was positively correlated with fine sand, silt and clay. Similar reports were obtained by Hire Kurabar (1998). There was also positive correlation between water soluble K with pH and EC of the soils.

Exchangeable Potassium

The exchangeable potassium in surface layer varied from 36.36 mg kg^{-1} in Profile III (Honnali taluk) to 219.72 mg kg^{-1} in Profile IV (Hirekerur taluk). Except Profile III all other soils contained higher amount of exchangeable potassium in surface horizons. A decreasing trend with depth was noticed in Profile IV (219.72 to 152.09 mg kg^{-1}). The mean values of profiles were in the order of Profile IV (178.41 mg kg^{-1}) > Profile I (151.45 mg kg^{-1}) > Profile II (95.57 mg kg^{-1}) > Profile III (75.70 mg kg^{-1}) (Table 5.2). The highest content of exchangeable K in surface layer may be due to intense weathering, vegetation, and release of labile K from other organic residues. The variation in the exchangeable K content may be attributed to differential release of K from non exchangeable and lattice K as well as variation in the lattice pool due to K fertilization. The exchangeable potassium was positively correlated with lattice K, total K. The positive correlation between exchangeable, non-exchangeable, lattice K signifies the fact of existence of dynamic equilibrium between the presence of difficulty replaceable form with easily replaceable form (Mishra and Shankar, 1970).

Non-Exchangeable K (Fixed Potassium)

The amount of fixed potassium in the surface horizon varied from 247.94 to 307.92 mg kg^{-1} (Table 5.2). The lowest and the highest profile mean values being in Shimoga taluk (Profile II) (222.95 mg kg^{-1}) and Honnali taluk (Profile III) (304.5 mg kg^{-1}) respectively. The value of fixed K in the soils are comparable to the findings of Ranganathan and Sathyanarayana (1980) in soils of Karnataka. The correlation studies indicated that non-exchangeable K was positively correlated with all the forms of K studied (Table 5.3). This implies that a dynamic equilibrium exists between all the forms of potassium in soil.

Table 5.2: Vertical Distribution of Different Forms of Potassium in Rice Growing Soils of Tunga Command Area

Name of the Pedon/Depth (cm)	*Water Soluble K (mg kg^{-1})*	*Exch. K (mg kg^{-1})*	*Non-exch. K (mg kg^{-1})*	*Mineral K (%)*	*Total K (%)*
		Profile I (Shimoga taluk)			
00–15	13.13	182.86	250.89	1.43	1.47
15–30	12.35	119.95	204.82	1.58	1.62
30–60	10.12	151.56	285.20	1.67	1.72
60–100	07.84	151.41	271.95	1.92	1.96
Mean	10.86	151.45	253.22	1.65	1.69
		Profile II (Shimoga taluk)			
00–15	I I. 76	098.49	247.94	1.98	2.01
15–30	10.12	138.18	200.90	2.02	2.06
30–60	08.43	077.32	213.15	2.32	2.35
60–100	07.64	068.30	229.82	1.88	1.91
Mean	09.49	095.57	222.95	2.05	2.08
		Profile III (Honnali taluk)			
00–15	04.90	036.26	272.44	1.05	1.09
15–30	04.12	118.37	230.31	0.99	1.03
30–60	03.53	114.06	313.61	1.03	1.09
60–100	03.14	034.10	401.80	1.33	1.37
Mean	03.92	075.70	304.54	1.10	1.15
		Profile IV (Hirekerur taluk)			
00–15	10.58	219.72	307.92	0.88	0.93
15–30	14.11	176.99	216.58	0.75	0.78
30–60	06.67	164.83	251.86	0.70	0.74
60–100	04.71	152.09	235.20	0.70	0.74
Mean	09.02	178.41	252.89	0.76	0.80

Mineral Potassium (Lattice K)

The mineral potassium in surface horizons (Table 5.2) varied from 0.88 per cent to 1.97 per cent. The mean values of K varied from 0.76 per cent in Profile IV to 2.05 per cent in Profile II. The positive and significant correlation between total K and mineral K (Table 5.3) indicated that most of the soil K was in the lattice form.

Total Potassium

Total K in surface horizon varied from 0.93 to 2.01 per cent (Table 5.2). The maximum and minimum total K was observed in surface horizons of Profile IV of Hirekerur taluk and Profile II of Shimoga taluk soils respectively. The mean total K content of profile varied from 0.80 per cent in Profile IV to 2.08 per cent in Profile II. Such a wide variation in total K may be attributed to predominance of insufficient of K bearing minerals, particle size and degree of weathering (Ranganathan and Sathyanarayana, 1980). The total K content of Profile I of Shimoga increased with depth and decreased with in Profile IV of Hirekerur taluk, while in remaining soil bodies, no definite pattern of distribution with depth was noticed, Singh *et al.* (1996) also reported no definite trend of distribution of K in soils. Total K content (Table 5.3) showed a highly positive and significant correlation with mineral K ($r = 0.999$**) indicating that most of soil K in mineral form. The correlation of positive nature had been obtained between total K and clay fraction indicating that more potassium was present on the finer fraction than the coarser fraction.

Table 5.3: Correlation Coefficient (r) Among the Forms of Potassium and Properties of Soils

	Water Soluble K	*Exch. K*	*Non-Exch. K*	*Lattice K*	*Total K*
Water soluble K					
Exchangeable K	0.268				
Non exchangeable K	0.209	0.627**			
Lattice K	0.241	0.191	0.248		
Total K	0.248	0.220	0.282	0.999**	
Clay	0.256	0.516*	0.570*	0.063	0.085
pH	0.055	0.529*	0.489*	–0.143	–0.121
EC	–0.109	0.257	0.278	–0.281	–0.267
OC	0.017	–0.136	–0.214	0.132	0.122
Exchangeable Acidity	–0.386	–0.309	–0.094	–0.304	–0.309
CEC	0.181	0.535*	0.575*	–0.135	–0.110

k*: Significant at 5 per cent level.

**: Significant at 1 per cent level.

The results discussed above clearly bring out that there is no definite pattern of distribution of potassium with depths. It was found that the positive and significant correlation between total K and mineral K indicated that most of the soils potassium was in the latice form. The results also indicated that soils of different profiles are well supplied with potassium and can support crop production without any deficiency in immediate future.

References

Chandel, A.S., Maharaj Singh and Singh, T.A., 1976. Forms of potassium in Mollisols of Nainital Tarai. *Bull. Indian Soc. Soil Sci.,* 10: 13–30.

Hirekurabar, B.M., 1998. Potassium dynamics in soils under cotton based Cropping system of North Karnataka. *Ph.D. Thesis,* Submitted to UAS, Dharwad.

Jackson, M.L., 1973. *Soil Chemical Analysis.* Prentice-Hall of India Pvt. Ltd.

Mishra, R.C. and Shankar, H., 1970. Potassium status of Uttar Pradesh soils. *J. Indian Soc. Soil Sci.,* 18: 319–325.

Pratt, P.F., 1982. Potassium in methods of soil analysis Part II: Chemical in microbiological properties. *Agronomy Monograph No. 9, 2nd edn.* American Soc. Agron. Madison, USA, pp. 225–246.

Ranganathan, A. and Satyanarayana, T., 1980. Studies on potassium status of soils of Karnataka. *J. Indian Soc. Soil Sci.*, 28(2): 148–153.

Singh, S.K., Das, K., Shyampura, R.L. and Singh, R.S., 1996. The status and release behaviour of potassium as influenced by soil moisture regime. *J. Indian Soc. Soil Sci.*, 44(3): 392–397.

Chapter 6

In vitro Efficacy of Indigenous Herbs Against *Trichosporon beigelii*

J. Das[1], *S. Goswami*[1], *R. Gupta*[1], *D.K. Jha*[2] *and R.B. Srivastava*[1]

[1]*Defence Research Laboratory, Post Bag No. 2, Tezpur – 784 001, Assam*
[2]*Gauhati University, Department of Botany, Guwahati – 781 014*

ABSTRACT

Twenty two extracts derived from leaves of six different plant species namely *Piper betle* (Type-Sathyavaram), *Allamanda cathartica, Ocimum sanctum, Ocimum gratissimum, Camelia sinensis* and *Alstonia scholaris* were tested for their efficacy against opportunistic fungal pathogen *Trichosporon beigelii* under laboratory conditions. Except *A. scholaris* all the extracts prepared from individual species exhibited a high degree of inhibition displaying varying inhibition zone ranged from 39.0 to 21.0 mm diameter. The extract prepared by combining of all the six species exhibited highest inhibitory, effect against *Trichosporon beigelii* causing inhibition zone of 42.6 mm diameter. It was also observed that the plant extracts were much more effective than commercial antimycotic drug clotrimazole which showed 12 mm inhibition zone diameter.

***Keywords**: Antifungal activity, Agar well diffusion method, Medicinal plant, Trichosporon beigelii.*

Introduction

Trichosporon beigelii is one of the most significant opportunistic fungal pathogen, normally inhabited in mouth, skin and nail and causes superficial and deep infections. White piedra, caused by *T. beigelii*, an infection of the hair, characterized by white nodules on hair shaft of axilla, moustache and beard is pre-dominant in tropical countries. Immuno-compromised hosts are particularly under risk to develop invasive infection, which usually progresses rapidly, involving various organs and systems including lungs, kidneys, spleens (Anonymous, 2006). It is capable of causing difficulty in breathing, distention of abdomen, and swelling of scrotum and lower limbs in children in spite of its lower incidence (Razik, *et al.*, 2000 and Singh, 2004). The infection poses a therapeutic problem, despite availability of several antimycotic drugs in the market. They are mostly fungistatic, besides side effect may become much worse than cure. Further more resistance to synthetic drugs is increasing at alarming rate.

To minimize all these ill effects in long run, scientific investigation for novel and safe antimycotic products is the need of the hour. Natural resources, especially plants are potential candidates to solve the problem. An increased interest in herbal medicines dominated the health scenario worldwide during the last quarter of the twentieth century. The research focus is turned onto plant products as alternatives, as they are seemed to fewer side effects, more effective and readily available.

Northeast region is one of the mega hotspots of biodiversity (Myers *et al.*, 2000) and the second richest center in plant diversity (Anonymous, 2005) in the world, particularly excellent reservoir of unique, endangered, endemic and high valued medicinal plants with wide diversity, contributing 53 per cent *i.e.*, more than 10000 plant species to the total Indian flora. WWF found an extraordinary number, 107 in a single 2000 sq meters of plot of North of Brahmaputra river in Assam, Himalayan foothills, parts of Arunachal Pradesh, North Bengal and Bhutan (Anonymous, 2005). Thus, the scope for developing effective antifungal products from the indigenous natural resources is immense.

Unfortunately in spite of tremendous bioresources, R&D works for exploitation of potential bioactivity of the valuable herbs are still in infant stage. Only less than 5 per cent of all plant species have been analyzed for potential medicines (Sujatha and Sandhya, 2005).

With this background the present investigation was undertaken to evaluate the activity methanol leaf extracts of six indigenous plant species namely *Piper betle* (type-Sathyavaram), *Allamanda cathartica, Ocimum sanctum, Ocimum gratissimum, Camelia sinensis* and *Alstonia scholaris* and their various combinations against *Trichosporon beigelii* in laboratory condition.

Materials and Methods

Preparation of Plant Extract

Fresh leaves of the plants were collected from their natural habitat of Sonitpur District (Assam). The leaves were washed thoroughly in running tap water, shade dried and ground into coarse powder. Each of the powdered materials was soaked in methanol for 7 days with vigorous shaking on every alternate day. The extracts were then filtered through Whatman filter paper No. 1 and the filtrates were air dried to remove the solvent. The dried crude extracts so obtained were considered as standard of 100 per cent concentration. From the standard, test extracts were prepared by dissolving the extract, initially in little amount of Dimethyle Sulphoxide (DMSO) and then adjusted the volume by adding double distilled water to obtain 20 per cent (w/v) solution and finally sterilized by filtering through Millipore filter (0.2 µm pore size). A total of 22 different test extracts were prepared, of which 6 extracts were obtained from the individual plant species, 15 extracts from mixture of 2 different plant extracts in all possible combinations and 1 extract was prepared by combining all the six extracts in equal proportion.

Preparation of Inoculum

Pure culture of the test fungus *T. beigelii* was obtained from school of Tropical medicine, Kolkata. The culture was maintained at 4°C on Saubauraud Dextrose Agar (SDA) slants and sub-cultured at regular intervals. The inoculum was prepared by adding of loop full of the stock culture in 50 ml of Saubauraud Dextrose Broth and incubated at 28±2°C for 48 hours.

Screening for Antifungal Activity

Agar well diffusion method was employed for determining the antifungal activity of the extracts (Grammar, 1976). With a sterile cotton swab 0.4 ml of inoculum was spread evenly on the surface of the petri dish containing solidified SDA. A single well of 7 mm diameter was made in the center of the inoculated agar plate. The wells were then loaded with 0.300 of the respective test extracts. The extracts were then allowed to diffuse at room temperature for 2 hours. The plates were then incubated at 28±2°C for 48 hours. The experiment was replicated thrice and the average results were recorded. A control set was maintained with DMSO. Clotrimazole in a concentration of 1000 µg/ml was used as reference standard. The antifungal activity was determined by reading the zone of inhibition around the well.

Activity Index Determination

The activity index of the extracts was determined by the following formula (Jain and Sharma, 2005).

$$\text{Activity Index} = \frac{\text{Zone of inhibition of extracts}}{\text{Zone of inhibition of standard}}$$

Percentage Inhibition

Percentage of inhibition was calculated according to the following formula (Vyas *et al.*, 2006).

$$\%\ \text{Inhibition} = \frac{\text{Inhibition zone in mm}}{\text{Control*}} \times 100$$

*: Growth zone is equal to plate diameter *i.e.*, 80 mm as growth occurs all over the agar plate.

Results and Discussion

The experimental findings in general, presented in Table 6.1, revealed a high degree of inhibitory effects by all the extracts except *A. scholaris.* Plant derivatives have been demonstrated to possess excellent antifungal activity (Parveen and Kumar, 2000; Shivpuri and Gupta, 2001, Premkumar and Shyamsundar, 2005, Lakshmi *et al.*, 2005, Bhattacharjee *et al.*, 2005). The inhibitory effect of *Terminalia chebula* dry fruit extract and oil of *Citrus limon* against *T beigelii* was reported (Dutta *et al.*, 2004, 2005). The antifimgal activity of leaf extract of *P. betle* and *A. cathartica* against plant pathogen *Fusarium oxysporum* was also described earlier (Das and Das, 2005). The extracts when tested alone displayed Inhibition Zone Diameter (IZD) ranging from 39.0 to 21.0 mm. The highest inhibition zone of 39.0 mm was exhibited by *P. betle* leaf extract, followed by *A. cathartica* (38.0 mm), O. *gratissimum* (35.0 mm) and O. *sanctum* (32.5 mm). C. *sinensis* was less effective with an inhibition zone of 21.0 mm.

The combined preparation of 2 different extracts showed almost similar results in terms of inhibition zone (39.0 to 23.0mm) as compared to the extract derived from single plant species. The inhibitory effect, however, was more pronounced when all the leaf extracts were used as composite preparation against *T. beigelii.* This showed an inhibition zone diameter of 42.6 mm with maximum activity index (3.55) and percentage inhibition (53.25 per cent). This study clearly indicated that combined effect of the leaf extracts of plant species namely *P. betle* (Type-Sathyavaram), *A. cathartica, O. sanctum, O. gratissimum, C. sinensis* and *A. scholaris* was much higher than the effect of the individual plant extract. This may be due to synergistic action of different compounds present in the plant extract, which together produced much greater effect than the individual extract. Another interesting finding was that except *A. scholaris,* the entire test extracts either singly or in combination with other extracts in crude form were more effective against the test pathogen than the commercial antifungal clotrimazole (12 mm). In control set no inhibition was observed.

Table 6.1: Effect of Extracts from Single Plant Species on *T. beigelii*

Plant Extract (20%)	*Inhibition Zone Diameter (mm)*	*Activity Index*	*Per cent Inhibition*
Piper betle (Sathyavaram) (Pb)	39.0	3.25	48.75
Allamanda cathartica (Ac)	38.0	3.17	47.50
Ocimum sanctum (Os)	35.0	2.92	43.75
Ocimum gratissimum (Og)	32.5	2.70	43.63
Canmelia sinensis (Cs)	21.0	1.75	26.25
Alstonia scholaris (As)	No zone	–	–

Table 6.2: Effect of Leaf Extracts in Different Combinations on *T. beigelii*

Plant Extract in Combination (20 %)	*Inhibition Zone Diameter (mm)*	*Activity Index*	*Per cent Inhibition*
Pb+Ac	39.0	3.25	48.75
Pb+Os	37.0	3.08	46.25
Pb+Og	36.0	3.00	45.00
Pb+Cs	36.0	3.00	45.00
Pb+As	35.0	2.92	43.75
Ac+Os	38.0	3.17	47.50
Ac+Og	37.0	3.08	46.25
Ac+Cs	37.0	3.08	46.25
Ac+As	37.0	3.08	46.25
Os+Og	30.0	2.50	37.50
Os+Cs	30.0	2.50	37.50
Os+As	30.0	2.50	37.50
Og+Cs	35.0	2.92	43.75
Og+As	34.0	2.80	42.50
Cs+As	23.0	1.92	28.75
Pb+Ac+Os+Og+Cs+As	42.6	3.55	53.25
Control (DMSO)	No zone		
Clotrimazole	12.0		

However, it is important to note that successful extraction of maximum bioactive components depends on various factors like solvent used for extraction, growing conditions of plants and collection and processing methods etc. The results of the present study are highly promising and highlight the new way for doing further studies to ascertain their minimum effective dose and activity in *in vivo* condition and also toxicity if any in order get utility out of these indigenous plants.

Acknowledgement

The authors are thankful to Dr. (Mrs) M. Begam, Head, Division of Biotechnology for her support and help throughout the study period.

References

Anonymous, 2005.

Anonymous, 2006.

Bhattacharjee, P.R., De, B., Datta, S., Das, A., Biswas, R. and Sangma, N.M., 2005. Antimicrobial and pharmacological evaluation of stem, seed and extracts of *M. paradisiaca* Linn. *Indian Drugs*, 42(4): 238–241.

Das, J. and Das, T.K., 2005. Effect of phytoextracts on inhibition of mycelial growth of *Fusarium oxysporum* Schlecht. *Environment and Ecology*, 23S(Spl–2): 362–364.

Dutta, B.K., Karmakar, S., Rahman, I. and Das, T.K., 2005. Antimycotic potential of an essential oil against dermatophytes. *Geobios*, 32(1): 103–104.

Dutta, B.K., Kannakar, S., Rahman, I., Das, J., Choudhury, K. and Das, T.K., 2004. Antidennatophytic property of dry fruit extract of *Terminalia chebula*. *Geobios*, 31(4): 293–295. .

Grammar, A., 1976. *Microbiological Methods*. Collins, C. H. and Lyne, P.M. Butterworths, London, p. 235.

Jain, N. and Sharma, M., 2005. Broad-spectrum antimycotic drug for the treatment of ringworm infection in human beings. *Current Science*, 85(1): 30–34.

Kamalakannan, A., Shanmugam, V., Surandran, M. and Srinivasan, R., 2001. Antifungal properties of plant extracts against *Pyricularia grisea*, the rice blast pathogen. *Indian Phytopath*., 54(4): 490–492.

Lakshmi, V., Gupta, P., Varshney, V., Kumar, R., Jain, P. and Khan, Z.K., 2005. Antifungal activity of *Petrosia nigricans*. *Indian Drugs*, 42(4): 213–216.

Myers, N., Russel, A.M., Cristina, G.M., Fonseca Gustavo, A.B. and Kent, J., 2000. *Nature*, 403: 853–858.

Parveen, S. and Kumar, V.R., 2000. Effect of extracts of some medicinal plants on the growth of *Alternaria triticina*. *J. Phytol. Res*., 13(2): 195–196.

Premkumar, V.G. and Shyamsundar, D., 2005. Antidermatophytic activity of *Pistia stratiotes*. *Indian J. Phamacol*., 37(2): 126–127.

Razik, M.A., Moawad, F. and Metawie, M., 2000. Prevalence of yeast fungi and related vaginal fungal infection among pregnant and unpregnant women at Ismailia. *Afr. J. Mycol. Biotech*., 8: 25–34.

Singh, K., 2004. Trichosporonosis in a previously healthy child in Nepal. *Indian Pediat*., 38: 667–678.

Shivpuri, A. and Gupta, R.B.L., 2001. Evaluation of different fungicides and plant extracts against *Sclerotinia sclerotiorum* causing stem rot of mustard. *Indian Phytopath*., 54(2): 272–274.

Sujatha, A. and Sandhya, B., 2005. Antimicrobial activity of *Syzygium aromaticum* (Clove) and *Millingtonia hortensis* (Indian cork tree) against human pathogens. *J. Curr. Sci*., 7(1): 123–127.

Vyas, Y.K., Bhatnagar, M. and Sharma, K., 2006. Antimicrobial activity of a herb, herbal based and synthetic dentrifrices against oral micro flora. *J. Cell and Tissue Research*, 6(1): 639–642.

Chapter 7

Allelopathic Effect of *Phargmites karka* on Free Floating and Submerged Freshwater Weeds

M.K. Saxena, J. Gupta and R.C. Meena

Department of Botany, University of Rajasthan, Jaipur-302004, India

ABSTRACT

Phragmites karka (Retz.) Trin. Ex. Steud. (= *P. vallatoria*) was screened for allelopathic potential on some freshwater weeds like *Eichhornia crassipes* (Mart.) So1m., *Salvinia molesta* Mitchell. (free-floating) *Hydrilla verticillala* L., *Ceratophyllum demersum* L. and *Utricularia flexuosa* (submerged). The free floating freshwater weeds were allowed to grow in experimental pots containing 3 per cent aqueous leachate (w/v) of above ground and belowground parts of *Phragmites* plant. The submerged plants were grown in 3 per cent aqueous leachate of aboveground parts of *P. karka*. *P. karka* was found to have high allelopathic potential to suppress the growth of both the investigated tree-floating weeds. *P. karka* killed *S. molesta* that died after 21 days. It also suppressed the growth of water hyacinth. The rate of loss of dry weight of water hyacinth in the shoot leachate of *P. karka* was higher (94 per cent) than the root leachate (75 per cent). However, *P. karka* had no effect on submerged plants.

These results confirmed that above ground parts of *P. karka* was found to be more toxic to the growth of Salvinia and water hyacinth than below ground and highlighted on the biological interactions of emergent and free floating weeds. This study also indicated the role of *P. karka* in integrated program for biocontrol of free floating obnoxious weeds.

***Keywords:** Allelopathy, Phytotoxins, Leachate, Water Hyacinth, Phragmites karka, Salvinia molesta.*

Introduction

Phragmites karka (Retz.) Trin. Ex. Steud commonly known as reed is a tall perennial macrophyte growing in shallow water, but also tolerates a wide range of water and soil regimes (Gopal 1990). It is found as monospecific stands in downstream areas of fresh water bodies and along the road sides in low-lying areas with very thick stand (4643–9731 g/m^{-2}). The growth of *phragmites communis* is easily suppressed by *Glyseria maxima* and *Agrostis* sp. (Szczepanski, 1977) whereas, it inhibits the seedling growth of some crops and vegetables (Kulshreshtha 1981). Gopal and Goel (1993) reported that vegetative multiplication of seedlings of *Typha aungustata* of an adverse effect of soil of *P. karka* stand. The soil collected from *P. karka* stand also inhibited the growth of seedlings of mustard (Gupta 1998). Besides, Sharma *et al.* (1990) have reported autotoxicity in this plant. All these studies indicate the potential of *Phragmites* plant to inhibit the growth of other plants. Szczepanski (1977) has specified the importance of allelopathy in biocontrol studies. Gopal and Goel (1993) have reviewed the role of competition and allelopathy in aquatic plant communities and thrown light on various aspects of allelopathy in aquatic systems.

Hence, study was conducted to determine the allelopathic effect of aqueous leachate of an emergent weed *P. karka* on the growth of some free floating and submerged fresh water weeds.

Materials and Methods

Collection of Plant Material

Fresh plants of *Phragmites karka* Trin. Ex. Steud (family Poaceae) water hyacinth (*Eichhornia crassipes* (Mart.) Solms., *Salvinia molesta* Mitchell, *Hydrilla verticillata* (L.F.) Royle, *Ceratophyllum demersum* L. and *Utricularia flexuosa* Vahi. were collected from a pond of 2.4 x 3.0 x 0.6 m in the Botanical Garden, University Campus, Jaipur.

Preparation of Leachate of *P. karka*

The above ground (shoots) and below ground parts (roots and rhizome) of *P. karka* were separated and cut in to 5 cm pieces. Thirty gram shoot were separately immersed in 1000 ml tap water (3 g/ 100ml=3 per cent) at room temperature (28–48°C) for 48 h. Tap water was used instead of distilled water in order to maintain more natural conditions. Thereafter, the mixture was passed twice through two layered cheese cloth and the filtrate was permitted to settle down for 8h. The upper portion of the filtrate was decanted and used for experiment. This leachate contained 3 per cent concentration of the shoot of *P. karka*. The same procedure was repeated by immerging thirty gram roots and rhizome in 1000 ml tap water (3 g/100 ml = 3 per cent) in tap water for obtaining 3 per cent root leachate of *P. karka*.

Simple growth experiments, were set up in plastic pots in an experimental chamber (3.0 x 2.4 x 1.8 m) made of an iron net (2.6 x 2.6 cm). Light intensity ranged from 1000–2500 Lux measured by Lux meter (model Aplab ML 4420). The temperature during these months ranged from 28–40°C. The following four experiments were performed.

Effect of Shoot/Root Leachate of *P. karka* on Free Floating Weeds Effect on Water Hyacinth

The experiment was conducted in plastic pots (2500 ml). Each of fifteen pots was filled with 2000 ml leachate obtained from shoot or root of *P. karka*. These were taken as the treated set. An equal number of pots filled with 2000 ml water served as the control set. Each pot had two water hyacinth plants.

Water hyacinth plant was consisting of a main plant body (1) without any off shoots connected by stolons. Each plant had a total of 4 leaves with maximum length of 8 cm. Each also had two leaf buds and roots of approximately of 11 cm. Initial dry weight of the above ground and below ground parts of hyacinth plants were 0.73±0.01 and 0.28±0.01g, respectively. Five pots (replicates) from each set were taken and the leaves were counted at weekly intervals. The senescence of the green parts of the water hyacinth was considered as complete death of the plant. On subsequent days 250 ml water was added daily to maintain water level during the experiment. The data for leaf number, dry matter, and relative growth rate was recorded at 7, 14 and 21 days. After being washed the shoots and roots were separated and dried in a hot air oven at 80°C to obtain dry weight.

Effect on Salvinia

Fresh plants of *S. molesta* were collected from a pond in the Botanical Garden, University Campus, Jaipur. An experiment was conducted in plastic disposable pots. The 3 per cent shoot and roots leachate was prepared as described above. Fifteen pots were filled with 200 ml (shoot/root) leachate obtained from *P. karka* plant. These were taken as treated set. Each pot had two small healthy *S. molesta* plant. Each plant had a total of 6 free floating leaves attached with submerged leaves with an apical bud having 0.08±0.005 g dry weight were grown in the shoot or root leachate of the *P. karka.* On subsequent days 250 ml water was added daily to maintain water level during the experiment. The senescence of the green parts of the *Salvinia* was considered as the complete, death of the plant. Plants were harvested after 21d. The number of the leaves and dry weight of *Salvinia* was measured. Relative Growth Rate (RGR) was calculated.

Effect on Submerged Macrophytes

Fresh twigs of *P. karka* were cut into small pieces and immersed in water at 30 g/liter for the duration of 24h. The leachate was filtered as described previously. Aqueous leachate of 3 per cent concentration (w/v) of fresh above ground parts of *P. karka* was used for the growth of different submerged plants. A total of 30 pots with five replicates (5 for each plants and control) were set up. Ten small twigs of submerged aquatic plants *viz. C. muricatum, U. flexuosa* and *H. verticillata* (having 7, 9 and 9 cm initial length with 2.28±0.02, 0.45±0.01, 1.20±0.02 g initial dry weight respectively) were allowed to grow in the aqueous leachate of the *P. karka*

On subsequent days 250 ml water was added daily to maintain water level during the experiment. The length and dry weight of the submerged plants (shoots) were measured at 7, 14 and 21 days.

Seed Germination Bioassay

The treatments consisted of 3 per cent aqueous leachate of shoot, root and field crops. [wheat (*Triticum aestivum* L. var. Kalyansona), mustard (*Brassica compestris* L. var. Varuna) and radish (*Raphanus sativus* L. Cv. 'HR–1']. The treatments were replicated five times. The bioassay with aqueous leachate (shoot, root) of *P. karka* was conducted in petriplates. The petriplates were sterilised in Autoclave at 1.06 kg per cm^2pressure for 20 min and after cooling, these were lined with 2 filter papers. Per petriplate, 20 seeds of either test crop (wheat, mustard and radish) were kept on filter paper. Thereafter, as per treatment 10 ml leachate or water (control) was added in each petriplate. The petriplates were kept in room at 28–40°C, in dark/light. On subsequent days 10 ml distilled water (1 ml per day) was added at 10 Days After Sowing (DAS). The seed germination, root and shoot length and dry matter were recorded at 10 DAS. The Germination Inhibition Rate (GIR) was calculated.

Statistical Analyses of the Data

Student's t test was applied to test the significance of difference between means of treated and control set following the method described by Allen (1989). Significance of these values at 5 and 1 per cent levels is shown in Tables 7.1–7.8. Anova was calculated to analyses the difference of mean values of each treated (shoot and root leachate) and control set, CD was calculated.

Table 7.1: Effect of 3 per cent Aqueous Leachate of *P. karka* on the Number of Leaves and Growth of *E. crassipes*

Phragmites karka	*Days*	*Control*	*Treated*	*% Loss of Control*	*CD*[b]
Shoot	7	4.33±0.21	1.83**±0040	–57.76	1.290
	14	5.33±0.33	1.0**±0045	–81–30	1.720
	21	6.67±0.21	0.16**±0.16	–97.60	0.542
Root	7	4.17±0040	3.83±0.31	–8.15	0.542
	14	5.66±0.21	3.5*±0.85	–38.16	2.041
	21	5.83±0.17	2.50**±0.50	–57.11	10434

a: Means of five replicates, b: Critical difference significant at 5 per cent level of probability.

Results

Effect of Shoot Leachate of *P. karka* on Water Hyacinth

The number of leaves and dry weight of water hyacinth was drastically decreased at each harvest in the 3 per cent shoot leachate of *P. karka,* whereas the same increased in the control. The dead leaves of water hyacinth were found floating on the surface of water in treated set after 21 days of plantation. Water hyacinth completely died at 30 per cent concentration of shoot leachate after 21 days. About 97.6 per cent inhibition in the number of the leaves, –99.26 per cent and –79.77 per cent in the dry weight of the above ground and under ground parts over control) was observed. Relative Growth Rate (RGR) values (the number of leaves per days) ranged –037.5 to –113.7 × 10^{-3} in the shoot leachate of *P. karka* (Table 7.2).

The shoot leachate of *P. karka* was found to be more toxic and caused complete death of water hyacinth.

Table 7.2: Effect of 3 per cent Aqueous Leachate of *P. karka* on the Relative Growth Rate x 10^{-3} of Water Hyacinth[a]

Phragmites karka	*Days*	*Control*	*Treated*	*CD*[b]
Shoot	7	5.0	–48.5	2.10
	14	13.0	–37.5	0.92
	21	14.0	–113.7	39.33
Root	7	2.6	–2.7	1.31
	14	19	–56	5.67
	21	1.80	–21	0.028

a: Means of five replicates, b: Critical difference significant at 5 per cent level of probability.

Effect of Root Leachate of *P. karka* on Water Hyacinth

In comparison to the shoot leachate, root leachate was found to be less inhibitory to the growth of free floating macrophytes. The number of the leaves of *E. crassipes* in the root leachate of *P. karka* gradually decreased (Table 7.1) about 57.11 per cent inhibition in the number of leaves were found after 21 days. Relative growth rate values (number of leaves per day) for water hyacinth ranged from –2.7 x 10.3 to 56 x 10^{-3}. The dry weight of the above ground and below ground parts of water hyacinth (Table 7.3) was inhibited –88.46 and –25.8 per cent over control after 21 days in the root leachate of *P. karka*.

Table 7.3: Effect of 3 per cent Aqueous Leachate of *P. karka* on the Dry Weight (g) of Water Hyacinth After 21 Days[a]

Phragmites karka	*Days*	*Control*	*Treated*	*% Loss of Control*
Shoot	AB	1.36±0.17	0.01[c]±0.01	99.26
	BG	0.89±0.03	1.60[c]±0.07	79.77
Root	AB	1.04±0.13	0.12[c]±0.04	88.46
	BG	0.31±0.001	0.23[b]±0.03	25.8

A: Means of five replicates; b and X: Significant at 5 & I per cent level of probability.

AB means above ground parts of water hyacinth; BG means below ground parts of water hyacinth.

Table 7.4: Effect of 3 per cent Aqueous Leachate of *P. karka* on the Number of Leaves and Growth of *Salvinia molesta*[a]

Phragmites karka	*Days*	*Control*	*Treated*	*% Loss of Control*	*CD[b]*
Shoot	7	50.33±05.04	19.33**±2.91	61.59	28.220
	14	40.66±2.01	1.33**±1.33	–96.70	5.172
	21	60.22±0.19	Death	–100	Death
Root	7	50.33±05.04	22.0*±4.04	–56.3	3.910
	14	40.66±2.01	10.66*±1.20	–73.78	6.362
	21	60.22±02.01	2.50**±01.04	–95.85	Death

a: Means of five replicates; b: Critical difference, significant at 5 per cent level of probability.

Table 7.5: Effect of 3 per cent Aqueous Leachate of *P. karka* on the Dry Weight (g) of *Savinia molestaa* After 21 Days

Phragmites karka	*Control*	*Treated*	*% Loss of Control*	*CD[b]*
Shoot	0.37±0.006	Death	–100.0	
Root	0.37±0.006	0.016*±0.001	–98.76	2.041

a: Means of five replicates; b: Critical difference significant at 5 per cent level of probability.

Effect of Shoot Leachate of *P. karka* on Salvinia

The shoot leachate of *P. karka* was found to be more toxic and caused complete death of *S. molesta*. The number of the leaves of *S. molesta* declined sharply after 7 and 14 days and –96.73 per cent inhibition over control was recorded in the treated set. The relative growth rate values ranged from –30 to –171 x 10^{-3} leaves per day in the shoot leachate of *P. karka*. The dry weight of the *S. molesta* (Table 7.6) was inhibited –99.08 per cent over control after 14 days, in the shoot leachate of *P. karka* and died thereafter.

Table 7.6: Effect of 3 per cent Aqueous Leachate of *P. karka* on the Relative Growth Rate x 10^{-3} of *Savinia molesta*[a]

Phragmites karka	*Days*	*Control*	*Treated*	*CD*[b]
Shoot	7	90.0	–30	2.10
	14	13.2	–166.1	0.92
	21	14.0	–171.0	39.33
Root	7	90.0	38.3	1.31
	14	13.2	–45.0	5.67
	21	14.0	–49.50	0.028

a: Means of five replicates; b: Critical difference significant at 5 per cent level of probability.

Effect of Root Leachate of *P. karka* on Salvinia

The root, leachate of *P. karka* was found less toxic but it also suppressed the growth and dry weight of *Salvinia*

The relative growth rate (Table 7.5) values ranged from 38.3 to 49.5 x 10^{-3} leaves per day in the root leachate of *P. karka*. The dry weight of the *S. molesta* (Table 7.6) was inhibited –98.76 per cent over control after 21 days in the root leachate of *P. karka*

Difference of mean values of each treated (shoot and root leachate) and control sets are greater (Tables 7.1–7.2) than the Critical Difference (CD) so it is concluded that *P. karka* significantly inhibited the growth of *E. crassipes* and *Salvinia molesta*.

Effect of Shoot Leachate of *P. karka* on Submerged Plant

The length of these macrophytes increased up to 113.19 in *C. muricatum*, 128.27 in *U. flexuosa* and 119.41 per cent in *H. verticillata* after 7 d in comparison to control.

The growth of *H. verticillata* was significantly (Table 7.7) promoted by the aqueous leachate of *P. karka* after 21 d at 5 and 10 per cent level respectively. However, *C. muricatum*, *U. flexuosa* were insignificantly affected.

Seed Germination and Seedling Growth Bioassay

The seed germination was observed 54, 93, and 96 per cent in wheat, mustard and radish (Table 7.8) respectively in the shoot leachate of *P. karka*. The Germination Inhibition Rate (GIR) was obtained to be 38.0, 1.1 and 3.1 in wheat, mustard and radish seeds respectively after 10 days. *P. karka* significantly reduced the growth of seedlings of wheat, mustard and radish (length of shoot and roots). The dry weight of the seedlings also decreased significantly in all the treated sets.

Table 7.7: Effect of Aqueous Leachate of *P. karka* on Seed Germination and Seedling Growth

Treatment	*Germination Per cent*	*GI/R*	*Length (cm)*		*Dry Matter (g)*
			Shoot	*Roots*	
			Wheat		
Control	87.00±03.00		10.15±00.89	10.60±00.65	3.29±01.09
Treatment	54.00±04.00	38.00	06.20±01.02	03.54±00.04	01.87±00.07
			Mustard		
Control	94.00±03.70		04.63±00.37	06.62±00.60	02.94±00.13
Treatment	93.00±02.0	01.10	01.56±00.23	00.76±00.15	01.56±00.07
			Radish		
Control	99.00±01.0		7.67±0.67	8.92±0.86	02.52±00.07
Treatment	96.00±01.9	3.1	5.11±0.85	2.29±0.27	

GIR: Germination inhibition rate, ±SE.

Table 7.8: Effect of 3 per cent Aqueous Leachate of *P. karka* on the Growth and Dry Weight (g) of Different Submerge Macrophytes

Species	*Length (cm)*				*Dry wt. (g)*		
	Treated	*Initial*	*7d*	*14d*	*21d*	*Initial*	*21d*
Ceratophyllum	Control	7.0±0.0	7.98±0.22	8.92±0.38	10.08±0.62	2.28±0.02	2.50±0.13
muricatum	Leaching	7.0	9.04[c]±0.23	9.4±0.40	9.89±0.45	2.28±0.02	2.18±0.08
Utricularia	Control	9.0±0.00	8.6±1.01	7.6±1.5	8.4±1.7	0.45±0.01	1.61±0.14
flexuosa	Leaching	9.0±0.0	11.01±1.01	11.5±1.3	10.8±1.5	0.45±0.01	1.85±0.08
Hydrilla	Control	9.0±0.0	12.5±0.3	13.56±0.41	16.6±0.5	1.2±0.02	2.34±0.1
verticillata	Leaching	9.0±0.0	14.93 [c]±0.50	17.50[c]±0.70	21.3 [c]±0.91	1.2±0.02	2.90[b]±0.12

a: Means of five replicates; b+c: Significant at 5 and 1 per cent level of probability.

Discussion

Allelopathy exhibits both promoting as well as inhibitory effects through the production of allelochemicals (Rice, 1984). There are a number of published information about allelopathic interactions among terrestrial plants (Rice, 1986) but much less is known about biological interactions among aquatic macrophyte (Szczepanska, 1971; Rice, 1984). Further developments in allelopathic studies in aquatic field have been reviewed. (Gopal and Goel, 1993).

Today, Allelopathy is well known in emergent macrophytes. McNaughton (1968) reported allelopathy for the first time in emmergent plant *Typha aungustifolia*. Later on, Szczepanska (1971) and Szczepanski (1977) also worked out on allelopathy of emergent macrophytes. They reported biological interaction among four macrophytes. Bonasera *et al.* (1979) investigated allelopathy in marshy emergent species. The most frequently studied emergent hydrophyte is a dwarf spikerush *Eleocharis* sp. which is reported to be allelopathic to the growth of pond weed (*Potamogeton* sp.) in laboratory experiments and under field conditions against the growth of *Potamogeton nodosus, P. pectinatus, Hydrilla verticillata,*

Elodea nuttalii and *Zannichellia palustris* (Ashton *et al.*, 1985, Frank and Dechoretz, 1980; Yeo and Thurston, 1984; Wooten and Elakovich, 1991; Sharma *et al.*, 1998).

The present study indicates the biological interactions between emergent and free floating weeds and confirms allelopathic nature of fast growing *P. karka* which killed the other two weeds water hyacinth and *Salvinia* within 21 days. The aqueous leachate of 3 per cent concentration was found to be highly toxic to the growth of *E. crassipes* and *S. molesta*. In earlier study, Saxena (1991) has reported the reduced length and breadth of leaf of water hyacinth (78.30 and 60.75 per cent in comparison to control) in the decomposing residue of aboveground parts of *P. karka*.

It is evidenced from the present study that the shoots of *P. karka* were more toxic than the roots. In our earlier study on the effect of 1 per cent extract of the leaves, stem and roots of *P. karka*, only leaf extract was found inhibitory (44 per cent reduction in the number of fronds) to the growth of duckweed (*Spirodela polyrrhiza*).

Thus, the present study confirmed the adverse toxic effect of *P. karka* to the growth of fee floating macrophytes *viz.*, water hyacinth (present study), *Salvinia molesta* (present study and Sengar and Sharma, 1993) and *Spirodela* polyrrhiza (Kulshreshtha, 1981).

Segar and Sharma (1993) however ambiguously reported the inhibition of *Salvinia* at 0.5 per cent concentration and promoting effect at higher concentration (1 and 3 per cent). Nevertheless, to mention here that allelochemicals behave like hormones, which promote the growth of other plants at lower concentration and inhibit the same at higher dose (Rice, 1986).

P. karka inhibited seedling growth (wheat, barley, pea and chickpea) of a number plants (Kulshreshtha, 1981). The 1 per cent extract inhibited 21, 29, 31 and 44 per cent length of radicles of these seedlings respectively over control and 25 and 4 per cent in the length of plumule in pea and barley. Besides, it was found to be inhibitory to the growth of wheat seedling (Konis, 1974). In the present study *P. karka* also inhibited the seedling growth of wheat, mustard and radish and confirmed the allelopathic nature of this emergent aquatic weed. It was reported that seedlings of *Typha aungustata* were not able to multiply vegetatively in the presence of soil of *P. karka* stand (Sharma *et al.*, 1990). Autotoxicity of this plant has also been worked out (McNaughton, 1968). Today allelopathy has been used as a means for biological control of aquatic weed (Szczepanski, 1977). However, an allelopathic interaction between submerged macrophytes has also been reported (Kulshreshtha and Gopal, 1983).

The importance of allelopathy in the distribution of the species in the community is well established *Phragmites karka* is known to form very thick persistent mono specific stand. The secretion and accumulation of toxic allelochemicals of *P. karka* may alter the surrounding aquatic vegetation. It was reported that seedlings of *Typha aungustata* were not able to multiply vegetatively in the presence of soil of *P. karka* stand (Gopal and Goel 1993). *P. karka* inhibited the germination growth of its own seedlings (autotoxicity) which was reported by Sharma *et al.*, 1990. Soil collected from *P. karka* stand also inhibited the seed germination and seedling growth of mustard (Gupta, 1998). Thus, there is a possibility for this plant to restrict the growth of other macrophytes in nature due to its allelopathic potential which needs to be investigated.

Today allelopathy has been used as a means for biological control of aquatic weed (Szezepanski, 1977; Saxena, 2000). Additionally, this plant promoted the growth of rice seedlings (Kumar *et al.*, 2004). As both are aquatic plants, the present investigation reveals weed (emergent–*P. karka*) to weed (free floating-water hyacinth/Salvinia) biochemical interaction in aquatic ecosystem. These observations on allelopathic potential of *P. karka* also highlight its importance in utilising this plant in biocontrol studies of free floating macrophytes in the field.

References

Allen, S.E., 1989. *Chemical Analysis of Ecological Materials,* 2nd edn. Blackwell Scientific Publications, Oxford London, Edinburgh Boston Melbourne, 565 pp.

Ashton, F.M., Tomason, J.M.D.I. and Anderson, L.W.A., 1985. Spikerush (*Eleocharis* spp): A source of allelopathics for the control of undesirable aquatic weeds. *Journal of Aquatic Plant Management,* 22: 52–56.

Bonasera, J., Lynch, J. and Leck, M.A., 1979. Comparison of the allelopathic potential of four marsh species. *Bull. Torry, Botanical Club,* 106: 217–222.

Frank, P.A. and Dechoretz, N., 1980. Allelopathy in dwarf Spikerush (*Eleocharis coloradoensis*). *Weed Science,* 28: 499–505.

Gopal, B. and Goel, U., 1993. Competition and allelopathy in aquatic plant communities. *Botanical Review,* 56: 156–210.

Gopal, B. (Ed.), 1990. *Ecology and Management of Aquatic Vegetation in the Indian Subcontinent.* Geobotany 16. Kluwer Academic Publishers, Dordrecht.

Gupta, J., 1998. Studies on allelopathic potential of some terrestrial and aquatic weeds. *Ph.D. Thesis,* University of Rajasthan, Jaipur.

Konis, E., 1974. On germination inhibitors. VI. The inhibiting action of leaf saps on germination and growth. *Palestinian Journal of Botany,* Jerusalem, Ser. IV. 4: 77 – 85.

Kulshreshtha, M. and Gopal, B. 1983. Allelopathic influence of *Hydrilla verticillata* (L.F.) Royle on the distribution of *Ceratophyllum* species. *Aquatic Botany,* 16: 207–209.

Kulshreshtha, M., 1981. Allelopathic influence of some macrophytes. *Actalimnologia Indica,* 1: 35–37

Kumar, R., Singh, S. and Saxena, M., 2004. *Phragmites karka* enhanced the growth of rice seedlings. International workshop on protocols and methodologies on Allelopathy (*IWPMA*), p. 32.

McNaughton, S.J., 1968. Autotoxic feedback in relation to germination and seedling growth in *Typha latifolia. Ecology,* 49: 367–369. .

Rice, E.L., 1986. Allelopathic growth stimulation. In: *The Science of Allelopathy,* (Eds.) Putnam, A.R. and Tang, C.S. Wiley Interscience, New York, pp. 23–43.

Rice, E.L., 1984. *Allelopathy,* 2nd Edition. Academic Press, New York.

Saxena, M.K., 1991. Effect of terrestrial litter inputs on the growth of a qua tic plants. In: *International Conference on Land-Water Interactions,* Abstract: National Institute of Ecology, New Delhi, p. 147,

Saxena, M.K., 2000. Aqueous leachate of *Lantana camara* kills water hyacinth. *Journal of Chemical Ecology,* 26(10): 2435–2446.

Sengar, R.S. and Sharma, K.P., 1993. Biological control of *Salvinia molesta* Mitchell with *P. karka* (Tetz.) ex. Steud. *Geobios,* 20: 267–268.

Sharma, K.P., Khuswaha, S.P.S. and Gopal, B., 1998. A comparative study of stand structure and standing crops of two wetland species, *Arundo donax* and *Phragmites karka,* and primary production in *A. donax* with observations on the effect of clipping. *Tropical Ecology,* 39(1): 3–14.

Sharma, K.P., Kushwaha, S.P.S. and Gopal, B., 1990. Autotoxic effect of *Phragmites karka* (Retz.) Trin. Ex. Steud. Plant on its seed germination. *Geobios,* 17: 287–288.

Szczepanska, W., 1971. Allelopathy among the aquatic plants. *Polish Hydrobiologia,* 18: 17–30.

Szczepanski, A.J., 1977. Allelopathy as a means of biological control of water weeds. *Aquatic Botany,* 3: 193–197.

Wooten, J.W. and Elakovich, S.D., 1991. Comparisons of potential of allelopathy of seven freshwater species of spikerush (*Eleocharis*). *Aquatic Plant Management,* 29: 12–15.

Yeo, R.R. and Thurston, J.R., 1984. The effect of dwarf spikerush (*Eleocharis coloradoensis*) on several submerged aquatic weeds. *Journal of Aquatic Plant Management,* 22: 52–56.

Chapter 8

Sodium and Chloride Ions Uptake and Translocation in *Sesbania sesban* (L). Merr. Dye to NaCl Stress

S.R. Harish, T. Gayathri and K. Murugan

Plant Biochemistry and Molecular Biology Lab, Department of Botany, University College, Thiruvananthapuram, Kerala – 695 034

ABSTRACT

Under salinity conditions, the increasing accumulation of Na^+ and Cl^- is of primary importance in physiological action of plants. Excess of these ions can be extremely toxic to plant cells. The restriction of the ion accumulation in roots and shoots is important in the salt-tolerance mechanisms. In the present study sodium and chlorine uptake and translocation from roots to shoots and percentage of their accumulation in roots of this plant was investigated. *Sesbania sesban* (L). Merr has a greater tolerance to NaCl and that could resist NaCl up to 200 mM. High salt tolerance of the species is probably related with the ability to restrict Na^+ and Cl^- content in roots and translocate higher amount of the ions to hold in shoots.

Introduction

Excess amount of salt in the soil adversely affects plant growth and development. Nearly 20 per cent of the world's cultivated area and nearly half of the world's irrigated lands are affected by salinity. Processes such as seed germination, seedling growth and vigour, vegetative growth, flowering and fruit set are adversely affected by high salt concentration, leading to poor economic yield and also quality of produce. Plants are classified as glycophytes or halophytes according to their capacity to grow on high salt medium. Most plants are glycophytes and cannot tolerate salt-stress. High salt concentrations decrease the osmotic potential of soil solution creating a water stress in plants.

Secondly, they cause severe ion toxicity, since Na^+ is not readily sequestered into vacuoles as in halophytes. Finally, the interactions of salts with mineral nutrition may result in nutrient imbalances and deficiencies. The consequence of all these can ultimately lead to plant death as a result of growth arrest and molecular damage. To achieve salt-tolerance, the foremost task is either to prevent or alleviate the damage, or to re-establish homeostatic conditions in the new stressful environment. Na^+ and Cl^- are of primary importance in physiological action of plants under saline stress. Excess of these ions can be extremely toxic to plant cells. The restriction of the ion accumulation in roots and shoots is important in the salt-tolerance mechanisms. Ion distribution within the plant may generally be described in two patterns. The first one, these plant have the ability to accumulate high levels of Na and CI in shoots without any detrimental effect (Murugan, 2005). In this pattern, ion translocation is an important factor for the tolerance of the plants. It is found that on average more than 90 per cent of absorbed Na^+ is translocation to shoots and at least 80 per cent accumulated in the leaves frequently in the old leaves, thus preventing any impact in young photosynthetically active leaves (Ghoulam *et al.*, 2002). In some plants, their leaves can avoid the injurious effect of salts by compartmentalization, succulence or excretion (Warwick and Halloran 1992). The second one, this plant group can basically tolerate salinity by including excessive salts in root cells. These plants have ability to restrict translocation of Na^+ and Cl^- from roots to shoots. Examples of these plants include rice and soybean (Rabie and Kumazawa 1988, Warwick and Halloran 1992). In the present study Sodium and chloride uptake and New File translocation from roots to shoots and percentage of their accumulation in roots of the plant was investigated under different concentrations of NaCl treatment.

Material and Methods

Seeds of *Sesbania sesban* were germinated and grown under different concentrations of NaCl ranging from 50–200 mM. When their seedlings reach the 3rd to 4th leaf stage, they were treated in solutions of higher concentrations, 50, 100, 150, and then 200 mM at two-day intervals. After reaching the final concentrations of NaCl, the seedlings were kept in a growth chamber. NaC1 solution containing nutrients were renewed every 4 days. Eighteen days after starting NaCl treatment, the seedlings were harvested for the following assays.

Bioassay

The test seedlings were divided into roots and shoots, and dried at 80°C for 48 h for the determination of dry weight.

Ions Analyses

The dried sample (0.1 g) of roots and shoots were ground into a fine powder for wet digestion and dry ashing. The solution obtained by wet digestion was analyzed for Na^+ content by a plasma atomic emission spectrophotometer. The solution prepared from dry ashing was analyzed for CI^- content by an ion chromatographic analyzer.

Results and Discussion

The effects of NaCl on shoots and roots dry weight are shown in Figure 8.1. Increase of NaCl concentration decreased shoots and roots dry weight in the species. At low concentrations (50 mM), NaCl had minimal effect on *S. sesban* dry weights. At the highest concentration (200 mM NaCl), although the shoot dry weight was reduced to 46 per cent of the control, the seedlings still survive. Figures 8.2a and b show the Na^+ and Cl^- distribution in *S. sesban*, respectively. In *S. sesban*, Na and Cl contents in roots and shoots were enhanced with increasing NaCl concentrations and those ion concentrations were higher in shoots than in roots. Translocation rates of Na^+ and Cl^- in *S. sesban* was shown in Table 8.1. The translocation rate of ions in *S. sesban* increased slightly with increasing NaCl concentration. At 50–200 mM NaCl treatment, 29–41 per cent of Na^+ and Cl^- remained in *S. sesban.*

Table 8.1: Uptake and Translocation of Na^+ and Cl^- in *S. sesban*

	Na^+ (μmol/plant)	*TR (%)*	*Cl^- (μmol/plant)*	*TR (%)*
NaCl (mM)				
0 Shoot	139	67.5	121	69.9
Root	67	32.5	52	30.1
Total	206		173	
50 Shoot	287	66.7	164	60.5
Root	143	33.3	107	39.5
Total	430		271	
100 Shoot	296	66.5	168	60.9
Root	149	33.5	108	39.1
Total	445		276	
150 Shoot	334	74.4	212	61.8
Root	165	36.7	131	38.2
Total	449		343	
200 Shoot	299	73	199	67.2
Root	110	26.9	97	32.8
Total	409		286	

The bioassay indicates that *S. sesban* has a greater tolerance to NaCl. *S. sesban* could resist NaCl up to 200 mM. During NaCl stress, roots of *S. sesban* responded to the stress by increasing Na^+ and Cl^- accumulation, but the quantity of these ions were greater in the shoots. Calculation of the translocation rate of Na^+ and Cl^- (Table 8.1) from roots to shoots and percentage of their remains in roots (Table 8.1) also confirmed that *S. sesban* has an ability to limit Na^+ and Cl^- accumulation in roots and translocate higher amount of the ions to shoots. The ability to absorb Na^+ and Cl^- from roots and transfer them to shoots is considered to be one of the mechanisms of salt tolerance in *S. sesban.* High accumulation of

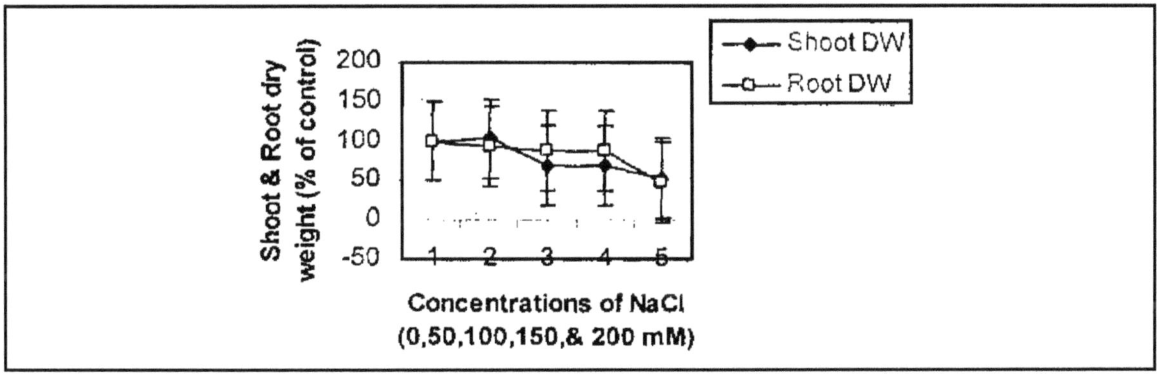

Figure 8.1: Dry Weight of Shoot and Root after NaCl Treatment under Different Concentrations

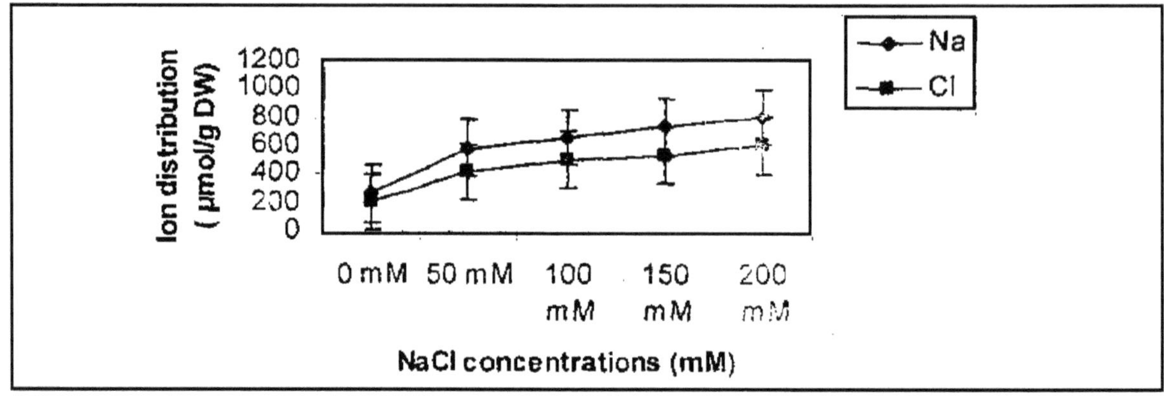

Figure 8.2a: Ion Distribution in Root Treated with NaCl in *S. sesban* after 18 Days (mM)

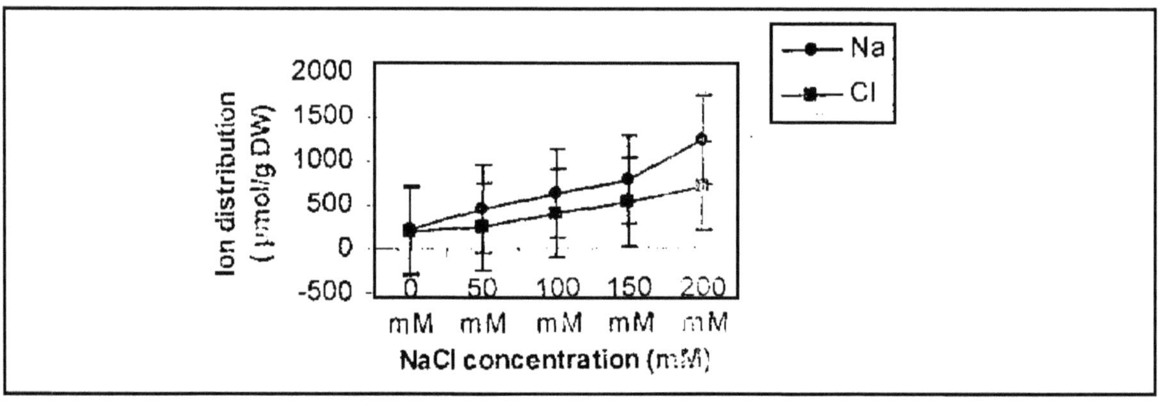

Figure 8.2b: Ion Distribution in Shoot Treated with NaCl in *S. sesban* after 18 Days

the excessive salts in shoots indicates that there have some mechanisms to reduce detrimental effect of the ions in the shoot cells. Halophytes resist salinity stress by accumulating high contents of Na^+ and Cl^- within the leaf cells. Localizing Na^+ and Cl^- in the vacuole and balancing by compatible solutes within the cytoplasm was reported as the important salt tolerant mechanisms at cellular level (Adams *et al.*, 1992; Bremberger and Luttge, 1992; Murugan and Sathish, 2005). The mechanisms of salt tolerance in the shoot cells of *S. sesban* need further clarification. The results obtained from this study are quite clear that the salt tolerance of *S. sesban* does not occur with ion inclusion in the roots but there is a true tolerance of high ion contents in the shoot cells. This is accepted as a characteristic of some halophytes as previously reported (Black, 1956; Scholander *et al.*, 1966; Rains and Epstein, 1967).

Conclusion

S. sesban has a greater tolerance to NaCl that could resist NaCl up to 200 mM. High salt tolerance of *S. sesban* is probably related with the ability to restrict Na^+ and Cl content in roots and translocate higher amount of the ions to hold in shoots. Further studies are warranted to analyze the impact of salinity on metabolism of the plant and also its biochemical salt tolerant characteristics.

References

Adams, P., Thomas, J.C., Vernon, D.M., Bohnert, H.J. and Jensen, R.G., 1992. Distinct cellular and organismic responses to salt stress. *Plant Cell Physiol.,* 33: 1215–1223.

Black, R.F., 1956. Effect of NaCl in water culture on the ion uptake and growth of *Atriplex hastata* L. *Aust. J. Biol. Sci.,* 9: 67–80.

Bremberger, C. and Luttge, U., 1992. Dynamics of tonoplast proton pumps and other tonoplast proteins of *Mesembryanthemum crystallinum* L. during the induction of Crassulacean acid metabolism. *Planta,* 188: 575–580.

Ghoulam, C., Foursy, A. and Fares, K., 2002. Effects of salt stress on growth, inorganic ions and proline accumulation in relation to osmotic adjustment in five sugar bee cultivars. *Environ. Exp. Bot.,* 47: 39–50.

Murugan, K., 2005. Mineral composition and salinity tolerance of mangroves in different habitats of Kerala. *Indian J. of Environ. and Ecoplan.,* 10: 391–396.

Murugan, K. and Sathish, D.K., 2005. Ameliorative effect of NaCl salinity stress by calcium related to praline metabolism in the callus of *Centella asiatica* L. *J of Plant Biochem. and Biotech.,* 14: 205–207.

Rabie, R.K. and Kumazawa, K., 1988. Effect of NaCl salinity on growth and distribution of sodium and some macronutrient elements in soybean Plant. *Soil Sci. Plant Nutr.,* 34: 375–384.

Rains, D.W. and Epstein, E., 1967. Preferential absorption of potassium by leaf tissue of the mangrove, *Avicennia marina:* An aspect of halophytic Competence in coping with salt. *Aust. J. Biol. Sci.,* 20: 847–857.

Scholander, P.F., Bradstreet, E.D., Hammel, H.T. and Hemmingsen, E.A., 1966. Sap concentrations in halophytes and some other plants. *Plant Physiol.,* 4: 529–532.

Warwick, N.W.M. and Halloran, G.M., 1992. Accumulation and excretion of sodium, potassium and chloride from leaves of two accessions of *Diplachne fusca* (L.) Beauv. *New Phytol.,* 121: 53–61.

Chapter 9

Physico-Chemical Analysis of Soil of Two Ponds of Bishunpur Sri Ram Village of Muzaffarpur (Bihar) in Relation to Pisciculture

Shishir Kumar[1], Rajeev Ranjan[2], Sachidanand Mishra[3] and Pankaj Kumar[4]

[1]Research Scholar, P.G. Department of Zoology, R.D.S. College, Muzaffarpur
[2]Research Scholar, M.P. Sinha Science College, Muzaffarpur
[3]Research Scholar, [4]Post Doctoral Research Scholar
P.G. Department of Zoology, B.R.A., Bihar University, Muzaffarpur – 842 001

ABSTRACT

Soil condition of two ponds of Bishunpur Sri Ram Village was completely alkaline, and presence of good amount of calcium and magnesium were somewhat similar to findings of Raheja (1968). Phosphate concentration was also higher. Bicarbonates were found in good amount but with carbonate. Therefore, soil condition of the present pond was found quite suitable for pisciculture.

Keywords*: Pond, Physico-chemical analysis of soil, Pisciculture.*

Introduction

Soil plays an important role in determining the fertility of fish ponds. Naturally, the constituents of soil have got bearing with the fishes. Soil constitutents have direct relationship with the growth of aquatic animals and plants. In India, different investigators (Saba *et al.*, 1971; Pillai and Sreeniwasam, 1975; Mandie and Moitra, 1975, Nasar, 1978) studied soil condition and ecological parameters of water bodies but report on their correlation is scarce.

Soil contains a variety of elements, chemical components and minerals, Calcium, Magnesium, Potassium, Hydrogen, Oxygen, Nitrogen, Carbon, Zinc, Sulphur, Molybdenum, Copper, Manganese, Boron, Iron, Aluminium, Silicon, Sodium and Chloride, which are found in soil of North Bihar. According to Boon, 1958, these elements are never found in the soil in free state but they combine with other elements to focus compounds such as nitrogen as nitrate and phosphorus as phosphate.

Materials and Methods

Soil samples were collected monthly, mixed thoroughly and dried under shades. The air dried soil sample were analysed for important chemical components by standard methods of Piper (1947) and Wilde *et al.* (1972). In the 2nd week of each months water sample were also collected on a clear day before noon in plastic jar. Water temperature, transparency, pH and dissolved oxygen concentration were measured on the spot using maximum-minimum thermometer, Sachchi disc, standard pH-paper, and Winkler's method (APHA), respectively. Total phosphorus, total Nirtogen, Magnesium and Sulphur content were estimated according to the method described by Adoni (1985).

Results

pH of Soil

The hydrogen ion concentration of Pond A (Table 9.1) was ranging from 7.0 to 7.8. The pH value of the soil of this pond was 7.5 during January to April, but during May to August, it was 7.8, which decreased during September to December and showed the lowest value of pH 7.0.

Table 9.1: Soil Contents of Bishunpur Sri Ram Pond–A

Sl.No.	Soil Constituents	January to April	May to August	September to December
1.	pH	7.5	7.8	7.0
2.	Phosphate (ppm)	15.5	107.00	84.00
3.	Organic carbon (per cent)	0.5	2.25	1.75
4.	Carbonate (mg/100gm)	00.000	00.000	00.554
5.	Ca^{+} Mg^{++} (mg/100gr)	0.4430	1.040	00.7730
6.	Bicarbonate (mg/l00gr)	1.75275	3.0415	0.88240
7.	Conductivity of soil ml/ (m^{hos}/cmat 25°C)	0.6500	0.8240	0.2400

The hydrogen ion concentration of Pond B was ranging more or less with in similar pattern (7.8 to 8.5), Table 9.2. The pH value was 8.0 during January to April.

During May to August it was 8.5, which showed highest value. It attained the minimum during the month of September to December (7.8).

Phosphate

Phosphate content of the soil varied from 15.5 to 84.0 ppm of the pond A. The phosphate value during January to April was 15.5ppm. The maximum value of phosphate content was observed during May to April (107.00 ppm). Now it decreased in. the subsequent months September to December to 84.00 ppm, Table 9.1.

In Pond B, the phosphate content of the soil varied from 17.00 to 85.00ppm. The phosphate value during January to April was recorded 17.00 ppm. It increased in the subsequent month *i.e.,* May to August to 1000.00ppm and it decreased during the months from September to December to 85.00 ppm, Table 9.2.

Table 9.2: Soil Contents of Bishunpur Sri Ram Pond–B

Sl.No.	Soil Constituents	January to April	May to August	September to December
1.	pH	8.00	8.5	7.8
2.	Phosphate (ppm)	17.00	100.00	85.00
3.	Organic carbon (per cent)	0.50	2.45	1.75
4.	Carbonate (mg/100gm)	0.5350	1.1450	0.8780
5.	Ca^{+} Mg^{++} (mg/100gr)	00.000	00.000	0.520
6.	Bicarbonate (me/100gr)	1.75375	3.02650	0.83300
7.	Conductivity of soil ml/ (m^{hos}/cmat 25°C)	0.600	0.775	0.2490

Organic Carbon

Organic carbon was ranging from 0.5 per cent to 1.75 per cent in Pond A. It was 0.5 per cent during the month or January to April. It increased during the month of May to August to 2.25 per cent. But it decreased during the months of September to December to 1.75 per cent. Thus the highest value of organic carbon in Pond A was in the month of May to August (Table 9.1).

In pond B, organic carbon was ranging from 0.5 per cent to 1.75 per cent. The minimum organic matter (0.5 per cent) was present during the months of January to April and maximum organic matter 2.45 per cent was available during the month of May to August, but during the months of September to December it was decreased to 1.75 per cent (Table 9.2).

Calcium and Magnesium

Very little quantity of calcium and magnesium, in soil was recorded from Pond A (Table 9.1). It ranged from 0.4430 me/gr to 1.040 me/gr. During January to April, it was 0.4430 me/100gr and it reached to maximum 1.040 me/100gr during May to August and a decreasing value was noted during the months of September to December (0.7730 me/100gr).

However, calcium and magnesium of the pond B, (Table 9.2), it ranged from 0.5350 me/100gr to 1.1450 me/100gr. During January to April, calcium and magnesium were 0.5350 me/100gr, which reached to maximum of 1.1450 me/100gr during May to August, and in the months of September to December, they were 0.8780 me/100gr.

Bicarbonate

Bicarbonate of the soil of Pond A ranged from 0.88240 me/100gr to 3.0415 me/100gr (Table 9.1).

The value of carbonate during January to April 1.75275 me/100gr. The highest value was observed during the month of May to August, 3.0415 me/100gr and minimum (0.88240 me/100gr) was observed during the months of September to December.

More or less similar observation of bicarbonate was found in the pond B (Table 9.2).

The value of carbonate ranged from 0.83300 me/100gr to 3.02650 me/100gr. During the month of January to April it was 1.75375 me/100gr, which increased during the months of May to August 3.02650 me/100gr. The minimum value was recorded during the months of September to December (0.83300 me/100gr).

Conductivity

The conductivity of soil ranged from 0.6500 to 0.8240 ml/nos of Pond A (Table 9.1). It was maximum during May to August (0.8240 ml/nos) and the minimum was during the month of January to April (0.6500), but it was 0.2400 ml/nos during the month of September to December. But the conductivity of soil ranged from 0.600 ml/nos to 0.7750 ml/nos of Pond B (Table 9.2). It was minimum during the months of January to April (.600 ml/nos), but maximum conductivity was during the months of May to August 0.7750 ml/nos. It was decreased to 0.2490 ml/nos during the months of September to December.

Discussion

Soil

The physico-chemical properties of the soil are responsible for the growth of the vegetation and determine the kind of organisms that can live. It forms the top cover of the earth. It plays, therefore, an important role in determining the fertility of a fishpond and other water bodies located in different environmental conditions.

Banerjee (1967) found following soil conditions necessary for a high productivity of any water body in relation to fishes:

1. Phosphate: 60 (greater than 60).
2. pH: 6.5 to 7.5.
3. Depth: more than 5 meter.
4. Temperature: 20°C.
5. Maximum temperature for Indian carp: 37.8°C
6. Organic Carbon: 1.5 to 2.5.
7. pH of water: 6.5 to 7.5 most favourable.
8. Presence of mud, deposit of organic matter derived from the decay and decomposition of animals and plants.

On the other hand soil contents of presently investigated pond during 2000 (Tables 9.1 and 9.2) showed complete alkaline nature, throughout the year. Calcium and Magnesium were also good in amount. Phosphate amount was also high. This might be due to degeneration of phytoplankton and deposition of shells and bones. Singh (1977) also reported high phosphate concentration while working on soil conditions of Motijheel lake of Motihari (Bihar).

Joseph *et al.* (1979), while working on the soil of Pulicut lake, correloated pH and phosphate of the soil but during the present study no such relationship was observed.

Reheza (1968) recorded pH of the soil of North Bihar, while M.S. Ahmed, E.N. Siddiqui and S.K. Khalid (1996) studied certain physico-chemical properties of soil of ponds of Darbhanga, Bihar are found these water bodies suitable for aquaculture. Further, Kumari *et al.* (1990), Jaiswal and Singh (1995), while working on Baswan lake of Motihari and Mushahari lake of Muzaffarpur arrive at the same conclusion and found the water bodies suitable for better fish productivity.

Acknowledgements

The authors are grateful to Prof. Y.K.P. Sinha, Head of the P.G. Department of Zoology and Dr. A.P. Mishra, Professor, P.G. Department of Zoology, B.R.A., Bihar University, Muzaffarpur for their incessant encouragement. First author is also thankful to Dr. C.M. Thakur and Dr. D.C. Baluni, P.G. Department of Zoology, R.D.S. College, Muzffarpur for their valuable suggestions.

References

Ahmad, M.S., Siddiqui, E.N. and Khalid, S.K., 1996. Studies on physico-chemical characteristics of water of two ponds at Darbhanga. *J. Indian. Bot. Soc.*, 75(96): 107–112.

Joseph, K.B., Raman, K. and Khan, P.M.A., 1979. Observation on the exchange of phosphorus between soil and water of lake Pulicut. In: *Proc. 66th Ind. Sci. Cong.*, 3: 46.

Kumari, Usha and Singh, S.K., 1990. Soil condition of Baswan lake, Motihari (Bihar). *Mendal*, 7(3–4): 381–383.

Raheza, P.C., 1968. Soil productivity and crop growth. *J. Central Arid. Zone Res. Inst.*, Jodhpur.

Singh, U.N., 1977. Ecological studies on aquatic insects of Kararia and Motijheel Lakes of East Champaran (Bihar). *Ph.D. Thesis*, M.U.

Chapter 10
Analysis of Stiffened Raft on Expansive Clays of Chennai

K. Premalatha

*Department of Civil Engineering, College of Engineering,
Guindy Campus, Anna University, Chennai – 600 025, Tamil Nadu, India*

ABSTRACT

A major part of Chennai city is underlined by potentially expansive clay. The choice of foundation for lightly loaded residential structure is based primarily on observation and local experience. The effect of neglecting soil characteristics in designing the foundation has also been reflected in the distress of numerous buildings in the city. The present study was carried out to examine the possibility of the use of stiffened mat for such structures, based on heave and edge moisture variation distance. An interactive analysis was carried out for two typical building plans for the profile of the ground surface considered. The interactive analysis was carried out using ANSYS (Release 5.6). The optimum size of beam and slab were obtained by satisfying the various permissible limits. The size of the foundation elements were also obtained based on BRAB (1968) and Lytton (1972) analyses and were compared.

Keywords: *Expansive soil, Loss of support, Heave, Edge moisture variation distance, Foundation elements, Serviceability criteria, Slab on mound.*

Introduction

Expansive soils and their problems have been the focus of research from the 1950's and several solution are found scattered in literature. These soils are mostly found in arid and semi-arid areas and exhibit volumetric changes (heave/shrinkage) in accordance with moisture changes due to seasonal variations and soil drainage conditions, beneath the foundation slab. The magnitude of vertical movements depends on a number of factors such as local climate, rainfall and drought occurrences, nature and density properties of soils etc. This process of heaving and shrinkage causes the development of a mound with loss of support around the periphery of the building Figure 10.1 shows the mound shape of the ground with peripheral loss of support.

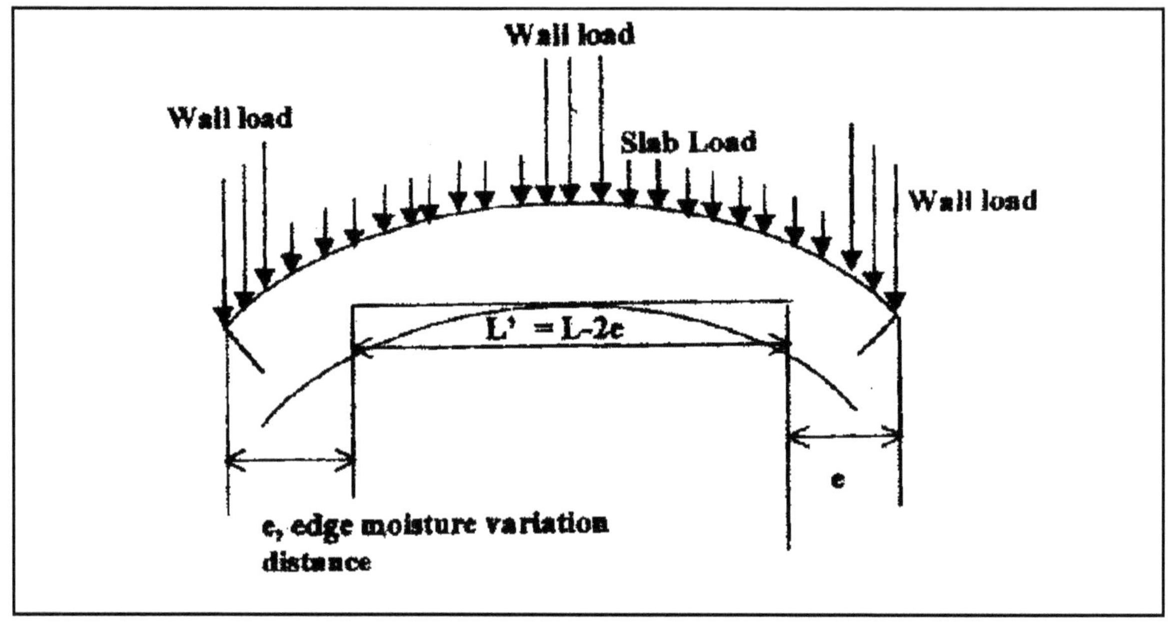

Figure 10.1: Mound Shape of the Ground with Peripheral Loss of Support

Potentially expansive clay underlies a major part of Chennai city. The choice of foundation for lightly loaded residential structures are based primarily on observation and local experience, often without adequate consideration of soil characteristics. Some of the foundations are unreinforced some are inverted Tee strip footings, isolated column footings, vierendeel frame foundations, uniform mat and under reamed piles of not more than 3 to 3.5 m etc. Structures founded on these types 0 foundations undergo mild to moderate distress. The presence of a soft clay layer, beneath the expansive stiff clay precludes the choice of under reamed piles in many locations. In such situations stiffened mat foundations are regarded as the most suitable and are widely used all over the world. Standard design procedures have been evolved to account for the differential ground movements (Building Research Advisory Board, 1968 and Lytton, 1972). The use of mat for low-rise structures as such is relatively limited in India. This paper discuss about the study that was carried out to examine the possibility of the use of stiffened mat for such structures, founded on the volumetrically active clays of Chennai city.

After arriving at realistic values of maximum heave and edge moisture variation distance, an interactive analysis was embarked to arrive at optimum sections of the foundation elements. The analysis of the stiffened mat, especially over an expansive soil bed is complex since factors influencing are too many. Here, the analysis of the problem is, a slab on mound, and is carried out through ANSYS (Release 5.6) package.

Materials and Methods

Model Idealisation

From the literature, it is observed that most of the methods of analysis are based on the condition of the beam or slab on a rectangular mound. The same idealization is modeled in this analysis with necessary input parameters defined. There are three components in the model, *viz.*, beam, slab, soil for which a suitable element has to be chosen. Solid 45 is used for soil, shell 63 is used for slab and beam 4 is used for beams.

Modelling of Soil Mass

Soil is modeled as a three dimensional homogeneous mass to a depth of 5 m. The length and width of soil mass are 2 m in addition to each side of the length and the width of the foundation. Structural shell element (SHELL63) is used to model the mat slab. Solid-45, 3D brick elements are used to formulate the soil mass soil properties like soil modulus, poisons ratio and density given as input material properties.

Modelling of Foundation

The mat foundation consists of beam and slab. Since the lightly loaded structures transfers the load through the load-bearing wall, beams are provided below the foundation to transfer the wall loading. Beam 4 is used to idealize the beams in footings. Elastic modulus, poison's ratio and density are applied as material properties. The cross-sectional area, the length and the depth of beam, the x-axis orientation and the moment of inertia about x, y and z directions are given through real constant sets.

Soil Structure Interaction

Common nodes of the structural shell elements and the soil elements (brick elements) on the surface just below the foundation ensure the interaction and integration between the soil and the structure. The shell element has six degrees of freedom and the brick element has three degrees of freedom that ensure that moments from the slab element are not transferred to the soil element.

Loads and Boundary Conditions

The uniformly distributed loads from the wall, are applied on the beam element. The live load and the dead load from the floor area are applied as pressure on the shell elements. Two typical building plans of light residential construction in Chennai city with the appropriate loading conditions are shown in Figures 10.2a and b

Analysis of the Problem

The validity of modeling in ANSYS is first checked by analyzing a simple problem using ANSYS and the results are compared. For analysis it is assumed that the building rests on a heaved up ground with loss of support all around the perimeter. The heave is characterized by a rectangular mound, with height as estimated heaves of 40 mm and 100 mm and by considering the EMVD around the periphery as 1.0 m and 1.2 m. Since the slab is placed upon the final heaved mound no swell pressure is given. An upper bound solution is then obtained by assuming the worst initial mound shape. To define the material constants of foundation elements, M20 grade concrete with E value 2.55 x 107 kN/m^2 and poison's ratio of 0.20 are used.

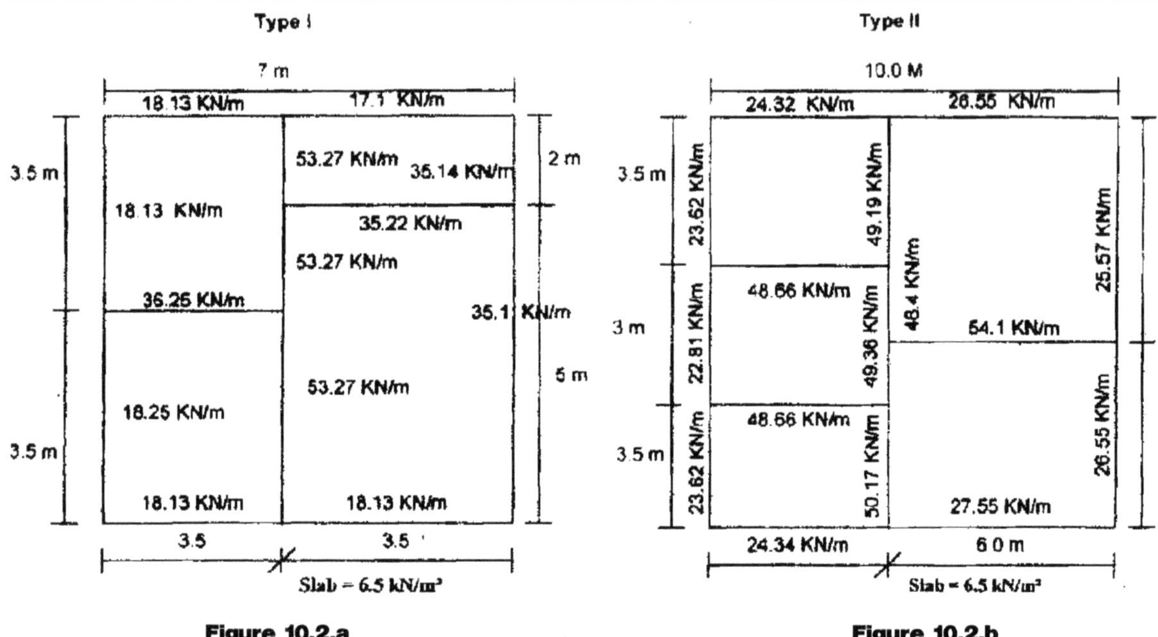

Figure 10.2.a **Figure 10.2.b**

Figures 10.2a and b: Typical Building Plan

Similarly to define the soil parameters, undisturbed soil samples are collected from different locations and the unconfined compressive strength test is carried out. Based on these results an average value of Eu = 2500 kN/m^2 (initial tangent deformation modulus) is determined for the subgrade in the analysis, with Poisson's ratio 0.48. The density of soil and concrete are taken as 15 kN/m^3 and 24 kN/m^3 respectively. The permissible bearing pressure and total settlement is also assumed as 100 kN/m^2 and 50mm as per I.S. code recommendations.

In the analysis, first the model is formulated and the loading and boundary conditions are applied. Analysis type is defined as static and the problem is analysed for current load steps. The results are reviewed, the settlement and the differential settlement of beam and slab are checked. Stresses in soil are also checked for allowable limits based on permissible bearing pressure. Section of beam and slab are revised until the settlement and stresses in the soil are within the permissible limits.

Serviceability Criteria

The literature pertaining to the serviceability criteria was reviewed, some of the guidelines and suggestions given by different authors are summarized below.

Skempton and MacDonald (1956) suggested the limiting criteria (angular distortion) as 1/300 for both independent footing and Raft foundation. Bjerrum (1963) suggested a limit of 1/150 for structural damage and 1/350 for safety against cracking of walls Grant *et al.* (1974) confirmed the value of 1/300 as a reasonable limit after studying the performance of 95 buildings. Meyerhoff (1953) earlier suggested values of 1/300 for the differential settlement between adjacent columns of open frame of encased steel and reinforced concrete at cracking.

Table 10.1: Size of Foundation Elements and Design Parameters Obtained from 3D Analysis

Building Type	*Heave mm*	*'e' Edge Moisture Variation*	*Size of Beam (mm)*	*Size of Slab (mm) Distance*	*Deflection of Beam*		*Max. Bending Moment (kN-m)*	*Beam Section Based on Moment (mm)*
					Max. (edge)	*Min (Centre)*		
Type I	40	1.0	300 x 700	12.5	28.9	19.0	179.4	300 x 600
	40	1.2	300 x 1000	15.0	43.3	30.1	352.5	300 x 840
	100	1.0	300 x 700	12.5	29.6	19.7	183.9	300 x 615
	100	1.2	300 x 1000	15.0	44.0	33.7	363.7	300 x 850
Type I	40	1.0	300 x 750	13.0	28	10.4	308.2	300 x 780
	40	1.2	300 x 1000	15.0	35.8	18.2	470.6	300 x 960
	100	1.0	300 x 750	13.0	28.7	10.6	316.2	300 x 790
	100	1.2	300 x 1000	15.0	30.4	18.6	475.6	300 x 960

Polshin and Tokar (1957) linked the onset of visible cracking to a specific tensile strain ($\in$ lim) dependant on the material involved. For brick work and block work set in cement mortor $\in$ lim lies between 0.03 and 0.05 per cent. In sagging mode the tensile effects are at the foundation soil interface while in hogging mode such restraints are absent on the upper surface. (The buildings are classified as framed type, load-bearing walls type undergoing sagging and hogging). The hogging mode is likely to occur at ratios about half of those in sagging mode). Further load-bearing wall, when subjected to hogging are susceptible to damage than framed buildings. Wilun and Starzewski (1975) recommends a ratio of 1/300 for statically indeterminate steel structures and load bearing brick work with reinforced concrete ring beam at every floor level, with longitudinally reinforced concrete strip foundation and with cross walls of at least 250 mm thickness and spaced at not more than 6 m centres.

For this analysis, based on the recommendation of Wilun and Starzewski (1975) and I.S.–456–2000. $\Delta L/L$ of 1/350 is considered. Table 10.1 shows the details of the sections (size of beam and slab) obtained along with the maximum design moment (beam) and deflections.

Results and Discussion

The interactive analysis optimized the beam and slab sections as 300 mm x 700 mm and 125 mm respectively for type-I building based on edge moisture variation distance value of 1.0 m and heave of 40 mm. Similarly for type-II building the corresponding dimension are 300 mm x 750 mm and 130 mm for beam and slab. Also from Table 10.1 it is understood that, the influence of heave on the size of foundation elements is relatively negligible, when compared to edge moisture variation distance. Hence the magnitude of edge moisture variation distance governs the analysis and the design parameters. The proposed sections from this analysis should therefore be adequate for satisfactory performance.

Comparison of Size Foundation Elements from Various Methods

Though the main existing design methods (Building Research, Advisory Board (1968), Lytton (1972), Walsh (1974,78), Frazer and Wardle (1975), and Post-Tensioning Institute, 1980, have resolved many issues, still a few doubts remain. Results obtained by the present analysis are compared with two well-established and published methods namely, the BRAB (1968) and Lytton (1972) procedures.

BRAB (1968) Method

This is an empirical method and based entirely on experience gained from observing the slabs-on-ground throughout the USA. The Type III. reinforced slab is recommended for use on expansive soils. The method may be outlined as follows:

1. The irregular shaped slab is divided into overlapping rectangles with long and short sides of length Land L′ respectively.
2. The total average load, (dead and live) w, is assumed to be uniformly distributed over the whole slab.
3. Various support modes such as center, edge, central support with cylindrical bending, edge support with cylindrical bending and corner support modes are considered.

Among the various support modes the center and edge support mode produces the most severe deflections. The centre support mode is the observed mode for this study.

One-dimensional analysis is carried out in both directions Land L′ to represent the two-dimensional case. This simplification produces smaller limiting moments but larger deflections, also it increases the factor of safety against extensive deflection and reduces the conservative safety margins against bending moment failure.

The allowable deflection ratio, $\Delta L/L$ recommended by BRAB for several different forms of construction are shown in Table 10.2 and these values need not necessarily represent a satisfactory level of serviceability.

Table 10.2: Allowable Deflection Ratios and Relative Rotations of BRAB (1968)

Type of superstructure	Allowable deflection ratio (ΔL/L) or relative rotation (β)
Wood frame and cladding	1/200 (ΔL/L)
Un reinforced load bearing wall	Un plastered 1/300 (ΔL/L); Plastered 1/360 (ΔL/L)

The design values for maximum one dimensional moment M_1, shear, V, and stiffness EI, considering the center heave conditions and the long direction of the slab, L, are given by

$$\left.\begin{aligned} M_1&: WL^2L'(1-C)\,\phi/L \\ V&: 4M_1/L \\ EI&: M_1L/6\,(\Delta L/L) \end{aligned}\right\} \qquad (1)$$

E is the conventional elastic modulus of concrete, I the second moment of inertia of the section and ϕ is a coefficient of reduction applied only in the long direction to account for the conservative assumptions of uniform contact pressure over the support area and the fact that the support index C, is independent of slab length.

For the two typical building plans (Figure 10.2) the bending moment shear force and EI values as per BRAB, Equation (1) are calculated. The calculations are made for uniform average load over the slab (Total load/total area). The details are listed in Table 10.3. Assuming M20 grade of concrete and Fe415 steel, based on limit state concept, the depth of beam is arrived at for the under-reinforced section and is listed in Table 10.4. The slab is designed as a conventional two-way slab and the thickness arrived is also shown in Table 10.4.

Table 10.3: Design Parameter (BRAB, 1968)

Building Type	*Loading Details*	*Max Bending Moment (kN-m)*	*Shear Force (kN)*	*EI (kN-m²)*
Type I (27 kN/m²)	Equivalent udl	97.24	55.6	0.324
Type II (27 kN/m²)	Equivalent udl	145.8	58.3	0.694

Table 10.4: Size of Foundation Elements BRAB (1968)

Building Type	*Loading Details*	*Size of Beam (mm)*	*Size of Slab (mm)*
Type I (27 kN/m²)	Equivalent udl	300 x 450	100
Type II (27 kN/m²)	Equivalent udl	300 x 550	125

Lyttons Method (1972)

Lytton (1972) proposed his design method, which is similar to BRAB in that initially the slab is analysed as a beam supported by one of the two modes such as central support with cylindrical bending and corner support with cylindrical bending. A major difference is that the loading is represented more realistically as line loads around the perimeter (P), and along the centre line of the slab (Q) and a uniformly distributed dead and live load of slab (q). The maximum moment (Q), is calculated in each direction, assuming both the soil and slab to be rigid, and then reduced by a correction term to account for soil compressibility. For the mode of current support (central support) conditions, the equation for the one-dimensional design moment, M_1 in the short direction is given by

$$M_1 = \frac{PLL'}{2} + \frac{QL'}{8} + \frac{qLL'}{8} - \frac{CTL'}{8} \tag{2}$$

where, T is the total load on the slab. The one-dimensional moment should be altered, if the centre doming and dishing modes are anticipated. However the maximum two-dimensional moment in the long direction. M_L is

$$M_L = M_1\,(1.4 - 0.4\,L/L') \geq (1.5 - C) \tag{3}$$

And in the short direction, Ms;

$$Ms = M_1\left[1 + 0.9\left(1.2C\left[\frac{L}{L'}\right] - 1\right)\right] \tag{4}$$

Design shear and stiffness are as that of BRAB (1968). The calculation of beam dimension uses conventional reinforced concrete design practice to ACI 318–636 but Lytton assumes an uncracked beam section.

For the two typical building types I and II (Figure 10.2) where the long and short spans are equal (bB), M. becomes the design moment. The maximum design moment and the size of the beam are shown in Table 10.5. The depth of the beam is arrived at based on the limit state concept for the under-reinforced section and the slab is designed as a conventional two-way slab.

Table 10.5: Design Parameters and Size of Foundation Elements (Lytton, 1972)

Building Type	Design Moment (kN-m)	Shear force (kN)	EI (kN-m²*)	Beam Section (mm)	Slab Section (mm)
Type I	451.1	257.8	1.51	300 x 940	100
Type II	505.8	202.3	2.41	300 x 990	125

* EI for $\left[\Delta L/L = \frac{1}{350}\right]$

Present Analysis

Results obtained through the present analysis are reproduced in Table 10.6.

Table 10.6: Design Parameters and Size of Foundation Elements of Present Analysis

Building Type	Maximum Bending Moment (kN-m)	Size of Foundation Element	
		Beam (mm)	Slab (mm)
Type I	179.4	300 x 700	125
Type II	308.2	300 x 750	130

Discussion

For a quantitative compression of the three different methods, BRAB (1968) Lytton (1972) and 3-D interaction analysis using ANSYS package (Release 5.6) the analysis is carried out and the various design parameters are listed in Table 10.7.

Table 10.7: Comparison of Foundation Elements Obtained in Various Methods

Building Type	Present Study		BRAB (1968)		Lytton (1972)	
	Beam (mm)	Slab (mm)	Beam (mm)	Slab (mm)	Beam (mm)	Slab (mm)
Type I	300 x 700	125	300 x 450	100	300 x 940	100
Type II	300 x 750	130	300 x 450	125	300 x 900	125

The raft dimensions and loadings used here are intended to be fairly representative of a small residential building. It is difficult to draw any conclusions from the design parameters such as bending moment and shear force because the analyses are entirely different, Even through BRAB (1968) recommended a cracked section and Lytton recommend an uncracked section, the size of foundation elements in the present study is obtained based on the limit state concept. The Lytton (1972) method indicates deeper beams than that from 3-D interactive analysis and BRAB (1968).

It is quite obvious, that the stiffness, obtained by Lytton (1972) is greater than that obtained by BRAB (1968), even for the same deflection ratio, since the stiffness is expressed as a function of the bending moment.

Conclusions

The present analysis is based on an analysis of the soil structure interaction of stiffened raft on observed elastic continuum mound profile and it is therefore reasonable to accept the relative magnitude of the design parameters and size of foundation elements. The major point of deviation appears to result from the influence of the choice of deflection ratio ($\Delta L/L$) and edge moisture variation distance. However it may be reasonably concluded based on both laboratory and field studies that $e \sim 1$ m and $\Delta L/L = 1/350$ represent realistic values for the climatic conditions, type of construction and serviceability requirements as evidenced at present in Chennai. Of course, continued monitoring of the performance of such building would be needed to verify the accuracy of this method of analysis and the size of foundation elements.

References

ANSYS Reference Manual, *SASIP*, Inc(c).

Bjerrum, L., 1963. Contribution to discussion session. In: *Proc. of European Conf. on Soil Mechanics and Foundation Engineering*. Wiesbaden II, pp. 135–137.

Building Research Advisory Board (BRAB), 1968. *Criteria for Selection and Design of Residential Slabs-on-Ground.* Publ. 1571, National Academy of Sciences Rep. No. 33 to Federal Housing Administration, NTIS No. PB, pp. 261– 551.

Fraser, R.A. and Wardle, K.J., 1975. The analysis of stiffened raft foundations on expansive soils. In: *Proc. Symp. Recent Developments in the Analysis of Soil Behaviour and their Application to Geotechnical Structures*, University of N.S.W., Sydney, Australia, pp. 89–98.

Grant, R., Christian, I.T. and Vanmarcke, E.H., 1974. Differential settlement of buildings. *J. Geotechnical Engineering Design*, ASCE, 100(GT9): 973–991

Lytton, R.L. and Woodburn, J.A., 1973. Design and performance of mat foundation on expansive of soils. *Haifa*, Israel, 1: 301–308.

Lytton, R.L., 1970. Design criteria for residential slabs and grillage rafts on reactive clay". *Rept. for Australian Commonwealth Scientific and Industrial Research Organization.* CSIRO, Melbourne, Australia, November.

Lytton, R.L., 1972. Design methods for concrete mats on unstable soils. In: *Third Inter. American Conf. on Materials Tech.*, Reo-de-Jenero, Brazil, pp. 171–177.

Meyerhoff, G.G., 1953. Some recent foundation research and its application to design. *The Structural Engineer*, 31: 35–69.

Polshin, D.E. and Tokar, R.A., 1957. Maximum allowable non-uniform settlement of structures. In: *Proc. IVth Int. Conf. on Soil Mechanics and Foundation Engineering*, London, 1: 402–405.

Premalatha, K., 2001. Prediction of heave using soil-water characteristic and analysis of stiffened raft on expansive clays. *Ph.D. Thesis*, Department of Civil Engineering, Anna university, Chennai.

Wilun, Z. and Starzewski, K., 1975. *Soil Mechanics and Foundation Engineering*, Vol. 2, 2nd edn. Survey University Press, London.

Chapter 11

On a New Species of the Genus *Anopheles* Meigen (Diptera : Culicidae) from India

B.P. Tingare and T.V. Sathe

Zoology Department, Shivaji University, Kolhapur – 416 004

ABSTRACT

A new species *Anopheles sangliensis* sp.nov. (Diptera : Culicidae) have been described for the first time form India. The female is 4.24 mm long, 1.6 mm broad; antenna 1.10 mm long; forewing 3.40 mm long, 0.64 mm broad; hindleg 6.38 mm long, brownish; thorax 1.54 mm long, 1.16 mm broad, brownish; pal pal bands alternately arranged in pal pi.

Flagellar formula: 1 L/W = 2.5, 13 L/W = 4, L1/13 = 0.83, W1/13 = 1.33.

Keywords: *Mosquito, Anopheles, Sangliensis sp. nov., Description.*

Introduction

The genus *Anopheles* have been erected by Meigen in 1818. The genus *Anopheles* is sub-divided into six subgenera *viz., Anopheles* Meigen, *Cellia* Theobald, *Kerteszia* Theobald, *Lophopodomyia* Antunnes, *Nyssorhynchus* Blanchard and *Stethomyia* Theobald. The present species comes under the sub-genus *Cellia*. From subgenus *Cellia* 197 species have been reported from the world and 34 species from India (Nagpal and Sharma, 1995). In past, Indian mosquitoes have been studied by Christhophers (1933), Barraud (1934), Foote and Cook (1959), Rao (1984), Nagpal and Sharma (1995), Sathe and Girhe (2001), Sathe and Girhe (2002), etc.

Materials and Methods

The species considered in this paper were collected from Sangli city of Maharashtra, India. The specimen were pinned in specimen tube. The dried specimen were kept in sterilized specimen tube by pinning inverted to wooden cork. Morphological study has been carried out with monocular microscope. Comparative measurements of body parts of specimen were made with ocular micrometer and calculated with the help of graduated mechanical stage. All measurements were made in millimeter. The mosquito species were identified consulting Barraud (1934), Rao (1984), Nagpal and Sharma (1995) and Sathe and Girhe (2002), etc.

Results

Anopheles (*Cellia*) *sangliensis* sp.nov

Female

4.24 mm long, 1.16 mm broad; antenna 1.10 mm long; forewing 3.40 mm long; 0.64 mm broad; hindleg 6.38 mm long; brownish; thorax 1.54 mm long, 1.16 mm broad, brownish.

Head

0.40 mm long, 0.54 mm broad, globular, brownish; small, rod like scale on vertex; eyes black, interocular space 0.20 mm; tempus semicircular; clypeus 0.20 mm long, brownish; proboscis 1.80 mm long, straight, brownish, covered with small scales; labium 1.66 mm long, cylindrical, scaly; labellum 0.14 mm long, brownish; maxillary palp five segmented, 1.90 mm long, yellowish brown, as long as proboscis, slender; mandible and maxillae well developed and tooted.

In male, palpi club shaped; proboscis straight line of body, with long blade like stylet; palpifer and palpus brownish; nape brownish, semicircular; antenna 1.10 mm long, 15 segmented, brownish, pilose; pedicel 0.08 mm long, 0.10 mm broad, brownish; flagellum 1.02 mm long, 13 segmented.

Flagellar Formula

1 L/W = 2.5, 13 L/W = 4, L1/13 = 0.83, L1/13 = 0.83, W1/13 = 1.33.

Thorax

1.54 mm long, 1.16 mm broad, brownish, undifferentiated, laterally compressed, thorax not covered with broad scales; scutum yellowish, shield shaped; scutellum half moon shaped, smooth; sternopleuron rectangular, brownish; pronotum without setae.

Forewing

3.40 mm long, 0.64 mm broad, yellowish, spotted with dark scales; scales 0.06 mm long, present on veins and as fringe, costa yellowish, four dark spots on costa; subcosta straight, no cross veins; apical scales longer than others.

Halter

0.18 mm long, 0.09 mm broad, triangular, yellowish brown, smooth, expanded at tip.

Hindleg

6.38 mm long, elongated, yellowish, slender, longer than body; coax 0.24 mm long, yellowish; trochanter 0.14 mm long, yellowish; femur 1.80 mm long, cylindrical, yellowish, without white spot but with tuft of white and black scales; tibia longer than femur, tibia 2.20 mm long, five segmented, yellowish, covered with small scales; first pretarsus longer than others; tip of hindtarsus not white, tarsi of front leg with broad pale bands; femora and tibia specked. claw simple, curved, yellowish, pulvillus and empodium densely hairy.

Other Legs

Special marks: similar.

Abdomen

2.30 mm long, 0.46 mm broad, brownish, densely hairy, brownish stripes on dorsal side; last abdominal segment narrow and shorter than others; anal cerci 0.12 mm long, 0.08 mm broad, hairy and brownish.

Colour

Brownish: Head, proboscis, antenna, thorax and abdomen

Yellowish: Wing, leg

Yellowish brown: Halter

Back: Eyes

Male

4.20 mm long, small, palpi club shaped, proboscis straight line of body, antenna elongated, plumose, brushy, phytophagus.

Host: Unknown.

Holotype

Female, India, Maharashtra, Sangli city, Coli Tingare B.P. 01-10-2005. Head, antenna, leg, abdomen, mounted on slide labelled as above.

Paratype

8 males; 20 females; sex ratio (M : F) = 1 : 2.5, Coli. Tingare, B.P., from September to November 2005, Sangli city; 2 males, 6 females; 15-10-2005

3 males, 10 females, Sangli city. 16-10-2005, 3 males, 4 females Sangli city 1-11-2005. Same data as above.

Discussion

This species runs close to *Anopheles* (*Cellia*) *stephensi* by having following characters.

1. Femora specked.
2. Female palp with broad apices and preapices pale band.
3. Tip of hind tarsi not white.

However, it differs from the above species by having following characters.

1. Tibia not specked completely but with yellowish dark banded.
2. Thorax without broad scales.
3. Terminal tip of palp is dark to blackish.
4. Flagellar formula:

   $1L/W = 2.5, 13L/W = 4, L1/13 = 0.83, W1/13 = 1.33$
5. Alternate black and white bands on palpi and alternate pattern on each palpi.

Acknowledgement

Authors are thankful to Shivaji University Kolhapur for providing facilities for this work.

References

Barraud, P.J., 1934. *The Fauna of British India including Ceylon and Burma*, V. Taylor and Francis, London. 1–463 pp.

Christophers, S.R., 1933. *The Fauna of British India including Ceylon and Burma, Diptera Vol. Family Culicidae, Tribe Anopheles.* Taylor and Francis, London, 1–371m pp.

Foote, Richard H. and Cook, David R., 1959. *Mosquitoes of Medical Importance.* U.S. Dept. Agric. Agric. Handbook No. 152, 158 pp.

Nagpal, B.N. and Sharma, V.P., 1995. *Indian Anopheles.* Oxford and IBH Publishing Co. Pvt. Ltd., 1–416 pp.

Rao, R.T., 1984. *The Anophelines* of *India.* Malaria Research Centre, Delhi ICMR, 518 pp.

Sathe, T.V. and Girhe, B.E., 2001. Biodiversity of mosquitoes in Kolhapur district, Maharashtra. *Riv. Oi. Parassitologia,* 18(67–3): 189–194.

Sathe, T.V. and Girhe, B.E. 2002. *Mosquitoes and Diseases.* Daya Publishing House, New Delhi, pp. 1–96.

Chapter 12

Eigen Value Analysis of Reinforced Concrete Silos: A Case Study

Jaspal Singh[1] and V.R. Sharma[2]

[1]*Associate Professor, Department of Civil Engineering, PAU, Ludhiana*
E-mail: jaspalsinghap@rediffmail.com
[2]*Dean, College of Agricultural Engineering, PAU, Ludhiana*
E-mail: vrsharma@rediffmail.com

ABSTRACT

In our country, a large amount of foodgrains are wasted due to the poor conditions of storage. Eventually this loss can be reduced by using proper storage structures such as bins (silos) which provide better protection against insects and rodents, pests and weather effects etc. Storage is one of the essential and vital stages between the marketing and consumption phases. The main purpose of silos is to store the commodity and to maintain quality. It is necessary to balance the supply and demand fluctuations and stabilize the prices. Silos and bunkers are made from different materials such as reinforced concrete silos or metal silos. Though metal silos have advantage over concrete silo in faster fabrication and erection, better quality control and air tightness, but in our country RCC silos are invariably used because of their ease of maintenance, superior architectural qualities and economical considerations. Practically RCC silos are maintenance free which is one of the consideration in overall cost.

RCC Cylinderical Silo can be flat bottom type or hopper bottom type. Though flat bottom silos can be built more easily than hopper bottom silos but it is desirable that bottom is self cleaning. It is because of this reason that hopper bottom silos are preferred. Given the various parameters such as height of silo, diameter of silo, properties of the material to be stored (density and angle of repose), grade of concrete, grade of steel and number of supports, various components are designed. Then for the designed components (thickness of wall, ring girder dimensions, column cross-section), mass matrix and influence coefficient matrix are generated. From mass and influence coefficient matrices, response parameters such as fundamental natural frequency and other frequencies along with mode shapes have been determined using Eigen value analysis. All the mode shapes have been checked for orthogonality condition implying that mode shapes obtained are correct. In this paper, the authors have carried out parametric study of RCC silos varying different parameters such as diameter, height and number of columns taking capacity of cylindrical portion as 500 m^3.

Introduction

The necessity of storing of materials and construction of large containers for coke, coal, cement, ores, grains (wheat, maize etc.) in the various steel plants and other industrial establishment cannot be over-emphasized. In cement factories as well as in construction projects, cement is stored in large silos. The structures for the foodgrain storage playa vital role in ensuring the regular supply of food grains and to build-up a buffer stock for emergency. Silos storing different materials are one among the important structures coming up in any industrial or organized storage complex. Reinforced concrete is an ideal structural material for building of permanent bulk storage facilities. RCC silos are fire proof, moisture proof, vermin proof and dust tight.

There are two main types of storage techniques *viz.*, bag storage and bulk storage structures. They are also known as godowns and vertical silos/bunkers respectively. The basic difference between the godown and silo/bunker is that godown is a horizontal, land using and labour using type while silo is a capital using but labour saving and land saving structure. Considering the advantages and disadvantages of both the systems, it can be concluded that silos are best suited for places where land is expensive, labour is dear and turnover is quick. Silos may be square, rectangular, circular or polygonal-single or in groups. Bulk storage structures are classified into two main categories *viz.*, Deep Bins/ Silos and Shallow Bins/bunkers. In shallow bins/bunkers, the plane of rupture of the material stored meets the top horizontal surface of the material before meeting the opposite sides of the structure.

In deep bins/silos, vertical walls are considerably taller than the lateral dimensions resulting in a tall structure. Consequently, the plane of rupture of the material stored meets the opposite sides of the structure before meeting the top horizontal surface of the material. The economy of deep bin (silo) lies in the proved phenomena that the major part of the weight of the fill materials is borne by the bin walls directly as a result of friction and arching action, and thus, it need not to be supported on the hopper/bottom, and hence the horizontal thrust due to filled material is comparatively small.

Silos are generally cylindrical in shape having an internal diameter not less than, what is required for the filling buckets etc. Their height is selected with respect to convenience in filling and emptying. It is always economical to have deep silos as the increase of pressure with depth in the lower portions of a silo is negligible. Flat bottom silos are seldom used since the undesirable dead storage reduces the efficiency of material flow. This is overcome by hopper forming concrete fill in the case of RCC silos. In the case of flat bottom silos, no ring beam is required as flat bottom does the function of a ring beam. Although flat bottoms can be built more easily than conical/hopper bottom ones, yet it is always desirable that the bottom is self cleaning. An angle of 45 degrees satisfies all the materials from self cleaning point of view and so in this study, hopper angle has been taken as 45 degrees. The silos are supported by a ring girder and which is further supported by columns. Normally, an even number of equally spaced columns are provided at the periphery.

The design process for silos comprises of Functional Design of Silo and Structural Design of Silo. Functional design must provide for adequate volume, proper protection of the stored materials, and satisfactory methods of filling and discharge. Structural Design constitutes conditions for stability, strength, and control (minimizing) of crack width and deflection. IS Code recommends Janssen' Theory to be used in structural design of a silo. To get the desired design components, the user has to feed the various parameters such as height of silo, diameter of silo, properties of the material to be stored (density and angle of repose), grade of concrete, grade of steel and number of supports. As prevalent in normal practice, the supports are assumed to be equally spaced along the periphery.

The designers have paid special attention to account for over pressures due to tilling and emptying of material. But loads caused by nature are not given due importance. In fact in contrast to vertical loads, lateral loads effects on silos are quite variable and are not properly documented. Unlike metal silos, RCC silos are very heavy and thus, are seldom subjected to wind loads when empty. It is in this context that they are more prone to earth quake failure. Hence in the present study, silos are designed taking into consideration both wind and earthquake into consideration.

For the designed components (thickness of wall, ring girder dimensions, column cross-section), depending upon height and diameter of the silos, mass matrix and influence coefficient matrix is generated. It is pertinent to mention that modern methods of structural analysis fall into two approaches *viz.*, system approach and finite element approach. System approach is suitable for hand computation whereas finite element approach is suitable for computers. The mass matrix and influence coefficient matrix have been obtained using element approach. Response parameters such as fundamental natural frequency and other frequencies, time periods in different modes and mode shape diagrams have been determined from Eigen value analysis. All the mode shapes have been checked for orthogonality condition implying that mode shapes obtained are correct.

Structural Design of Components

Taking capacity of cylindrical portion of RCC silo as 500 ml, diameters were determined for different heights say 12 m, 14 m and 16 m. For each case, number of supports have been varied in multiple of 2, starting from 4 say 4, 6, 8, 10, 12 etc. As is the normal practice, the supports are assumed to be equally spaced along the periphery. In all, 30 cases have been were analyzed (15 for silo full condition and 15 for silo empty condition). It is recommended that to ensure gravity flow, in hopper bottom, it shall have at least 10 degree more than angle of repose of the material to be stored. Normally this angle is kept as 45 degree as it suits all the materials.

Depending upon various parameters such as height of silo, diameter of silo, properties of the material to be stored (density and angle of repose), grade of concrete, grade of steel and number of supports, the various components such as thickness of wall, top slab, top ring beam, bottom ring beam and diameter of the supports are determined satisfying both silo full and silo empty conditions under earthquake and wind loading. RCC silos fall in the category of very heavy structures and thus are seldom subjected to wind loads when empty. It is in this context that they are more prone to earth quake failure. It has been observed that for all the cases, earthquake governs the design.

Generation of Element Mass Matrix and Element Influence Coefficient Matrix

Height has been varied in multiple of 2 depending upon diameter of the silo and satisfying the conditions:

1. $H > 1.5 \sqrt{A}$ (Given by Dishinger)
2. $H > 1.5 D$ (Soviet Code)

where, H is height of the cylindrical portion of silo, A is the cross-sectional area and D is the diameter of the silo.

Elements have been taken at every 2 m of height of the cylindrical portion to determine the mass matrix in each of the cases. To generate element Influence Coefficient Matrix, again elements at every 2 m of cylindrical portion are considered. So the number of levels considered are 7, 8 and 9 corresponding to 12, 14 and 16 m (respectively) as height of cylinderical portion. That is to say, the order of the matrices corresponding to 12, 14 and 16 m is respectively 7x7, 8x8 and 9x9.

Determination of Mode Shapes

The mode shapes have been determined for all the cases with the aid of a computer program developed. Table 12.1 gives values of normalized Eigen vector values in silo full and silo empty conditions. It is pertinent to mention that in all the cases orthogonality conditions are satisfied implying that mode shapes obtained are correct.

Determination of Frequencies

Knowing the element mass matrix and Influence Coefficient Matrix, Fundamental frequency and hence Fundamental Time Period for different cases has been determined. The second and third mode frequencies have also been determined corresponding to all cases. Table 12.2 indicates the fundamental and other frequencies (cycles/second) for the cases considered. The code gives us certain formulae for calculating the fundamental time period in case of buildings. Those formulae are not applicable for special structures such as silos. For special structures, it is recommended that Eigen value analysis is carried out to determine the frequencies from which time period in different modes can be computed.

The range of fundamental frequencies for diameter 7.28 m and height 12 m; diameter 6.74 m and height 14 m and diameter 6.31 and height 16 m is respectively 1.467–1.652 cycles/see, 1.265–1.401 cycles/see and 1.087–1.180 cycles/see when the silo is filled whereas the range of fundamental frequencies for diameter 7.28 m and height 12 m; diameter 6.74 m and height 14 m and diameter 6.31 and height 16 m is respectively 2.628–2.953 cycles/sec, 2.284–2.519 cycles/see and 1.963–2.127 cycles/see when the silo is empty.

Conclusions

It is recommended to carry out Eigen value analysis (frequency analysis) in case of special structures such as silos.

Average fundamental frequencies in first mode in case of silos having height 12 m, 14 m and 16 m are respectively 1.557 cycles/sec, 1.330 cycles/sec and 1.136 cycles/sec in silo full condition.

Average fundamental frequencies in first mode in case of silos having height 12 m, 14 m and 16 m are respectively 2.790 cycles/sec, 2.395 cycles/sec and 2.051 cycles/sec in silo empty condition.

Average frequencies in second mode in case of silos having height 12 m. 14 m and 16 m are respectively 9.230 cycles/sec, 7.799 cycles/sec and 6.356 cycles/sec in silo full condition.

Average frequencies in second mode in case of silos having height 12 m, 14 m and 16 m are respectively 15.268 cycles/sec, 13.055 cycles/sec and 10.908 cycles/sec in silo empty condition.

Average frequencies in third mode in case of silos having height 12 m, 14 m and 16 m are respectively 27.017 cycles/sec, 20.411 cycles/sec and 18.148 cycles/sec in silo full condition.

Average frequencies in third mode in case of silos having height 12 m, 14 m and 16 m are respectively 46.916 cycles/sec, 37.150 cycles/sec and 29.405 cycles/sec in silo empty condition.

Table 12.1: Normalized Mode Shaped Values for Silo having Capacity of Cylindrical Portion = 500 m³

Diameter (m)	Height (m)	Number of Columns	Level 1	Level 2	Level 3	Level 4	Level 5	Level 6	Level 7	Level 8	Level 9
Silo	Full	Case									
7.28	12	4	4.3326	3.7364	3.1316	2.5465	1.9985	1.4646	1.000		
			–1.3197	–0.5051	0.0365	0.5816	0.8793	1.0843	1.000		
			1.4037	–0.8599	–0.3594	–0.5537	–1.1021	0.5270	1.000		
		6	4.4991	3.8660	3.2221	2.603	2.0287	1.4726	1.000		
			–1.3493	–0.4917	0.0669	0.6181	0.9155	1.1086	1.000		
			1.3902	–0.9463	–0.3262	–0.4235	–1.1367	0.5725	1.000		
		8	4.5816	3.9297	3.2674	2.6308	2.0443	1.4764	1.000		
			–1.3572	–0.4847	0.0740	0.6340	0.9271	1.1190	1.000		
			1.3864	–0.9985	–0.2655	–0.3915	–1.1777	0.5969	1.000		
		10	4.4675	3.8412	3.2052	2.5927	2.0238	1.4713	1.000		
			–1.3390	–0.4853	0.0690	0.6169	0.9113	11.1058	1.000		
			1.3682	–0.9414	–0.3080	–0.4215	–1.1126	1.6036	1.000		
		12	4.3736	–3.7678	3.1540	2.5604	2.0063	1.4672	1.000		
			–1.3209	–0.4855	0.0625	0.6002	0.8958	1.0935	1.000		
			1.3162	–0.8639	–0.3342	–0.4230	–1.0519	0.5503	1.000		
6.74	14	4	5.5132	4.7861	4.0798	3.3756	12.7025	2.1038	1.4961	1.000	
			–1.3306	–0.7733	–0.0001	0.4921	1.9215	0.9627	1.744	1.000	
			2.5611	–1.0535	–0.6389	–2.1074	–1.4239	1.8485	0.5745	1.000	
		6	5.7794	4.9987	4.2422	3.4883	2.7717	2.1439	1.5056	1.000	
			–1.3433	–0.7764	0.0301	0.5277	0.9657	0.9856	1.1937	1.000	
			2.7779	–1.3075	–0.6086	–2.2373	–1.4857	2.1977	0.6086	1.000	
		8	5.9260	5.1153	4.3314	3.5501	2.8092	2.1659	1.5102	1.000	
			–1.3509	–0.7768	0.0426	0.5441	0.9824	0.9975	1.2047	1.000	
			2.8662	–1.4146	–0.5808	–2.3000	–1.5033	2.3199	0.6109	1.000	

Contd...

Table 12.1–Contd...

Diameter (m)	Height (m)	Number of Columns	Level 1	Level 2	Level 3	Level 4	Level 5	Level 6	Level 7	Level 8	Level 9
Silo	Full	Case									
		10	5.7248	4.9550	4.2090	3.4651	2.7576	2.1360	1.5035	1.000	
			–1.3365	–0.7719	0.0279	0.5285	0.9602	0.9840	1.1926	1.000	
			2.7610	–1.3188	–0.6036	–2.2027	–1.4494	2.1848	0.6129	1.000	
		12	5.5589	4.8227	4.1 081	3.3952	2.7151	2.1111	1.4978	1.000	
			–1.3244	–0.7675	0.0174	0.5107	0.9405	0.9734	1.1820	1.000	
			2.7388	–1.3120	–0.5632	–2.1938	–1.4660	2.1786	0.6048	1.000	
6.31	16	4	6.9291	6.0925	5.2436	4.4205	0. 3.6136	2.8519	2.1817	1.5244	1.000
			–1.6937	–0.7871	–0.2240	0.5000	0.9084	1.2234	1.2719	1.2505	1.000
			1.0192	0.1729	–0.4897	–0.8552	–0.7701	–0.3229	–0.0179	0.8093	1.000
		6	7.3594	6.4491	5.5254	4.6326	3.7593	2.9401	2.2302	1.5357	1.000
			–1.7357	–0.7761	–0.2032	0.5497	0.9570	1.2708	1.3114	1.2684	1.000
			1.0125	0.1697	–0.5299	–0.8502	–0.7459	–0.2729	0.0063	1.8562	1.000
		8	7.5210	6.5833	5.6316	4.7121	3.8140	2.9731	2.2485	1.5400	1.000
			–1.7496	–0.7732	–0.1971	0.5649	0.9757	1.2893	1.3226	1.2747	1.000
			0.9989	1.1677	–0.5216	–0.8415	–0.7508	–0.2727	1.0254	0.8621	1.000
		10	7.2748	6.3791	5.4699	14.5907	3.7303	2.9226	12.2206	1.5334	1.000
			–1.7324	–0.7779	–0.2032	0.5462	0.9557	1.2699	1.3075	1.2673	1.000
			0.9929	0.1746	–0.5281	–0.8373	–0.7367	–0.2736	0.0316	0.8529	1.000
		112	7.0913	6.2264	5.3495	4.5001	13.6684	2.8853	2.2002	1.5289	1.000
			–1.7178	–0.7766	–0.2095	0.5309	0.9398	1.2522	1.2950	1.2594	1.000
			0.9705	0.2003	–0.5316	–0.8221	–0.7489	–0.2873	0.0548	0.8518	1.000
Silo	Empty	Case									
7.28	12	4	4.4070	3.7861	3.1641	2.5638	12.0015	1.4668	1.000		
			–0.9533	–0.2880	0.2049	0.6629	0.9474	11.0844	1.000		
			1.1899	–1.5343	–1.3107	–1.5420	–1.8965	0.1652	1.000		

Contd...

Table 12.1–Contd...

Diameter (m)	Height (m)	Number of Columns	Level 1	Level 2	Level 3	Level 4	Level 5	Level 6	Level 7	Level 8	Level 9
Silo	Empty	Case									
		6	4.5938	3.9294	3.2635	2.6256	12.0331	1.4753	1.000		
			–0.9464	–0.2497	0.2523	0.7099	0.9920	11.1103	1.000		
			1.1132	–1.5294	–1.2489	–1.3497	–1.7801	1.2455	1.000		
		8	4.6852	3.9993	3.3131	2.6556	2.0493	1.4794	1.000		
			–0.9433	–0.2359	0.2648	0.7284	1.0058	1.1206	1.000		
			1.0914	–1.5295	–1.2050	–1.3284	–1.7546	0.2737	1.000		
		10	4.5654	3.9075	3.2491	2.6169	2.0294	1.4746	1.000		
			–0.9242	–0.2278	0.2665	0.7202	0.9964	1.1107	1.000		
			1.0348	–1.4462	–1.1540	–1.2428	–1.6393	0.3193	1.000		
		12	4.4657	3.8309	3.1959	2.5841	2.0126	1.4704	1.000		
			–0.9079	–0.2239	0.2631	0.7087	0.9839	1.1015	1.000		
			0.9777	–1.3326	–1.1134	–1.1763	–1.5389	0.2825	1.000		
6.74	14	4	5.6167	4.8641	4.1304	3.4070	2.7189	2.1242	1.4981	1.000	
			–1.0032	–0.4758	0.1533	0.5935	0.9528	0.9478	1.1635	1.000	
			1.5628	–1.0185	–1.7284	–2.6304	–2.0553	1.0623	0.2879	1.000	
		6	5.9098	5.0972	4.3067	3.5287	2.7930	2.1696	1.5081	1.000	
			–0.9891	–0.4552	0.2015	0.6432	1.0054	0.9721	1.1855	1.000	
			1.5438	–1.2022	–1.6557	–2.5637	–1.9432	1.3170	0.3799	1.000	
		8	6.0685	5.2230	4.4020	3.5944	2.8326	2.1938	1.5129	1.000	
			–0.9881	–0.4482	0.2174	0.6614	1.0235	0.9831	1.1962	1.000	
			1.5476	–1.2581	–1.6339	–2.5789	–1.9177	1.3734	0.3941	1.000	
		10	5.8574	5.0555	4.2756	13.5072	2.7804	2.1620	1.5063	1.000	
			–0.9751	–0.4437	1.2114	0.6528	1.0091	0.9739	1.1863	1.000	
			1.4797	–1.2119	–1.5627	–2.4467	–1.8109	1.3295	0.4184	1.000	
		12	5.6823	4.9166	4.1708	3.4352	2.7370	2.1355	1.5008	1.000	
			–0.9647	–0.4404	0.2050	0.6407	0.9953	0.9661	1.1782	1.000	
			1.4288	–1.1828	–1.4806	–2.3600	–1.7574	1.3134	0.4219	1.000	

Contd...

Table 12.1–Contd...

Diameter (m)	*Height (m)*	*Number of Columns*	*Level 1*	*Level 2*	*Level 3*	*Level 4*	*Level 5*	*Level 6*	*Level 7*	*Level 8*	*Level 9*
Silo	*Empty*	*Case*									
6.31	16	4	17.0735	6.2005	5.3220	4.4708	3.6440	2.8676	2.1795	11.5260	1.000
			–1.2278	–0.4850	0.0246	0.6152	0.9704	1.2298	1.3078	11.2372	1.000
			1.0405	–0.3714	–0.7829	–1.4435	–1.1755	–0.7055	–0.6934	0.6612	1.000
		6	7.5403	6.5845	5.6241	4.6965	3.7986	12.9606	2.2285	1.5380	1.000
			–1.2393	–0.4481	0.0685	0.6810	1.0314	11.2853	1.3552	1.2568	1.000
			0.9828	–0.3955	–0.7775	–1.3740	–1.0652	–0.5813	–0.6365	0.7321	1.000
		8	7.7145	6.7283	5.7374	4.7807	3.8561	12.9951	2.2468	1.5425'	1.000
			–1.2435	–0.4381	0.0799	0.7004	1.0521	11.3046	1.3687	1.2631	1.000
			0.9671	–0.4050	–0.7589	–1.3549	–1.0507	–0.5655	–0.6177	0.7476	1.000
		10	7.4550	6.5143	5.5688	4.6550	3.7700	2.9435	2.2196	1.5359	1.000
			–1.2358	–0.4439	0.0738	0.6877	1.0381	1.2913	1.3572	1.2573	1.000
			0.9378	–0.3855	–0.7488	–1.3149	–1.0076	–0.5414	–0.5805	0.7430	1.000
		12	7.2636	6.3562	5.4449	4.5627	3.7074	2.9061	2.2002	1.5316	1.000
			–1.2288	–0.4412	0.0695	0.6782	1.0278	1.2795	1.3479	1.2512	1.000
			0.9024	–0.3558	–0.7280	–1.2716	–0.9831	–0.5089	–0.5447	0.7526	1.000
		10	7.4550	6.5143	5.5688	4.6550	3.7700	2.9435	2.2196	1.5359	1.000
			–1.2358	–0.4439	0.0738	0.6877	1.0381	1.2913	1.3572	1.2573	1.000
			0.9378	–0.3855	–0.7488	–1.3149	–1.0076	–0.5414	–0.5805	0.7430	1.000
		12	7.2636	6.3562	5.4449	4.5627	3.7074	2.9061	2.2002	1.5316	1.000
			–1.2288	–0.4412	0.0695	0.6782	1.0278	1.2795	1.3479	1.2512	1.000
			0.9024	–0.3558	–0.7280	–1.2716	–0.9831	–0.5089	–0.5447	0.7526	1.000

Remarks:

Level 1 corresponds to the tip of the silo.

Other Levels are chosen at an interval of 2 m below the tip.

Last level corresponds to the level at the junction of cylindrical wall and hopper.

For different levels corresponding to each case; first, second and third value of each row corresponds to model, mode 2 and mode 3 respectively.

Table 12.2: Frequencies in Cycles/Sec in Different Modes in Silo Full and Silo Empty Conditions

Diameter (metre)	*Height of Cylin Portion (metre)*	*No. of Supports*	*Silo Full*			*Silo Empty*		
			1st	*2nd*	*3rd*	*1st*	*2nd*	*3rd*
7.28	12	4	1.467	8.953	26.398	2.628	14.515	45.607
7.28	12	6	1.595	9.285	27.051	2.854	15.270	46.820
7.28	12	8	1.652	9.422	27.413	2.953	15.560	47.602
7.28	12	10	1.571	9.308	27.074	2.818	15.543	47. 198
7.28	12	12	1.499	9.181	27.147	2.693	15.452	47.354
6.74	14	4	1.268	7.860	20.273	2.284	12.444	36.441
6.74	14	6	1.357	7.782	20.377	2.442	13.093	37.084
6.74	14	8	1.401	7.891	20.533	2.519	13.317	37.458
6.74	14	10	1.339	7.786	20.484	2.414	13.259	37.433
6.74	14	12	1.283	7.676	20.386	2.316	13.161	37.336
6.31	16	4	1.087	6.207	17.786	1.963	10.497	28.488
6.31	16	6	1.156	6.399	18.139	2.086	10.961	29.407
6.31	16	8	1.180	6.460	18.334	2.127	11.099	29.677
6.31	16	10	1.143	6.386	18.252	2.064	11.017	29.62
6.31	16	12	1.114	6.330	18.228	2.014	10.967	29.825

The fundamental frequency, second and third mode frequencies in silo empty condition are more in all the cases as compared to silo full condition.

For a number of supports, with increase in H/D ratio, frequencies in both silo full and silo empty conditions decrease because the silo (structure) tends to become more flexible as the height increases with respect to the diameter.

All the mode shapes satisfy orthogonality condition implying that mode shapes obtained are correct.

References

Adidam S., Sai, R., Roy, B.N. and Kumar, Veerendra, 1989. Plastic analysis of silos. *Journal of Structural Engineering,* 16(2): 35–42.

Ayuga, F., Guaita, M. and Aguado, P., 2001. Static and dynamic silo loads using finite element models. *Journal Agricultural Engineering Research,* 78(3): 299–308.

Ayuga, F., Guaita, M., Aguado, P.J. and Couto, A., 2001. Discharge and the eccentricity of the hopper influence on the silo wall pressures. *Journal of Engineering Mechanics,* 127(10): 1067–1074.

Lal, Bansal Mohan, 2001. Analysis of circular steel silos for wind loading. *Ph.D. Dissertation,* Punjab Agricultural University, Ludhiana, India.

Demetres, Briassoulis and James, Curtis, 1985. Design and analysis of silos for friction forces. *Journal of Structural Engineering, ASCE,* 111(6): 1377–1397.

Chandrasekaran, A.R. and Jain, P.C., 1968. Effective live load of storage materials under dynamic conditions. *Indian Concrete Journal,* p. 364–365.

Phirke, P.S. and Bhole, N.G., 1993. Development of prediction equation for dynamic wall strain in deep bins for wheat. *Journal of Institution of Engineers (Civil Engineering Division),* 74: 26–31.

Rahal Mohammed, A. and Vuez Alain, R., 1998. Analysis of settlement and pore pressure induced by cyclic loading of silo. *Journal of Geotechnical and Geoenvironmental Engineering, ASCE,* 124(12): 1208–1210.

Sayed George Abdel, Monasa Frank and Siddall Wayne, 1985. Cold-formed steel farm structures. *Journal of Structural Engineering ASCE,* 3(10): 2065–2089.

Singh, Jaspal and Sharma, V.R., 2006. Eigen value analysis for RCC hopper bottom silos. In: *Proceedings National Conference on Civil Engineering: Meeting the Challenges of Tomorrow,* Ludhiana, p. 385–389.

Singh, Jaspal and Sharma, V.R., 2006. Knowledge based systems in civil engineering. In: *Proceedings National Conference on Civil Engineering: Meeting the Challenges of Tomorrow,* Ludhiana, p. 662–669.

Zhao, Y. and Teng, J.G., 2001. Buckling experiments on cone-cylinder intersections under internal pressure. *Journal of Engineering Mechanics,* 127(12): 31–39.

Chapter 13

Food and Feeding Habits of Two Ponds of Bishunpur Sri Ram Village of Muzaffarpur (Bihar)

***Shishir Kumar*[1], *Sachidanand Mishra*[2], *Rajeev Ranjan*[3], and *Pankaj Kumar*[4]**

[1]*Research Scholar, P.G. Department of Zoology, R.D.S. College, Muzaffarpur*
[3]*Research Scholar, M.P. Sinha Science College, Muzaffarpur*
[2]*Research Scholar,* [4]*Post Doctoral Research Scholar*
P.G. Department of Zoology, B.R.A., Bihar University, Muzaffarpur – 842 001

ABSTRACT

On the basis of findings carried out on the ponds of Bishunpur Sri Ram Village, it is possible to divide broadly the six fishes of the presently studied pond into (*a*) surface feeders-which feed on surface food organisms *i.e., C. catla* (*b*) mid-feeders-which feed on sub-surface food organisms *i.e., L. rohita, W. attu* and (*c*) bottom feeders-which feed on bottom plants and animals *i.e.,* C *mrigala, H. fassilis* and *C. punctatus.*

Another interesting feature that emerges out of this study is that a definite correlation has been established between the plankton population and the food and feeding habits of fishes. When phytoplankton reaches its maximum peak, all plankton feeding fishes have shown high percentage of phytoplankton in their stomach. Similar findings have been made in regard to zooplankton.

The above findings are also supported by the studies of Srivastava (1959) and Singh (1977).

Keywords: *Feeding habits, Gut contents, Fishes, Ponds.*

Introduction

Most of the biologists agree that the fishes have some scientific affinities to different zones of the water-bodies. Some fishes are completely plankton feeder and some fishes are bottom fauna feeders throughout their life, and some others are omnivorous at certain stage of their life cycles.

Schaperclans (1933), classified natural food of fishes under the groups:

1. 'Main food' or the natural food which the fish prefers under favourable conditions and on which it thrives best.
2. 'Occasional food' or the natural food that is well liked and consumed as and when available.
3. 'Emergency food', which is ingested when the preferred food items are not available and on which the fish is just able to survive.

Alikunhi (1952) recorded certain microscopic planktonic crustaceans group and rotifers as 'Main food' of spawn and fry of the Indian Major Carps and majority of other culturable sps., and phytoplankton as *"Emergency food"*.

Nikolski (1963) divided food of fishes into four groups:

1. "Main or Basic food" which is the natural food and is usually consumed by fish under normal conditions.
2. The 'Secondary food' or 'Occasional food' which is consumed by the fish in small quantities when available.
3. 'Incidental food', which only rarely enters the gut.
4. 'Obligatory food', or 'Emergency food', which is consumed at the time of emergency for survival of fishes.

Nikolski (1963) further categorised fishes according to the extent of variation in types of food consumed by them, such as:

1. Europhagic, feeding on variety of foods.
2. Stenophagic, feeding on a few selected types of food.
3. Monophagic, feeding on a single type of food.

On the basis of the works of Khan (1934), Hora (1943), Das and Moitra (1955) and Singh (1977) and others, fishes prefer to feed in the area which they occupy in a water-body. In this way, they can be divided into three groups: (*a*) Plankton eating surface feeders such as *Catla;* (*b*) Mid feeders such as, *Labeo rohita* and *Labeo batao;* (*c*) Bottom feeders, such as *Cirrhinus mrigala, C. reba* and *Labeo calbasu.*

On further study of different works on food and gut content of fishes Chaeiko and Kuriyan (1948 and 1949), Das and Srivastava (1955) have classified fishes on the basis of their feeding habits into (*i*) Herbivorous–Fishes feed mainly on plant food consisting of Algae and higher aquatic plants. (*ii*) Carnivorous–Fishes feed more on animal food. (*iii*) Omnivorous–Fishes feed, both on plant and animal food. It has been reported that seasonal fluctuations occur and more marked in herbivorous and carnivorous fishes. It is probably due to difference in the availability of food. However, the omnivorous show little variation because of their adaptability of both types of food. Due to availability of proper food, seasonal variation in growth rate of fishes has also been observed.

No work on the above lines has been done on fishes of Bishunpur Sri Ram Ponds and it is imperative that one should have a clear picture regarding the food habits of fishes on aquaculture so that the artificial food can be supplemented to increase fish productivity.

Materials and Methods

The fishes of both the ponds were collected at intervals with the help of local fisherman. For the collection of fish fauna from these water-bodies following fishing gears (nets) were used *i.e.*, Cast Net, Drag Net and Scoop Net.

The fishes collected were first preserved in 8 per cent formalin for 48 hours. After that the fishes were transferred in 5 per cent formalin and preserved, for detailed study and identification in the laboratory. The identification and classification of fishes were made with the help of Days *"Fish Fauna of British India"* and *"The classification of the fishes, Present and Extinct"* of Leo S. Brig. respectively.

Results

Food and Gut Content of Six Food Fishes of Bishunpur Sri Ram Pond A and Pond B

Monthly collection of fish were made in Bishunpur Sri Ram Pond A and Pond B for a year and out of all the catches six types of fishes were brought to the laboratory for investigation namely

1. *Labeo rohita*
2. *Wallago attu*
3. *Heteropneustes fossilies*
4. *Channa punctatus*
5. *Cirrhina mrigala*
6. *Catla catla*

They were measured and weighted. The abdomen of each fish was cut open and fishes were kept in 10 per cent formalin for further investigation of the gut contents. The stomach and intestine were carefully removed from the body of the fishes. These organs were cut and the gut contents were washed with water into a petridish. All animals and plants were identified and counted under the high and low powers of microscope on a counting side.

The percentage composition of gut contents were estimated by the method followed by Hynes (1950). The results were calculated in Tables 13.1 and 13.2.

Gut Contents of *Labeo rohita*

While analysing the gut contents of *Labeo rohita* of Pond A, the following unicellular, multicellular algae (Desnides, Diatoms, filamentous algae, microcystes, *Spirogyra and Ulothrix*), aquatic weeds, crustaceans, rotifers, insects and larvae were noticed along with vegetables and debris. The unicellular and multicellular algae were present in a little quantity but aquatic weeds shared more (50.40 per cent) of the total gut contents. The contents of plankton (uicellular and multicellular algae), crustaceans, rotifers and insects larvae were much lesser in quantity (6.25 per cent, 5.40 per cent, 10.20 per cent, 6.70 per cent and 3.45 per cent) respectively along with 15.50 per cent vegetable and mud and 2.10 per cent miscellaneous (Table 13.1).

Gut content of *Labeo rohita* of Pond B were analysed. The aquatic weeds shared more than half (61.50 per cent) of the total gut contents. Thus, the total herbivorous diet was constituted about 76.20

Table 13.1: Average Percentage of Gut Contents of Six Fishes of Bishunpur Sri Ram Pond A

Name of Fishes	Gut Contents											
	Unicellular Algae	Multicellular Algae	Aquatic Weeds	Crustaceans	Rotifers	Insects and Larvas	Fish Fry and Fingerlings	Molluscs	Shrimps	Fish Fin and Scales	Veg. Debris and Mud	Misc.
L. rohita	6.25	5.40	50.40	10.20	6.70	3.45	–	–	–	–	15.50	2.10
W. attu	–	–	–	28.50	7.70	14.30	48.2	–	–	–	–	1.3
H. fossilis	–	–	–	26.8	8.5	19.50	25.5	2.0	3.7	8.2	4.3	1.5
C. punctatus	–	–	–	20.50	7.60	36.30	15.30	10.20	6.20	2.20	–	1.7
C. mrigala	10.9	13.3	59.8	–	–	–	–	–	–	–	14.0	2.00
C. catla	26.7	25.3	16.3	27.2	3.0	–	–	–	–	–	–	1.5

Table 13.2: Average Percentage of Gut Contents of Six Fishes of Bishunpur Sri Ram Pond B

Name of Fishes	Gut Contents											
	Unicellular Algae	Multicellular Algae	Aquatic Weeds	Crustaceans	Rotifers	Insects and Larvas	Fish Fry and Fingerlings	Molluscs	Shrimps	Fish Fin and Scales	Veg. Debris and Mud	Misc.
L. rohita	7.50	7.20	61.50	4.70	4.70	–	–	–	–	–	12.00	2.30
W. attu	–	–	–	42.7	11.3	19.2	24.9	–	–	–	–	1.9
H. fossilis	2.3	4.2	4.3	7.3	4.2	20.6	56.4	–	–	–	–	0.7
C. punctatus	–	–	–	15.8	7.3	45.7	15.3	10.1	1.5	2.2	1.1	1.1
C. mrigala	8.9	14.1	65.5	–	–	–	–	–	–	–	10.3	1.2
C. catla	21.3	17.8	20.7	27.5	10.7	–	–	–	–	–	–	2.00

per cent while carnivorous diet was 9.40 per cent. Besides, there was about 12 per cent of vegetable debris and mud with 2.30 per cent miscellaneous diet (Table 13.2).

Gut Contents of *Wallago attu*

In the gut content of *Wallago attu* of Pond A, only carnivorous food was found and there was no trace of herbivorous food. The carnivorous food consists of plankton, crustaceans, rotifers, insects and their larvae along with small fishes and fingerlinges. The total percentage of carnivorous food was 98.7 per cent and 1.3 per cent miscellaneous (Table 13.1).

The gut contents of same fish of pond B also consisted of only carnivorous diet having plankton, rotifers, crustaceans, insects and their larvae and small fishes, thus, making about 98.1 per cent carnivorous food and 1.9 per cent miscellaneous. In this, there was no sign of herbivorous food in this fish also (Table 13.2).

Gut Contents of *Heteropneustes fossilis*

The gut contents of *Heteropneustes fossilis* also were analysed in Pond A. There was no trace of herbivorous food (Table 13.1) but only carnivorous food was noticed, consisting of planktons, crustaceans, rotifers, insects, insects larvae, fishes, molluses, shrimps, fish-fins and scales. Thus, total carnivorous diet was 94.2 per cent. Besides, there was about 4.3 per cent debris and mud along with 1.5 per cent miscellaneous in the alimentary canal tract.

While analysing the gut contents of *Heteropneustes fossilis* of Pond B it consisted of unicellular, multicellular algae and aquatic weeds in a very little quantity (10.8 per cent), wereas, the carnivorous diet was 88.5 per cent. There was no sign of vegetable debris and mud. The miscellaneous was about 0.7 per cent (Table 13.2).

Gut Contents of *Channa punctatus*

In pond A, the gut contents of *Channa punctatus* consisted of only carnivorous food (crustaceans, rotifers, insects and their larvae, shrimps, fishes and their scales). This constituted about 98.3 per cent with 1.7 per cent miscellaneous food (Table 13.1).

In pond B, also the gut contents of *Channa punctatus* were analysed. There was no trace of herbivorous food. The carnivorous food consisted of plankton, crustaceans, rotifers, insects and their larvae, molluscs, shrimps along with fins, fries and fingerlings, fish-fin and scales. The percentage of carnivorous food was 97.9 per cent and 1.0 per cent miscellaneous. The vegetable debris and mud were present 1.1 per cent (Table 13.2).

Gut Content of *Cirrhinus mrigala*

While analysing the gut contents of *Cirrhinus mrigala,* the unicellular algae and aquatic weeds were recorded, but there was no traces of crustaceans and rotifers in the gut. In pond A, the total herbivorous food consists about 84.0 per cent while debris and mud about 14.0 per cent with 2 per cent miscellaneous food (Table 13.1).

In pond B, in the food analysis of *Cirrhinus mrigala* only herbivorous diet was recognised which was composed of unicellular and multicellular algae and aquatic weeds along with little amount of vegetable debris and mud. Thus herbivorous food constituted about 88.5 per cent while vegetable debris and mud shared 10.3 per cent with 1.2 per cent miscellaneous (Table 13.2).

Gut Content of *Catla catla*

In the gut contents of *Catla catla* of pond A, the unicellular and multicellular algae and aquatic weeds were found as herbivorous food about 68.3 per cent while plankton, crustaceans and rotifers were 30.2 per cent along with 1.5 per cent miscellaneous (Table 13.1).

The gut content of *Catla catla* of pond B was analysed. The unicellular and multicellular algae and aquatic weeds were again found as herbivorous food which shared about 59.8 per cent of the total gut contents, while crustaceans, rotifers were about 38.2 per cent along with 2.0 per cent miscellaneous (Table 13.2).

Discussion

Observation and examination of gut contents of *Labeo rohita, Wallago attu, H. fossilis, C. punctatus, C. mrigala* and *C. catla* showed that they fed on a particular type of food available in different zones of the water bodies. The gut content of:

1. *Labeo rohita*–Consisted of a few quantities of unicellular and multicellular algae but a large quantity of aquatic weeds and some traces of adult crustaceans, insects, vegetables debris and mud were also noticed. Thus, it is clear that *Labeo rohtia* is omnivorous. Since most of the food organisms, recovered from gut contents, are distributed more in the middle layer of the pond, hence the fish is mid feeder.
2. *Wallago attu*–Consisted of adult crustaceans, insects and a number of small fishes. Thus, it is evident that *Wallago attu* is completely carnivorous. Mishra (1953), Das and Srivastava (1959), suggested that *Wallago attu* is mid feeder as they feed on food organisms from the mid-surface zone.
3. *C. mrigala*–Consisted of unicellular and multicellular algae with good amount of aquatic weeds along with vegetables debris and mud. Thus, *C. marigala* is herbivorous. It takes food from bottom layers of the water body and so they are considered as bottom feeder.
4. *C. Catla*–Consisted of unicellular and multicellular algae, aquatic weeds, crustaceans and rotifers. Since it feeds on the food, available on the surface zone of the water, so it is surface feeder.
5. *H. fossilis*–Consisted poor percentage of unicellular and multicellular algae with some amount of aquatic weeds. It takes crustaceans, rotifers, insects and their larvae and molluscs as their food. Thus, it is reported as omnivorous and considered as bottom feeder.
6. *C. punctatus*–Consisted crustaceans, rotifers, insects and their larvae, fry and fingerlings, molluscs and shrimps as their food and therefore, they are carnivorous in feedings habits. As this fish depends on the benthic organisms, it is considered as bottom feeders.

Acknowledgements

The authors are grateful to Prof. Y.K.P. Sinha, Head of the P.G Department of Zoology and Dr. A.P. Mishra, Professor, P.G Department of Zoology, RRA, Bihar University, Muzaffarpur for their incessant encouragement. First author is also thankful to Dr. C.M. Thakur and Dr. D.C. Baluni, P.G Department of Zoology, R.D.S. College, Muzffarpur for their valuable suggestions.

References

Chacko, P.I. and Kuriyan, T.G.K., 1948. On a fish survey of the Cavery. *Proc. Ind. Sci. Cong.*, 35(3): 205.

Chacko, P.I. and Kuriyan, G.K., 1949. Feeding and breeding habits of the common carps of South India. *Proc. Ind. Sc. Cong.*, 36(3): 167.

Das, S.M. and Moitra, S.K., 1963. Studies on the food and feedings habits of some freshwater fishes. Part IV. A Review on the food and feeding habits with general conclusions. *Ichthyologica*, 2(1–2): 107–115.

Das, S.M. and Srivastava, V.K., 1955. Quantitative studies on freshwater plankton. I. Plankton of a fish tank of Lucknow, *India. Nat. Acad. Sci.*, 26: 83–89.

Hynes, H.R.N., 1950. The food of freshwater stickle lakes with a review of the methods used in the studies of the food and fishes. *J. Anim. Ecol.*, 19: 36–58.

Joseph, K.R., Raman, K. and Khan, P.M.A., 1979. Observation on the exchange of phosphorous between soil and water of lake Pulicut. In: *Proc. 66th Ind. Sc. Cong.*, 3: 46.

Khan, M.H., 1934. Habits and habitats of food fish of Punjab. *J. Bombay Nat. Hist. Soc.*, 37: 655–668.

Mishra, R.N., 1953. On the gut contents of *Labeo rohita* (Ham.), *Cirrhinus marigala* (Ham.), *Catla catla* (Ham.). In: *Proc. 40th Ind. Sc. Congr.* Abs., p. 210.

Raheza, P.C., 1968. Soil productivity and crop growth. *J. Central Arid Zone Res. Inst.*, Jodhpur.

Singh, U.N., 1977. Ecological studies on aquatic insects of Kararia and Motijheel lakes of East Champaran (Bihar). *Ph.D. Thesis*, M.U.

Chapter 14

Characterisation of Anhydrous Blended Cement Sample

G. Sivakumar[1], *K. Mohanraj*[2] *and S. Barathan*[2]

[1]*Centralised Instrumentation and Service Laboratory,* [2]*Department of Physics, Annamalai University, Annamalainagar – 608 002, Tamil Nadu, India*

ABSTRACT

In order to cope up to the necessity and demand each manufacturer is releasing varieties of grade in cement. This investigation characterizes the market available Blended Cement (BC) and compare it with Ordinary Portland Cement (OPC) using SEM micrographs with EDS analysis and FTIR spectra. A consumer has to identify the component responsible for Blended Cement is the objective of the present study.

Keywords: *OPC, SEM, EDS, FTIR.*

Itnroduction

Cements may be defined as adhesive substances capable of uniting fragments or mass of solid matter to a compact whole. Its plays a vital role in modern construction. It is one of the agglomerated material and most commonly used for construction work in modern world (Lea, 1970). The main phases of cement is tricalcium silicate (C_3S), dicalcium silicate (C_2S), tricalcium aluminate (C_3A) and tetra calcium aluminoferrite (C_4AF). Blended Cements (BC) are produced by intimately and uniformly inter grinding or blending two or more types of fine materials. The primary materials are Ordinary Portland Cement (OPC) and other blended materials such as Ground Granulated Blast Furnace Slag (GGBFS), Flyash (FA), Silicafume (SF) (Shetty, 2004). For several decades, the concrete and cement industry has routinely incorporated flyash into cement as a partial replacement. Flyash improves concrete properties, lowers the cost of concrete production and is ecologically beneficial (Malhotra,

1996). The characterization of anhydrous samples can well be explained through the spectroscopic techniques like FTIR, X-ray, NMR, Raman and SEM analysis. The annual production of cement in our country is large amounts by cement industries. Due to this large production, manufactures place advertisements like quick setting, high strength etc. to promote their sales in the market. Hence it is of author's interest to carryout this investigation regarding the characteristics of cement component (OPC and BC) through FTIR and SEM.

Experimental Studies

The Fourier Transform Infrared Spectroscopy is one of the powerful techniques normally used for molecular characterization as it contains both phases and structural information. The sample is mixed with spectra grade KBr in the proportion 1 : 30 ratio and made to a pellet after a piolet study.

SEM with EDS is a very powerful non-destructive tool, which covers the observation from the fine structures on a surface of specimen to elemental analysis on a micro area. The sample is dispersed in specimen stubs using double-sided adhesive carbon tape and is coated with the help of gold coater. SEM micrographs are recorded from the sample using JEOL-SEM model, JSM-5610 LV.

The collected samples were individually subjected to chemical analysis for their constituents using standard procedure suggested by American Society of Testing Materials (ASTM-CI14-58) and are reported given below.

Samples	*CaO*	SiO_2	Al_2O_3	Fe_2O_3	SO_3	*MgO*	*LOI*	*Others*
OPC	63.15	20.70	5.40	3.12	2.90	2.70	1.56	0.47
BC	62.60	21.41	4.75	3.40	2.70	2.85	1.80	0.49

Results and Discussion

FTIR spectra of anhydrous Blended Cement (BC) and Ordinary Portland Cement (OPC) were recorded (4000–400 cm^{-1}) and are shown in Figures 14.1(*a*) and (*b*) respectively.

The small shallow at 3630 cm^{-1} is due to OH asymmetric stretching ($_{v3}$) vibration from residual (Harchand *et al.*, 1980). Broad band centered at 3413 cm^{-1} is due to symmetric stretching ($_{v1}$) vibrations

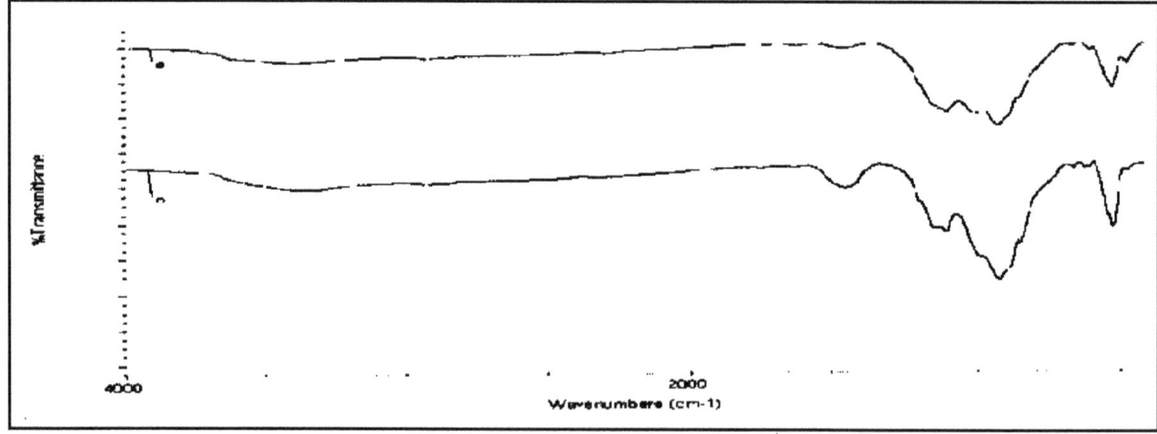

Figure 14.1(*a*) BC and (*b*) OPC

of H_2O from gypsum. Band at 1460–1420 cm^{-1} is attributed to ($_{v3}$) vibration of carbonate (Mollah *et al.*, 1993; Hassaan and Hakeem, 1989) formed by CO_2 attack and shallow at 874 cm^{-1} is due to out of plane bending ($_{v4}$) vibration.

Weak doublet observed in the region 1150–1090 cm^{-1} is attributed to $_{v3}$ vibrations of sulphates or gypsum. Bands at 668 cm^{-1} and 601 cm^{-1} are due to out of plane ($_{v4}$) and in plane bending ($_{v2}$) vibration of sulphates. The sulphate bands are weak due to their low concentration.

A strong band at 920 cm^{-1} is due to $_{v3}$ vibration of silicate (Ghosh and Chatterjee, 1975) of C_3S. Strong bands at 520 cm^{-1} and 465 cm^{-1} are attributed to $_{v4}$ and $_{v2}$ bending vibrations of silicates. Band in the region 1100–990 may be attributed to higher frequency $_{v3}$ vibration of silicates of flyash incorporated during the time of production (Hanna *et al.*, 1995).

The band assignments of *OPC* are almost same as that of *BC*. However, they observed some differences. Bands in the region 1150–1090 cm^{-1} are stronger owing to relatively higher sulphates in BC. The silicate bands occurring in the region 990–1100 cm^{-1} for BC absented itself due to absence of flyash in OPC. SEM micrograph (X 2000) of Figures 14.2(*a*) BC and (*b*) OPC shows dispersed anhydrous samples.

Figure 14.2(*a*) Micrograph of BC (X 2000)

Figure 14.2(*b*) Micrograph of OPC (X 2000)

BC cement particles are extremely angular and spherical in nature except for tricalcium silicate (C_3S) and dicalcium silicate (C_2S) respectively (Dalgleish *et al.*, 1982) Both C_3S and C_2S appear to be embedded in a fine-grained matrix such as tricalcium aluminate (C_3A) and a variable composition of tetracalcium alumino ferrite (C_4AF) (Maximilienne Bishop, 2003). In this figure, the larger particles may be monomineralic, but the majority is likely to be composites. The C_3S particles are about 25 µm in size and angular in shape and the C_2S particles are smaller about 10 µm in size. The gypsum, which is relatively soft compared with cement phases, occurs in the fine grained matrix phases. The inter ground flyash particles comprise approximately 10–20 per cent of volume. On contrast to cement particles, the shape of the fly ash particles is a smooth surfaced sphere. Figure 14.2(*c*) shows, the flyash particles are poorly graded and the large particles are (less than 10 µm) smaller than that of the Portland cement particles. The various particle sizes flyash are clearly seen in figure. The diameter of the various particles ranges between 0.5 and 10 µm. The biggest particles are usually rich in calcium and silicon (Papadakis and Vagelis, 1999).

The SEM micrograph of *OPC* is similar to that of *DC* except the absence of flyash particles. The micrograph 14.2(*d*) represents the higher magnification (X3500) and the cement particle (C3S) is about 25 µm in size.

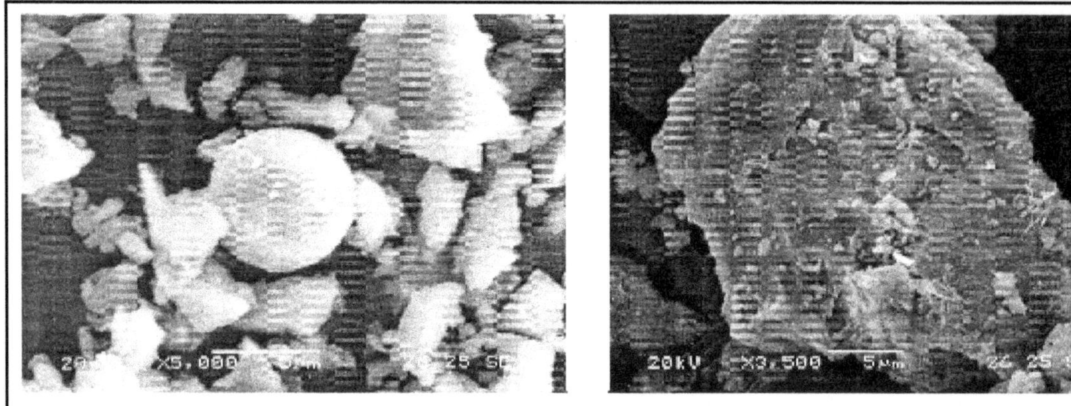

Figures 14.2(*c*): Micrograph of BC (X 5000) **Figures 14.2(*d*): Micrograph of OPC (X 3500)**

The anhydrous BC and OPC have been analysed by EDS technique for qualitative and semi-quantitative information. Since EDS is not purely a surface technique and has penetration greater than 1 μm, below the data given the composition of the sample in the bulk of the matrix rather than on the surface.

Samples	*Ca*	*Si*	*Al*	*Fe*	*S*	*K*	*Na*	*Mg*	*Mn*	*Others*
OPC	62.81	14.52	3.16	2.75	2.06	1.53	0.81	0.94	0.86	10.66
BC	60.86	18.56	2.36	3.24	1.72	0.32	0.54	0.98	0.97	10.45

Conclusion

The commercially available blended cement has been characterized using FTIR, SEM and compare with ordinary portland cement OPC. It is observed that the BC is found to contain flyash about to 10-20 per cent of volume in addition to OPC. The IR and SEM are found to be an effective tool in characterisation of cement.

References

Dalgleish, B.J., Pratt, P.L. and Toulson, E., 1982. Fractographic studies of microstructral development in hydrated Portland cement. *J. Mater. Sci.*, 17: 2199–2207.

Ghosh, S.N. and Chatterjee, A.K., 1974. Absorption and reflection Infrared spectra of major cement minerals, clinkers and cement. *J. Mater. Sci.*, 10: 1574–1584.

Hanna, R.A., Barrie, P.J., Cheeseman, C.R., Hills, C.D., Buchler, D.M. and Perry, R., 1995. Solid state 29Si and 27 Al NMR and FTIR study of cement pastes containing industrial wastes and organic. *Cem. Concr. Res.*, 25(7): 1435–1444.

Harchand, K.S., Vishwamittar and Chandra, K., 1980. Infrared and Mossbauer study of two Indian Cements. *Cem. Concr. Res.*, 10: 243–252.

Hassaan, M.Y. and Abdel Hakeem, N., 1989. Study of anhydrous and hydrated Portland cement containing alkalai ions by Infrared spectroscopy. *J. Mater. Sci. Lett.*, 8: 578–580.

Lea, F.M., 1970. *The Chemistry of Cement and Concrete.* Edward Arnold (Publishers) Ltd., London.

Malhotra, P.K. and Mehta, 1996. *Pozzolanic and Cementitious Materials.* Gordon and Breach, The Netherlands.

Maximilienne Bishop, Simon G. Bott and Barron, Andrew R., 2003. A new mechanism for cement hydration inhibition: Solid state chemistry of calcium Nitrilotris (methylene) triphosphate. *Chem. Mater.*, 15: 3074–3088.

Mollah, M.Y.A., Hess, Thomas R., Tsai, Yung-Nian and Cocke, D.L., 1993. An FTIR and XRD investigations of the effects of carbonation on the solidification/stabilization of cement based systems–Portland type V with zinc. *Cem. Caner. Res.*, 23: 773–784.

Mollah, M.Y.A., Yu, Wenhong, Schennach, Robert and Cocke, David L., 2000. A fourier transform infrared spectroscopic investigation of the early hydration of Portland cement and the influence of sodium lignosulfonate. *Cem. Concr. Res.*, 30: 267–273.

Papadakis, V.G., Pederson, E.J. and Lindgreen, 1999. An AFM-SEM investigation of silica fume and flyash on cement paste microstructure. *J. Mater. Sci.*, 34: 683–690.

Shetty, M.S., 2004. *Concrete Technology*. S. Chand and Co., Ltd, Delhi.

Chapter 15

A Review on Biofertilizers and Biocides: A Best Alternative of Chemical Fertilizers and Pesticides

Deepali and Kamal K. Gangwar

Department of Zoology and Environmental Sciences,
Gurukul Kangri University, Haridwar – 249 404

In India, due to growing human population it is necessary to increase crop production and land productivity to fulfill their food requirements. It is estimated that up to 2020, total production required by country will become 321 million tones of foodgrains and for their proper growth nutrient requirement will become 28.8 million tones but only 21.6 million tones nutrient will be available from a deficit about 7.2 million tones. To increase crop production use of chemical fertilizers and plant protection materials is also increased. Increasing use of chemical fertilizers in agriculture make country self sufficient in food production but it pollute environment and cause slow deterioration of living beings (Saxena and Joshi, 2002a).

Plant requires essential nutrients like nitrogen (N), phosphorus (P), potassium (K), and several other minerals for their growth and receives them from soil (Arya, 2000). Nitrogen and phosphate relationship is very important and shows a direct impact on productivity of soil (Hatchinson and Richards, 1921). N.P.K. fertilizers used for the crop improvement are not completely utilized by crops but excess of fertilizers washed away from land by rainwater and causes water pollution. A large amount of nutrients that are washed away in water bodies during rainy season can cause eutrophication and the whole stretch of water many become chocked.

The liberal use of chemical fertilizers especially in paddy fields may affect the growth inhabiting microorganisms (Bishara, 1978, 79; Konar and Sarkar, 1983). These fertilizers, reach into water bodies, also affect fishes (Palanichamy, 1985) and other organisms living their. In agriculture, chemical fertilizers are used extensively but they are costly and also have various adverse effects on soils *i.e.*, depletes water holding capacity, soil fertility and disparity in soil nutrients. Due to insufficient uptake of these fertilizers by plants results in the leaching away from soil. Hence, it is necessary to develop some low cost fertilizers which work without disturbing nature.

Microbial Fertilizers/Biofertilizers/Biomanure

From the 1950s to 1970s considerable number of nitrogen fixing bacteria were found to be associated with crop. Several soil microbiologists suggests that nitrogen fixing bacteria associated with the plants may be the source of agronomically significant nitrogen inputs to the sugarcane crop in Brazil (Ray and Handerson, 2001). A number of microorganisms (bacteria, fungi and algae) are considered as beneficial for agriculture and used as biofertilizers. Microbial consortia are inoculated in the field for the improvement and supply of nitrogen, phosphorus, potassium and other essential elements which are necessary for the proper growth of plants. Microorganisms produce a range of extra cellular enzyme which has the potential to mediate utilization of organic sources of nitrogen and phosphorus in soil (Saxena and Joshi, 2002b).

Biofertilizers are supposed to be a safe alternative to chemical fertilizers to minimize the ecological disturbance. These are natural, organic, non-pollutant, cheap products that are required in a small dose (Gahukar, 2005–06). Microorganisms can fix atmospheric nitrogen, and solubilize insoluble phosphates, produce growth promoting substances for plants *e.g.*, vitamins and hormones (Yojana, 1992). Many workers have reported that the uses of biofertilizers are beneficial for soil, as well as for crops. It is the safest method to maintain soil fertility. Biofertilizers are also called as "Microbial inoculants" (Subha Rao, 1982) due to use of many nitrogen fixing bacteria and cyanobacteria. Microbial inoculants are considered as new feature of agricultural system that facilitate or enhance the microbial process in the soil. Common biofertilizers such as *azotobacter, azospirillum, rhizobium*, blue green algae etc. are good nitrogen fixer and *Bacillus megatherium var phosphaticum, Aspergillus awamori, Penicillium digitatum* etc. are good phosphate solubilizers.

Nitrogen fixing organisms such as *azotobacter, Azotomonas, Azotococcus, Biejerinckia, cyanobacteria* (*Anabaena* spp.) can fix nitrogen under aerobic condition and can be utilized for nitrogen deficient soil, while some facultative anaerobes such as *Bacillus, Klebsiella, Rhodopseudomonas* and some species of the anaerobic genus *Clostridia* fix nitrogen under anaerobic condition (Emerich and Wall, 1985). Nitrogen fixation by nitrogen fixing microorganism is catalyzed by the enzyme 'nitrogenase' which is inhibited by free oxygen (Emerich and Wall, 1985). The reduction of atmospheric nitrogen to ammonia by the nitrogenase enzymes is expressed as follows:

$$N_2 + 8H^+ + 6e^- \rightarrow 2NH_4^+$$

Microbes are effective in inducing plant growth as they secrets plant growth promoters (auxins, abscisic acid, gibberellic acid, cytokinins, ethylene) and affects seed germination and root growth (Ramarethinum *et al.*, 2005). They also play considerable role in decomposition of organic materials and enrichment of compost.

Biofertilizers have 75 per cent moisture and it could be applied to the field directly. Biofertilizers contained 3.5 per cent–4 per cent nitrogen, 2 per cent–2.5 per cent phosphorus and 1.5 per cent potassium. In terms of N : P : K, it was found to be superior to farmyard manure and other type of manure (Mukhopadhyay, 2006).

Table 15.1: N : P : K Composition of Different Field Manure

Manure Type	Nitrogen per cent	Phosphorus As P_2O_5, per cent	Potassium as K_2O per cent
Farmyard manure	0.40	0.20	0.40
Urban compost	0.60	0.50	0.60
Green manure	0.600.70	0.100.20	1.25
Biofertilizer (50 per cent day)	1.802.40	1.001.20	0.600.80

Biofertilizers can be produced in the anaerobic digester used for production of biogas as a co-product of biogas. It consists of remaining residual solids after draining bioliquid. Excess sludge can perhaps serve as a fertilizers additive due to nitrogen fixing capability (Kargi and Ozmihci, 2004).

At present, annual production of biofertilizers is estimated as around 7000 tones from nearly 70 units and expected consumption of biofertilizers is approximately 6000 tones. In India nearly 3.5 lakh small and large biogas units have been installed during 1981–1991 (Makhopadhyay, 2006). Now Central Government also provides some financial assistance for setting up biofertilizer units.

Biofertilizers are an alternative to the conventional approach as they have lower cost than the chemical fertilizers and when they are required in bulk can be generated at the farm itself hence; these are economically attractive for the farmers (Venkatramani, 1996). The extent of cultivation depends on the market type and its proximity to the processing facility.

Transportation costs and distance to markets will affect product value and its potential use (Saxena and Joshi, 2002b).

Nitrogen Fixing Bacteria

Rhizobia

Legume plants have root nodules, where atmospheric nitrogen fixation is done by bacteria belonging to genera, *Rhizobium* (fast growing rhizobia), *Bradyshzodium* (slow growing rhizobia), *Sinorhizobium, Azorhizobium* and *Mesorhizobium* collectively called as rhizobia (Jordan, 1984; Chimote and Kashyap, 2001).

Rhizobium is a free living, gram negative, non-sporulating, aerobic, motile and rod shaped bacterium which occur in soil. It occurs in the roots of leguminous plants and forms nodules where it fixes nitrogen in the presence of leghaemoglobin. Leghaemoglobin promotes O_2 utilization in bacteroids and favours nitrogen fixation and converts atmospheric nitrogen into ammonia in presence of phosphorus and molybdenum (Gahukar, 2001). In the absence of leghaemoglobin nitrogen fixation can not takes place. When rhizobial culture is inoculated in field, pulse crops yield can be increased due to rhizobial symbiosis (Dubey, 2001). *Rhizobium* resides in roots of beans, grams, groundnut and soyabean etc. *Rhizobium* is a crop specific inocultant. It can fix 15–20 kg N/ha and increase crop yield upto 20 per cent. Higher Mg content due to inoculation of *Rhizobium* was observed by Kumudha (2005). Bhaskar and Kashyap (2004) reported that wild type strain of *R. ciceri* 18–7 mutant M126 and complemented mutant M126 (C_4) were characterized for symbiotic properties (*viz.*, acetylene reduction assay, total nitrogen content, nodule number and fresh and dry weight of the injected plants) and nitrogenase activity).

Azorhizobium is a stem nodule forming bacteria and for fixes nitrogen symbionts of the stem nodule also produce large amount of IAA that promotes plant growth (Ghose and Basu, 1997). They have isolated *Azorhizobium caulinodans* from the stem nodule of the leguminous emergent hydrophyte *Aeschynomene aspera* and their purpose of study was to check the ability of *Azorhizobium* spp. for EPS (extracellular polysaccharides) production for its importance in nodule symbiosis and in industry. EPS acts as determinants of host plant specificity and play a role in the first step of root hair infection in N_2 fixation (Olivares *et al.*, 1984). Ghosh and Basu (2001) also reported the plant *Aeschynomene aspera* possess many sessile stem nodule containing high amount of IAA.

Bradyrhizobium is also reported us good nitrogen fixer by various workers. Mathew *et al.* (2003) reported that when Mucunna seed were applied with an effective isolate of *Bradyrhizobium* in rubber (*Hevea brasiliensis*) Plantation, prior to souring it increased plant growth and consequently plant biomass, reduction in the weed population and increased soil microbial population were reported. It was also reported that *Bradyrhizobiuin* strain inoculation causes improvement in total organic carbon, N_2, available phosphorus and potassium in the soil.

Diazotrophs

These are aerobic chemolithotrophs and anaerobic photoautotrophs. These are non-nodule forming bacteria. They include members of the families:

Azotobacteracae *e.g., Azotobacter*

They are the free living aerobic, photoautotrophic, non-symbiotic bacteria. They secretes vitamin-B complex, gibberellins, napthalene, acetic acid and other substances (Jain, 1998 and Gohukar, 2001) that inhibit certain root pathogens and improves root growth and uptake of plant nutrients. It was also reported that Azotobacter inoculation is effective in soil only in the presence of a native *Azotobacter* population. It was also reported that *Azotobacter* inoculation is effective in soil only in the presence of a native *Azotobacter* population. It occurs in the roots of *Paspalum notatum* (tropical grasses) and other spp. of this genus or other genera (Dobereiner, 1970). It adds 15–93 kg N/ha/annum on *P. notatum* roots (Dobereiner *et al.*, 1973). *Azotobacter indicunl* occurs in acidic soil in sugarcane plant roots. It can apply in cereals, millets, vegetables and flowers through seed, seedlings soil treatment.

Spirillaceae–*e.g., Azospirillum, Herbaspirillum*

These are gram negative, free living, associative symbiotic and non-nodule forming, aerobic bacteria. *Azospirillum* is a wide spread bacterium, occurs in the roots of dicots and monocot plants *i.e.*, corn, sorghum, wheat etc. (Tarrand *et al.*, 1978 and Elmerich, 1984), and responsible for nitrogen fixation in association with several cereals (Balandreau, 1983 and Vose, 1983).

It is easy to culture and identify. *Azospirillum* is found to be very effective in increasing 10–15 per cent yield of cereal crops and fixes N_2 upto 20–40 per cent kg/ha. Different *A. brasillense* strains inoculation in the wheat seed causes increase in seed germination, plumule and radicle length (Tien *et al.*, 1979; Gunasekaran and Purushothaman, 1980). *Azospirillum* inoculations cause plant growth. It may be due to either by nitrogenase activity or by its ability to produce plant growth promoters (Okon and Labandera-Gonzalez, 1994). Ray *et al.* (2004) have conducted a field experiment to assess the response of winter and autumn rice varieties to *Azospirillim brasillense*, strain Sp7 and *Azospirillum lipoferum*, strain C_2 with and without inorganic nitrogen. They have observed that there was an increase in the grain yield of all the 4 varieties as well as dry matter. They have also reported that *Azospirillim brasillense* (Sp7) inoculation gave higher per cent grain and straw yield than *A. Lipoferum* (C_2).

Herbaspirillum species occurs in roots, stems and leaves of sugarcane and rice. They produce growth promoters (IAA, Gibberillins, Cytokinins). That promotes root development and enhances uptake of plant nutrients (N, P, K). In legume plants, *Azospirillum* fixes atmospheric nitrogen.

Acetobacter Diazotrophicus

Another diazotroph is *Acetobacter diazotrophicus* occurs in roots, stem and leaves of sugarcane and sugar beat crops as nitrogen fixer and applied through soil treatment. It also produces growth promoters *e.g.*, IAA. That helps in nutrients uptake, seed germination, and root growth. This bacterium fixes nitrogen upto 15 kg/ha/year and enhance upto 0.5–1 per cent crop yield (Gahukar, 2005–06). It releases various organic acids (Mahesh Kumar, 1999), succinate, tarterate, citrate and gluconate etc. which generated H^+ ion that dissolve mineral phosphate and make it easily available to plants. Mowade and Bhattacharya (2000) have reported that organism have the capacity to solubilising insoluble phosphate in Pikovskay's medium and Sperber's medium and tested for its response to different antibiosis under in vitro condition.

Cyanobacteria (Blue Green Algae)

Nostoc, Anabaena, Oscillatoria, Aulosira, Lyngbya etc. are the prokaryotic organisms and phototropic in nature. They play an important role in enriching paddy field soil by fixing atmospheric nitrogen and supply vitamin B complex and growth promoting substance (Sharma, 1986) which makes the plant grow vigorously. In paddy field there is no need of cyanobacterial inoculation. They also convert insoluble phosphorus into soluble form by excreting organic acids. Cyanobacteria fixes 20–30 kg/N/ha and increase 10–15 per cent crop yield when applied at 10 kg/ha. Cyanobacteria oxygenate the water impounded in the field. The cyanobacterial mat in paddy fields also reduces loss of moisture from the soil. Blue green algae can be used in the form of flakes or these flakes can be powdered and mixed with farmyard manure and soil then can be applied in the field. Algal flakes are dried and mixed at the rate of 10^{15} kg/ha, after one week rice is planted (Arya, 2000).

Azolla-Anabaena Symbiosis

It is a free floating, aquatic fern found on water surface having a cyanobacterial symbiont *Anabaena azollae* (heterocystous) in their leaves. Azolla fixes atmospheric nitrogen in association with nitrogen fixing cyanobacteria in paddy field and excrete organic nitrogen in water during its growth and also immediately upon trampling. Azolla is used as biofertilizer in India, USA, Sri Lanka, China, Indonesia, Bangladesh and many other countries. Azolla fixes 40, 60 kg N/ha in a month and increases 10–20 per cent yield cif paddy crops (Kumar, 2004).

Azolla contributes nitrogen, phosphorus (15–20 kg/ha/month), potassium (20–25 kg/ha/month) and organic carbon etc. and when applied in rice field also suppresses weed growth. It is also susceptible to high temperature (>40°C) and scarcity of water (Gahukar, 2005–06). Azolla also absorbs traces of potassium from irrigation water. Azolla can be used as green manure before rice planting. Azolla spp. are metal tolerant hence, can be applied near heavy metal polluted areas.

Phosphate Solubilising Bacteria

Pseudomonas fluorescens, Bacillus megatherium var. phosphaticum, Acrobacter acrogens, Nitrobacter spp., *Escherichia freundii, Serratia* spp., *Pseudomonas striata, Bacillus polymyxa* are the bacteria have phosphate solubilising ability.

'Phosphobacterin' are the bacterial fertilizers containing cells of *Bacillus megatherium var. phosphaticum* prepared firstly by USSR scientists. They increased about 10 to 20 per cent crop yield (Cooper, 1959).

It can be applied for low land and upland rice. They dissolve soil phosphate and make it available for crops. A large number of soil microorganisms have the ability to solubilize inorganic phosphates through their metabolic activity directly or indirectly (Arya, 2000). They produces plant growth promoting hormones *e.g.*, IAA, GA and various organic acids *e.g.*, lactic acid, citric acid, fumaric acid, succinic acid etc. These organic acids help in phosphate solubilising activity of soil.

Phosphate Solubilizing Fungi

Some fungi also have phosphate dissolving ability *e.g.*, *Aspergillus niger*, *Aspergillus awamori*, *Penicillium digitatum* etc.

Plant Growth Promoting Rhizobacteria (PGPR)

Recently ability of plant growth promoting rhizobacteria is also studies for microbial control. They are also called as microbial pesticides *e.g.*, *Bacillus* spp. and *Pseudomonas fluorescens*. Sipirin (2000) reported that the growth and survival or vetiver is possible without nitrogen and phosphorus application especially in the infertile soil with the help of diazotrophs including the genera of *Pseudomonas*. Chakraborti *et al.* (2003) reported that four bacterial spp. *Serratia marcescens* TR10, *Ochrobactrum ocnthropi* TR9, *Bacillus pumilus* TR24 and Bacillus spp. TR16, isolated from rhizosphere of tea plant have antagonistic ability against the root pathogens of the tea plant. They have also reported that *Srratia* spp. and *Ochrobactrum* spp. are able to promote growth of plants. It was founded by Paul *et al.* (2003) that *Pseudomonas fluorescens* application to the black pepper rhizophere resulted in easy mobilisation of the essential nutrients in the rhizosphere microcosm and resulted in enhanced uptake of nutrients, which reflected in increased plant biomass.

A study was carried out by Joseph *et al.* (2003) in the Kerala from 10 rubber growing areas and isolated 17 fluorescent rhizobacterial strains and founded that all the 17 isolates shows antagonism against 5 pathogens of rubber. They also observed that these organism also improve the growth of *H. brasiliensis* and associated crop cover *i.e.*, shoot weight, root weight, nodulation, nitrogenase activity.

Mycorrhiza

Mycorrhizas are developed due to the symbiosis between some specific root inhabiting fungi and plant roots. This means these are fungi grow on roots of the plants. There are specific fungi for vegetables, fodder crops, flowers, trees etc. Mycorrhizas are used as biofertilizers (Kumudha, 2005) and biocides. They absorb nutrients such as manganese, phosphorus, iron, sulphur, zinc etc. from the soil and pass it on to the plant. They also show higher tolerance to high soil temperature. Mycorrhizal fungi increase the yield of crop fields by 30–40 per cent. Mycorrhiza also produces plant growth promoting substances. It was also reported that seedlings having mycorrhizal fungi growing on their roots grow faster after inoculation. Mycorrhiza occurs in low land and upland rice and it mobilize phosphorus required by rice.

Commonly mycorrhizas are of two types: Ectomycorrhiza and Endomycorrhiza. Both are different in structure and systematic position of fungi.

Ectomycorrhiza

They occur in the roots of the higher plants *e.g.*, gymnosperms and angiosperms. These are the fungi requires pH 5–6 for their proper growth. Excessive use of inorganic fertilizers and shading supresses the development of ectomycorrhizal fungi. The poor development of mycorrhizal fungi in the roots of plants causes stunted growth and chlorosis of leaves.

Endomycorrhiza (VAM Fungi)

They occur commonly in the roots of crop plants. These fungi penetrate the cortical cells and get established them intracellularly by secreting extracellular enzymes. VAM fungal hyphae enhance the uptake of phosphorus and other nutrients that is a responsible for plant growth stimulation including roots and shoot length (Hayman, 1980). VAM also enhances the growth of black pepper, protects from *Phytophthora capsid, Radopholus similis* and *Melvidogyne incognita* (Anandraj *et al.*, 2001). It was reported by Kumudha (2005) that VAM inoculation was found to be effective in chlorophyll pigmentation. Trappe and Fogel (1977) reported that the increased nitrogen uptake in VAM might also be due to the increased phosphorus uptake which in turn might enhances the activity of NAD dependent enzyme which might contribute to nitrate reductase activity. VAM fungi enhance water uptake in plants and also provide heavy metals tolerance to plants.

Benefits from Biofertilizers

They are environmentally sound, pollution free and of low cost. They increase crop yield upto 10–40 per cent and fix nitrogen upto 40–50 kg. After using 3–4 years continuously, there is no need of application of biofertilizers because parental inoculums are sufficient for growth and multiplication. They improve soil texture, pH, and other properties of soil. They produces plant growth promoting substances *e.g.*, IAA amino acids, vitamins etc.

Precautions

1. They are living organism hence, handling should be careful.
2. Biofertilizers should be used before expiry date.
3. They are species specific hence, particular biofertilizer should be used for a particular crop plant.
4. Recommended dose of biofertilizer should be applied.
5. They should not be exposed to direct sunlight.

References

Arya, A., 2000. Biofertilizer for sustainable plant development. *Everyman's Science*, 35(1): 19–25.

Anandraj, M., Venugopal, M.N., Veena, S.S., Kumar, A. and Sharma, Y.R., 2001. Eco-friendly management of disease of spices. *Indian Species*, 38(3): 28–31.

Bhaskar, V.V. and Kashyap, L.R., 2004. Azide resistance in *Rhizobium ciceri* linked with superior symbiotic nitrogen fixation. *Ind. J. Exp. Biol.*, 42: 1177–2285.

Bishara, N.F., 1978. Fertilizing fish ponds II growth of *Mugil cephla* in Egypt by pond fertilization and feeding. *Aquaculture*, 13: 361–367.

Bishara, N.F., 1979. Fertilizing fish ponds III growth of *Mugil apito* in Egypt by pond fertilization and feeding. *Aquaculture*, 16: 47–55.

Balandreau, J., 1983. Microbiology of the association. *Can. J. Microbiol.*, 29: 851–859.

Chakraborty, V., Chakraborty, B.N., Roy Chawdhary, N., Tongden, C. and Basnet, M., 2003. Investigations on plant growth promoting rhizobacteria of tea rhizosphere. 6th *International PGPR Workshop*, 5–10 October, Calicut, India.

Chimote, V. and Kashyap, L.R., 2001. Lipochito oligosaccharides and legume rhizobium symbiosis: A new concept. *Indian J. Exp. Biol.*, 39: 401–409.

Cooper, 1959. *Soil Fertilizers*, 22: 227–233.

Dobereiner, J., 1970. *Zent. Bakteriol* II. 124: 224–230.

Dobereiner, J., Day, J.M. and Dart, P.J., 1973. *Perg. Agr Pe. Bras*, 8: 153–157.

Dubey, R.C., 2001. *A Textbook of Biotechnology*. S. Chand and Company Ltd., New Delhi.

Elmerich, C., 1984. *Biotechnology*, 2: 967–978.

Emerich, D.W. and Wall, J.D., 1985. Nitrogen fixation In: *Comprehensive Biotechnology*, Vol. 4, (Ed.) M. Moo-young. Pergamon Press, N.Y., pp. 73–106.

Gahukar, R.T., 2001. *Kisan World*, 28(3): 25–27.

Gahukar, R.T., 2005–06. Potential and use of biofertilizers in India. *Everyman's Science*, 40(5): 354–361.

Ghosh, A.C. and Basu, P.S., 1997. Culture growth and IAA production by a microbial diazotropic symbiont of stem-nodule of the legume *Aeschynomene aspera. Folia. Microbiol.*, 42: 595.

Ghosh, A.C. and Basu, P.S., 2001. Extracellular polysaccharide production by *Azorhizobium caulinodans* from stem nodules of Leguminous emergent hydrophyte, *Aeschynomene aspera. Ind. J. Exp. Biol.*, 39: 155–159.

Gunasekaran, S. and Purushothaman, D., 1980. Nitrogen fixation by Azopirillum in the rhizophere of cotton. In: *National Symposium on Biological Nitrogen Fixation in Relation to Crop Production*, T.N. Agric University, Coimbtore, pp. 27.

Hayman, D.S., 1980. Mycorrhiza and crop production. *Nature*, London, 287: 487–488.

Hutchinson, H.B. and Richards, E.H., 1921. *J. Ministry of Agriculture*, 28: 398.

Jain, R., 1998. *Millennium Biofertilizer Guide*. Nargi Printer, New Delhi, pp. 52.

Jordan, D.C., 1984. *Bergeys Manual of Systematic Bacteriology*, pp. 234.

Joseph, K., Sunju, Y., George, J., Mathew, J. and Kunivilla Jacob, C., 2003. Plant growth promoting Rhizobacteria in Rubber (*Hevea brasiliensis*) plantations. In: 6th *International PGPR Workshop*, 5–10 October, Calicut, India.

Konar, S.K. and Sarkar, S.K., 1983. Acute toxicity of agricultural fertilizers to fish. *Geobios*, 10: 6–9.

Kargi, F. and Ozmihci, S., 2004. Batch biological treatment of nitrogen deficient synthetic wastewater using Azotobacter supplemented activated sludge. *Bioresource Technology*, 94: 113–117.

Kumar, N., 2004. *Indian Farmer's Digest*, 37(2): 7–9.

Kumudha, P., 2005. Studies on the effect of biofertilizers on the germination of *Acacia nilotica* Linn. *Seeds Ad. Plant Sci.*, 18(11): 679–684.

Mahesh Kumar, K.S. *et al.*, 1999. *Curr. Sci.*, 76: 874–875.

Mathew, J., Joseph, K., Lakshmanan, R., Jose, G., Kethandaraman, R. and Kuruvilla Jacob, C., 2003. Effect of Bradyrhizobium inoculation of *Mucuna bracteala* and its impact on the properties of soil under Hevea. In: 6th *International PGPR Workshop*, 5–10 October, Calicut, India.

Mowade, S. and Bhattacharya, P., 2000. Resistance of P-solubilosing Acetobacter diazotrophicus to antibiotics. *Curr. Sci*, 79(II).

Mukhopadhyay, S.N., 2006. Eco-friendly products through process biotechnology in the provision of biotechnology economy: Recent advances. *Technorama*, A. Supplement to IEI News, March.

Okon, Y. and Labandera-Gonzalez, C.A., 1994. Agronomic application of Azospirillum: An evaluation of 20 years worldwide field inoculation. *Soil Biol., Biochem.*, 26: 1591–1601.

Olivares, J., Bedman, E.J. and Martinez, M.E., 1984. Infertility of *Rhizohium melilotias* affected by extracellular polysaccharides. *J. Appl. Bacteriol.*, 56: 389.

Palanichamy, S., Seeniammal, K. and Arunachalam, S., 1985. Effects of agricultural fertilizers complex on food consumption and growth in the fish *Sarotherodon mossambicum* (Trewaves). *J. Env. Biol.*, 6(2): 71–76.

Paul, D., Srinivasan, V., Anandraj, M. and Sarma, Y.R., 2003. *Pseudomonas fluorescens* mediated nutrient flux in the black pepper Rhizosphere microcosm and enhanced plant growth. In: *6th International PGPR Workshop*, 5–10 October, Calicut, India.

Ramarethinam, S., Murugusen, N.V. and Rajalakshmi, N., 2005. *Pestology*, 20(4): 12–14.

Ray, N., Baruah, R. and Goswami, S.R., 2004. Effect of seedling bacterization of wetland rice with Azospirillum. *Ind. J. Environ. and Ecoplan.*, 8(3): 843–848.

Ray, R.N. and Handerson, G., 2001. Endophytic nitrogen fixation in sugarcane: Present knowledge and future applications, Technical expert meeting on increasing the use of biological nitrogen fixation (BNF) in agriculture, FAO, Rome.

Saxena, P. and Joshi, N., 2002a. A comparative study of compost formation by the *Mucor* sps. and *Penicillium* sps. on the soil. *Him. J. Env. Zool.*, 16(1): 83–96.

Saxena, P. and Joshi, N., 2002b. In: Role of microorganisms in the decomposition of organic wastes matter. *Ph.D. Thesis*, Gurukul Kangri University, Haridwar.

Sipirin, S., Thinathorn, A., Pintarak, A. and Aibcharoen, P., 2000. Effect of associative nitrogen fixing bacterial inoculation on growth of vetiver grass. A poster paper presented at ICV–2.

Subha Rao, N.S., 1982. In: Biofertilizers. In: *Advances in Agriculture Microbiology* (Ed.). Oxford and IBH Publication, New Delhi, pp. 219–242.

Sharma, V.K., 1986. A review of recent work on pesticide studies on the nitrogen fixing algae. *J. Env. Biol.*, 7(3): 171–176.

Tarrand, J.J., Kreig, N.R. and Dobereiner, J., 1978. A taxonomic study of the spirillum lipoferum groups, with description of a new genus, *Azospirillum gen. novo* and two sps. *Azosperillum brasiliense* sp novo. *Can. J. Microbiol.*, 24: 967–980.

Tien, T.M., Gaskins, M.H. and Hubbeli, D.H., 1979. Plant growth substances produced by *Azospirillim brasilense* and their effect on the growth of pearl millet (*Penniselum americanum* L.) *Appl. Environ. Microbiol.*, 37: 1016–1024.

Trappe, J.M. and Fogel, R.S., 1977. In: Ecosystem function of mycorrhizae in the below ground ecosystem: A synthesis of plant associated process, (Ed.) J.K. Marshall. Colarado State University, Range, Sci, Deptt. Sci. Services, No. 26 Fur Collins.

Venkatramani, G., 1996. In: *Hindus Survey of Indian Agriculture*, p. 29.

Vose, P.B., I983. Development in non-legume N_2 fixing systems. *Can. J. Microbial.*, 29: 837–850.

Yojana, 1992. In: Biofertilizers. In: *Advances in Agricultural Microbiology*, (Ed.) N.S. Subba Rao. Oxford and IBH Pub. Co., New Delhi, pp. 219–242.

Chapter 16

Impact of Monocrotophos on Major Nutrient Molecules in Different Tissues of Albino Mice

***P. Madhaveelatha*[1], *K. Sankaraiyah*[1], *R. Jayakumar*[2] and *K. Jayantha Rao*[1]**

[1]*Department of Zoology, Dr. S.R.K. Government Arts College, Yanam – 533 464*
[2]*Department of Zoology, S.V. University, Tirupati – 517 502*

ABSTRACT

Levels of major nutrient molecule such as carbohydrates, proteins, tree amino acids and lipids were determined in brain, liver, kidney and intestine tissues of albino mice exposed to single and multiple doses of monocrotophos. Carbohydrates, proteins and lipids showed a decrease whereas tree amino acids increased in all the tissues of experimental animals. The change in the major nutrient molecules content showed dose dependency.

Keywords: *Carbohydrates, Proteins, Free amino acids, Lipids, Monocrotophos.*

Introduction

Since pesticides impose stress condition in exposed animals, they may induce change in the major nutrient molecules such as carbohydrates, proteins amino acids and total lipid levels. Elumalai, *et al.* (1999) reported decreased tree sugar levels in rats exposed to monocrotophos, Sharma, *et al.* (2000) noticed gluconeogenesis and increased energy demand in rats. Sublethal exposure of monocrotophos and fenvalerate mixture induced a decrease in total carbohydrates level in *Labeo rohita* (Tilak *et al.*, 2001). Monocrotophos treated albino rats showed decreased protein content (Elumalai, *et al.*, 1999) and increased tree amino acid levels (Venkataswamy, 1991).

In the present investigation, an attempt was made to asses the toxic effect of sub-lethal concentration of monocrotophos on major nutrient molecules level in different tissues of albino mice after exposure to single and multiple doses of monocrotophos.

Material and Methods

The animals were obtained from Indian Institute of Science, Bangalore and mice colony was maintained in the laboratory at 27±2°C with 12 h light and 12 h darkness. Mice were fed on standard diet and water supplied *ad libitum*. A stock solution (1 mg/10 ml) of technical grade monocrotophos was prepared and required dilutions were made from the stock. As the acute LD_{50} value of monocrotophos through intra peritoneal injection was 11.2 mg/kg body weight sub-lethal concentration (1.2 mg/kg body weight) was selected for the study. Healthy animals weighing 30±2 gm were divided into 3 groups. The first set of animals was administered with single dose of monocrotophos (*i.e.* on first day) intra peritonially. Similarly multiple doses (*i.e.*, on 1st day, 3rd day, 5th day and 7th day) were given to the second set of animals.

The animals of third set served as corresponding controls. The control and treated animals were sacrificed on 9th day by cervical dislocation. The tissues were isolated and used for the estimations. Carrol *et al.* (1956), Lowry *et al.* (1951), More and Stein (1954) and Folch *et al.* (1957) methods were used to estimate carbohydrates, proteins, free amino acids, and total lipids.

Results and Discussion

The total carbohydrate content of different tissues showed great reduction (Table 16.1). The decrease was more in the tissues of animals exposed to multiple doses showing dose dependent action of monocrotophos. It may be due to higher energy demand under monocrotophos stress and their decreased synthesis of carbohydrates as a consequence of toxic stress. It signifies its utilization probably to meet high energy demands and decreased rate of gluconeogenesis (Sharma *et al.*, 2000) under monocrotophos stress. The same response was reported by several workers (Elumalai *et al.*, 1999; Tilak *et al.*, 2001) with different pesticides. Carbohydrates that are convertible (or related) to glycogen are being trapped extensively to meet the energy demand under insecticide stress (Sivaprasada Rao, 1980). The decrease may also be due to stimulation of hormones that accelerates carbohydrate utilization or inhibition of other factors which contribute to the carbohydrate synthesis under monocrotophos intoxication. The pesticide in general causes either anoxia or hypoxia in non-target organisms, with damage in respiratory system; it may be another reason for the decreased level in carbohydrate content.

A significant decrease in total proteins and an elevation in total free amino acids were observed in all the tissues of experimental mice (Table 16.1), when compared to controls. The decrease in protein may be due to their break down into amino acids, leading to increased amino acid levels in different tissues. The depletion of protein might be due to deceased protein synthesis as a consequence of pesticide stress. Active transamination and decreased utilization of amino acids synthesized from other sources like glucose and fatty acids might cause their elevation in different tissues of experimental animals under monocrotophos stress. The animals may elevate their amino acid levels to maintain the osmoregulation (Sambasiva Rao, 1984). These results are in accordance with the findings of Zaid *et al.* (1990), Rajamnnar and Manohar (1998), Elumalai *et al.* (1999), Das and Mukherzee (2000). The rise in free amino acid levels should lead to increased proteins synthesis. But monocrotophos stress might disturb the protein synthesis leading to their degradation.

Table 16.1

Sl.No.	Tissue	Carbohydrates (in mg/gm wet weight of tissue)			Proteins (in mg/gm wet weight of tissue)			Amino Acid (in mg/gm wet weight of tissue)			Total Lipids (in mg/gm dry weight of tissue)		
		Control	Single Dose	Multiple Doses	Control	Single Dose	Multiple Doses	Control	Single Dose	Multiple Doses	Control	Single Dose	Multiple Doses
1.	Brain	8.04 ±0.36	6.05 ±o.26 (−24.84)	4.03 ±29 (−49.81)	118.18 ±1.11	100.92 ±1.0 1 (−14.61)	87.16 ±1.66 (−26.25)	11.55 ±0.496	12.23 ±0.79 (5.91)	14.38 ±0.50 (24.53)	140.36 ±0.89	124.83 ±0.78 (11.05)	110.18 ±1.05 (−21.49)
2.	Liver	40.01 ±0.26	34.67 ±0.60 (−13.33)	27.90 ±0.64 (−30.26)	135.81 ±1.02	110.75 ±1.67 (−18.45)	70.82 ±1.49 (−47.85)	14.28 ±0.48	14.77 ±0.70 (3.38)	16.83 ±0.80 (17.85)	185.49 1.32	170.49 ±1.17 (−8.08)	150.25 1.41 (−19.0)
3.	Kidney	27.90 ±0.44	20.0100 ±0.26 (−28.28)	17.11 ±0.32 (−38.68)	143.66 ±1.46	123.50 ±1.45 (−14.03)	101.35 ±1.40 (−29.45)	17.05 ±0.65	18.02 ±0.61 (5.65)	22.5 5 ±1.12 (32.23)	150.30 ±1.3 1	130.52 ±1.1 7 (−13.15)	120.83 ±1.35 (−19.60)
4.	Intestine	11.72 ±0.68	10.0500 ±0.46 (−14.2)	6.26 ±0.60 (−46.53)	121.43 ±1.54	110.43 ±l.31 (−9.06)	96.68 ±2.79 (−20.38)	13.83 ±0.69	15.55 ±0.51 (12.41)	18.78 ±0.65 (35.78)	170.45 ±1.05	160.62 ±1.1 0 (−5.76)	150.19 ±1.27 (−11.88)

Values in the parenthesis are per cent change over control each value is mean of six individual observations. ±SD.

An increase in the free amino acid content in the tissues indicates metabolic adaptation to overcome the effect of monocrotophos toxicity. It appears that free amino acid pool may form a possible source of energy to meet the energy requirement. The changes in liver are predominant to other tissues indicating that liver was the most affected organ. The elevation of free amino acids may be due to active transamination and decreased utilization of amino acids from other sources like glucose and fatty acids. The elevation of amino acids may also be due to the damage of tissues (Dikshith *et al.*, 1975; Jayantha Rao *et al.*, 1983). Another reason for the elevation of amino acids is that they may be increased to maintain the osmoregulation of the animal, whose physiological capabilities are disrupted under pesticide stress (Sambsiva Rao, 1984). The same trend was reported by a number of workers (Baig, 1988; Venkataswamy 1991) in different tissues of animals exposed to various pesticides.

A significant decrease in total lipid content was noticed in all tissues of monocrotophos exposed albino mice (Table 16.1) suggesting liposysis in all most all tissues. Bhatia and Venkatasubramanian (1972) and Peter *et al.* (1973) advocated that pesticides affect the lipids, besides the carbohydrates, Lipids might have been channeled to meet the metabolic demand for extra energy that is needed to mitigate the toxic stress. The decrease in the total lipid content of the tissues supports the observations of Siva Prasada Rao (1980) and Sivaprasad and Rao (1981). The decrease may be due to increased lipolysis or due to reduction in the fatty acid synthesizing enzymes. Reduced cholesterol levels, triglycerides, phospholipids and increased lipid peroxidation might cause the decrease in total lipid content of experimental animals.

References

Baig, Md, A., 1988. Effect of heptachlor, an organochlorine, insecticide on functionally different muscles of fresh water edible fish *Channa punctatus*. *Ph.D. Thesis*. S.V. University, Tirupati, India.

Bakthavathsalam, R., 1987. Protein metabolism during disastar exposure in *Anabas testudineus* (Bloch.). *Poll. Res.*, 6(1): 1–4.

Bhatia, S.C. and Venkata Subramania, T.A., 1972. Mechanism of dieldrin induced fat accumulation in rat liver. *J. Ag. Food. Chem.*, 20: 993–996.

Carrol, M.V., Langley, R.W. and Row, J.M., 1956. Glycogen determination in liver and muscle by use of anthrone reagent. *J. Biol. Chem.*, 22: 583–593.

Das, B.K. and Mukherjee, S.C., 2000. Chronic toxic effects of quinolphos on some biochemical parameters in *Lobeo rohita* (Ham.). *Toxicol. Lett.*, 119(1–3): 11–18.

Dikshith, T.S.S., Behari, J.R., Datta, K.K. and Mathur, A.K., 1975. Effect of diazinan in male histopathological and biochemical studies. *Environ. Physiol. Biochem.*, 5: 293–299.

Elumalai, M., Jayakumar, R. and Balasubramanian, M.P., 1999. Impact of monocrotophos on protein and carbohydrate metabolism in different tissues of albino rat. *Cytobios*, 98(389): 131–136.

Folch, J., Lees, M. and Sloane G.m Syanley, 1957. A simple method for the isolation and purification of total lipids from animal tissues. *J. Biol. Chem.*, 226: 497–509.

Jayantha Rao, K., Madhu, Cb. and Murthy, V.S.R., 1983. Histopathology of malathion effect on gills of freshwater teleost *Tilapia mossambica* (Peters). *J. Environ. Biol.*, 4(1): 19–23.

Lowry, O.H., Rosebrough, N.J., Farr, A.L. and Randall, R.J., 1951. Protein measurement with Folin-phenol reagent. *J. Biol. Chem.*, 193: 265–275.

Moores and Stein, W.R., 1954. A modified ninhydrin reagent for the photometric determination of amino acids and related compounds. *Biol. Chem.*, pp. 221, 907.

Peters, H.D., Selherst, P., Linnendhal, V., Helm, K.V. and Sehoellhofer, P.S., 1973. *Arch. Toxicol.*, 30: 139.

Sambasiva Rao, K.B.S., 1984. Toxic impact of carboryl-1 and phenthoate on some metabolic aspects of a murrel [*Channa punctatus* (Block)]: A synergistic study. *Ph.D. Thesis.* S.V. University, Tiropati, India.

Sharma, Neeta Raj, Kusbwah, Aneeta, and Kushwah, H.S., 2000. Effect of dietary protein against monocrotophos toxicity. *J. Environ. Sci. Health,* B. 2004; 39(5–6): 805–818.

Siva Prasada Rao, K., 1980. Studies on some aspects of metabolic changes with emphasis on carbohydrate utilization of cell tree systems of the freshwater teleost *Tilapia mosambica* (Peters) subjecte to methylparathion exposure. *Ph.D. Thesis*, S.V. University, Tirupati, India.

Sivaprasad, K.S.R and Rao, K.V.R., 1981. Lipid derivative in the tissue of the freshwater teleost. *Tilapia mosambica* (Peters). Effect of methyl parathion. *Ind. Natl. Sci. Acad.*, p. B47.

Tilak, K.S., Veeraiah, K., Laksbmi, and Vijaya (2001). Biochemical changes induced in freshwater fish *Labeo rohita* (Hamilton), exposed to pesticide mixture. *Asian Journal of Microbiology, Biotechnology and Environmental Sciences*, 3(4): 315–319.

Venkataswamy, K., 1991. Neurochemical studies during the development of behavioral tolerance to organophosphate toxicity in Albino rats. *Ph.D. Thesis,* S.V. University, Thirupati, A.P., India.

Zaid, S.A., Singh, S., Palni, L. and Singh, V.S., 1990. Biochemical alterations in the levels of DNA, RNA and Protein in discrete areas of rat brain following nuvacron toxicity. *J. Pak Med. Assoc.*, 40(11): 261–263.

Chapter 17

Impact of Fly Ash-Soil Amendment on Vegetable Production

***N. Tripathi*[1], *P.K. Mishra*[2], *R.S. Singh*[1] and *P.S.M. Tripathi*[3]**

[1]*Central Mining Research Institute, Barwa Road, Dhanbad – 826 001, Jharkhand, India*
[2]*V.B. University, Hazaribagh, Jharkhand, India*
[3]*Central Fuel Research Institute, P.O. Digwadih, Dhanbad – 826 001, Jharkhand, India*

ABSTRACT

The present paper discusses the prospects of utilization of fly ash in agriculture sector to get enhanced yields of vegetable crops in the degraded soils. The physico-chemical properties, such as pH, EC, porosity, Bulk Density (BD), Water Holding Capacity (WHC), Cation Exchange Capacity (CEC), dehydrogenase activity and texture of the soil, after mixing with fly ash were found to improve the soil properties greatly, making the amended soil richer in essential nutrient elements and more fertile. Amendment of fly ash enhances the growth performance of selected vegetables *viz.*, carrot (*Daucus carota*–Family Umbelliferae) and radish (*Raphanus sativus*–Family Cruciferae). The pot experiments were carried out with mixed soil, fly ash dose being 5 per cent, 10 per cent and 20 per cent by weight, in addition to the usual application of recommended doses of NPK fertilizer. The vegetables grown in fly ash amended soil have also shown higher yield compared to non-amended soil. The fly ash-soil amendment technology will thus, go a long way in greatly enhancing and improving the agricultural production and economy of the state.

Keywords: *Fly ash amendment, Carrot, Radish, NPK fertilizer, Yield.*

Introduction

Sobriqueted as the mineral bowl of India, the newly created state of Jharkhand accounts for about 48 per cent of the country's coal production. Opening of 50 new coal mines in Jharkhand area is on the anvil to achieve the target production of 417 million tons of coal per year. To meet the present as well as future energy need and to bridge the gap between the demand and supply of power in the state, establishment of many more big coal-fired thermal power plants as also several captive power generation plants is well in the offing. The coming of several new coal-fired power plants along with the existing operational power plants in the state will simultaneously give rise to another serious problem, *i.e.*, generation of unmanageable quantities of fly ash (current production of fly ash alone being ca. 110 million tonnes per annum) (C.F.R.I. Report, 2002; C.E.G.B., 1973], for the storage and disposal of which quite a big area of precious agricultural land will be occupied near the power plants, making the available cultivable land in the state shrink. It has already assumed alarming proportions, which calls for taking urgent remedial measures for proper and bulk gainful utilization of coal ashes, fly ash in particular, in an eco-friendly manner. Particularly, with reference to fly ash utilization in agriculture, it has been found that fly ash can be used as a soil amendment and soil-coal ash admixture (20–30 per cent) is optimum in terms of soil health, plant growth, fertility status and yield, where heavy metals in crop produce were within the permissible limits (Mandai and Saxena, 1999). The results indicated the prospect of safe disposal and utilization of fly ash in agriculture for retaining productivity of problem soils, reduce the usage of costly chemical lenilizer, bring greater economy in cultivation and minimize environmental problems (Mitra *et al.*, 2003).

In the present paper the effect of amendment of red lateritic soil with different doses of fly ash on the physico-chemical characteristics of soil and yield of root vegetables, carrot and radish has been discussed.

Materials and Methods

Collection and Characterization of Samples

The soil samples from Sudamdih area of Dhanbad (Jharkhand, India) and fly ash samples from DPPL, Durgapur (West Bengal, India) were collected for pot experiment and analyzed for various physico-chemical properties, following prescribed standard methods.

The physical properties, *viz.*, colour, texture, Bulk Density (BD), porosity, Water Holding Capacity (WHC) of fly ash and soil sample with and without the additions of fly ash were determined following the prescribed methods (Piper, 1950 and Black, 1965). Chemical properties, *viz.*, pH, electrical conductivity, total and available nutrients (N, P, and S) of fly ash and soil were also estimated following the methods as described by Black (1965), Jackson (1973) and Tandon (1995).

Pot experiments were done using acidic soil (pH 5.0). After proper mixing of fly ash with the soil with pre-decided doses treatments, *i.e.*, 100, 300 and 400 t/ha (equivalent to 5, 10 and 20 per cent fly ash w/w, respectively), 4 kg each of amended soils was added in their respective pots. The data were compared with the control soil (*i.e.*, without fly ash amendment). Recommended doses of NPK fertilizers (1:2: 1) were applied in both control soil and fly ash amended soil. The pots were properly lined with the polythene sheets to check the leaching of trace and heavy metals, if any, from fly ash-soil amended system under green house conditions. Radish and Carrot were grown in pots with different treatments of fly ash and their yield (root) data were collected after four months compared with control soil pot.

Results and Discussion

The analytical results of soil samples before the start of experiment and after harvest of each crop are shown, in respect of physico-chemical characteristics and yield of the vegetable crops grown in fly ash amended soil in Tables 17.1–17.3.

Table 17.1: Initial Physico-chemical Characterization of Soil and Fly Ash

Parameter	Soil	Fly Ash
pH	5.01	8.92
Electrical Conductivity	0.05	0.12
Bulk Density (glee)	1.63	0.81
Water Holding Capacity	30.92	54.35
Porosity (per cent)	37.31	65.81
Texture		
Sand (per cent)	66.1 0	33.50
Silt (per cent)	21.60	50.10
Clay (per cent)	12.30	16.40
Textural Class	Sandy loam	Silty loam
Nitrogen (per cent)	0.0515	BDL
Phosphorus (per cent)	0.0140	0.29
Potassium (per cent)	0.77	0.84
Calcium (per cent)	0.86	1.09
Magnesium (per cent)	0.56	0.79
Sulphur (per cent)	0.052	0.026
Zinc (per cent)	44.10	71.72
Copper (per cent)	42.90	78.20
Iron (per cent)	229	293
Manganese (per cent)	284.40	212.80
Lead (per cent)	22.60	52.0
Nickel (per cent)	228.30	134.0
Chromium (per cent)	179.0	225.0
Cobalt (per cent)	32.05	51.20

Physico-chemical Properties of Soil

The comparative physico-chemical characteristics of the fly ash and soil show (Table 17.1) that the soil is acidic (pH = 5.01), while the fly ash is alkaline (pH = 8.92) in nature. Bulk density (BD), water holding capacity and porosity values of the fly ash are 0.81 g/cc, 54.34 per cent and 65.81 per cent, respectively, while those of the soil are 1.63 g/cc, 30.92 per cent and 37.31 per cent, respectively. The alkaline fly ash promoted the nutrient supplying capacity of the soil by raising pH level from 5.01 to 5.48 in soil. Addition of fly ash in soil resulted into the decreased bulk density and sand content of soil and increased pH, WHC, porosity, silt and clay contents of soil, besides increasing the available nutrients like P, K, Ca, Mg, Zn and Cu. There was an increase of 8 per cent, 11 per cent, 12 per cent, 14 per cent and 3 per cent in pH, WHC, porosity, silt and clay contents, respectively, while the EC, bulk density and sand content were declined by 4 per cent, 7 per cent and 6 per cent, respectively in fly ash amended soil compared to control soil.

Table 17.2: Physico-chemical Characteristics of Soil After Fly Ash Amendment

Parameter	Fly ash-amended soil
pH	5.48
EC (d s/m)	0.03
BD (glee)	1.52
WHC (per cent)	34.82
Porosity (per cent)	42.35
Mechanical Composition	
Sand (per cent)	61.95
Silt (per cent)	25.30
Clay (per cent)	12.75
Textural Class	Sandy loam
Nitrogen (per cent)	0.0156
Phosphorus (per cent)	0.001 08
Potassium (per cent)	0.01004
Calcium (per cent)	0.00288
Magnesium (per cent)	0.00206
Sulphur (per cent)	0.00420
Zinc (ppm)	2.36
Copper (ppm)	2.70
Iron (ppm)	44.10
Manganese (ppm)	15.62
Lead (ppm)	3.66
Nickel (ppm)	0.61
Chromium (ppm)	BDL
Cobalt (ppm)	0.05

The results have shown that the available nutrients, such as P, K, Ca, Mg, Zn and Cu were increased in fly ash amended soil in comparison to control soil by 12 per cent, 2 per cent, 16 per cent, 10 per cent, 11 per cent and 13 per cent, respectively.

The results of the physical characteristics of soil and fly ash indicate that due to its alkaline nature, fly ash can be suitably applied to the acidic soils to neutralize their acidity, thereby acting as a liming agent for the acidic soils. Jala and Goyal (2005) reported a positive impact of fly ash on micro ecology and chemistry of soil along with physical characteristics, such as bulk density, water holding capacity and soil structure. Similarly, Water Holding Capacity (WHC) and porosity of the soil are significantly improved by fly ash addition into the soil. Chang *et al.* (1977) reported that an addition of Fly Ash (8 per cent by weight) increased the water holding capacity of soil. This improvement in water-holding capacity is beneficial to the plants especially under rainfed agriculture. Thus, fly ash can be potentially used and is quite effective to act as an efficient soil modifier/conditioner.

Table 17.3: Yield of Carrot and Radish After Amendment with Different Doses of Fly Ash (on harvest and oven dry basis)

Yield (g/pot)	Carrot (after 4 months)	Radish (after 4 months)
	Harvest basis	
Control	89.6	124.6
*T_1	108.8	145.0
*T_2	120.9	156.8
*T_3	141.6	174.6
CD (5 per cent)	0.8	1.6
	Oven dry (60°C) basis	
Control	8.10	6.59
*T_1	9.92	7.55
*T_2	10.9	8.32
*T_3	12.5	9.22

T_1, T_2, T_3 are the treatments at different doses of fly ash (5 per cent, 10 per cent and 20 per cent, respectively).
Level of significance ($P > 0.05$).

The reduction in bulk density of amended soil was because of lower bulk density of fly ash (0.81 g/cc) as compared to that of soil (1.63 g/cc). Optimum bulk density, in turn, improves the soil porosity, the workability of the soil, the root penetration and the moisture retention capacity of the soil. The potential benefits of fly ash addition to soil include improved structure for fine- and coarse-textured soil (Ghodrati *et al.*, 1994). Capp (1978) also suggested that both sandy and clayey soils tend to become loamy in texture after fly ash addition. Sharma *et al.* (1989) reported the average silt content in U.S. fly ash to be about 63.2 per cent, while in India this content ranges from about 16 per cent to 45 per cent. In the present study, the silt content in fly ash is 50 per cent.

Major and Secondary Nutrients in Soil

The results (Table 17.3) reveal that the application of maximum dose of fly ash results in the enrichment of the available major and secondary plant nutrients (N, P, K, S, Ca and Mg) status of the soil, which has contributed to the improvement of soil fertility. The concentration of major and secondary plant nutrients and also micronutrients (such as Cu, Zn, Mn, Fe) increased progressively with the increasing dose of fly ash in comparison to that in the control. The presence of essential plant nutrients in fly ash has also been reported by several other workers (Ashokan *et al.*, 1995; Neelima *et al.*, 1995). Saxena *et al.* (2005) observed that in a paddy sunflower cropping sequence, there was an increase in the concentration of the micronutrients, such as Cu, Zn, Mn and Fe increased with fly ash amendment @ 10 per cent and 20 per cent in the soil.

Similar observation with respect to the impact of fly ash application on micronutrient status was made by Bhoyar and Matte (2005), where they found an increase in available micro nutrients with increasing dose of fly ash.

Crop Yield

The results of the yield of all the crops grown with different fly ash doses are included in Table 17.3, The yield data of carrot (Table 17.3) grown in the fly ash amended soil with different doses of fly ash show that in control soil the yield, on harvest basis, was 89.6 g/pot, while 5 per cent, 10 per cent and 20 per cent amendments have shown the yield to be 108.8, 120.9 and 141.6 g/pot, respectively. On oven dried (60°C) basis the yield was 8.1 g/pot in control soil, while 9.92, 10.9 and 12.5 g/pot, respectively in the soils with 5 per cent, 10 per cent and 20 per cent fly ash additions.

In radish, the yield (Table 17.3), on harvest basis, was 124.6 g/pot in control, while it was 145.0, 156.8 and 174.6 g/pot in the soils with 5 per cent, 10 per cent and 20 per cent fly ash additions. On oven dried (60°C) basis the yield of radish was 6.59 g/pot in control soil, while 7.55, 8.32 and 9.22 g/pot, respectively in the soils with 5 per cent, 10 per cent and 20 per cent fly ash amendments.

The percent increase in carrot and radish yield is found to be 18.34–35.2 and 12.7–28.5, respectively. The increase in root crop yield could possibly be due to the improved physico-chemical properties and fertility status of soil after fly ash amendment. Similar to our findings, Ashoka (2005) has also reported an increase in growth and yield parameters after amendment of soil with fly ash @20 t/ha and 30 t/ha, either alone or in combination with vermicompost. Iyer and Scott (2001) attributed the greater yield to the higher water holding capacity of the fly ash. In yet another study. Patil *et al.* (1999) have also found an increase in yield of groundnut, sunflower and maize by 75 per cent, 25 per cent and 15 per cent, respectively. Mittra *et al.* (2003) suggested that the increase in pod yield may be due to increased nodule number. They postulated that the application of fly ash to soil reduced its bulk density, which, in turn, helped in better pegging and pod formation.

More interestingly, the penetration of the roots of such crops was also found to be better in fly ash-amended soil than in control, obviously owing to the better workability. Several other workers (Kukier *et al.*, 1994; Dosskey *et al.*, 1993; Bilske *et al.*, 1995) have also reported a modification in the soil characteristics and increase in the crop yield due to fly ash addition in soil. Mittra *et al.* (2005) observed that the application of fly ash in acidic lateritic soil@10 t/ha, in combination with organic wastes, such as paper factory sludge, Farmyard Manure (FYM), crop residue and chemical fertilizers for growing rice-peanut crop, resulted in an increase in yield. Sajwan *et al.* (1996) found that the soils amended with fly ash generally improved plant growth and consequently resulted in enhanced yield, at the fly ash dose of 25 tons/acre, and decreased at higher fly ash doses. The variation in the increase in the crop yield of different crops could be attributed to the selective uptake of the nutrients by the plants and, to some extent, also the higher uptake of zinc. The decrease in yield at higher application rate may be due to the accumulation of high levels of Band Zn which are phytotoxic and/or elevated levels of inorganic dissolved salts. Thus, our study is in consonance with the general findings that the amendment of soil with fly ash results in higher yields of crops.

Statistical Analysis

The statistical analysis was done using the statistical software SPSSwin (1993). There is a significant difference in the physico-chemical properties of control and those of amended soil. The most significant difference has been found at 20 per cent amendment ($P<0.05$). The biochemical properties of the vegetables grown in fly ash amended soil have shown a significant increase ($P<0.05$) as compared to those of the control; the greatest difference found at 20 per cent amendment.

Conclusion

Fly Ash has potential for use in agriculture because it contains almost all macro as well as micro nutrients except organic carbon and nitrogen. It can be used in conjunction with the recommended doses of chemical fertilizer for different crops to increase the yield of various agricultural crops. Although, Fly Ash contains moderate quantities of trace and heavy metals, radioactive elements, but their concentrations are too low to have any harmful effect and their effects on ground water, soil health and uptake by plants are probably negligible. Thus, the soil-fly ash amendment improves various physico-chemical properties of soil making rather unproductive land soil more fertile, without having any adverse environmental impact. Therefore. it is imperative that concomitant with the plans for more coal mining and power generation, blue print should also be carefully prepared for the environmentally-safe management and gainful utilization of fly ash in different ways.

Acknowledgement

Council of Scientific and Industrial Research (CSIR), New Delhi is acknowledged.

References

Ashoka, J., 2005. Effect of fly ash application on growth and yield and mulberry and subsequent effect on mulberry silkworm. In: *Proceedings of Fly Ash India, 2005, Fly Ash Utilization Programme (FAUP)*, TIFAC, DST, New Delhi.

Ashokan, P., Saxena, M., Bose, S.K.J. and Khazenchi, A.C., 1995. *Proceedings of Workshop on Fly ash Management in the State of Orissa*, RRL, Bhubaneshwar, April 11, p. 64–75.

Bhoyar, S. and Matte, D.B., 2005. Effect of fly ash application on some physical properties and available micronutrients status of black soil. In: *Proceedings of Fly Ash India, 2005, Fly Ash Utilization Programme (FAUP)*, TIFAC, DST, New Delhi.

Bilski, J.J., Alva, A.K. and Sajwan, K.S., 1995. *Soil Amendments and Environmental Quality*.

Black, C.A., (Ed.). 1965. *Methods of Soil Analysis*, Parts 1 and 2. American Society of Agronomy Inc. Publisher, Wisconsin, USA.

Cappo, J.P., 1978. Power plant fly ash utilization for land reclamation in the Eastern United States. In: *Reclamation of Drastically Disturbed Lands*, (Eds.) Schaller, F.W. and P. Sutton. Soil Sci. Soc. of America Madison, WI, p. 339.

C.F.R.I. (Central Fuel Research Institute), 2002. Utilization of fly ash in agriculture, at Bakreshwar Thermal Power Project, Bakreshwar (West Bengal). *CF.R.I. Dhanbad (India) Report No. TR/CF.R.1./3*, 03/2001–2002.

Central Electricity Generating Board, 1973. The uptake of minerals by plants growing on pulverized fuel ash. CEGB, London, U.K. *Report No. SSD. MID./R.2/73*, pp. 30.

Chang, A.C., Lund, L.J., Page, A.L. and Warneke, J.E., 1977. Physical properties of fly ash amended soils. *Journal of Environmental Quality*, 6(3): 267.

Dosskey, M.G. and Adrino, D.C., 1993. *J. Soil Biol. Biochem.*, 25: 1547–1552.

Ghodrati, M., Sims, J.T. and Vasilas, B.L., 1994. Evaluation of fly ash as a soil amendment for the Atlantic coastal plain. I. Soil Hydraulic properties and elemental leaching. *Water, Air and Soil Pollution*, 81: 349–361.

Lyer, R.S. and Scott, J.A., 2001. Power station fly ash: A review of value-added utilization outside of the construction industry. *Resources, Conservation and Recycling*, 31(3): 217–228

Jackson, M.L., 1967. *Soil Chemical Analysis*. Prentice Hall of India Pvt. Ltd., Delhi.

Jala, S. and Goyal, D., 2005. Fly ash as a soil ameliorating agent in forestry plantations. In: *Proceedings of Fly Ash India, 2005, Fly Ash Utilization Programme (FAUP)*, TIFAC, DST, New Delhi.

Kukier, U., Summer, M.E. and Miller, E.P.J., 1994. *Environ. Quality*, 23: 596–603.

Mandal, S. and Saxena, M., 1999. Fly ash: A boon for clay and sandy soil. *J. Env. Res.*, 9(1): 19–24.

Mittra, B.N., Karmakar, S., Swain, D.K. and Ghosh, B.C., 2003. Fly ash: A potential source of soil amendment and a component of integrated plant nutrient supply system. In: *Proceedings of International Fly Ash Utilization Symposium*, Centre for Applied Energy Research, University of Kentucky, Paper No. 28.

Mittra, B.N. Karmakar, S., Swain, D.K. and Ghosh, B.C., 2005. Fly ash: A potential source of fly ash amendment and a component of integrated plant nutrient supply system, in proceedings of *Fly Ash India, 2005, Fly Ash Utilization Programme (FAUP),* TIFAC, DST, New Delhi.

Reddy, Neelima M., Khandaul, S., Tripathy, A. and Sahu, P.K., 1995. *Proceedings of Workshop Oil Fly Ash Management in the State of Orissa,* RRL, Bhubaneswar, April 11, 1995, p. 76–87.

Patil, C.Y., Prakash, S.S., Yeledhalli, N.A. and Rajkumar, G.R., 1999. Characterization of fly ash for utilization in agriculture. In: *Proceedings of the National Seminar on Fly Ash Characterization and its Geotechnical Applications,* Indian Institute of Science, Bangalore, pp. 151–156.

Piper, C.S., 1950. *Soil and Plant Analysis.* University of Adelaide, Adelaide, Australia.

Saxena, M., Murali, S., Asokan, P. and Yadav, B., 2005. Bulk utilization of pond ash in agriculture: A review on technology demonstration by RRL, Bhopal. In: *Proceedings of Fly Ash India, 2005, Fly Ash Utilization Programme* (*FAUP*), TIFAC, DST, New Delhi.

Sharma, S., 1989. Fly ash dynamics in soil-water systems. *Critical Reviews in Environmental Control,* 19(3): 251–275.

SPSSwin, 1993.

Tandon, L. (Ed.). 1995. *Methods of Analysis of Soils, Plants, Water and Fertilizers*. Fertilizer Development and Consultation Organization, New Delhi.

Chapter 18

On the Validity of the Species *Procamallanus* (*monospiculus*) *devendrii* Sinha and Sahay, 1966

***Rajendra Prasad Singh*[1] *and Umapati Sahay*[2]**

[1]*Marwari College, Ranchi*
[2]*Former University Professor and HOD of Zoology, Ranchi University*

ABSTRACT

Out of eleven gubernaculate *Procamallanids Procamallanus* (*monospiculus*) *devendrii* Sinha and Sahay, 1966 is one of them. This species was synonimised with *P. mathurai* by Soota, 1983. It seems that Soota is misinformed about the details of *P.* (*m*) *devendrii*. The authors have reasons to believe that *P.* (*m*) *devendrii* is valid.

Keywords: *P. devendrii, P. mathurai and Revalidation.*

Introduction

In the year 1960 Yeh wrote on the reconstruction of the genus *Camallanus* in which he divided the family *Camallanidae* into two sub-families namely *Procamallaninae* and *Camallaninae*. Under the sub-family *Procamallaninae*, the genera *Procamallanus* (Baylis, 1923) and *Spirocamallanus* (Olsen, 1952) were included while under the subfamily *Camallaninae* the following genera were included *Procamallanus* (Yorke and Maplestone, 1926); *Piscilania* (Yeh, 1960); *Serpinema* (Yeh, 1960); *Camallanus* (Railliet and Henry, 1915). Sahay in the year 1976–77 added another genus *Neo-zeylanema* in the sub-family *Camallaninae*.

Ali, 1956 splitted the genus *Procamallanus* into 3 sub-genera *viz.*, *Monospiculus*, *Isospiculus* and *Procamallnus* and later in 1960, he added yet another species and genus *Aspiculus*.

This splitting has been supported by (Sahay, 1966); (Bilquees, Khanum and Jehan, 1971); (Gupta and Duggal, 1973); (Dhar and Fotedar, 1980). Though (Campana Rouget, 1961); (Pande *et al.*, 1963); Fernando and Furtado (1963); (Agarwal, 1966); (Sood, 1967); and (Akram, 1975) did not accept this splitting.

(Chabaud, 1975); (Petter, 1979); (Arya, 1980); (Naidu and Murher, 1980) and (Soota, 1983) supported Campana Rouget.

(Petter, 1979) however, separated all the species of the sub-family *Procamallaninae* into four genera

1. *Procamallanus*: Distinguished by smooth inner wall of buccal capsule.
2. *Onchocamallanus*: Distinguished by incomplete transverse crest on the inner wall and the base of buccal capsule.
3. *Spirocamallanus*: Distinguished by the spiral crest on the inner wall of the buccal capsule.
4. *Mallayocamallanus*: Distinguished by longitudinal crest in the inner wall of the buccal capsule.

It seems that controversy exists regarding the genus *Spirocamallanus* and some other genera. However, the existence of the genus *Procamallanus* is undisputable.

The present paper deals with the validity of the species *Procamallanus* (*monospiculus*) *devendrii* Sinha and Sahay, 1966. The arguments in favour of this validity are as follows:

Materials and Methods

Original paper of *P.* (*m*) *devendrii* Sinha and Sahay, 1966 and *P. mathurai* Pande *et al.*, 1983 have been consulted.

Discussion

So far the authors are aware nearly eleven species of the genus *Procamallanus* are provided with gubernaculum in male. These are:

1. *Procamallanus ahiri* Karve, 1952
2. *P. saccobranchi* Karve 1951
3. *P. spiculogubernaculus* Agrawal, 1958
4. *P. daccai* Gupta, 1952
5. **Neocamallanus heteropneusti* Chakravorty *et al.*, 1961
6. *P. confusus* Fernando and Furtado, 1963
7. *P. mathurai* Pande *et al.*, 1963
8. *P.* (*monspiculus*) *devendrii* Sinha and Sahay, 1966
9. *P. mahendri* Singh, 1970

----

* When Dr. Judith, M. Humphrey of Beltsville Parasitological Laboratory Maryland, USA communicated to Dr. Chakravorty the preoccupation of the name for Chakravorty's genus *Neocamallanus* founded by (Ali, 1956) a new name was proposed by him as *Indocamallanus heteropneusti.* (Fernando and Furtado, 1963) proposed a new name for Chakravorty's worm as *Procamallanus chakravortyii.* Ignorant of the literature of (Fernando and Furtado, 1963), Sinha and Sahay gave a new nomenclature to Chakravorty's worm as *Procamallanus* (*monspiculus*) *heteropneusti* nom.nudum.

10. *P. ramteki* Naidu and Murhar, 1980
11. *P. guptai* Gupta and Naiyar, 1999.

For considering the validity, the following characters have been chosen to be of significance.

Gubernaculum

P. (monspiculus) devendrii (Sinha and Sahay, 1966) and *P. mathurai* (Pande *et al.*, 1963) are provided with Y-shaped gubernaculum with little variation in sclerotization, however, the differences in the shape of gubernaculum is well marked in *P. (monspiculus) devendrii* (Sinha and Sahay, 1966) in which case the two anterior limbs have been shown to be unequal but bifurcated at their tips (longer right lateral measures 0.024–0.028 mm while the left lateral is 0.01–0.014 mm. The median limb however, is slightly curved and measures 0.02–0.024 mm. Such a condition is not met in *P. mathurai* (Pande *et al.*, 1963) where Y-shaped gubernaculum in its total length measures 0.06–0.07 mm but the longer arm is 0.036–0.04 mm and the shorter arm along with median arm is 0.044–0.15 mm. Pande *et al.*, 1963 did not give the measurement of shorter arm independently, neither they have mentioned any bifurcation at the tip of the anterior limbs. Therefore, the structural difference do exist and they do not point towards their being synonyms.

Noble and Noble, 1974 mentions "transfer of sperms to the female worm is aided by a pair of spicules in many species of the round worms. These long hardened structures may be thrust out through cloaca, may serve the additional function of facilitating as sensory organs. Another male structure, the gubernaculum is a sclerotized thickening of cuticle in some species of the worm. This organ is formed from the spicular pouch, it lies on the dorsal side of cloaca and probably helps to guide the spicule as they are thrust out".

Sahay and Sahay, 1999 opine "spicules dilate the vulvular opening and the gubernaculum guides this operation and therefore, spicules and gubernaculum are different structures. This structural configuration of gubernaculum therefore, is of paramount importance in guiding and manouvering the spicules".

Muscular and Glandular Oesophagus

In the genus *Procamallanus,* the oesophagus is divided into two parts, the anterior muscular and posterior glandular part, which is always longer than the anterior one. The ratio between muscular and glandular oesophagus in male and female worms of *P. (m) devendrii* Sinha and Sahay, 1966 is 1 : 1. 42–1.57 and 1 : 1.35–1.356, while in males of *P. mathurai* (Pande *et al.*, 1963), this ratio is 1 : 1.13 and 1 : 1.81–1.83 in female. The difference in ratio also does not point towards their synonymy.

Spicules

It is of great importance to a taxonomist but very week scerotization sometimes may lead a taxonomist to a state of confusion and should be dealt cautiously says Yeh, 1960. The spicule measures in *P. (m) devendrii* Sinha and Sahay, 1966 to be 0.288–0.3 mm and in *P. mathurai* Pande *et al.*, 1963 it is 0.25–0.27 mm. The measurements do not show much difference but other measurements do exist which are of more importance.

Position of Vulva

It is generally post-equatorial in the family *Camallanidae.* In *P. (m) devendrii* (Sinha and Sahay, 1966), the position of vulva from anterior end in worms ranging from 6.4–6.8 mm, is 3.17–3.62 mm and the ratio of body length versus vulva is 1:1.84–1.87. This however, in *P. mathurai* (Pande *et al.*, 1963) is at 2.46–3.52 mm from posterior end but its ratio from body length is 1:2.03–2.2. This data also shows that the species are dissimilar and they are not synonyms.

Body Length Versus Length of Buccal Capsule

In *P.* (*m*) *devendrii* (Sinha and Sahay, 1966) the ratio between length of body and length of buccal capsule in male and female worms is 1 : 43.75–100 and 1 : 133.33–136 respectively, while in the male of *P. mathurai* (Pande *et al.*, 1963) the ratio is 1 : 66.66–80.39 and in female 1 : 92.59–121.8.

None of these ratios fall in the ratio range of *P.* (*m*) *devendrii* (Sinha and Sahay, 1966). This also does not support their synonymy.

Body Length Versus Breadth of Buccal Capsule

The ratio between the body length and the breadth of buccal capsule in *P.* (*m*) *devendrii* is 1 : 92.10–114.28 in male and 1 : 212.5–213.33 in female but in *P. mathurai* (Pande *et al.*, 1963) this ratio is 1 : 1 00–107.89 in male and 1 : 125–195 in female. Though the ratio range of male worms of *P. mathurai* (Pande *et al.*, 1963) fall in the ratio range of *P.* (*m*) *devendrii* (Sinha and Sahay) but the ratio in females do not.

Caudal Papillae

In *P.* (*m*) *devendrii* (Sinha and Sahay, 1966) the number of caudal papillae has been show to be 13 pairs (8 pairs preanal and 5 postanal papillae) but in *P. mathurai* (Pande *et al.*, 1963) the caudal papillae are 12 pairs (7 pairs preanal, 1 adanal and 4 pairs postanal). The aforesaid species differ remarkably in the number and disposition of the papillae.

Tail Versus Body Length

This has been shown in *P.* (*m*) *devendrii* (Sinha and Sahay, 1966) to be 1 : 1.06–87.5 in male and 1 : 80–82.92 in the female, but this value in *P. mathurai* (Pande *et al.*, 1963) is 1 : 55.55–69.64 in female. The ratio in case of male is not possible to give as the authors of *P. mathurai* forgot to mention the tail length.

Never Ring Versus Body Length

In *P.* (*m*) *devendrii* (Sinha and Sahay, 1966) the nerve ring is situated at a distance of 0.074–0.076 mm in male and 0.062 mm in female from the anterior end, but in *P. mathurai* it is at 0.12–0.13 mm in male and 0.14–0.16 in female worms. The ratio between length of body and nerve ring being 1 : 43.24–46.05 in male and 1 : 103.2–109.67 in female worms in *P.* (*m*) *devendrii* (Sinha and Sahay, 1966) but it is 1 : 20.66–31.54 in male 1 : 35.714–40.75 in female worms of *P. mathurai* (Pande *et al.*, 1963).

It appears, therefore, that in most of the details the differences exist in the two aforesaid species except the spicules, which can safely be ignored in preference to dissimilarities. The authors think that *P.* (*m*) *devendrii* (Sinha and Sahay, 1966) is a valid species; and revalidates it.

Acknowledgements

The authors are thankful to late Prof. P.N. Mehrotra, the then Head of the department of Zoology, Ranchi University and Prof. K.N. Dubey, present Head of the department of Zoology, Ranchi University for library facilities.

References

Agarwal, S.C., 1958. On a new species of *Procamallanus* Baylis, 1923. *Curr. Sci.*, 27: 348–349 (W.L. 7021 N).

Agarwal, V., 1966. On a new neamtode *Procamallanus muelleri* n. sp from the stomach of freshwater fish, *Heteropneustes fossilis*. In: *Proc. Helminth. Soc. Wash.*, 33(2): 204–208.

Akram, M., 1975. A preliminary review of the genus *Procamallanus* Baylis, 1923 with the description of a new species from the marine fish of Karanchi Coast. *Biologia* (Lahore), 21(2): 93–100.

Ali, S.M., 1956. Studies on the nematode parasites of fish and birds found in Hyderabad state. *Ind. Jour. Helminth.*, 8: 1–88.

Arya, S.N., 1980. A new nematode of the genus *Goezia* Zeder, 1800 from a marine fish of India. *Ind. Jour. Helmith.*, 30: 96–99.

Baylis, H.A., 1923. Report on a collection of parasitic nematode mainly from Egypt. Part. I: *Ascaridae* and *Heterekidae;* Part II: *Oxyruidae;* Part III: *Camallanidae* etc. with a note on *Prostmayria* and appendix on acanthocephala. *Parasitology,* 15(1): 3, 14–23–38.

Baylis, H.A. and Doubney, R., 1922. Report on the parasitic nematodes in the collection of Zoological Survey of India. *Mem. Ind. Mus.*, 7(4): 263–347.

Bilquees, F.M., Khanum, Z. and Jehan, Q., 1971. Marine fish nematodes of West Pakistan. I. Description of seven new species of Karanchi Coast. *J. Sci. Karachi,* 1(1): 175–184.

Campana Rouget, Y., 1961. Exploration hydrobiologique des lacs Kivu, Eduard et Albert. Nematodes de Poissons. *Inst. Roy. Sc. Nat. Belg.*, III, fasc, 4: 3–61.

Chabaud, A.G., 1975. Keys to the genera of the oder *Spiruida.* Part I. *Camallanoidea, Drcunculoidea, Gnathostomatoidea, Physalopteroidea* and *Thelazoidea.* CIH keys to the nematode parasites of vertebrates. Edit. Anderson Chaboud and Wilmott, Commonwealth Agricultural Bureaux Farnham Royal, Bucks, England, pp. 277.

Chakravorty, G.K. and Majumdar, G., 1960. On the classification of the nematode family *Camallanidae* Railliet and Henry, 1915. *Ind. Jour. Helminth.*, 12(2): 93–94.

Chakravorty, G.K. and Majumdar, G., 1962. New nematode parasites from birds and fish. *Proc. Zool. Soc. Calcutta.*, 15: 21–26.

Chakravorty, G.K., Majumdar, G. and Sain, S.K., 1963. The nematode genus *Indocamallanus* (nom. novo pro. *Neocamallanus*) in *Heteropneustes fossilis. Sci. Cult.*, 29: 415–416.

Dhar, R.L. and Fotedar, D.N., 1980. On *Procamallanus* (*monospiculus*) *kashmirensis* sp. novo from freshwater fish *Wallago attu* from Jammu, India. *Ind. Jour. Helminth.*, 31: 128–134.

Fernando, C.H. and Furtado, J.I., 1963. A study of some helminth parasites of freshwater fishes in Ceylon. *Z. Parasitkunde.*, 23: 141–163.

Gupta, N.K. and Duggal, C.L., 1973. On a new and one already known species of the sub-genus *Procamallanus* (Baylis, 1923) Ali, 1956 (Nematoda: *Camallanidae*) from the freshwater fish and a key to species of the sub-genus. *Riv. Parasit.*, 34(4): 295–304.

Gupta, S.P., Nematode parasites of vertebrates of East Pakistan. III. *Camallanidae* from fish, amphibia and reptiles. *Can. J. Zool.*, 39: 771 –779.

Gupta, V. and Naiyar, N., 1990. On a new nematode *Procamallanus guptai* sp. nov from the intestine of freshwater fish *Heteropneustes fossillis* Block from Lucknow. *Indian J. Helminthol.*, 42(1): 67–71.

Karve, J.N., 1952. Some parasitic nematodes of fishes. III. *J. Univ.*, Bombay, 21(3): 1–14.

Karve, J.N. and Naik, G.G., 1951. Some parasitic nematodes of fishes. II. *J. Univ.*, Bombay, 19(5): 1–37.

Naidu, T.S.V. and Murher, B.M., 1980. Two new species of the genus *Procamallanus* Baylis, 1923 (Nematoda : *Camallanidae) from Saccobranchus fossilis* (B1) of Ramtek (M.S.) India. *Riv. Parasit.*, 41(1): 105–111.

Noble, E.R. and Noble Glen, A., 1974. *The Biology of Animal Parasites*. Lea and Febiger, Philadelphia.

Olsen, L.S., 1952. Some nematodes parasitic in marine fishes. *Pub. Inst. Mar. Sci. Univ. Texas.*, 11(2): 171–215.

Pande, B.P., Bhatia, B.B. and Rai, P., 1963. On the Camallanid genus *Procamallanus* Baylis, 1923 in two of the freshwater fishes. *Ind. Jour. Helminthol.*, 15(2): 105–118.

Petter, A.J., 1979. Essai. De classification de la sous–famille des *Procamallaninae* (Nematoda: *Camallanidae*). *Bull. Mus. Natn. Hist. nat. Paris.*, ser 4(1) Section A(1): 219–239.

Rai, P., 1969. On some of the heither to known and unknown nematodes parasitic in some freshwater siluroid fishes. *Ind. Jour. Helminthol.*, 21(2): 94–108.

Railliet, A. and Henry, A., 1915. Sur les nematode du genere *Camallanus* Railliet, A. and Henery, 1915 (*Cucullanus* auct., no Mueller, 1777). *Bull. Soc. Path. Exot.*, p. 117–119.

Sahay, Umapati, 1966. On a new key to the genus *Procamallanus* with a historical review. *Jap. Jour. Med. Sci. Biol.*, 19(3): 165–170.

Sahay, P. and Sahay, Umapati, 1999. On a new key to the genus *Procamallanus* (*Camallanidae : Procamallaninae*). *Trans. Zool. Soc. East India*, 3(1): 27–42.

Sinha, D.P. and Sahay, Umapati, 1966. On a new species of *Procamallanus* (*Camallanidae* : Nematoda) with a discussion on the validity of the genus *Indocamallanus* Chakravorty *et al.*, 1961. *Zool. Anz.*, 176(5): 384–388.

Singh, S.S., 1970. *Ph.D. Thesis*, Patna University, Patna.

Sood, M.L., 1967. On some species of the genus *Procamallanus* Baylis, 1923 from freshwater fishes of India. *Proc. Nat. Acad. Sci. India.*, 37(B): 291–308.

Soota, T.D., 1983. Studies on nematode parasites of Indian vertebrates. I. Fishes. *Rec. Zool. Survey India.* (Occasional Pub. No. 54): 1–352.

Srivastava, Hem and Sahay, Umapati, 2004. On a new species of the genus *Procamallanus* Baylis, 1923 from *Heteropneustes fossilis*. *Uttar Pradesh J. Zool.*, 24(3): 251–254.

Yeh, L.S., 1960. On a collection of Camallanid nematodes from freshwater in Ceylon. *J. Helminth.*, 34(1–2): 107–116.

Yeh, L.S., 1960. On a reconstruction of the genus *Camallanus* Railliet, et Henery, 1915. *J. Helminth.*, 34(1–2): 117–124.

Yorke, W. and Maplestone, P.A., 1926. *The Nematode Parasites of Vertebrates*. J. and A. Churchill Editt, London. 536 pp.

Chapter 19

Effect of Micronutrients on Seed Production in Okra

R.V. Patil, S.V. Kolase and K.G. Kadam

Breeder Seed Production (Vegetable), Seed Cell, MPKV, Rahuri – 413 722, Maharashtra

ABSTRACT

The experiment was carried out at Breeder Seed Production (Vegetable) farm, MPKV, Rahuri during Kharif 2004–05 to find out the effect of micronutrients on seed production in Okra. It was found that the seed yield of okra can be increased significantly by three sprays of mixture of all nutrients (*i.e.*, Boric add, Zinc sulphate, Ammonium molybdate, Copper sulphate, ferrous sulphate, Manganese sulphate) at 100 ppm, first spray was given before flowering, second spray after third picking and third 15 days after second spray. The maximum seed germination, seed vigor and seed yield was recorded 86.33 per cent, 6.78 and 4.53 q/ha respectively, followed by commercial formulation (Multiplex), while the minimum germination percentage 86.33 per cent, seed vigour 6.78 and seed yield 4.53 q/ha respectively was recorded in the treatment Control (water spray).

Keywords: *Okra, Micronutrients, Seed yield.*

Introduction

Seed is the most important basic input in vegetable production. Micronutrients play important role in plant growth and metabolism. Okra is a very important vegetable crop grown in Maharashtra. Flowering starts 50–55 days after sowing seeds. It is palatable and it has high protective value. The experiment was conducted on the Breeder Seed Production (Vegetable) farm, MPKV, Rahuri on medium black soil, The objective of the present investigation was to study the effect of different micronutrients as a foliar spray on seed production in Okra.

Materials and Methods

The field trial consisting of nine treatments with three replications in randomized block design was laid out during Rabi 2004–05 with variety Arka Anamika. The gross pot size was 5.0 m × 3.0 m. The seeds of okra dibbled at spacing of 45 × 10 cm on ridges and furrows on both the sides. Application of 50 kg N, 50 kg P and 50 kg K as basal dose and 50 kg N per hectare after thirty days of seed dibbling. The foliar spray was given for three times first before flowering, second after third pickings and third fifteen days after second spray with the concentrations 100 ppm.

Results and Discussion

Form the present studies, the data presented in Table 19.1 it was observed that, by the application of mixture of all micronutrients (*i.e.,* Boric add, Zinc sulphate, Ammonium molybdate, Copper sulphate, Ferrous sulphate, Manganese sulphate) the maximum germination percentage 86.33 per cent, seed vigour 6.78 and seed yield 4.53 q/ha respectively followed by the treatment commercial formulation (Multiplex)–100 ppm with germination percentage 84.6 per cent, seed vigour 6.64 and seed yield 4.47 q/ha. The minimum germination percentage (67.33 per cent), seed vigour (5.24) and seed yield (3.67 q/ha) was observed in the treatment of control (water spray). These results are in agreement with the findings of Patil (2002), Sontake *et al.* (1996), Suryanarayana and Rao (1981), Tosh, Choudhary and Chattarjee (1980) in Okra.

Table 19.1: Effect of Micronutrients on Seed Production in Okra

Sl.No.	*Treatments*	*Seed Yield q/ha*	*Germination %*	*Seed Vigour Index*
1.	Control (Water spray)	3.67	67.33	5.24
2.	Boric acid–100 ppm	3.95	79.33	6.19
3.	Zinc sulphate–100 ppm	4.25	80.33	6.28
4.	Ammonium molybdate–100 ppm	4.07	79.67	6.22
5.	Copper sulphate–100 pprn	4.00	72.67	5.66
6.	Ferrous sulphate–100 ppm	3.87	76.67	5.98
7.	Manganese sulphate–100 ppm	4.45	82.00	6.41
8.	Mixture of all–100 ppm	4.53	86.33	6.78
9.	Commercial formulation (Multiplex)–100 ppm	4.47	84.67	6.64
	S.E±	0.012	0.425	–
	C.D. at 5 per cent	0.035	1.273	–

References

Panse, V.G. and Suknatma, P.V., 1985. *Statistical Methods for Agricultural Workers.* ICAR, New Delhi.

Patil, K.B., 2002. Effect of micronutrients on yield of Okra [*Abelmoschus escantulus* (L) Moench). *Orissa J. of Horticulture,* 29(2): 118.

Sontake, M.B., Pillamari, V.D., Mandage, A.S. and Shinde, N.N., 1996. Effect of N levels on yield of okra. *J. Mah. Agril. Univ.,* 21(2): 292–293.

Suryanarayana, V. and Rao, K.V.S., 1981. Effect of growth regulators and nutrient sprays on the yield of okra. *Vegetable Science*, 8(1): 12–14.

Tosh, S., Choudhary, M.A. and Chattarjee, S.K., 1980. Effect of some micronutrients on growth, development and fruit yield of okra [*Abelmoschus escantulus* (L) Moench]. *Science and Culture*, 46(7): 271–274.

Chapter 20

Effect of Micronutrients on Seed Production in Brinjal

R.V. Patil, S.V. Kolase and K.G. Kadam

Breeder Seed Production (Vegetable), Seed Cell, MPKV, Rahuri – 413 722, Maharashtra

ABSTRACT

The experiment was carried out at Breeder Seed Production (Vegetable) farm, MPKV, Rahuri during Kharif 2004–05 to find out the effect of micro-nutrients on seed production in Brinjal. It was found that the seed yield of Brinjal can be increased significantly by three sprays of mixture of all nutrients (*i.e.*, Boric add, Zinc sulphate, Ammonium molybdate, Copper sulphate, ferrous sulphate, Manganese sulphate) at 100 ppm, first spray was given before flowering, second spray after third picking and third 15 days after second spray. The maximum seed germination, seed vigor and seed yield was recorded 84.00 per cent, 12.09 and 4.52 q/ha respectively, followed by commercial formulation (Multiplex), where the minimum germination percentage 82.67 per cent, seed vigor 9.51 and seed yield 4.31 q/ha respectively was recorded in the treatment Control (water spray).

Keywords: *Brinjal, Micronutrients, Seed yield.*

Introduction

Micronutrients when applied as foliar sprays exert a profound influence on yield of potato and tomato. Brinjal is one of the popular vegetable grown extensively in all over the country. It is heavy yielder and high remunerative crop but sometimes growers suffer with recurring economic loss due to poor plant vigour and low fruit setting. The experiment was conducted on the Breeder Seed Production (Vegetable) farm, MPKV, Rahuri on medium black soil, The objective of the present investigation was to study the effect of different micronutrients as a foliar spray on seed production in Brinjal.

Materials and Methods

The field trial consisting of nine treatments with three replications in randomized block design was laid out during Kharif 2004–05 with variety Manjri Gota. The gross plot size was 5.0 mx3.0 m. The plants of Brinjal were transplanted at space 90x60 cm on ridges and furrows. Application of 50 kg N, 50 kg P and 50 kg K as basal dose and 50 kg N per hector after thirty days of transplanting of Brinjal plants. The foliar spray was given for three times first before flowering, second after third pickings and third fifteen days after second spray with the concentrations 100 ppm.

Results and Discussion

In the present studies, the data presented in Table 20.1, it was observed that, by the application of mixture of all micronutrients (*i.e.*, Boric add, Zinc sulphate, Ammonium molybdate, Copper sulphate, Ferrous sulphate, Manganese sulphate) the maximum germination percentage 84.00 per cent, seed vigour 12.09 and seed yield 4.52 q/ha respectively followed by the treatment commercial formulation (Multiplex)–100 ppm with germination percentage 82.67 per cent, seed vigour 9.51 and seed yield 4.31 q/ha. The minimum germination percentage (66.67 per cent), seed vigour (6.00) and seed yield 93.38 q/ha) was observed in the treatment of control (water spray). The above results are in line with the findings of Gedez (965) in potato, Jyothi and Shanmugavelu (1985) in brinjal and Sycharaivo (1965) in tomato.

Table 20.1: Effect of Micronutrients on Seed Production in Brinjal

Sl.No.	Treatments	Seed Yield q/ha	Germination %	Seed Vigour Index
1.	Control (Water spray)	3038	66067	6000
2.	Boric add–100 ppm	3070	72.33	7002
3.	Zinc sulphate–100 ppm	3072	75.33	8.44
4.	Ammonium molybdate–100 ppm	3079	74.33	7028
5.	Copper sulphate–100 ppm	3053	69000	6.49
6.	Ferrous sulphate–100 ppm	3063	70.33	6.68
7.	Manganese sulphate–100 ppm	4.06	76.00	8.59
8.	Mixture of all–100 ppm	4.52	84.00	12.09
9.	Commercial formulation (Multiplex)–100 ppm	4.31	82.67	9.51
	S.Em±	0.648	0.709	–
	CD. at 5 per cent	1.942	2.126	–

References

Gedez, S.M. (1965). Effect of Mn, B, and Cu on some physico-chemical processes of metabolism of potato plants, yield of tubers and its quality. In: *Application of Trace Elements in Agriculture*. Navkova Domeka Keiv, p. 73–81.

Jyothi, K. Uma and Shanmugavelu, K.G., 1985. Studies on the effect of tricontanol, 2,4-D and boron on the yield of Brinjal Cv MDU.1.

Panse, V.G. and Suknatma, P.V., 1985. *Statistical Methods for Agricultural Workers*. ICAR, New Delhi.

Sycharaivo, KCh., 1965. Effect of trace elements on the yield and quality of green house grown tomatoes. In: *Application of Trace Elements in Agriculture*. Navkova Domeka Keiv, p. 134–138.

Chapter 21

Influence of Day Length on the Ovarian Development in a Strongly Photoperiodic Murrel, *Channa gachua* (Teleostei, Ophiocephaliformes)

Tanuja Sharma and Amitabh Hore

Department of Zoology, Ranchi University, Ranchi – 834 008

ABSTRACT

Sequence of changes in GSI, HSI, frequency of oocyte stages in ovary, protein, cholesterol, glucose and calcium in blood serun, hepatic and ovarian tissues has been observed month wise in four groups of *C. gachua* kept in different lighting regimes. Group I, the control, revealed the existence of a precise annual ovarian cycle in the fish, divisible in four phases–the preparatory phase (January to March), the pre-spawning phase (April, June), spawning phase (July, August) and spent phase (September to December) based on ovarian histology showing fluctuation in the frequency of oocyte stages. Changes in the biochemical parameters, protein, cholesterol, glucose and calcium also showed annual cyclicity associated with the ovarian histology. Gr. II fishes kept in 12L : 12D lighting regime during the preparatory phase (January to March) produced sequence of changes similar to the controls in all parameters studied. Group III fishes were kept in short photoperiod (16D : 8L) during the preparatory phase (January to March) and the onset of pre-spawning and spawning phases in these fishes were delayed. The frequency of atretic follicles was high in this group. Biochemical parameters studied also showed similar shift in their cyclicity. Group IV fishes, exposed to long day length (16L : 8D) during January through March, showed early matuirity of the oocytes. The biochemical parameters pointed towards an early vitellogenesis.

The result indicate that the variation in photoperiod alone or in association with some other physical factor(s) like water temperature may be involved in the maturation the ovary in this fish. These observations provide a basis for further experimental studies on the specific role of photoperiods in ovarian maturation in this murrel.

Keywords: *Day length, Ovarian development, C. gachua.*

Introduction

Seasonal reproduction is an artful stratagem employed by most temperate animals to ensure propagation of the species (Reiter, 1980). The majority of fish species in the temperate region exhibit an annual cycle of reproductive development which is maintained for as long as the animal is reproductively competent (Breder and Rosen, 1966; Bye, 1984). In order to produce offspring at a time when the chance of survival is highest, animals need to sense environmental cues telling them what season it is (Berg *et al.*, 2004).

Temporal organization of seasonal breeding is a species-specific phenomenon, which may be self-sustained circannual rhythmic function (Bromage *et al.*, 2001) or may be influenced by one or more meteorological factors (Vivien-Roels, 1985). The most important of external cues is the photoperiod, although other environmental factors such as temperature are also involved in the control of seasonal reproduction of some fishes (Borg *et al.*, 2004). It is long known that the reproduction (gonadal growth) and sexual maturity in the three spined stickleback (*Gastrosleus aeuleatus*) can be manipulated by changing photoperiods (Baggerman, 1980; Borg, 2004). Some fishes like cyprinids and sticklebacks spawn when days are lengthening (long-day breeders), whereas, others, such as most salmonids, spawn when days are shortening (short-day breeders). In both the cases, experimental studies have demonstrated effects of photoperiod on the timing of reproduction (Duston and Bromage, 1987: Awaji and Hanyu, 1988). However, the effect of photoperiod on the reproductive cycle has been investigated in very few Indian freshwater teleosts (*Mystus tengara*–Guraya *et al.*, 1976; *Channa punctatus*–Garg and Jain, 1985; Srivastava and Singh, 1991; *Heteropneustes fossilis*–Sundararaj and Sehgal, 1970; Vasal and Sundararaj, 1976; Chaube and Joy, 2002; *Clarias batrachus*–Acharia *et al.*, 2000; *Cirrhina reba*–Verghese, 1975; *Catla catla*–Dey *et al.*, 2004, 2005). In the present communication attempts have been made to find out the relation of biometeorological parameters and the annual reproductive cycle in female *C, gachua,* if any, by using artificial photoperiodic regimens.

Material and Methods

Female specimens of C. *gachua,* 10–12 cm in length and 100–125 gm in weight, collected from ponds in and around Ranchi (23°20′N lat. and 85°30′N long.) were used for the present study. More or less identical length and weight of fish was taken as an indication of their similar age. Immediately after collection (during September–October 2004) the fish were brought to laboratory, treated with 0.5 per cent $KmnO_4$ solution and kept in glass aquaria with artificial aeration to get acclimatized to the laboratory conditions. Commercial fish food as well as pieces of goat liver was given to the fish on alternate days.

The fish were kept in three different groups (Group 2, Group 3 and Group 4) for experimental studies. Group 1 fish (controls) were directly collected from wild every month. Group 2 experimental fish were kept in a 12L : 12D lighting regime during the preparatory phase of its reproductive cycle (January through March 2005). Group 3 and Group 4 fish were kept in 16L : 8D and 8L : 16D lighting regimes respectively during the preparatory phase *i.e.*, January through March 2005.

Ten fish from each experimental group (2, 3 and 4) and ten collected from the wild (*Gr.* 1) were sacrificed on 15th of every month after weighing. The blood was collected immediately, pooled and the serum preserved at –4°C for biochemical studies. Ovary and liver were taken out, extra tissue removed and weighed separately. While the ovary alone was subjected to histological and histochemical techniques, both the liver and the ovary were processed biochemically for the quantitative estimation of calcium, cholesterol, protein and glucose.

The ovary was fixed in aqueous Bouin's. Paraffin blocks of the tissue were prepared after dehydration through graded alcohol, sections cut at 5μ and stained with Mallory's triple for histological studies (after McMannus and Mowry, 1960).

Serum value of calcium was estimated after Anderegg (1954) and cholesterol after Zak and Henley (Oser, 1976). Protein and glucose was estimated by Biuret's method (as described in Oser, 1976) and after Kemp and Heijninger (1954) respectively.

Significance of the results was calculated by students' 't' test at 5 per cent confidence limit.

Observations

The mean value of monthly day length as well as different indices of ovarian and liver activity has been shown in Table 21.1.

Controls (Group I)

The month-wise ovarian and liver activity show time bound changes in rise and fall during January through December 2005. The gono- somatic index (GSI) (weight of the gonad/weight of the fish x 100) shows a progressive rise during April (3.1±0.9) through August (10.2±2.1), which is significant. From September onwards there is fall in the value of GSI (Figure 21.1). The hepato somatic

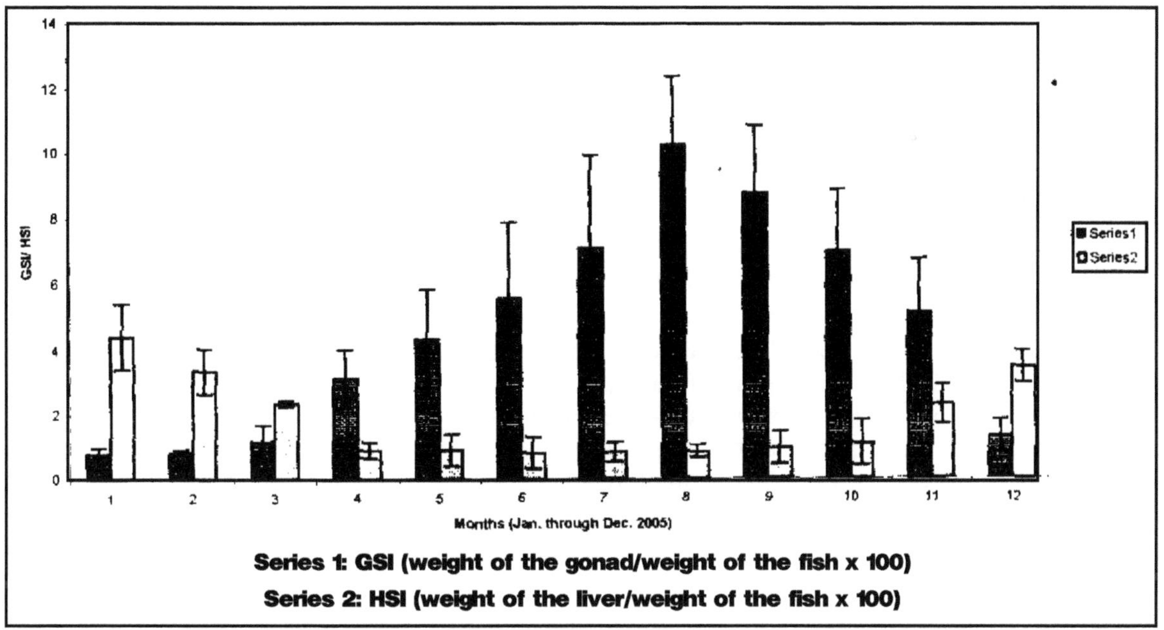

Series 1: GSI (weight of the gonad/weight of the fish x 100)
Series 2: HSI (weight of the liver/weight of the fish x 100)

Figure 21.1: Bar Diagram Showing Monthly GSI and HSI in Control

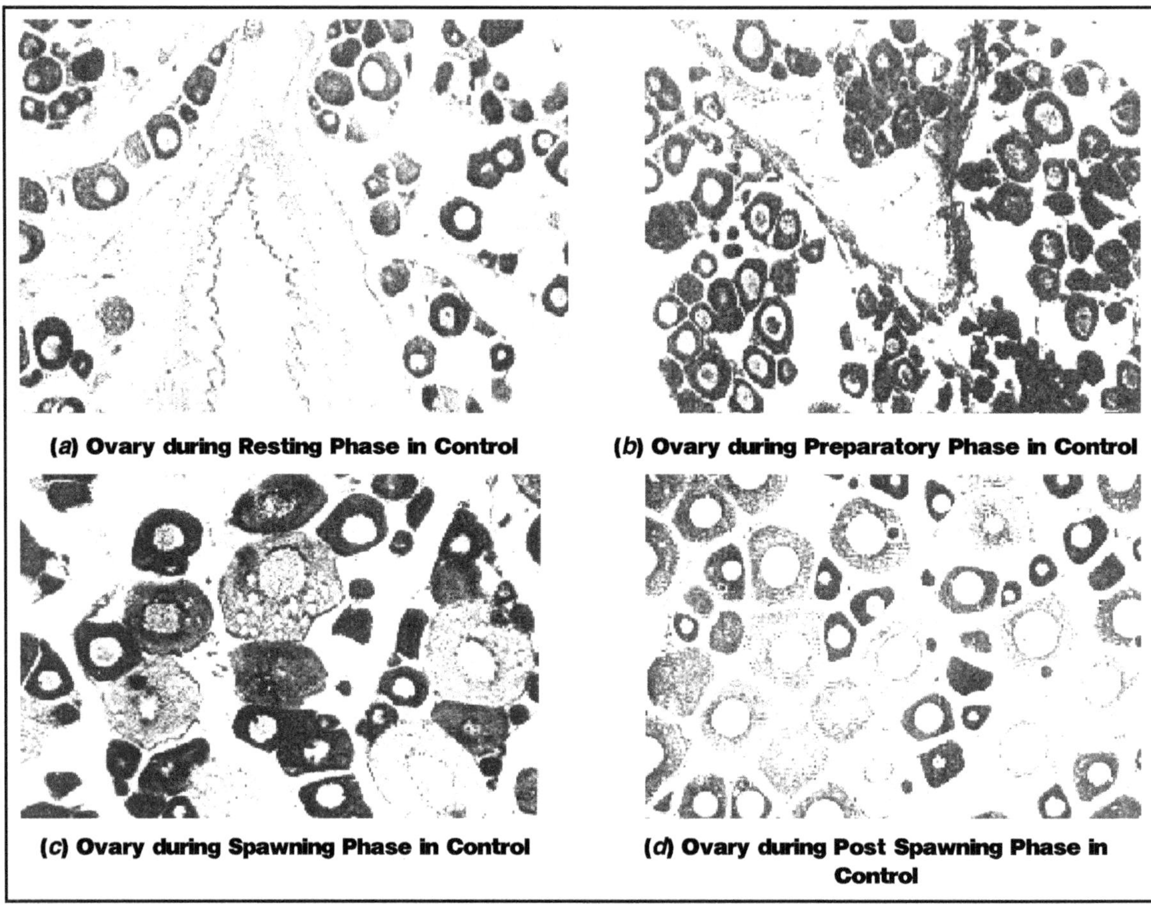

Plate 21.1: Showing Transverse Section of Ovary during Different Reproductive Phases in Control. Photo taken at 450 X. Mallory's triple.

index (HIS) (weight of the liver/weight of the fish x 100) shows an inverse relationship with GSI. HIS shows a progressive fall in its value from January (4.4±1) through June (0.81±0.5) after which its value rises.

The ovary in *C. gachua* is a hollow paired organ. It consists of oogonia, oocytes in different stages of development and their surrounding follicular cells, supporting connective tissue or stroma and vascular and nervous tissue. In microscopic observations the sections show the presence of one or more stages of growing oocytes, the relative percentage of which depends on the month of observation (Table 21.1; Figure 21.2). The classification of the oocyte stages was done after Wallace and Selman (1981) and Poortenaar *et al.* (2001).

Oogonia: The most primitive type of ovarian germ cells remain attached to ovigerous sac does not exceed 50µ in diameter.

Oocyte II: These are primary oocytes ranging between 50–200µ in diameter, with homogenous basophilic cytoplasm with a large nucleus containing 3–6 nucleoli.

Table 21.1: Monthly Variation in GSI, HSI, Stages of Growing Oocytes (%), Amount of Serum Calcium (mg/100 ml), Serum Cholesterol (mg/100 ml), Protein (mg/gm) and Glucose (mg/gm) in Ovarian as well as Hepatic Tissue in Control Fish (Gr. I)

2005	*Average Day-Length**	*GSI*	*HSI*	*Oogonia (%)*	*Oocyte II (%)*	*Oocyte III (%)*	*Oocyte IV (%)*	*Atretic Follicle*	*Serum Calcium*	*Serum Choles-terol*	*Protein Ovary*	*Protein Liver*	*Glucose Ovary*	*Glucose Liver*
Jan	10.49	0.81 ±0.2	4.4 ±1	98 ±1.7	1.3 ±.7	0	0	0	14.50 ±2.6	22.50 ±2.3	119.81 ±2.7	407.75 ±5.1	2.25 ±0.65	25.7 ±3.2
Feb	11.10	0.83 ±0.1	3.35 ±0.7	97 ±1.5	2.3 ±1.2	0	0	0	15.25 ±1.7	29.50 ±1.9	204.45 ±3.8	399.50 ±3.7	3.45 ±1.35	20.9 ±2.5
Mar	11.37	1.2 ±0.5	2.34 ±0.1	94.5 ±1.7	3.1 ±1.5	2.2 ±1.2	0	0	16.74 ±2.3	50.75 ±3.7	251.27 ±4.3	374.95 ±4.3	3.6 ±0.5	19.9 ±4.1
Apr	12.50	3.1 ±0.9	0.91 ±.25	74.5 ±1.5	4.4 ±2.4	12.4 ±1.3	0	0	18.15 ±1.9	75.45 ±2.7	330.25 ±2.9	307.90 ±3.8	4.5 ±.75	17.5 ±2.6
May	13.27	4.3 ±1	0.88 ±0.5	46.7 ±3.6	5.2 ±3.7	18.6 ±1.5	24.8 ±3.5	4.1 ±2.1	25.15 ±1.3	105.24 ±2.5	467.83 ±3.7	260.53 ±2.5	5.6 ±.37	8.5 ±4.5
Jun	13.48	5.5 ±2.3	0.81 ±0.5	25 ±2.2	5.6 ±3.7	25.9 ±4.2	40.4 ±2.6	4.5 ±1.7	30.65 ±2.3	126.35 ±3.1	505.55 ±5.1	230.15 ±3.8	6.7 ±1.5	4.2 ±3.7
Jul	13.35	7.0 ±2.9	0.83 ±0.3	22.8 ±1.4	5.7 ±2.3	15.7 ±3.5	51.7 ±4.7	12.7 ±1.7	20.75 ±3.4	131.70 ±3.9	500.15 ±4.3	244.15 ±4.3	6.5 ±2.4	4.9 ±2.9
Aug	12.50	10.2 ±2.1	0.85 ±0.2	13.5 ±1.6	5.5 ±2.7	0	45 ±2.6	31.5 ±3.6	10.70 ±1.8	85.85 ±3.8	470.63 ±3.7	278.90 ±3.4	4.8 ±1.7	5.7 ±2.2
Sep	12.37	8.7 ±2.1	0.95 ±0.5	48.5 ±1.7	3.1 ±2.1	0	0	23.5 ±3.5	10.55 ±2.7	65.28 ±4.5	390.5 ±4.1	299.75 ±5.1	4.0 ±1.5	7.6 ±3.1
Oct	11.51	6.9 ±1.9	1.10 ±0.7	96 ±1.9	0	0	0	3.5 ±2.3	11.50 ±3.4	61.75 ±3.4	325.67 ±2.8	315.68 ±4.3	3.75 ±0.9	14.5 ±3.2
Nov	10.55	5.1 ±1.6	2.28 ±0.6	98.5 ±1.2	0	0	0	0	12.25 ±1.9	41.82 ±3.5	254.75 ±2.7	345.55 ±3.5	3.65 ±1.8	21.9 ±3.9
Dec	10.42	1.3 1±0.5	3.44 ±0.5	100	0	0	0	0	13.42 ±2.4	26.55 ±4.3	174.75 ±45	390.45 ±2.3±	2.45 ±1.25	23.5 ±4.2

*: In hours and minutes; other values are mean±SEM (n=5).

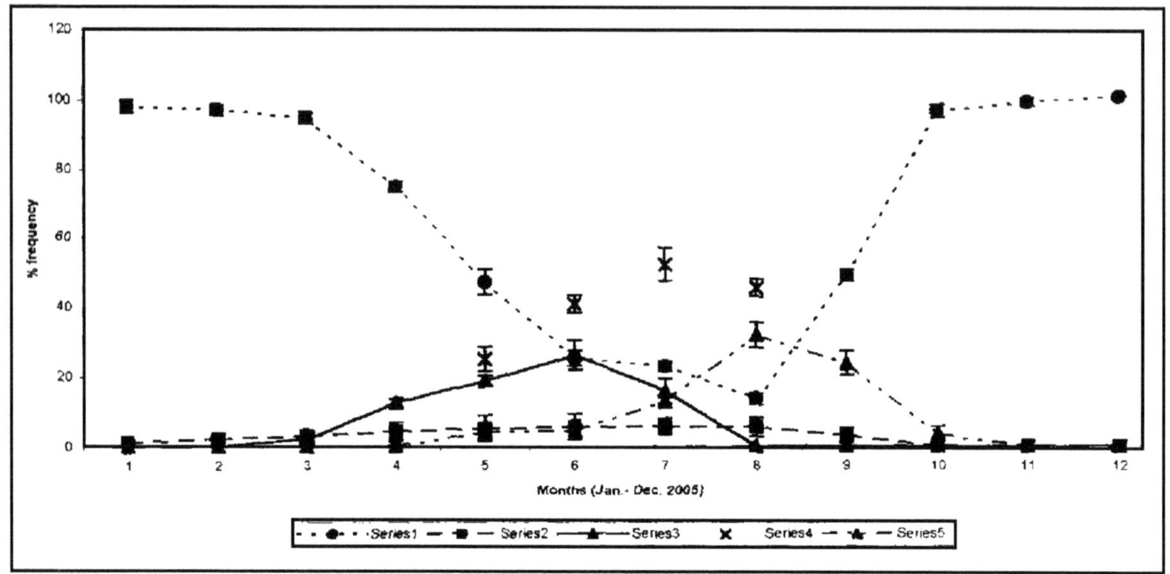

Figure 21.2: Monthly Per cent of Oocyte Stages
Series 1: Oogonia; Series 2: Oocyte 2; Series 3: Oocyte 3; Series 4: Oocyte 4; Series 5: Atretic follicle.

Oocyte III: These are primary oocyte between 200–350µ in diameter and identified by the presence cortical alveoli or vacuoles considered to be the indication of the start of vitellogenesis.

Oocyte IV: These stages are characterized by the presence of yolk in the oocyte cytoplasm. They range in size from 400–700µ.

Atretic follicles–oocytes in any stage of development show atresia and degenerate. None of the experimental groups (Grs. 2, 3 and 4) spawned in captivity and showed a greater per cent of atretic follicle than the control (Gr. 1).

Monthly variations in protein content in hepatic as well as ovarian tissue in control show a progressive rise and fall inversely related to one another (Table 21.1; Figure 21.3). In ovarian tissue the maximum protein content found was in the months of June and July (505.55±5. 1 mg/100 gm and 500.15±4.3 mg/100 gm) after which it starts falling reaching the minimum in the month of January (119.81±2.7 mg/100 gm). Protein in liver showed an opposite behavior. The minimum was recorded in June (230.15±3.8 mg/100 gm) and starts increasing thereafter reaching the maximum in January (407.75±5.1 mg/100 gm).

In ovary the lowest value of glucose was recorded in January (2.25±0.65 mg/100 gm). It showed a progressive increase then onwards reaching the maximum in June (67±1.5 mg/100 gm). After June the glucose content of ovarian tissue gradually declined. The glucose in hepatic tissue showed an opposite behavior. The maximum was recorded in January (25 7±3.2 mg/100 gm) and minimum in June (4.2±3.7 mg/100 gm) with progressive waxing and waning (Table 21.1; Figure 21.4).

The serum cholesterol and serum calcium also shows a progressive rise and fall during the year both reaching the maximum during June and July (Table 21.1; Figure 21.5).

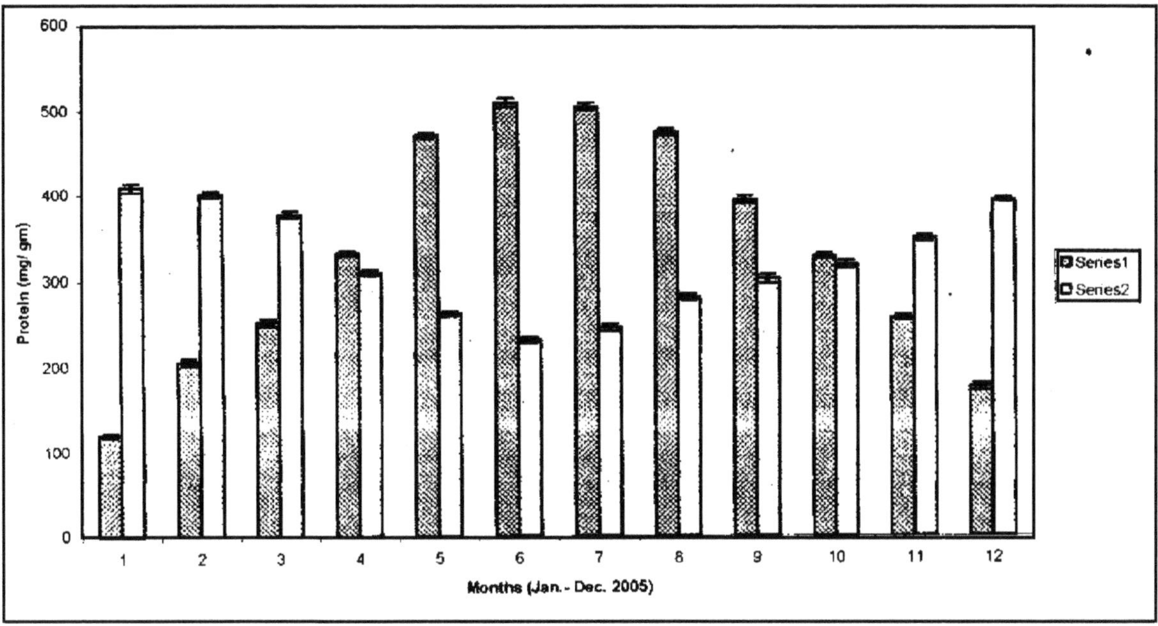

Figure 21.3: Monthly Variation in Protein Content of Ovarian and Hepatic Tissue in Control Series 1: Ovarian tissue; Series 2: Hepatic tissue

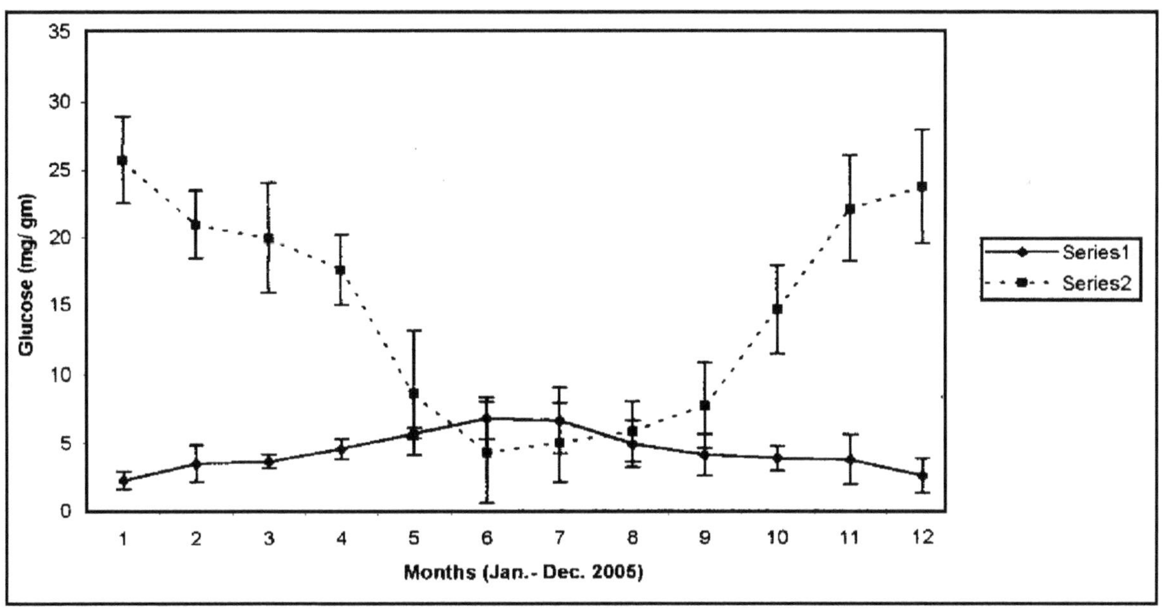

Figure 21.4: Monthly Variation in Glucose Content of Ovarian and Hepatic Tissue in Control Series 1: Ovarian tissue; Series 2: Hepatic tissue

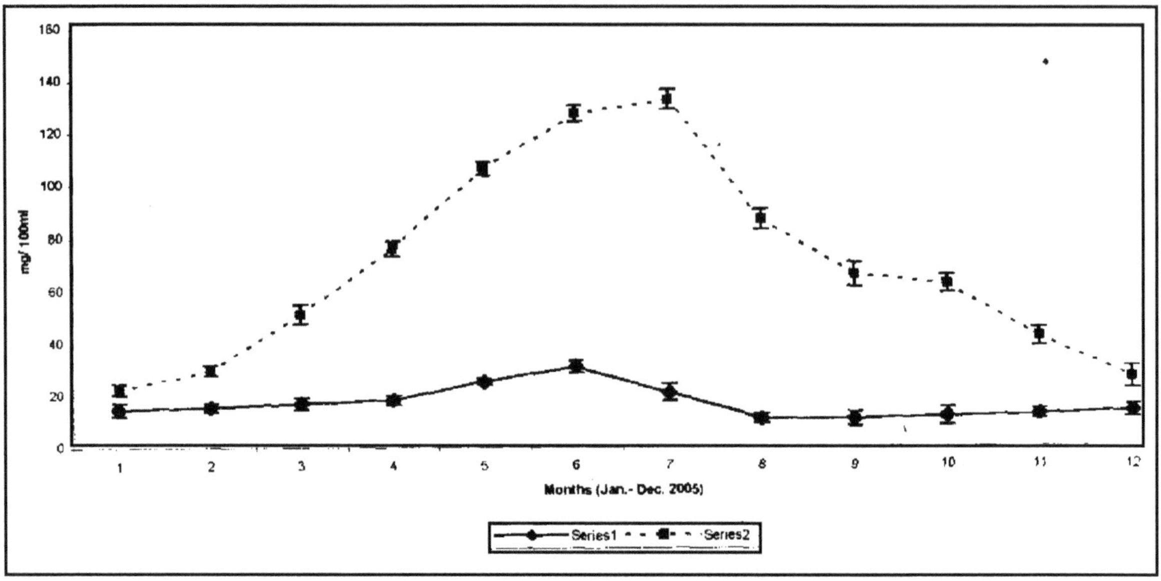

Figure 21.5: Monthly Variation in Serum Calcium and Cholesterol in Control
Series 1: Serum calcium; Series 2: Serum cholesterol

Ovarian Cycle in Experimental Groups (Groups II, III, IV)

In group II (exposed to 12L : 12D during January through March) the GSI and Ovarian Cycle was similar to that of the control. In group III (exposed to 16L : 8D during January through March) the increase in GSI took place earlier than that of the control, whereas in group IV (exposed to 8L : 16D during January through March) GSI increased later than the control. In both the II, III and IV experimental groups the GSI did not fall abruptly after September (the end of spawning period) as in control (Figure 21.6).

Discussion

The present results clearly indicate the presence of an annual breeding cycle in *C. gachua*. The GSI as well as the oocyte stages in control shows a free running reproductive cycle in this fish which can be divided in four phases; the preparatory phase (January–March), the pre-spawning phase (April–June), the spawning or reproductive phase (July–September) and the post-spawning or resting phase (October–December). Qayyum and Qasim (1964) studied the spawning cycle of the common murrel *Ophiocephalus* (*Channa*) *punctatus*. They reported the fish attaining the peak of gonadal maturity in May/June and spawning during July-mid September. The annual cycle of gonad weight clearly indicated the spawning cyclicity (Qayyum and Qasim, 1964). Monthwise study of the variation in protein, glucose and lipid of both liver and the ovary of *Ophiocephalus* (*Channa*) *punctatus* has shown the existence of a annual breeding cycle in the fish (Verma *et al.*, 1985). Most of the teleosts found in the freshwaters of our country have an annual breeding cycle comparable to *C. gachua* presently studied. Statistical analysis of the oocyte stages in *Catla catla* has shown that the fish attains sexual maturity at a particular time of the year when the mature/ripe stages of oocyte are found. This temporal pattern coincides with the annual reproductive cycle of the fish (Dey *et al.*, 2004).

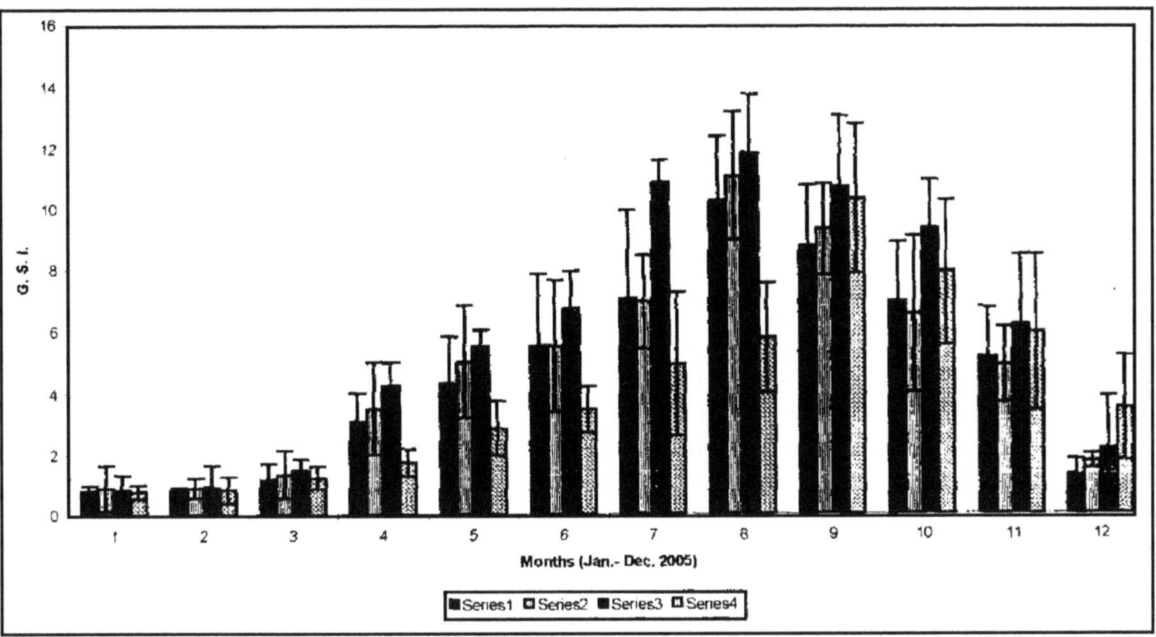

Figure 21.6: Monthly Variation of GSI in Experimental Groups

In teleosts, which reproduce by external fertilization, the ovulatory process marks a distinct and critical transition in female reproduction. In the pre-ovulatory period, growing oocytes accumulate high energy yolk stores (vitellogenin) to provide nutrients for future larvae. Following completion of vitellogenesis, oocytes are then maintained within the ovary for a variable period until a series of endocrine events stimulates their final maturation [migration and breakdown of the nucleus (germinal vesicle) accompanied by completion of meiosis] and ovulation (Jalabert, 1976). At this point of time, ovulated oocyte must be oviposited and fertilized within a relatively brief period if maximal viability of larvae is to be achieved (Stacey, 1984).

Liver of fish is known as one of the important organs, which reportedly stored glucose, protein and/or lipid for utilization during the reproductive cycle of the ovary (Verma *et al.*, 1985). Glucose level of the liver and the ovary in *C. gachua* appears to be inversely related. Considering the present observations it can be visualized that the carbohydrate content of the ovary of this fish is perhaps synthesized from the monomers depleted from liver during the late pre-spawning and reproductive (spawning) phases. Since perhaps the ovarian glucose is very little during the resting phase, the glucose level of the liver is maximum. As regards the protein level of the liver and the ovary, it appears to be inversely related. It can be envisaged that the liver possibly synthesizes and contributes its protein-containing yolk precursor for vitellogenesis (exogenous) in C. *gachua*. Similar observations have also been made by Idler and Bitners (1960) in *Salmo*, by Peute *et al.* (1978) in the zebrafish *Brachydanio*, by Yagamoto and Egami (1974) in the medaka *Olyzias*, by Aida *et al.* (*1973*) in the ayu, *Plecoglossus*, by van Bohemen *et al.* (1981) in the rainbow trout *Salmo* and by Verma *et al.* (1985) in *Channa punctatus* and *Heteropneustes fossilis*. The serum Calcium and cholesterol values as an index of vitellogenesis show similar behavior. Their basal levels were found till April and the increase in their circulating levels coincided with vitellogenesis probably under the influence of growing levels of estrogen. *Bromage et al.* (1982) have made similar observations.

The timing of reproduction in female teleosts may be viewed as the product of numerous biotic and abiotic stimuli which exert both long-term effects on ovarian growth and short term effects on final maturation and ovulation of the oocytes (Bye, 1984). Environment has a definite influence on gonadal activity in fish (Lam, 1983) and reproduction styles in every fish form a distinct pattern (Breder and Rosen, 1966). However, the probable environmental synchronizer of the cyclic activity is not definitely known. To find out the influence of photoperiod in gonadal ripening the present experiment was designed. It was found in the control that the actual photoperiod was less significant than the direction of photoperiod- in other wards, the photoperiod in January (*i.e.*, towards a long day regime) was stimulatory whereas the same in September (*i.e.*, towards a short day regime) was not stimulatory. The group III fish showed an early maturation of gonads as they were subjected to a longer than natural photoperiod during the photo-inducible phase and in the group IV fish the gonadal maturation was delayed as a shorter than natural photoperiod during the photo-inducible phase was available to the fish. Although it is apparent that the photoperiod is the most important environmental factor which is responsible for timing the gonadal growth many more environmental as well as limnological factors *viz.*, temperature, pH of water, amount of dissolved oxygen and dissolved carbon dioxide may have a role independently or in association (Borg *et al.*, 2004; Dey *et al.*, 2004). Moreover, it has to be further elucidated how the information regarding the environmental cues is picked up and finally processed to the reproductive axis for its synchronization (Dey *et al.*, 2005).

References

Acharia, K., Lal, B., Singh, T.P. and Pati, A.K., 2000. Circadian phase dependent thermal stimulation of ovarian recrudescence in Indian catfish, *Clarias batrachus. Biol. Rhythm Res.*, 31: 125–135.

Aida, K., Hirose, K., Yokote, M. and Hibiya, T., 1973. Physiological studies on gonadal maturation of fishes. II. Histological changes in the liver cells of ayu following gonadal maturation and estrogen administration. *Bull. Jap. Soc. Sci. Fish,* 39: 1107–1115.

Andeegg, C.H., 1954. *Practical Clinical Biochemistry,* 5th edn. Vol. I, (Eds.) Varley, H., Gowenlock, A.H. and M. Bell. William Heinman Medical Books Ltd., London.

Awaji, M. and Hanyu, I., 1988. Effects of water temperature and photoperiod on the beginning of spawning season in the orange-type medaka. *Zool. Sci.,* 5: 1059–1064.

Baggerman, B., 1980. Photoperiodic and endogenous control of the annual reproductive cycle in teleost fishes. In: *Environmental Physiology of Fishes,* (Ed.) M. Ali. Plenum Press, New York, pp. 533–567.

Borg, B., Bornestaf, C., Hellqvist, A., Schmitz, M. and Mayer, I., 2004. Mechanism in the photoperiodic control of reproduction in the stickleback. *Behaviour,* 141: 1521–1530.

Breder, C.M. and Rosen, D.E., 1966. In: *Modes of Reproduction in Fishes.* Natural History Press, Garden City, New York.

Bromage, N.R., Porter, M.J.R. and Randall, C.F., 2001. The environmental regulation of maturation in farmed finfish with special reference to the role of photoperiod and melatonin. *Aquaculture,* 197: 63–98.

Bromage, N., Whitehead, C., Elliott, J., Breton, B. and Matty, A., 182. Investigations into the importance of day length on the photoperiodic control of reproduction in the female rainbow trout. In: *Proc.*

Int. Symp. Repr. Physiol. Fish. Wagenillgen, The Netherlands. Compilers Richter, C.J.J. and Goos H.J.Th., pp. 233–236.

Bye, V.J., 1984. The role of environmental factors in the timing of reproductive cycles. In: *Fish Reproduction, Strategies and Tactics,* (Eds) Potts, G.W. and R.J. Wootton. Academic Press, New York, pp. 187–205.

Chaube, R. and Joy, K.P., 2002 Effect of altered photoperiod and temperature, serotonin affecting drugs, melatonin on brain tyrosine hydrolase activity in female catfish, *Heteropnelistes fossilis:* A study correlating ovarian activity changes. *J. Exp. Zool.,* 293: 585–593.

de Vlaming, V.L., 1972. Environmental control of teleost reproductive cycles: A brief review. *J. Fish. Biol.,* 4: 131– 40.

Dey, R., Bhattacharya, S. and Maitra, S.K., 2004. Temporal pattern of ovarian activity in a major carp *Catla catla* and its possible environmental correlate in an annual cycle. *Biol. Rhythm Res.,* 35(4/5): 329–353.

Dey, R., Bhattacharya, S. and Maitra, S.K., 2005. Importance of photoperiods in regulation of ovarian activities in Indian major carp *Catla catla* in an annual cycle. *J. Biol. Rhythms,* 20(2): 145–154.

Duston, J. and Bromage, N., 1987. Constant photoperiod regimes and the entrainment of the annual cycle of reproduction in the female rainbow trout (*Salmo gairdneri*). *Gen. Comp. Endocrinol.,* 65: 373–384.

Garg, S.K. and Jain, S.K., 1985. Effect of photoperiod and temperature on ovarian activity in the Indian murrel, *Channa (Ophicephalus) punctatus* (Bloch). *Can. J. Zool.,* 63: 834–842.

Guraya, S.S., Saxena, P.K. and Gill, M., 1976. Effects of long photoperiod on the maturation of ovary of catfish, *Mystus tengara* (Ham). *Acta. Morphol. Neerl. Scand.,* 14: 331–338.

Idler, D.R. and Bitners, I., 1960. Biochemical studies on sockeye salmon during spawning migration. IX Fat, protein and water in the major internal organs and cholesterol in the liver and gonads of the standard fish. *J. Fish. Res. Board Can.,* 17: 113–122.

Jalabert, B., 1976. *In vitro* oocyte maturation and ovulation in rainbow trout (*Salmo gairdneri*), northern pike (*Esox lucius*) and goldfish (*Carassius auratus*). *J. Fish. Res. Board Can.,* 33: 974–988.

Kemp, A. and van Heijninger, K., 1954. A colorimetric micromethod for determination of glycogen in tissues. *Biochemical Journal,* 56: 646.

Lam, T.J., 1983. Environmental influences on gonadal activity in fish. In: *Fish Physiology,* (Eds.) Randall, D. J. and E.M. Donaldson. Academic Press, New York, 9B: Chapter 2.

Oser, B.L., 1976. *Hawk's Physiological Chemistry,* 14th edn. Tata McGraw-Hill Publishing Company Ltd., New Delhi.

McMannus, J.F.A. and Mowry, R.W., 1960. *Staining Methods, Histologic and Histochemical.* Hoeber, New York.

Peute, J., van der Gaag, M.A. and Lambert, J.G.D.J., 978. Ultrastructure and lipid content of the liver of the zebra fish, *Brachidanio rerio* related to vitellogenin synthesis. *Cell Tissue Res.,* 186: 297–308.

Poortenaar, C.W., Hickman, R.W., Tait, M.J. and Giambartolomei, F.M., 2001. Seasonal changes in the ovarian activity of New Zealand turbot (*Colistium mudipinnis*) and Brill (*C. guntheri*). *New Zealand J. Mar. Freshwater Res.,* 35: 521–529.

Qayyum, A. and Qasim, S.Z., 1964. Studies on the biology of some freshwater fishes. Part 1. *Ophicephailis punctatus* Bloch. *J. Bombay Nat. Hist. Soc.*, 61(1): 74–98.

Srivastava, S.J. and Singh, R., 1991. Effects of constant photoperiod-temperature regimes on the ovarian activity during during the annual reproductive cycle in murrel, *Channa punctatus* (Bloch). *Aquaculture*, 96: 383–391.

Stacey, N.E., 1984. Control of the timing of ovulation by exogenous and endogenous factors. In: *Fish Reproduction, Strategies and Taclics,* (Eds.) Potts, G.W. and R.J. Wootton. Academic Press, New York, pp. 207–221.

Sundararaj, B.I. and Sehgal, A., 1970. Effects of long or an increasing photoperiod on the initiation of ovarian recrudescence during the preparatory period in the catfish, *Heteropneustes fossilis* (Bloch). *Biol. Reprod.*, 2: 413–423.

Sundararaj, B.I. and Vasal, S., 1976. Photoperiod and temperature control in the regulation of reproduction in the female catfish, *Heteropneustes fossilis. J. Fish. Res. Biol., Can.*, 33: 989–973.

van Bohemen, C.G., Lambert, J.G.D. and Peute, J., 1981. Annual changes in plasma and liver in relation to vitellogenesis in the female rainbow trout, *Salmo gairdneri. Cell. Compo Endocrinol.*, 44: 94–107.

Verghese, P.U., 1975. Internal rhythm of sexual cycle in a carp *Cirrhina reba* (Ham.) under artificial conditions of darkness. *J. Inland Fish. Soc., India*, 7: 182–188.

Verma, G.P., Sahu, K.C., Mohapatra, S., Mohapatra, S. and Das, C.C., 1985. A comparative histo-biochemical study of vitellogenesis in teleosts, *Ophiocephalus punctalus* and *Heleropneustes fossilis.* In: *Recent Advances in Zoology,* (Eds.) Srivastava, C.B.L. and S.K. Goel. Rastogi and Company, Meerut, pp. 45–60.

Vivien-Roels, B., 1985. Interactions between photoperiod, temperature, pineal and seasonal reproduction in non-mammalian vertebrates. In: *The Pineal Gland: Current State of Pineal Research,* (Eds.) Mess, B., Ruzsas, C.S., Tima, L. and P. Pevet. Elsevier Sciences, Budapest, Hungary, pp. 187–209.

Wallace, R.A. and Selman, K., 1981. Cellular and dynamic aspects of oocyte growth in teleosts. *Amer. Zoologist,* 21: 325–343.

Yagamoto, K. and Egami, N., 1974. Sexual differences and age changes in the fine structure of hepatocytes in the medaka, *Oryzias lalipes. J. Fac. Sci. Univ. Tokyo,* 13: 199–210.

Chapter 22

Investigation on the Pattern of Distribution of Macrophytes in Hidenkompat Lake, Bishnupur (Manipur)

N. Pinky Devi and B. Manihar Sharma

Ecology Laboratory, Department of Life Sciences, Manipur University, Canchipur – 795 003

ABSTRACT

Hidenkompat is a freshwater saucer shaped natural lake. It is situated at Keinou Village in the Bishnupur District at about 22 km due South west of Imphal with an area of 0.2461 sq. km. It is situated at an altitude of 775 above mean sea level. A total of 32 macrophytic species were recorded during the study period which were categorized into submerged (4 species), rooted with floating leaves (3 species), free floating (5 species), and emergent (20 species). Maximum number of macrophytes (24 species) have been recorded in site III followed by site I, II and IV (23 species each). Significant variations were observed in the distribution pattern of the macrophytes in the lake.

Keywords: *Hidenkompat lake, Macrophytes, Morphometry, Bathymetry, Eutrophication.*

Introduction

Freshwater macrophytes determine the overall ecosystem physiognomy and plays a significant biotype of the aquatic ecosystem and determine the trophic levels of the food chain from aquatic to the terrestrial life Aquatic macrophytes respond to the change in water quality and have been used as bio-indicator of pollution (Tripathi and Shukla, 1991). They have significant ecological and economic importance as they contribute to the productivity of an aquatic ecosystem and mobilize mineral nutrients from the bottom sediment. Pearsall (1921) studied the dynamic relationship between the macrophytic plant communities and their environment in the lake. One of the important approaches in better interpreting the complex relationship between the composition and distribution pattern of aquatic macrophytes and the environment variables of a lake includes the ecological vegetation mapping (Kuechler, 1988). Floristic survey of a community provides valuable background information for analysing the structural diversity and dynamics of the plant species.

In India floristic studies of macrophytic vegetations were initiated quite earlier by Biswas and Calder (1936). The North Eastern region of India which constitutes one of the 25 biodiversity Hot spots of the world (Myers *et al.*, 2000) are found much degraded. The situation is more acute in Manipur as most of few existing lakes including the Loktak lake (a Ramsar site) Waithou, Ikop, Utrapat and Sanapat lakes have been seriously degraded due to eutrophication. In the present study a comprehensive floristic survey of Hidenkompat lake has been undertaken for understanding the ecological role of macrophytic plants in maintaining the freshwater ecosystem.

Study Sites

The total surface area of Hidenkompat lake is 0.2461 sq. km. It lies between 24°40′18.39″ and 24°40′49.5″N latitude and 93°48′11.14″ 93°48′29.25″E longitude. The depth varies from 0.84 cm to 1.45 cm during the different seasons. The morphometry and bathymetric characters of the lake are given in Table 22.1.

Table 22.1: Morphometry and Bathymetry of Hidenkompat Lake, Manipur

Altitude (m MSL)	775
Latitude	24°40'18.39"N–24°40'49.5"N
Longitude	93°48'11.14"E–93°48'29.25"E
Area (Km^2)	0.2461 sqkm
Maximum Depth (m)	1.45
Minimum Depth (m)	0.84
Mean depth (m)	1.15
Mean depth/Maximum depth ratio	0.79
Basin shape	Saucer shape
Basin slope	Gentle slope
Bottom texture	Silted bottom
Mean Maximum temperature (°C)	21–30
Mean Minimum temperature (°C)	20–29

Materials and Methods

Regular periodical survey and sampling of the vegetation was done at monthly intervals at four different sites. These sites are located within a short distance from each other. The plants were identified according to the latest International code of Botanical Nomenclature. The growth habits were assigned following Mueller-Dombois and Ellenberg (1974).

Results and Discussions

Lake Morphometry

Morphometric and bathymetric characters of the lake revealed saucer shaped basin with gentle slope and silted bottom which are indicative of eutrophic nature of the lake. The bathymetric features mainly the nature and slope of the basin have been emphasized by the ecologists in arriving at the trophic structure of a lake (Russel-Hunter, 1970, Moss, 1989). Moss (1989) reported that the ratio of the mean depth to maximum depth has been found to be correlated to the contributions of the aquatic plants. The ratio of the mean depth to maximum depth (0.79) is indicative of the luxuriant growth of the macrophytes with high magnitude of production.

Table 22.2: Floristic Composition of the Four Study Sites.
The presence and absence of a species is indicated by '+' and '–' respectively.

Sl.No.	Name of Species	Site–I	Site–II	Site–III	Site–IV
1.	*Alternanthera philoxeroides* Griseb	+	+	+	+
2.	*Atylosia scarabaeoides* (Linn) Benth	+	+	+	–
3.	*Azolla pinnata* R. Br.	+	+	+	+
4.	*Bracharia mutica*	–	–	–	+
5.	*Carex cruciata* Wahlenb	+	+	–	+
6.	*Ceratophyllum demersum* Linn	+	+	+	+
7.	*Colocassia esculenta* Schott	–	–	+	+
8.	*Commelina* sp.	+	–	+	+
9.	*Echinochloa stagnina* (Retz) P. Beauv.	+	+	+	–
10.	*Eichhornia crassipes* (Mart) Solms	+	+	+	+
11.	*Enhydra fluctuans* Lour	–	+	+	+
12.	*Euryale ferox* Salisb	–	+	–	–
13.	*Hydril/a verticil/ata* (Linn. f) Royle	+	+	+	–
14.	*Hygroryza aristata* (Retz). Nees	+	+	+	+
15.	*Ipomoea aquatica* Forsk	+	+	+	+
16.	*Leersia hexardra* Swartz	+	+	+	+
17.	*Ludwigia adscendens* (Linn) Hara	+	+	+	+
18.	*Marsilea quadrifoliata* Linn	–	+	–	+
19.	*Mikania micrantha* Kenth	–	–	–	+
20.	*Nymphoides indicum* (Linn) O. Kuntz	+	+	+	+
21.	*Oenanthe javanica* (Bl.) DC	–	+	–	–
22.	*Oryza minuta* Pres 1	+	–	+	–
23.	*Oryza rufipogon* Griff	+	–	+	–
24.	*Polytoca* sp.	+	–	+	–
25.	*Pistia straNotis* Linn.	–	–	–	+
26.	*Pseudoraphis minuta* Pilger	+	+	+	+
27.	*Salvinia cucullata* Roxb.	+	+	+	+
28.	*Salvinia natans* Linn.	+	+	+	+
29.	*Trapa natans* Linn Bispinosa (Rox)	–	+	+	–
30.	*Utricularia aurea* (Lour)	+	+	+	+
31.	*Vallisneria spiralis* Linn.	–	–	–	+
32.	*Zizania latifolia* (Griseb) Stapf.	+	+	+	+

Floristic Composition

During the study period, a total of 32 macrophytes were recorded from the lake out of which 24 species were in site III and 23 species each in sites I, II and site IV. Out of 32 species recorded, 14 species comprising *Alternathera philxeroides, Azolla pinnata, Ceratophyllum demersum, Enhydra fluctuans, Eichhornea crassipes, Hygroryza aristata, Ipomoea aquatica, Nymphoides indicum, Pseudoraphis minuta, Salvinia cucullata, Salvinia natans, Utricularia aurea, Ludwigia adscendens* were found throughout the different seasons in all the four different study sites. *Genathe javanica, Euryale ferox* were found only in site II. *Bracharia mutica, Mikania micrantha, Pistia stratiotis, Vallisneria spiralis* were found only in site IV. The floristic distribution of plant species in the study area is given in Table 22.2. The plants were categorized into different groups: *viz.*, Emergent (20 species) Rooted with floating leaves (3). Free floating (5) and Submerged (4 species). The groupwise distribution of macrophytic species have been set in Table 22.3.

Table 22.3: Groupwise Distribution of Macrophytes in the Four Study Sites

Sl.No.	Name of the Growth Habit	Site–I	Site–II	Site–III	Site–IV	Whole Lake
1.	Emergent	15	13	14	14	20
2.	Rooted with floating leaf	2	3	3	1	3
3.	Free floating	3	4	4	5	5
4.	Submerged	3	3	3	3	4

The present findings are in conformity with the findings of Devi, Ch. B. (2001) who reported 34 species from the Sanapat Lake, Manipur of which 7 were submerged, 6 free floating and rooted with floating leaves and 15 emergents. Okram *et al.* (1996 and 1999) recorded 18 emergents, 4 free floating, 6 rooted with floating leaves and 6 submerged species out of 34 species in Waithou Lake, Manipur. In the Ikop lake, Devi, Ch. N. and Sharma, (2003) reported 29 species of which 2 species were found in submerged, 5 in rooted with floating leaved species, 5 in free floating and 12 in emergent group. Devi (1998) reported 26 macrophytes consisting 6 species in the submerged, 3 rooted floating leaved, 3 in free floating and 11 species in emergents in Utrapat lake, Manipur. Shah and Abbas (1979) reported 28 macrophytic species in Ganga river, Bhagalpur out of which 22 species were emergent, 4 submerged species and 2 floating species. Purushothama *et al.* (2005) in Kanale tank, Karnataka reported 19 species of which 2 free floating, 5 rooted floating 6 submerged, 1 emergent and 5 semi-aquatic Mishra and Tripathi (2004) also reported 12 species, in the unpolluted site of Ganga river in Varanasi.

A comparatively higher number of macrophytic species were reported by Devi (1993) with 86 species in the Loktak lake, Manipur, out of which 73 species were reported from the non-phumdi zone of which 6 species were submerged, 4 species free floating and 3 rooted with floating leaved species. Handoo and Kaul (1982) reported 58 species in the wetlands of Kashmir out of which 23 species were in the emergent group, 21 in the ground layer, 7 species in floating leaved and 7 species were in the submerged group. Daisy and Gupta (2006) also reported 56 species in Keibul Lamjao National Park, Manipur. Recently, Devi and Sharma (2007) observed 36 macrophytic species (20 emergents, 8 submerged, 4 rooted with floating leaves and 4 free floating species) in A wangsoipat lake (Bishnupur), Manipur. Kumar and Pandit (2005) recorded 46 species in the wetlands of Hokarsar, Kashmir of which 30 were emergent, 7 rooted with floating leaves, 2 free floating and 7 submerged. Crowder *et al.* (1977) also reported heterogenous composition of submerged vegetation in lake Opinicon, Canada. Seshavatharam *et al.* (1982) also reported intermixed mats of macrophytes in Kolleru lake, Andhra pradesh.. Similarly in the present study a clear cut zonation of floating, emergent and submerged species could not be observed and hence all such communities were found in intermixed mats.

Table 22.4: List of the Species in Hidenkompat Showing Family and Growth Habit

Sl.No.	Name of Species	Family	Habit
1.	*Alternanthera philoxeroides* Griseb	Amaranthaceae	Emergent
2.	*Atylosia scarabaeoides* (Linn) benth	Papilionaceae	Emergent
3.	*Azolla pinnata* R. Br.	Salvinaceae	Free floating
4.	*Bracharia mutica*	Poaceae	Emergent
5.	*Carex cruciata* Wahlenb	Cyperaceae	Emergent
6.	*Ceratophyllum demersum* Linn	Ceratophyllaceae	Submerged
7.	*Colocassia esculenta* Schott	Araceae	Emergent
8.	*Commelina* sp.	Commelianaceae	Emergent
9.	*Echinochloa stagnina* (Retz) P. Beauv.	Poaceae	Emergent
10.	*Eichhornia crassipes* (Mart) Solms	Pontederiaceae	Free floating
11.	*Enhydra fluctuans* Lour	Asteraceae	Emergent
12.	*Euryale ferox* Salisb	Nymphaeaceae	Rooted with floating leaves
13.	*Hydrilla verticillata* (Linn. f) Royle	Hydrocharitaceae	Submerged
14.	*Hygroryza aristata* (Retz). Nees	Poaceae	Emergent
15.	*Ipomoea aquatica* Forsk	Convolvulaceae	Emergent
16.	*Leersia hexardra* Swartz	Poaceae	Emergent
17.	*Ludwigia adscendens* (Linn) Hara	Onagraceae	Emergent
18.	*Marsi/ea quadrifoliata* Linn	Asteraceae	Emergent
19.	*Mikania micrantha* Kenth	Menyanthaceae	Rooted with floating leaves
20.	*Nymphoides indicum* (Linn) O. Kuntz	Apiaceae	Emergent
21.	*Oenanthe javanica* (B1.) DC	Apiaceae	Emergent
22.	*Oryza minuta* Presl	Poaceae	Emergent
23.	*Oryza rufipogon* Griff	Poaceae	Emergent
24.	*Polytoca* sp.	Poaceae	Emergent
25.	*Pistia stratiotis* Linn.	Araceae	Free floating
26.	*Pseudoraphis minuta* Pilger	Poaceae	Emergent
27.	*Salvinia cucullata* Roxb.	Salvinaceae	Free floating
28.	*Salvinia natans* Linn.	Salvinaceae	Free floating
29.	*Trapa natans* Linn Bispinosa (Rox)	Trapaceae	Rooted with floating leaves
30.	*Utricularia aurea* (Lour)	Lentibulariaceae	Submerged
31.	*Vallisneria spiralis* Linn.	Hydrocharitaceae	Submerged
32.	*Zizania latifolia* (Griseb) Stapf.	Poaceae	Emergent

Out of the total 18 different families recorded in the lake, 9 species belonged to Poaceae family while 3 species belonged to the family *Salvinaceae.* 2 species each were contributed by *Araceae, Asteraceae,*

Hydrocharitaceae. Each of the remaining species were represented by families like *Amaranthaceae, Apiaceae, Ceratophyllaceae, Commelinaceae, Cyperaceae, Convolvulaceae etc.*

Conclusion

The unique morphometric and bathymetric observation indicates the cutrophic nature of the lake in the present investigation. The lake has got luxuriant growth of vegetation with rich Biodiversity. The rich growth of emergent species indicate the eutrophic nature of lake. This has resulted serious threats to the ecosystem stabilization and the very existence of the lake. Hence relevant conservation and management measures of this valuable fresh water lake is needed to save the important species from extinction and conserve the lake from further degradation.

References

Biswas, K. and Calder, C.C., 1936. *Handbook of Common Water and Marsh Plants of India and Burma*. Government of India Publication, Calcutta.

Crowder, A.A., Briston, J.M., King, M.R. and Vander Kloet, S., 1977. Distribution seasonality and biomass of aquatic macrophytes in lake Opinicon. *Naturalists Can.*, 104: 441–456.

Daisy Angom and Gupta, Asha, 2005. Biological spectrum of Keibul Lamjao National Park, Manipur N.E. *Indian J. Curr. Sci.*, 9(1): 125–132.

Devi, Ch. Bebika, 2001. Variations in species distribution and primary production of the macrophytes in Sanapat lake Manipur. *Ph.D. Thesis*, Manipur University, Manipur.

Devi, Ch. Nivanonee and Sharma, B.M., 2003. Ecological study of the macrophytes of Ikop lake, Manipur. Morphometry and Qualitative analysis. *Indian J. Environ. and Ecoplan.*, 7: 243–250.

Devi, K. Indira, 1998. Ecological studies of freshwater macrophytes in Utrapat lake, Manipur. *Ph.D. Thesis*, Manipur University, Manipur.

Devi, L. Geetabali and Sharma, B.M., 2007. Studies on the diversity of the Macrophytes in Awangsoipat lake (Bishnupur), Manipur, India. In: *Biodiversity Conservation and Legal Aspects*, (Eds.) A.K. Kandya and Asha Gupta. Aavishkar Publishers, Distributors, Jaipur, pp. 62–71.

Devi, N. Beenakumari, 1993. Phytosociology, primary production and nutrient status of macrophytes of Loktak lake, Manipur. *Ph.D. Thesis*, Manipur University, Manipur.

Handoo, J.K. and Kaul, V., 1982. Phytosociological and standing crop studies in Wetlands of Kashmir. In: *Wetlands Ecology and Management*, (Eds.) B. Gopal, R.E. Turner, R.G. Wetzel and D.F. Whigham. National Institute of Ecology and International Scientific Publication, India, pp. 187–195.

Kuechler, A.W., 1988. Ecological vegetation maps and their interpretation. In: *Vegetation Mapping*, (Eds.) Kuechler, A.W. and I.S. Zooneveld. Kluwer Academic Publishers, The Hague, pp. 469–481.

Kumar, R. and Pandit, A.K., 2005. Community architecture of macrophytes in Hokarsar Wetland, Kashmir. *Indian J. Environ. and Ecoplan.*, 10(3): 565–573.

Mishra, B.P. and Tripathi, B.D., 2004. Distribution of macrophytes and phytosociology of *Hydrilla verticillata* Casp, and *Lemna minor* Linn in lotic and lentic aquatic ecosystem. *Ecol. Env and Cons.*, 10(1): 37–41.

Moss, B., 1989. *Ecology of Freshwaters: Man and Medium.* Blackwell Scientific Publication, Oxford.

Mueller-Dombois, D. and Ellenberg, 1974. *Aims and Methods in Vegetation Ecology*. John Wiley and Sons, New York.

Myers, N., Mittermeler, R.A., Mittermeler, C.G. da Fonseca, G.A.B. and Kent, J., 2000. Biodiversity hot spots for conservation priorities. *Nature*, 3: 853–852.

Okram, I.D., Sharma, B.M. and Singh, E.J., 1996. Ecological study of Waithou lake, Manipur: Morphometry and qualitative analysis of macrophytic vegetation. *J. Freshwater Biology*, 8: 177–189.

Pearsall, W.H., 1921. A suggestion to factors influencing the distribution of free floating vegetation. *J. Ecol.*, 9: 243–253.

Purushothama, R., Naryana, J., Kiran, B.R. and Harish Kumar, K., 2005. Nutrient status of Kanale tank at Sagara taluk in Karnataka with reference to diversity of aquatic macrophytes. *J. Curr. Sci.*, 7: 161–164.

Russel Hunter, W.D., 1970. *Aquatic Productivity: An Introduction to Some Basic Aspects of Biological Oceanography and Limnology*. Macmilan Publishing Co. Inc., New York.

Seshavatharam, V., Dutta, E.S.M. and Venu, P., 1982. An ecological study of the vegetation of the Kolleru lake. *Bull. Bot. Surv., India*, 24: 70–75.

Shah, J.D. and Abbas, S.G., 1979. Seasonal variations in freshwater, density, Biomass and rate of production of some aquatic macrophytes of the river Ganges at Bhagalpur (Bihar). *Trop. Ecol.*, 20: 127–134.

Tripathi, B.D. and Shukla, S.C., 1991. Biological treatment of Wastewater by selected aquatic plants. *Environ. Poll.*, 6: 69–78.

Chapter 23

Water Scarcity Zones of Jamshedpur, Jharkhand

Gouri Suresh[1], Neena Sharma[2], G. Rajalakshmi[2] and Rajni Kumari[2]

[1]Reader in Botany, Jamshedpur Women's College, Jamshedpur
[2]Guest Faculty, Environment and Water Management, Jamshedpur Women's College, Jamshedpur

ABSTRACT

The present paper is based on a survey conducted by Jamshedpur Women's College Eco-Club, of which the author is the moderator. The city of Jamshedpur was divided into 9 segments and questionnaires were distributed to households in a random fashion in all the segments. Based on the information collected, a water scarcity map of Jamshedpur was prepared; this map will definitely be of use to the city planners. It was also found that awareness activities are essential in the city.

Keywords: *Water source, Water quality, Awareness, Water scarcity map, Water conservation.*

Introduction

Jamshedpur is a picturesque city situated in the state of Jharkhand, at the confluence of two rivers, Subarnarekha and Kharkhai, and surrounded by the rugged hills of the Chota Nagpur plateau. 64 sq.kms in area, the city is located at 532 metres above sea level. The Mapping of Water Scarcity Zones of Jamshedpur was a project, sponsored by the Ministry of Environment and Forests, Government of India, undertaken by the Eco-Club of Jamshedpur Women's College. Eco-Clubs of different schools of Jamshedpur were also involved in this massive venture.

Materials and Methods

Jamshedpur was divided into 9 zones of approximately equal areas based on the main roads separating these zones. Copies of a questionnaire prepared were distributed among the members of the Eco-Clubs of our college and also the participating schools. The members were divided into 9 groups based on the localities they lived in and a leader was selected for each group. Thereafter, the groups fanned out into the 9 zones and collected answers for the questions from the local people. A very large quantity of data was obtained by this method.

The questionnaire:

1. Name of the zone:
2. Information collected from:
3. Main source of water–Supply Water/Surface Water/Groundwater
   Any specification:
4. If groundwater source water is available, at what depth:
   Maximum depth Minimum depth
5. Quality of water:
6. Is there any scarcity of water in your area? Yes/No
   If yes, throughout the year/for a few months only
   Specifications, if any:
7. Cleaning procedure practised:
   Self/Community/Company or Municipality
   Details:
8. Awareness regarding water problems among the local people:
9. Is there any water body in your zone? Lakes/Ponds/River/Any other
10. Any other water related problem faced by the local people of the zone:
11. Suggestions to improve the prevailing condition:
    (*a*) At individual level
    (*b*) At community level
    (*c*) At Government level
12. Suggestions by the Eco-Club members.

Results and Discussion

Based on the collected data, a detailed map of Jamshedpur with regard to its Water Scarcity Zones was prepared. This map is sure to help the authorities in chalking out solutions to the problem of water scarcity in parts of Jamshedpur.

From the map of the city, it is clear that Tata Steel Works is situated almost in the centre of the city, towards the southern part and it occupies a considerable area. This area was excluded for the purpose of the study. The 9 zones studied are all outside the company area.

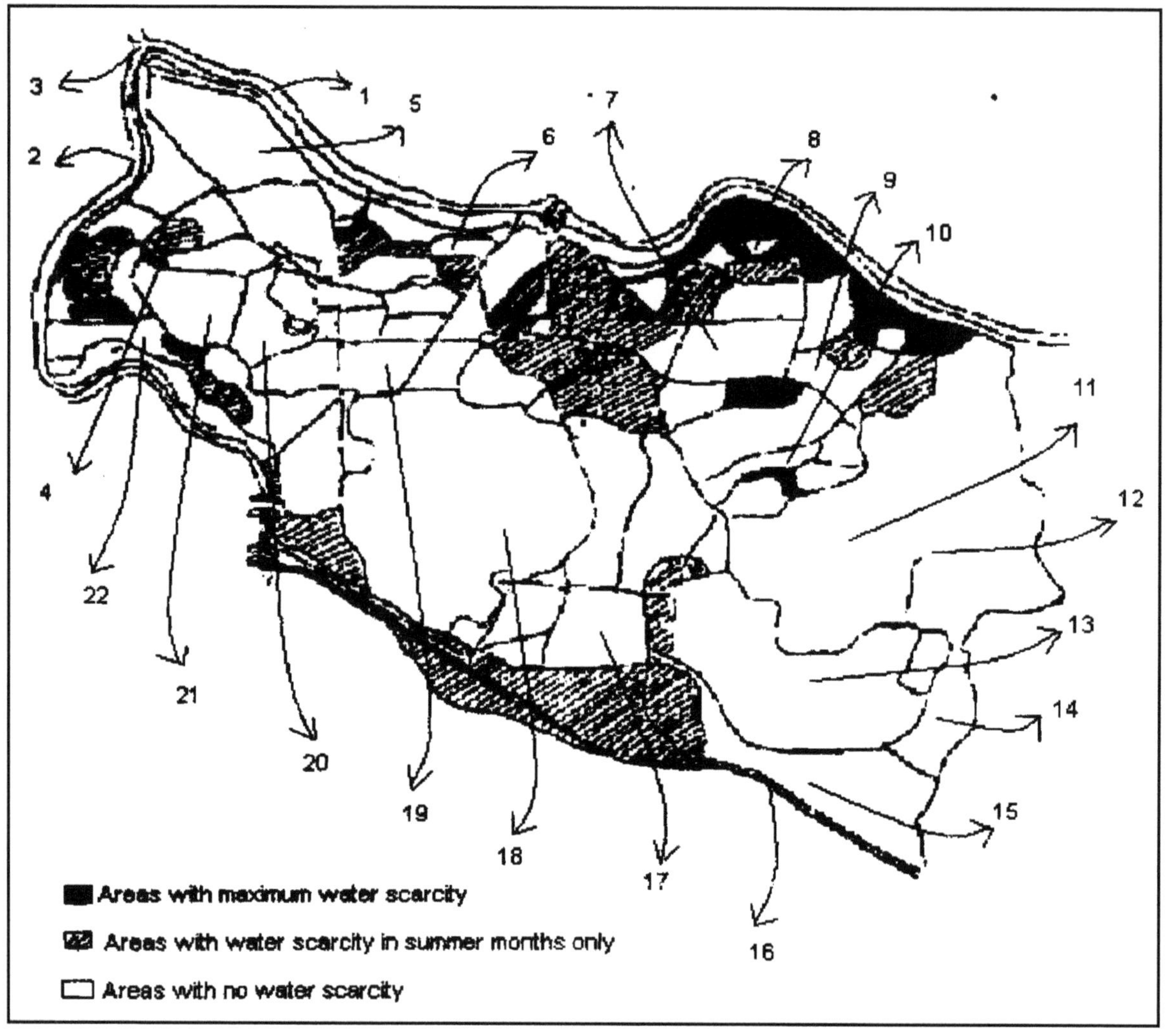

Figure 23.1: Water Scarcity Zones of Jamshedpur, Jharkhand

1: River Subarnarekha; 2: River Kharkhai; 3: Rivers Meet (Do Muhani); 4: Uliyan; 5: Sonari; 6: Jubilee Lake (Jayanti Sarovar); 7: Sidgora; 8: Fly Ash Pond; 9: Old Baridih; 10: New Baridih; 11: Telco Colony; 12: TRF Colony; 13: Tata Engineering Works; 14: Hudco; 15: Lafarge Cement; 16: Railway Line; 17: Tubes Division; 18: Tata Steel Works; 19: Northern Town; 20: B.H. Area; 21: Farm Area; 22: Kadma.

Tata, the biggest company of Jamshedpur, does maintenance of most parts of the city through JUSCO (Jamshedpur Utility Services Company). In these areas, there is ample water supply. However, some other areas (about 20 per cent of the total) face water shortage in summer months from April to July. Yet other such areas (about 10 per cent of the total) face the problem throughout the year.

Some parts of the city where there is acute water shortage all the year round are located away from water bodies; however, quite contrary to expectations, some such areas are very near to either River

Subarnarekha or River Kharkhai. To know the exact reason for this, hydro-geological studies will have to be conducted.

The survey has also thrown up some facts regarding the approach of the public in dealing with water supply and crisis. A majority of the people feel that supplying water is purely the responsibility of the government. It is true that water, like land and forests, is a state subject and it is the duty of the state governments to enforce legislations for surface and subsurface water. Yet people's participation is a must for the redressing of any local scarcity problem. The present survey showed that local people in the affected' areas are either non-committal or ignorant about water conservation methods. From this, it is obvious that there is an urgent need of educating the people and creating awareness in them regarding different methods of water conservation.

In many suburban areas of Jamshedpur, where acute water scarcity was seen, clusters of residential complexes have been built after draining wetlands. Inland wetlands help to replenish groundwater sources (Miller, Jr., 1996). Therefore draining them is suicidal to water conservation efforts and not many people know this.

It has also been observed that the public is using some water bodies as dumping grounds. These can be cleaned and maintained by the same public what is needed is motivation. Success stories like that of how such a dump was transformed back to the original 9,000 square meter *jheel* of clear water in Kolkata (Dasgupta, 2003) can be an inspiration.

In areas outside the command of JUSCO, people use mainly hand pumps for all their water needs. But bore wells would yield water only if they are recharged over a long period of time (Kumar, 2001). Therefore people must realise that as many types of water harvesting techniques as possible must be implemented. The 3 main methods are roof harvesting, runoff harvesting and floodwater harvesting (Suresh, 2004).

Of the several avenues to raising water productivity, the key is pricing water at its market value (Brown, 2001). However, there is a contrary view that privatisation of water resources implies that water is an economic commodity and scarcity is a business opportunity (Jayaraman, 2003). Thus making the local populace realise the essentiality of practising water conservation measures is the need of the hour. Motivating them to do so is the duty of all those who are aware of the grimness of the situation.

Acknowledgement

The authors are thankful to the Ministry of Environment and Forests, Government of India for sponsoring the project and to Gram Vikas Kendra, Jamshedpur Regional Resource Agency for forwarding the same to the ministry. Thanks are due to all the Eco-Club members who participated in the project. The authors are also grateful to the Principal, Jamshedpur Women's College for providing the facilities needed.

References

Kumar, Dinesh M., 2001. Drinking water: A scarce resource. *Survey of the Environment*, (Ed.) N. Ravi. *The Hindu*, pp. 93–101.

Miller, Jr. G. Tyler, 1996. *Living in the Environment: Principles, Connections and Solutions*. Wadsworth Publishing Company, New York, pp. 160–162

Brown, Lester R., 2001. *State of the World.* W.W. Norton and Company, New York, London, pp. 54–55.

Jayaraman, Nityanand, 2003. Privatisation of water. *Survey of the Environment*, 2003, (Ed.) N. Ravi. *The Hindu*, pp. 83–86.

Dasgupta, Rajashri, 2003. Kolkata: Clean waters again. *Survey of the Environment*, 2003, (Ed.) N. Ravi. *The Hindu*, pp. 165–170.

Suresh, R., 2004. *Soil and Water Conservation Engineering.* Standard Publishers Distributors, Delhi, pp. 523–527.

Chapter 24

Yield Maximisation through Plant Density and N Levels on Baby Corn (*Zea mays* L.) var. COBC1

S. Ravi, S. Ramesh, K. Sathyamoorthi[1] and S.D. Sundar Singh[2]

[1]Associate Professor, [2]Registrar
Tamil Nadu Agricultural University, Coimbatore – 641 003, Tamil Nadu, India

ABSTRACT

Field experiments were conducted at eastern block farm, Tamil Nadu Agricultural University, Coimbatore during *Kharif* 2001(August to October) to study the yield maximization through plant density and N levels on baby com variety COBC 1. The treatments consisted of two nitrogen levels in the main plots *viz.*, 150 kg ha^{-1} (N_1) and 200 kg ha^{-1} (N_2) and four plant densities levels *viz.*, 45 × 30cm (D_1), 45 × 25cm (D_2), 45 × 20cm (D_3) and 45 × 15cm (D_4) were assigned to sub-pot in a Factorial Randomized Block design. The trials were replicated thrice. Growth characters *viz.*, plant height, LAI and DMP; and yield attributes *viz.*, length of cob and com, weight of the cob and com, and green cob yield were significantly higher at 45 × 20cm (D_3) spacing with combination of 200 kg N ha^{-1} recorded higher yield of 8,426 kg ha^{-1}. This study may help to optimize plant density and nitrogen levels.

Keywords: *Baby com, Plant density, Nitrogen level, Cob yield.*

Introduction

Maize or Indian com (*Zea mays L.*) is one of the most important cereals in the global agricultural economy both as food for man and feed for animal. It is a miracle crop, has very high yield potential, there is no cereal crop on the earth that has shown immense potential and so called queen of cereals.

The physiological maize is rated as the most effective species domesticated by man (Singh, 1998). During the late 1980's and early 90's people started consuming immature young com harvested when the silks have either not emerged or just emerged and no fertilization has taken place (Bar Zur and Saddi,1990). These young ears are de-husked and eaten as vegetables, whose delicious sweet flavour and crispness or much flavoured in western countries and USA. Such young delicious vegetable is called Baby com. Importantly baby com in free from pest and diseases and its nutritional value is comparable to cauliflower, cabbage, tomato, brinjal and cucumber (Yodpet, 1979). Its by-products such as tassel, young husk, silk and green stalks provide good cattle feed. The sweet succulent and delicious baby com is a medium plant type and provides green ears within 65–75 days after sowing. As it is a new plant type, there is an urgent need to find out suitable agro-techniques for higher production and higher income of farmers.

Nitrogen is a essential constituent of chlorophyll and the synthesis of chlorophyll is highly limited by the availability of nitrogen (Pandey and Sinha, 1998). As like maize, baby com is an exhaustive crop and prefers light soil types for its cultivation, which further aggravates the nutrient deficiency, as they are lost quickly after fertilizer application in the field. The nitrogen management therefore is of paramount important in the baby crop production.

Crop geometry is an important factor to achieve higher production by better utilization of moisture and nutrients from the soil (root spread) and with above soil (plant canopy) by harvesting the maximum possible solar radiation and in turn better photosynthates formation through the spacing requirement of grain and fodder maize has been standardized, the information on the influence of spacing on yield and quality of baby com is yet to be optimized under Coimbatore condition.

Materials and Methods

Field experiments were conducted during *Kharif* 2001 (August to October) at eastern block, Tamil Nadu Agricultural University, Coimbatore. The experiment site is located at 11°N latitude, 77°E longtitude with an altitude of 427 above MSL. The soil of the experimental area was sandy clay loam (*Typic Ustropept*) with alkaline pH, low in organic carbon (0.31 per cent) and available N (223.6 kg ha^{-1}), medium in available P (12.9 kg ha^{-1}) and high in available K (421.6 kg ha^{-1}) during *Kharif* 2001. Baby com composite COBC 1 was chosen for the study.

The experiments were laid out in Factorial Randomized Block design with three replication on a net plot size of 4 × 5 m. In main plots two nitrogen levels *viz.*, N_1: 150 kg ha^{-1} and N_2: 200 kg ha^{-1} and in sub-plots four levels of plant densities *viz.*, D_1: 45 × 30cm; D_2: 45 × 25 cm; D_3: 45 × 20cm and D_4: 45 × 15 cm were assigned. Nitrogen as urea, Phosphorus (60 kg ha-I) as single super phosphate and potassium (40 kg ha^{-1}) as muriate of potash were applied as per the treatment schedule. 50 per cent of N and K along with full dose of P were applied as basal. Remaining half of the N and K were applied as top dressing at 25 days after sowing. All the agronomic practices were carried out uniformly to raise the crop.

To record various bio-metric observations on baby com, a sample consisting of five plants was selected at random. Plant height was measured from base to tip of the terminal of the main stem and expressed in cm. Leaf Area Index was measured by using the formula. The uprooted plants were dried under shade and oven dried at 75°C till a constant weight was obtained and expressed in kg ha^{-1}. Length, diameter and weight of cobs from the tagged plants were measured and weight were calculated. Ratio between green cob weights to baby com weight was worked out and expressed in percentage. Harvested cobs from the net plot were weighed and cob yield was recorded from individual plots and expressed in kg ha^{-1}. After harvest of the cobs, the baby com stalk were harvested from net plot area, weighed and expressed as green fodder yield (t ha^{-1}). The data subjected to statistical analysis.

Results and Discussion

Growth Attributes

Nitrogen Levels

During the study, two levels of nitrogen *viz.*, 150 and 200 kg ha^{-1} were evaluated and during *Kharif* 2001, the application of 200 kg N ha^{-1} had registered significantly higher plant height, LAI and DMP. Highest plant and LAI recorded under 200 kg ha^{-1} might be ascribed to the fact that the treatment enhanced the availability of N which is required for the vegetative growth, throughout the crop period. Increase in plant height resulted on account of cumulative effect of need based, gradual and continuous supply of N at the time of its critical requirement by the crop, efficient utilization of supplied N (Gaur *et al.*, 1992). The increase was due to better utilization of applied nutrients, increased sink capacity and higher nutrient uptake of crops. Increase in DMP due to 200 kg ha^{-1} N application is due to the increase in plant height and leaf area index and continuous availability of the fertilizer nitrogen and less nutrient losses which might have resulted in higher vegetative growth and thereby enhanced the total drymatter production. The yield potential of baby com is decided by growth and yield components as reported in the present study. Higher yield of baby com under the higher level of applied nitrogen was also reported by Thakur *et al.*, 1995.

Crop Density

Crop geometry, a non-monetary input play major role upon introduction of new plant types. Optimum space requirement of the crop to utilize the available resources such as moisture and nutrients and to harvest the maximum possible solar radiation for the formation and translocation of photosynthates is one of the pre-requisites to attain higher productivity.

The effect of crop geometry on growth characters of baby com was significant. The experimental results revealed that the different treatments maintenance of plant density 45 × 20cm (D_3) spacing (1, 11, 111 plants ha^{-1}) and 200 kg N ha^{-1} (N_2) combination have a profound influence on the growth and yield attributes of baby com at all the stages of crop growth *viz.*, 25th, 40th, 50th and harvest stage. The crop density had significant influence on growth parameters of baby com during *Kharif* 2001. Baby com raised at 45 × 20cm (D_3) noticed taller plants compared to D_4 and this might be due the competition for light due to closer plant to plant space. Paulpandi *et al.* (1998) opined that wider row spacing had taller plants due to better availability of resources. Taller plants under wider spaced planting were also documented by Mazaheri and Akbary (1998). Significantly higher LAI at respective stages of the investigation were noticed under 45 × 20cm compared to 45 × 15 cm, 45 × 25 cm and 45 × 30cm. Wider space availability between the rows and the closer intra-rows might have increased the root spread which eventually utilized the resources such as water, nutrients, space and light very effectively as compared to 25 and 30cm intra-row spacing. Better utilization of available resources must have increased the functional leaves and in turn enhanced the LAI. Similar results were also reported by Abo-shetaia *et al.*, 2002 and Sahoo and Panda (1999). The crop geometry 45 × 20cm registers significantly higher DMP as compared to other spacing at all the stages of baby corn. The utilization of available resources in greater extent favoured higher plant height and LAI which combinely caused the increase of DMP at 45 × 20cm spacing as compared to other intra-row spacing. The results are in conformity with reports of Lucas (1986), Tetio-Kagho and Gardner (1988) and Jiotode *et al.* (2002).

Yield Attributes and Yield

Nitrogen Levels

The application of 200 kg ha^{-1} produced significantly higher yield and yield parameters than lower level of 150 kg N ha^{-1}. It may be due to higher response of baby corn to N levels since it is a exhaustive crop and the nitrogen utilization pattern increased from seeding to knee-high stage and reaches to the peak at the tasseling stage during which the baby corn removes 4–5 kg ha^{-1}day^{-1}. It is interesting to note that the response of applied nitrogen is highest in poor fertility condition. In case of baby corn composite and hybrids, the economical yield of baby corn has been obtained even up to the application of 200 kg ha^{-1} and lower levels of N (50 and 100 kg ha^{-1}) did not responsed well on the yield attributes and yield as reported by Thakur *et al.* (1995). This results fall in line with the findings of Rajendran and Sunder Singh (1999) who reported that application of 180 kg N ha^{-1} produced significantly higher green corn yield than lower levels of 120 kg ha^{-1}. With respect to green fodder yield, the spacing of 45 × 15cm with higher population in combination with application of 200 kg N ha^{-1} registered the highest stalk yield. This results corroborates the finding of Sukanaya *et al.*, 1999.

Crop Geometry

During the present investigation, the number of cobs per plant did not differ due to the alteration of crop geometry. Since the number of cons planr1 is the genetic make-up, it could not easily be altered by the agronomic management practices. In general the yield attributes such as length and weight of cobs and corns were significantly higher in the treatment 45 × 20cm due to the alteration of crop density (Table 24.1). The enhanced yield components might be due to increased LAI, leading to higher photosynthetic rate and accumulation of more assimilates which in turn increased the sink size. During the present investigation, the results revealed that the influence of different spacing in baby com opined that yields were higher with 45 × 20cm where as 45 × 15cm produced the maximum green

Table 24.1: Effect of N Levels and Plant Density on the Growth Attributes of Baby Corn Var. Co. BC-1

Treatments	*Kharif 2001*		
	Plant Height (cm)	*LAI*	*DMP (kg ha^{-1})*
N_1D_1	169.70	3.20	6845.0
N_1D_2	192.30	3.65	7769.0
N_1D_3	195.70	3.62	7690.0
N_1D_4	185.70	3.40	7299.0
N_2D_1	178.00	3.40	7024.0
N_2D_2	200.70	3.71	7894.0
N_2D_3	209.00	3.87	8267.0
N_2D_4	193.00	3.46	7404.3
SEd (N)	0.48	0.01	13.52
CD 5 per cent (N)	1.46	0.02	41.11
SEd (D)	0.68	0.01	19.12
CD (0.05) (D)	2.06	0.02	58.14
SEd (ND)	0.96	0.01	27.04
CD (0.05) (ND)	2.91	0.03	82.22

Table 24.2: Effect of N Levels and Plant Density on the Yield Attributes of Baby Corn Var. Co. BC-1

Treatments	*Kharif 2001*				
	Cob Length (cm)	*Corn Length (cm)*	*Cob Weight (g)*	*Corn Weight (g)*	*Number of Cob Plant^{-1}*
N_1D_1	22.70	12.61	47.25	9.77	2.21
N_1D_2	24.40	13.68	51.22	10.54	2.26
N_1D_3	26.20	14.30	51.37	11.12	2.33
N_1D_4	23.67	13.00	48.32	10.46	2.22
N_2D_1	25.30	13.22	49.25	10.11	2.30
N_2D_2	27.37	14.52	52.43	11.68	2.42
N_2D_3	28.50	14.93	55.82	12.22	2.51
N_2D_4	26.60	14.48	51.44	11.31	2.35
SEd (N)	0.06	0.04	0.07	0.02	0.01
CD 5 per cent (N)	0.18	0.12	0.21	0.07	0.02
SEd (D)	0.09	0.06	0.10	0.03	0.01
CD (0.05) (D)	0.26	0.17	0.30	0.10	0.03
SEd (ND)	0.12	0.08	0.14	0.05	0.01
CD (0.05) (ND)	0.37	0.24	0.43	0.15	0.04

Table 24.3: Effect of N Levels and Plant Density on the Yield and Cob-Corn Ratio of Baby Corn Var. CO. BC-1

Treatments	*Kharif 2001*		
	Cob Yield (kg ha^{-1})	*Fodder Yield (t ha^{-1})*	*Cob-Corn Ratio*
N_1D_1	6803.0	27.50	5.11
N_1D_2	7318.0	30.30	5.18
N_1D_3	7458.0	32.40	5.19
N_1D_4	6944.0	28.90	5.14
N_2D_1	7383.0	30.80	5.10
N_2D_2	7607.7	32.80	5.21
N_2D_3	8426.0	34.10	5.28
N_2D_4	7525.0	35.50	5.14
SEd (N)	28.05	0.17	0.01
CD 5 per cent (N)	85.27	0.53	0.01
SEd (D)	39.66	0.24	0.01
CD (0.05) (D)	120.60	0.74	0.01
SEd (ND)	56.09	0.35	0.01
CD (0.05) (ND)	170.55	1.05	0.02

fodder yield. The increase was due to effective utilization of applied nutrients, increased sink capacity. The yield potential of baby corn is decided by the growth and yield components. This was reflected in the present study. The strong positive and significant correlation of LAI and DMP at different stages, weight of cobs and corns with cob yield was also noted. Similar results were obtained by Thakur *et al.*, 1995, Sahoo and Ponda, 1999. Turger (2000) reported similar result of response to intra row spacing with 20 cm recording the higher cob yield than 15, 20 and 30 cm intra-row spacing. Similar results of higher monetary return and B : C ratio with narrow row spacing as compared to wider spacing was reported by Singh *et al.* (1992).

Conclusion

The present investigation revealed that raising baby com in an optimum spacing of 45 × 20cm with the application of 200 kg N ha^{-1} recorded the highest yield which in turn resulted in higher monetary returns as compared to other crop geometry and N levels.

References

Abo-Shetaia, A.M., El-Gowad, A.A.A., Mohamad, A.A. and Abdel-Wahab, T.T., 2002. Yield dynamics in four Yellow maize (*Zea mays* L.) hybrids. *Arab Univ. J. Agric. Sci.*, 10(1): 205–219. [Cited: Field Crop Abstract, 2002. 55(10): 1258].

Bar-Zur, A.R. and Saadi, H., 1990. Prolific maize hybrids for baby corn. *J. Hort. Sci.*, 65(1): 97–100.

Gaur, B.L., Mansion, P.R. and Gupta, D.C., 1992. Effect of nitrogen levels and their splits on yield of winter maize (*Zea mays* L.). *Indian J. Agron.*, 37(4): 816–817.

Jiotode, D.J., Lamb, D.L. and Dhawad, C.S., 2002. Growth parameters and water use studies of maize influenced by irrigation levels and row spacings. *Crop Res.*, 24(2): 292–295.

Lucas, E.D., 1986. Effect of density and nitrogen fertilizer on growth and yield of maize in Nigeria. *J. Agric. Res.*, 107: 573–578.

Mazaheri, D. and Akbary, G.A., 1998. Effect of plant density and different amount of nitrogen and potassium fertilizers on vegetative growth and forage yield of maize. *Seed Plant*, 14(1): 32–49.

Pandey, S.N. and Sinha, B.K., 1998. *Plant Physiology*. Vikas Publishing House Pvt. Ltd., New Delhi, p. 105.

Paulpandi, V.K., Solaiyappan, U. and Palaniappan, S.P., 1998. Effect of plant geometry and fertilizer levels on yield and yield attributes in irrigated sorghum. *Indian J. Agric. Res.*, 33(2): 125–128.

Rajenderan, K. and Singh, S.D. Sundar, 1999. Effect of irrigation regimes and nitrogen levels on yield, water requirements, water use efficiency and quality of baby corn (*Zea mays*, L.). *Agric. Sic. Digest.*, 19(3): 159–161.

Sahoo, S.C. and Panda, M.M., 1999. Determination of optimum plant geometry for baby corn (*Zea mays*). *Indian J. Agric. Sci.*, 69(9): 664–665.

Singh, C., 1998. *Modern Techniques of Raising Field Crops*. Oxford and IBH Publishing Co. Pvt. Ltd., New Delhi, pp. 74–94.

Singh, K.K., Bojai, R.P. and Sisodia, R.J., 1992. Effect of sowing dates and plant population on yield of maize. *Curr. Res.*, 21(2): 24–25.

Sukanya, T.S., Nanjappa, H.V. and Ramachandarappa, K., 1999. Effect of spacing on growth, development and yield of baby corn (*Zea mays* L.) Varieties. *Karnataka J. Agric. Sci.*, 12(1–4): 10–14.

Thakur, D.R., Kharawra, A.C. and Prakash, Om, 1995. Effect of Nitrogen and plant spacing on growth, development and yield of baby corn (*Zea mays* L.). *Himachal J. Agric. Res.*, 21(1–2): 5–10.

Tetio-Kagho, F. and Gardner, F.P., 1988. Response of maize to population density I. Canopy development, light relationship and vegetative growth. *Agron. J.*, 80: 930–935.

Turget, I.I., 2000. Effect of plant population and nitrogen doses on fresh ear and yield components of Sweet com (*Zea mays* saccharata Sturt.) growth under bursa conditions. *Turkish J. Agric. For.*, 24(3): 341–347.

Yodpet, 1979. Studies on sweet com as potential cob corn (*Zea mays* L.). *Ph.D. Thesis*, University of the Philippines, Philippines.

Chapter 25

Analysis of Weibull Deteriorating Inventory with Quadratic Demand Rate

Sudhir K. Sahu[1] *and Sirish K. Dash*[2]

[1]*P.G. Department of Statistics,* [2]*P.G. Department of Mathematics*
Sambalpur University, Burla – 768 019, Orissa, India

ABSTRACT

This paper investigates inventory production systems where items follow Weibull deterioration. The objective is to develop an optimal policy that minimizes total average cost. The quadratic demand technique is applied to control the problem in order to determine the optimal production policy, holding cost and cost of deterioration. Sensitivity analysis is conducted to study the effect of the cost parameters on the objective function.

***Keywords**: Production, Inventory, Deterioration, Shortage, and Replenishment.*

Introduction

Goods deteriorate and their value reduces with time. Electronic products may become obsolete as technology changes. Fashion tends to depreciate the value of clothing over time. Batteries die out as they age. The effect of time is even more critical for perishable goods such as foodstuff and cigarettes. The effect of deterioration and time/age is that the classical inventory model has to be readjusted (Heng and Labban, 1991).

In general, deterioration is defined as decay, damage, spoilage, evaporation, obsolesce, pilferage, loss of utility or loss of marginal value of a commodity that results in decrease of usefulness from the original one. The decrease or loss of utility due to decay is usually a function of the on-hand inventory. It is reasonable to note that a product may be understood to have lifetime, which ends when utility reaches zero.

The continuously decaying/deterioration of items is classified as age-dependent ongoing deterioration, and age-independent ongoing deterioration. Blood, fish, strawberry are some of the examples of the former while alcohol, gasoline and radioactive chemical and grain products are examples of the latter (Wee, 1993).

Haiping and Wang (1990) developed an economic policy model for deteriorating items with time proportional demand. Donaldson (1977) derived an analytical solution to the problems of obtaining the optimal number of replenishments and the optimal replenishment times of an EOQ model with a linearly time dependent demand pattern, over a finite time horizon. Zangwill (1966) developed a discrete-in-time dynamic programming algorithm to solve an inventory model by allowing the inventory levels to be negative where the demand pattern is time dependent. Following the approach of Donaldson (1977), Murdeshwar (1988) has tried to derive an exact solution for a finite horizon inventory model to obtain the optimal number of replenishments, optimal replenishment times and the optimal times at which the inventory level falls to zero, assuming the demand rate to be linearly time dependent and shortages. Hamid (1989) presented a heuristic model for determining the ordering schedule when inventory items are subject to deterioration and demand changes linearly over time and obtained an optimal replenishment cycle length. Goswami and Chaudhury (1991) presented an EOQ model for deteriorating items with shortage and linear trend in demand. Bradshaw and Erol (1980), published a paper in which they derived unbounded control policies for a class of linear time invariant production inventory systems.

This paper investigates inventory-production systems where items follow Weibull deterioration. The objective is to develop an optimal policy that minimizes the cost associated with inventory and production rate. The quadratic demand technique is applied to control the problem in order to determine the optimal production policy. Sensitivity analysis is conducted to study the effect of the cost parameters on the objective function.

Assumptions and Notations

1. The demand rate is assumed to be $R(t) = a + bt + ct^2$, a, b and c being constants.
2. The production rate Say $k = rR(t)$, where $r > 1$.
3. A fraction $z(t)$ of the on-hand inventory deteriorates per unit time, where $z(t) = \alpha\beta t^{b-1}$, $0 < \alpha < 1$, $t > 1$ and $\beta \geq 21$.
4. The lead-time is zero and shortages are not allowed.
5. Unit holding cost C_1 per unit time and unit deterioration cost C_3 per unit time are known and constants.
6. C is the total average cost for the production cycle and S is the stock level reached in the cycle.
7. The set up cost is not considered in this model because it is taken to be fixed for the whole cycle time.
8. Planning horizon is finite.

Mathematical Formulation and Solution

Let q be the inventory level at any time $t(0 \leq t \leq t_2)$. The differential equations governing the system in the interval $(0, t_2)$ are

$$\frac{dq}{dt} + z(t)q = k - R(t), \quad 0 \leq t \leq t_1 \tag{1}$$

$$\frac{dq}{dt} + z(t)q = -R, \quad t_1 \leq t \leq t_2 \tag{2}$$

The stock level initially is zero. Production begins just after $t = 0$, continues upto $t = t_1$ and stops as soon as the stock level becomes S. Then the inventory level decreases due to demand and deterioration both till it becomes zero at $t = t_2$. The cycle then repeats itself. Our objective is to determine the optimum values of S, C, t, and t_2. The intensity of deterioration is very low initially but it increases with time. However, it remains bounded for $t >> 1$

U sing the values of $R(t)$ and $z(t)$, the two equations (1) and (2) take the form

$$\frac{dq}{dt} + \alpha\beta t^{\beta-1}q = (r-1)(a + bt + ct^2), \quad 0 \leq t \leq t_1 \tag{3}$$

and
$$\frac{dq}{dt} + \alpha\beta t^{\beta-1}q = -(a + bt + ct^2), \quad t_1 \leq t \leq t_2 \tag{4}$$

The solution of equation (3) with initial conditions is

$$qe^{\alpha t^{\beta}} = (r-1)\int_0^t (1 - \alpha t^{\beta}) + (a + bt + ct^2)dt$$

or
$$q = (r-1)\left(at + \frac{bt^2}{2} + \frac{ct^3}{3} + \frac{a\alpha t^{\beta+1}}{\beta+1} - \frac{b\alpha\beta t^{\beta+2}}{2(\beta+2)} - \frac{c\alpha\beta t^{\beta+3}}{3(\beta+3)}\right) \tag{5}$$

Neglecting the powers of α greater than 1.

Similarly, the solution of equation (4) also is (neglecting the powers of α greater than 1)

$$q\exp(\alpha t^{\beta}) = C' - \int (a + by + ct^2)\exp(\alpha t^{\beta})dt$$

or
$$q = C'(1 - \alpha t^{\beta}) - \left(at + \frac{bt^2}{2} + \frac{ct^3}{3} - \frac{a\alpha t^{\beta+1}}{\beta+1} - \frac{b\alpha\beta t^{\beta+2}}{2(\beta+2)} - \frac{c\alpha\beta t^{\beta+3}}{3(\beta+3)}\right) \tag{6}$$

for $t = t_1, q = S$

$$\therefore S = C'(1 - \alpha t_1^{\beta}) - \left(at_1 + \frac{bt_1^2}{2} + \frac{ct_1^3}{3} - \frac{a\alpha t_1^{\beta+1}}{\beta+1} - \frac{b\alpha\beta t_1^{\beta+2}}{2(\beta+2)} - \frac{c\alpha\beta t_1^{\beta+3}}{3(\beta+3)}\right) \tag{7}$$

From (6) and (7) we get the relation

$$q = S(1-\alpha t_1^{\beta} - \alpha t^{\beta}) + a\left(t_1 - t + \frac{\alpha\beta t^{\beta+1}}{(\beta+1)} - \frac{\alpha t\beta t_1^{\beta+1}}{2(\beta+2)} - \alpha t_1 t^{\beta}\right)$$

$$+ b\left(\frac{(t_1^2 - t^2}{2} + \frac{\alpha\beta t^{\beta+2}}{2(\beta+2)} + \frac{\alpha t_1^{\beta+2}}{\beta+2} - \frac{\alpha t_1^2 t^{\beta}}{2}\right)$$

$$+ c\left(\frac{(t_1^3 - t^3}{3} + \frac{\alpha\beta t^{\beta+3}}{3(\beta+3)} + \frac{\alpha t_1^{\beta+3}}{\beta+3} - \frac{\alpha t_1^3 t^{\beta}}{3}\right) \tag{8}$$

Using the condition $q = 0$ for $t = t_2$ in equation (8), we get

$$s = a\left(t_2 - t_1 + \frac{\alpha t_1^{\beta+2}}{(\beta+1)} + \frac{\alpha t_1^{\beta+1}}{(\beta+1)} - \alpha t_1^{\beta} t_2\right) + b\left(\frac{(t_2^2 - t_1^2}{2} + \frac{\alpha t_1^{\beta+2}}{(\beta+2)} + \frac{\alpha t_1^{\beta+2}}{(\beta+2)} - \frac{\alpha t_1^{\beta} t_2^2}{2}\right)$$

$$+ c\left(\frac{(t_2^3 - t_1^3}{3} + \frac{\alpha\beta t_2^{\beta+3}}{(\beta+3)} + \frac{\alpha t_1^{\beta+3}}{(\beta+3)} - \frac{\alpha t_1^{\beta} t_2^3}{3}\right) \tag{9}$$

Now the average holding cost becomes

$$H(t) = \frac{C_1}{t_2}\left(\int_0^{t_1} q(t)dt + \int_{t_1}^{t2} q(t)dt\right)$$

$$= \frac{C_1}{t_2}\left[\int_0^{t_1}(r-1)\left(at + \frac{bt^2}{2} + \frac{ct^3}{3} - \frac{a\alpha t^{\beta+1}}{\beta+1} - \frac{b\alpha\beta t^{\beta+2}}{2(\beta+2)} - \frac{c\alpha\beta t^{\beta+3}}{3(\beta+3)}\right)dt\right.$$

$$+ \int_{t_1}^{t_2}\left(S(1+\alpha t_1^{\beta} - \alpha t^{\beta}) + a\left(t_1 - t + \frac{\alpha\beta t^{\beta+1}}{(\beta+1)} + \frac{\alpha\beta t_1^{\beta+1}}{(\beta+1)} - \alpha t_1 t^{\beta}\right)\right.$$

$$\left.\left. + b\left(\left(\frac{(t_1^2 + t^2)}{2} + \frac{\alpha\beta t^{\beta+2}}{2(\beta+2)} + \frac{\alpha t_1^{\beta+2}}{\beta+2} - \frac{\alpha t_1^2 t^{\beta}}{2} + c\left(\frac{(t_1^3 + t^3)}{3} + \frac{\alpha\beta t^{\beta+3}}{3(\beta+3)} + \frac{\alpha t_1^{\beta+3}}{\beta+3} - \frac{\alpha t_1^3 t^{\beta}}{3}\right)\right)\right)dt\right]$$

Now substituting the value of S from (9) and simplifying we get

$$H(t) = \frac{C_1}{t_2}\left(\frac{a}{2}\left(rt_1^2 - \frac{2\alpha\beta r t_1^{\beta+2}}{(\beta+1)(\beta+2)} + t_2^2 - 2t_1t_2 + \frac{2\alpha\beta t_2^{\beta+2}}{(\beta+1)(\beta+2)} - \frac{2\alpha t_2^{\beta+1}t_1}{(\beta+1)} + \frac{2\alpha t_1^{\beta+1}t_2}{(\beta+1)}\right)\right.$$

$$+\frac{b}{6}\left(rt_1^3+2t_2^3-3t_1t_2^2+\frac{6\alpha\beta_2^{\beta+2}}{(\beta+1)(\beta+3)}-\frac{6\alpha t_1t_2^{\beta+2}}{(\beta+2)}+\frac{3\alpha t_1^{\beta+1}t_2^2}{(\beta+1)}-\frac{3\alpha\beta rt_1^{\beta+3}}{(\beta+2)(\beta+3)}\right)$$

$$+\frac{c}{12}\left(rt_1^4+3t_2^4-4t_1t_2^3+\frac{12\alpha\beta_2^{\beta+4}}{(\beta+1)(\beta+4)}-\frac{12\alpha t_1t_2^{\beta+3}}{(\beta+3)}+\frac{4\alpha t_1^{\beta+1}t_2^3}{(\beta+1)}-\frac{4\alpha\beta rt_1^{\beta+4}}{(\beta+4)(\beta+3)}\right)\Bigg) \quad (10)$$

The average cost due to deterioration in the total cycle time is

$$d(t)=\frac{C_3}{t_2}\left[r\int_0^{t_1}(a+bt+ct^2)dt-\int_{t_1}^{t_2}(a+bt+ct^2)dt\right]$$

$$=\frac{C_3}{t_2}\left[a(rt_1-t_2)+\frac{b}{2}(rt_1^2-t_2^2)+\frac{c}{3}(rt_1^3-t_2^3)\right] \quad (11)$$

From (10) and (11) the total average cost of the inventory is

$$C=\frac{C_1}{t_2}\left(\frac{a}{2}\left(rt_1^2-\frac{2\alpha\beta rt_1^{\beta+2}}{(\beta+1)(\beta+2)}+t_2^2-2t_1t_2+\frac{2\alpha\beta t_2^{\beta+2}}{(\beta+1)(\beta+2)}-\frac{2\alpha t_2^{\beta+1}t_1}{(\beta+1)}+\frac{2\alpha t_1^{\beta+1}t_2}{(\beta+1)}\right)\right.$$

$$+\frac{b}{6}\left(rt_1^3+2t_2^3-3t_1t_2^2+\frac{6\alpha\beta_2^{\beta+2}}{(\beta+1)(\beta+3)}-\frac{6\alpha t_1t_2^{\beta+2}}{(\beta+2)}+\frac{3\alpha t_1^{\beta+1}t_2^2}{(\beta+1)}-\frac{3\alpha\beta rt_1^{\beta+3}}{(\beta+2)(\beta+3)}\right)$$

$$+\frac{c}{12}\left(rt_1^4+3t_2^4-4t_1t_2^3+\frac{12\alpha\beta_2^{\beta+4}}{(\beta+1)(\beta+4)}-\frac{12\alpha t_1t_2^{\beta+3}}{(\beta+3)}+\frac{4\alpha t_1^{\beta+1}t_2^3}{(\beta+1)}-\frac{4\alpha\beta rt_1^{\beta+4}}{(\beta+4)(\beta+3)}\right)\Bigg)$$

$$+\frac{C_3}{t_2}\left[a(rt_1-t_2)+\frac{b}{2}(rt_1^2-t_2^2)+\frac{c}{3}(rt_1^3-t_2^3)\right] \quad (11)$$

By putting $t_1=zt_2$ (where $0<z<1$) in equation (12), we get

$$C=\frac{C_1}{t_2}\left(\frac{a}{2}\left(rz^2t_2^2-\frac{2\alpha\beta rz^{\beta+2}t_2^{\beta+2}}{(\beta+1)(\beta+2)}+t_2^2-2zt_2^2+\frac{2\alpha\beta t_2^{\beta+2}}{(\beta+1)(\beta+2)}-\frac{2\alpha zt_2^{\beta+1}t_1}{(\beta+1)}+\frac{2\alpha z^{\beta+1}t_2^{\beta+2}}{(\beta+1)}\right)\right.$$

$$+\frac{b}{6}\left(rz^3t_2^3+2t_2^3-3zt_2^3+\frac{6\alpha\beta_2^{\beta+3}}{(\beta+1)(\beta+3)}-\frac{6\alpha zt_2^{\beta+3}}{(\beta+2)}+\frac{3\alpha z^{\beta+1}t_2^{\beta+3}}{(\beta+1)}-\frac{3\alpha z^{\beta+3}t_2^{\beta+3}}{(\beta+2)(\beta+3)}\right)$$

$$+\frac{c}{12}\left(rz^4t_2^4+3t_2^4-4zt_2^4+\frac{12\alpha\beta t_2^{\beta+4}}{(\beta+1)(\beta+4)}-\frac{12\alpha zt_2^{\beta+4}}{(\beta+3)}+\frac{4\alpha z^{\beta+1}t_2^{\beta+4}}{(\beta+1)}-\frac{4\alpha\beta rz^{\beta+4}t_2^{\beta+4}}{(\beta+4)(\beta+3)}\right)\Bigg)$$

$$+\frac{C_3}{t_2}\left[a(rzt_2-t_2)+\frac{b}{2}(rz^2t_2^2-t_2^2)+\frac{c}{3}(rz^3t_2^3-t_2^3)\right] \tag{13}$$

For calculating the optimum value of e we differentiate it partially *w.r.t.t2* and equate them to zero. Thus we get the following equation.

$$\frac{dC}{dt_2}=C_1\left(\frac{a}{2}\left(rz^2-\frac{2\alpha\beta z^{\beta+2}t_2^{\beta}}{(\beta+2)}+1-2z+\frac{2\alpha\beta t_2^{\beta}}{(\beta+2)}-2\alpha zt_2^{\beta}+2\alpha z^{\beta+1}t_2^{\beta}\right)\right.$$

$$+\frac{b}{6}\left(2rz^3t_2+4t_2-6zt_2+\frac{6\alpha\beta(\beta+2)t_2^{\beta+1}}{(\beta+1)(\beta+3)}-6\alpha zt_2^{\beta+1}+\frac{3\alpha z^{\beta+1}(\beta+2)t_2^{\beta+1}}{(\beta+1)}-\frac{3\alpha\beta z^{\beta+3}t_2^{\beta+1}}{(\beta+3)}\right)$$

$$\left.+\frac{c}{12}\left(3rz^4t_2^2+9t_2^2-12zt_2^2+\frac{12\alpha\beta(\beta+3)t_2^{\beta+2}}{(\beta+1)(\beta+4)}-12\alpha zt_2^{\beta+2}+\frac{4\alpha z^{\beta+1}(\beta+3)t_2^{\beta+2}}{(\beta+1)}-\frac{4\alpha\beta rz^{\beta+4}t_2^{\beta+2}}{(\beta+4)}\right)\right)$$

$$+C_3\left[\frac{b}{2}(rz^2-1)+\frac{2c}{3}(rz^3-1)_{t_2}\right] \tag{14}$$

This equation gives us the optimum value of Iz which, when substituted equation (13), give the minimum total average cost, provided $\frac{d^2C}{dt_2^2}>C$. Equation (14) is highly non-linear in t_2 and cannot be solved analytically. This equation, therefore, can be solved by some suitable numerical method like Newton-Raphson, and optimal value of t_2 can be obtained. This optimal value of t_2 gives the minimum cost of the system in question. We have solved this equation on computer for a set of values of the parameters with the help of Newton-Raphson method. A numerical example is given below as an illustration.

Example 1

Let $\alpha = 0.0001$, $\beta = 2.0$, $r = 2.0$, $z = 0.7$, $C_1 = 5.0$, $C_3 = 60$, $a = 250.0$, $b = 10$, $c = 10$ in suitable units. The solution for optimal values of t_1 and t_2 is $t_1^* = 3.2827$, $t_2^* = 4.6896$, which gives minimum average cost $C^* = 6706.7988$.

Following are a number of tables representing the optimal values of t_1, t_2 and C as also the no-production interval $t_2 - t_1$.

Table 25.1
Values of t_1, t_2 and C for different values of α taking $\beta = 2.0$, $r = 2.0$, $z = 0.7$, $C_1 = 5.0$, $C_3 = 60$, $a = 250.0$, $b = 10$, $c = 10$

α	t_1	t_2	C	$t_2 - t_1$
0.00010	3.2827	4.6896	6706.7988	1.4069
0.00015	3.2808	4.6869	6714.9443	1.4061
0.00020	3.2788	4.6841	6723.0590	1.4053
0.00025	3.2769	4.6814	6731.1474	1.4045
0.00030	3.2750	4.6787	6739.2070	1.4036

Table 25.2
Values of t_1, t_2 and C for different values of β taking $\alpha = 0.0001$, $r = 2.0$, $z = 0.7$, $C = 5.0$, $C_3 = 60$, $a = 250.0$, $b = 10$, $c = 10$

β	t_1	t_2	C	$t_2 - t_1$
1.0	3.2853	4.6934	6691.4291	1.4080
1.5	3.2842	4.6918	6695.0712	1.4075
2.0	3.2827	4.6896	6706.7988	1.4068
2.5	3.2814	4.6878	6740.8789	1.4063

Table 25.3
Values of t_1, t_2 and C for different values of r taking $\alpha = 0.0001$, $\beta = 2.0$, $a = 0.7$, $C_1 = 5.0$, $C_3 = 60$, $a = 250.0$, $b = 10$, $c = 10$

r	t_1	t_2	C	$t_2 - t_1$
1.7	1.0369	1.4813	3011.3706	0.4444
1.8	1.4532	2.0761	4165.8305	0.6228
1.9	2.0527	2.9325	5376.1621	0.8797
2.0	3.2827	4.6896	6706.7988	1.4068
2.2	4.1199	5.8857	9868.3427	1.7658

Table 25.4
Values of t_1, t_2 and C for different values of a taking $\alpha = 0.0001$, $\beta = 2.0$, $z = 0.7$, $C_1 = 5.0$, $C_3 = 60$, $r = 2.0$, $b = 10$, $c = 10$

α	t_1	t_2	C	$t_2 - t_1$
230.0	2.8007	4.0010	6094.0712	1.2003
240.0	3.0253	4.3219	6397.3735	1.2966
250.0	3.2827	4.6896	6706.7988	1.4069
260.0	3.5975	5.1394	7024.1030	1.5419
270.0	4.0339	5.7628	7353.4199	1.7289

Table 24.5
Values of t_1, t_2 and C for different values of C_1 taking $\alpha = 0.0001$, $\beta = 2.0$, $z = 0.7$, $a = 250$, $C_3 = 60$, $r = 2.0$, $c = 10$, $b = 10$

C_1	t_1	t_2	C	t_2-t_1
4.0	2.0257	2.8938	6386.6791	0.8681
5.0	3.2827	4.6896	6706.7988	1.4069
6.0	4.7006	6.7152	7400.9125	2.0146
7.0	5.7526	8.2180	8668.0664	2.4654
8.0	7.0350	10.0500	11078.3515	3.0150

Table 24.6
Values of t_1, t_2 and C for different values of C_2 taking $a = 0.0001$, $\beta = 2.0$, $z = 0.7$, $a = 250$, $C_1 = 5$, $r = 2.0$, $b = 10$, $c = 10$

C_2	t_1	t_2	C	t_2-t_1
60.0	3.2827	4.6896	6706.7988	1.4069
65.0	2.6380	3.7687	7104.9282	1.1307
70.0	2.2799	3.2570	7535.8500	0.9771
75.0	2.0257	2.8939	7983.3491	0.8682

Table 24.7
Values of t_1, t_2 and C for different values of b taking $\alpha = 0.0001$, $\beta = 2.0$, $z = 0.7$, $a = 250$, $C_1 = 5$, $r = 2.0$, $C_3 = 60$, $c = 10$

b	t_1	t_2	C	t_2-t_1
08.0	3.1437	4.4910	6689.7036	1.3473
10.0	3.2827	4.6896	6706.7988	1.4069
12.0	3.4539	4.9342	6726.3891	1.4803
14.0	3.6927	5.2754	6749.9902	1.5827
16.0	4.1103	5.8719	6782.2915	1.7616

Table 24.8
Values of t_1, t_2 and C for different values of c taking $\alpha = 0.0001$, $\beta = 2.0$, $z = 0.7$, $a = 250$, $C_1 = 5$, $r = 2.0$, $C_3 = 60$, $b = 10$

c	t_1	t_2	C	t_2-t_1
08.0	3.6306	5.1867	6868.5703	1.5561
10.0	3.2827	4.6896	6706.7988	1.4069
12.0	2.3296	3.3280	6662.1054	0.9984
14.0	1.8543	2.6490	6600.9335	0.7947
16.0	1.5458	2.2083	6543.3886	0.6625

A sensitivity analysis could also be employed for this model to see the effect of changes in the various parameters on the average total cost of the system. However we have given a tabular representation.

We have discussed here the model with deterioration of items. The shortages are not taken into consideration. The deterioration rate is time dependent but it is not taken to be linear function of time. However, for $\beta = 2$ it becomes directly proportional to time t and for $\beta = 1$ it becomes constant. The production rate is taken to be dependent on demand rate, which in turn is itself a quadratic function of time. That is, with the increase in time the demand increases and so also there is a proportional increase in the rate of production. The case of backlogging of shortages can be studied further.

Conclusion

1. Increase in the values of either of the parameters α, β, a decrease the values of t_1, t_2 but increases the value of C.
2. Increase in the value of r increases the value of t_1, t_2 and C.
3. Increase in the value of holding cost C_1 increases the value of the cost C, t_1, t_2.
4. Increase in the value of deterioration cost C_3 increases the value of the cost C.
5. However the values of t_1, t_2 decrease.
6. Increase in the values of a and b increases the value of C, t_1, t_2. Increase in the value of c, decreases the value of C, t_1, t_2.
7. Keeping these variations in mind of the decision maker of the inventory system can control the parameters so as to optimize the objective function. The decision maker may control particularly the holding cost and the cost of deterioration for minimizing the total average cost.

References

Bradshaw, A. and Erol, Y., 1980. Control policies for production inventory systems with bounded input. *Int. J. Sys. Sci.*, 11: 947–959.

Donaldson, W.A., 1977. Inventory replenishment policy for a linear trend in demand: An analytical solution. *Opl Res. Q.*, 28: 663–670.

Goswami, A. and Chaudhuri, K., 1991. An EOQ model for deteriorating items with shortages and a linear trend in demand. *J. Opl. Res. Soc.*, 42: 1105–1110.

Haiping, U. and Wang, H., 1990. An economic ordering policy model for deteriorating items with time proportional demand. *Eur. J. Oper. Res.*, 46: 21–27.

Hamid, B., 1989. Replenishment schedule for deteriorating items with time proportional demand. *J. Opel. Res. Soc.*, 40: 75–81.

Heng, K., Labban, J. and Linn, R., 1991. An order level for deteriorating items with partial back ordering. *Comput. Ind. Eng.*, 20: 187–197.

Murdeshwar, T.M., 1988. Inventory replenishment policy for linearly increasing demand considering shortages: An optimal solution. *J. Opl. Res. Soc.*, 39: 687–692.

Wee, H., 1993. Economic production lot size model for deteriorating items with partial back ordering. *Comput. Ind. Eng.*, 24: 449–458.

Zangwill, W.I., 1966. A deterministic multi-period production scheduling model with backlogging. *Mgmt. Sci.*, 13: 105–119.

Chapter 26

Effect of Different Levels of K on Dry Matter Yield and Uptake Efficiency of K by Different FCV Tobacco Varieties Grown in Southern Transition Zone of Karnataka

K.T. Gurumurthy, H.C. Prakasha, N.S. Marvakar and C.J. Sridhara

Department of Soil Science and Agricultural Chemistry, College of Agriculture, Navile, Shimoga – 577 204, Karnataka, India

ABSTRACT

A pot culture experiment was conducted at College of Agriculture, Navile, Shimoga. The results indicated that the variety Thrupthi (KST-19) recorded significantly higher dry matter accumulation (68.33 g plant^{-1}) followed by Bhavya (68.16 g plant^{-1}). The potassium uptake (2.9 g plant^{-1}) by Thrupthi (KST-19) variety was also highest among the four varieties studied. The potassium uptake efficiency tended to decrease with increased potassium supply from 44.50 to 23.50 mg day^{-1} plan^{-1}.

Keywords: *Potassium uptake efficiency, Flue cured tobacco (FCY), and Dry matter.*

Introduction

Potassium plays a vital role in increasing the tobacco yield, leaf size, specific leaf weight, colour and improve the quality by affecting the biochemical processes which determine the chemical constituents such as alkaloids, organic acids, amino acids and sugars. However tobacco plants is highly sensitive to even light changes in the environment (Mc Cants and Woltz, 1967). Apparently, the need of the hour is to find genotypes, which can take up potassium efficiently even under adverse environmental situations. It is satisfying to note that there is considerable variation among FCY tobacco varieties in potassium content, uptake and translocation under field condition (Janardhan *et al.*, 1996). Despite higher heritability values for potassium content, uptake and translocation, the expected genetic advance of environment. Thus, field evaluation for potassium uptake efficiency appears to be very little relevance. It was suggested that studies under controlled conditions are needed to establish genotypic differences unequivocally for potassium uptake and utilization (Janardhan *et al.*, 1996). In view of this we have made an attempt to establish differences in potassium uptake and dry matter yield among the FCY tobacco varieties.

Materials and Methods

A pot culture experiment was conducted at College of Agriculture, Navile, Shimoga with four varieties of FCV tobacco, *viz.*, Kanchana, Thrupthi (KST-19), Rathna and Bhavya grown under four levels of potassium (0, 40,80, 120 kg K_2O ha^{-1}). The soil used to study belong to the sandy loam in texture, had pH 6.50, CEC 6.8 cmol (p+) kg^{-1}, exchangeable Ca, Mg 3.1, 1.2 cmol (p+) kg^{-1}, organic carbon 3.6 g kg^{-1} and available N, P, K contents 169.84, 20.92 and 210 kg ha^{-1} respectively. Treatments were imposed using Sulphate of potash as a source of potassium after applying recommended dose of N and P *viz.*, 40 : 30 kg ha^{-1} in the form of calcium ammonium nitrate, single super phosphate. All the recommended package of practices was followed in the nursery, as well as during pot culture to raise a good crop.

Seventy days after planting tobacco leaves were primed as and when they matured, cleaned in water, dried and kept in cupboard for 2–3 days and then subjected to air drying and weights were recorded. After the final priming of tobacco leaves, stems and roots were dried in air and oven at 60°C. The powdered materials were used for chemical analysis. Plant samples were digested by wet oxidation method and potassium was determined flame photometrically. Potassium concentration in both leaf and stem of tobacco plant was multiplied with the corresponding dry matter yield to obtain nutrient uptake by leaf and stem of tobacco plants respectively. Taking in to account, the uptake of leaf and stem were computed for total nutrient uptake by tobacco plant.

The uptake efficiency of applied potassium at different levels in different varieties of FCV tobacco was calculated by the method as described by Singh and Chaudary (1979)

$$\text{K uptake efficiency (mg/plant/day)} = \frac{\text{Total K uptake in plant}}{\text{No. of days} \times \text{Dry matter yield}}$$

Results and Discussion

The results on dry matter yields showed that the yield of Thrupthi (KST-19) variety (68.3 g $plant^{-1}$) was on par with the dry matter yield of Bhavya variety (68.2 g $plant^{-1}$) and the yields of both these varieties were significantly higher than the yields of other two varieties of Kanchan (57.9 g $plant^{-1}$) and Rathna (61.7 g $plant^{-1}$) (Table 26.1). With regard to the response of the tobacco varieties to K fertilization, all the four varieties have responded to K application significantly up to 80 kg K_2O ha^{-1} level with no further positive response to K application at 120 kg K_2O ha^{-1}. The interaction effects between varieties and K levels were non-significant. Significant increase in the dry matter yield of FCV tobacco varieties due to K application may be attributed to the enhanced leaf area, leaf weight, root weight and root length. Potassium had greater effect on leaf expansion accounting for enhanced yield in tobacco and other crops (Rama Rao, 1986)

Table 26.1: Dry Matter Accumulation of Different Varieties of FCV Tobacco at 70 Days After Planting as Influenced by Levels of K Application

K Levels (kg K_2O/ ha)/Variety	Dry Matter (g $plant^{-1}$)				
	Thrupthi (KST-19)	Kanchan	Bhavya	Rathna	Mean
0	57.00	51.35	57.66	52.96	54.74
40	64.66	55.33	67.99	61.35	62.33
80	79.33	63.99	75.32	68.33	71.74
120	72.32	61.30	71.66	64.33	67.40
Mean	68.33	57.99		61.74	
			SEm±	CD at 5%	
Variety			0.80	2.32	
K levels			0.80	2.32	
Variety x K level			1.60	0.05	
CV per cent			4.3	n.s.	

Potassium content in leaf of Thrupthi (KST-19) (4.7 4 per cent) was significantly superior to that of all varieties studied. The interaction between variety and K levels of application was significant. Treatment mean values indicated that Potassium content in tobacco leaf increased from 3.18 per cent in control to 5.22 per cent at 80 kg K_2O ha^{-1} (Table 26.2) The potassium uptake (2.97 g $plant^{-1}$) by

Table 26.2: Effect of Different Levels of K Application on K Content and K Uptake of FCV Tobacco Varieties

K Levels (kg K_2O/ha)/ Variety	Potassium (%)					Total K Uptake (g $plant^{-1}$)				
	Thrupthi (KST-19)	Kanchan	Bhavya	Rathna	Mean	Thrupthi (KST-19)	Kanchan	Bhavya	Rathna	Mean
0	3.18	3.19	3.13	3.23	3.18	1.66	1.57	1.73	1.48	1.61
40	4.91	3.58	4.21	4.58	4.32	2.91	1.88	2.62	2.58	2.49
80	5.52	5.24	5.46	4.64	5.22	3.88	3.14	3.42	2.92	3.34
120	5.36	5.00	5.25	4.46	5.02	3.41	2.81	3.26	2.68	3.04
Mean	4.74	4.25	4.51	4.45		2.97	2.35	2.75	2.41	
				SEm±	CD at 5%			SEm±	CD at 5%	
Variety				0.009	0.02			0.03	0.08	
K levels				0.009	0.02			0.03	0.08	
Variety x K level				0.019	0.05			0.06	0.17	
CV per cent				2.8				4.00		

Thrupthi (KST-19) was also highest among the four varieties studied. The concentration and uptake of potassium in tobacco plant increased due to the application of different levels of K. The potassium content and its uptake increased significantly up to 80 kg K_2O ha^{-1} level. Further, increase in K level did not influence the potassium uptake (Table 26.2).

The K uptake efficiency (mg/day/plant) was highest for Thrupthi (KST-19) followed by Bhavya (Figure 26.1). Further it was also noted that K uptake efficiency tended to decrease with increased K supply from 44.5 to 23.5 mg/day/plant. It is generally known that under limited nutrient supply also the plant delivered certain mechanisms to accumulate some amount of nutrient in their tissue. In the present study all most all the varieties were found to be accumulated more K at 80 kg K_2O ha^{-1} applied. Similar, findings were reported by Nataraju *et al.*, 2002). Overall, it is to be noted that Thrupthi (KST-19) and Bhavya were significantly superior in utilizing the applied K. The higher dry matter yield recorded by these varieties was due to the higher K uptake efficiency.

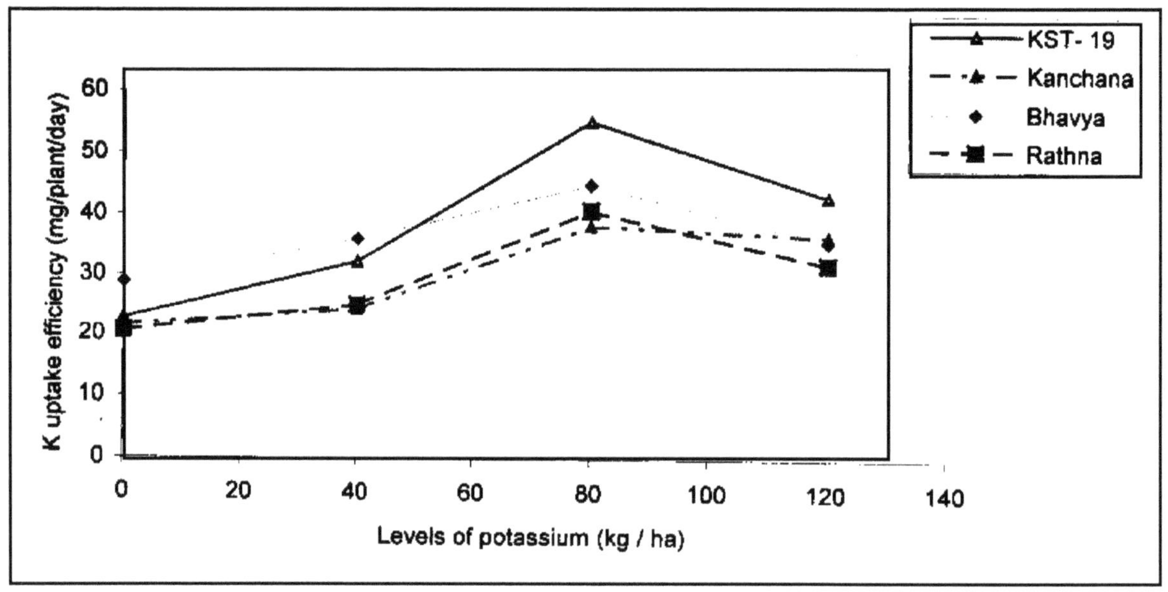

Figure 26.1: K Uptake Efficiency of Different FCV Tobacco Varieties

References

Janardhan, K.V., Nataraju, S.P., Shetty, Y.V., Gurumurthy, B.R. and Bhojaraja, R., 1996. Genetic variability and possible genetic improvement of potassium nutrition in FCV tobacco. *Tob. Res.*, 22(2): 80–87.

McCants, C.B. and Woltz, W.G., 1967. Growth and mineral nutrition of tobacco. *Advances in Agronomy*, 19: 215–265.

Nataraju, S.P., Jayadeva, H.M., Janardhan, K.V. and Ashoka, 2002. Screening of flue cured tobacco Genotypes for potassium uptake efficiency. *Karnataka J. Agri. Sci.*, 15(1): 18–23.

Rama Rao, N., 1986, Potassium requirement by growth and its related processes determined by plant analysis in wheat. *Plant and Soil*, p. 1–9.

Singh, R.K. and Chaudhary, B.D., 1979. *Biometrical Studies in Quantitative Genetic Analysis*. Kalyani Publishers, pp. 304.

Chapter 27

Stratified Scavenging System with Exhaust Gas Recirculation for Two Stroke Spark Ignition Engine

P. Srinivasa Rao[1] *and K. Raja Gopal*[2]

[1]*Department of Mechanical Engineering, Chaitanya Bharathi Institute of Technology, Gandipet, Hyderabad – 500 075, Andhra Pradesh*

[2]*Jawaharlal Nehru Technological University, Hyderabad – 500 072, Andhra Pradesh*

ABSTRACT

From the beginning, two stroke engines have suffered from high emissions and poor fuel economy compared to the larger, heavier but more efficient four stroke engines. Over the last century, many innovative ideas have been designed and patented with the intent or improving these two critical short comings. Because of stringent emission regulations introduced in the last two decades, designers have now given impetus to revive, revise and apply technologies to meet these requirements.

Among the most successful technological improvements, stratified scavenging and charging is the most important one which reduced the two stroke engine emissions by as close as 40 per cent. However the major research is only concentrated either on low capacity engines as 60cc, 65cc and 75cc or as high as 250cc and 350cc engines. But in Indian subcontinent, major part or the two wheelers is of either 125cc or 150cc engines, which emit dangerous pollutants causing health hazards to human beings apart from environmental disorders. The suitability of technology as stratified charging or scavenging to such engines is not validated and substantiated so far. This chapter presents the results and conclusions from a conceptual design study for a stratified charging concept for a higher cubic capacity engines.

Introduction

The civilization of any county depends on the number of vehicles used by the public. For heavy duties, diesel engines are preferred, while for individual transport, a small duty, two-stroke petrol engines are being employed. However, from the beginning, two stroke engines have suffered from high emissions and poor fuel economy compared to the larger, heavier but more efficient four stroke engines. The major pollutants emitted from these two-stroke spark ignition engines are carbon monoxide and un-burnt hydro carbons. Breathing of these emissions causes detrimental effects on human, animal and plant life besides environmental disorders (Sharma, 1996; Fulekar, 1999). Hence globally, stringent regulations are made fix permissible levels of pollutants in the exhaust of 2 and 4 stroke spark ignition engines. A two-stroke engine is recently receiving a renewed interest in the automotive industry due to its lower weight and volume resulting in a more compact body and twice in power compared to four stroke engines of same cubic capacity. The short circuiting loss of fresh charge has been known ever since two stroke engines were first made more than a century ago by Sir Dugald Clerk in 1879 equally old are all the innovative technologies and concepts that have been proposed or attempted in order to circumvent the short circuiting loss of fresh charge, as stratified scavenging through air head, exhaust gas recirculation, air assisted fuel injection, Compressed Wave Injection (CWI), direct or indirect fuel injection etc. The charge stratified engine discovered by Heiko Rosskemp (2003), G.P. Blair (1981) and B.W. Hill (1983) are in another interesting design disclosed by Huber (1943) stratification occurs in three phases air, air fuel mixture and air which improves trapping of fresh charge. In the year 1901 patent (Huber, 1943) Woolf used poppet valve in a unitlow crankcase scavenged and carbureted two stroke engine. In another interesting design more or less similar to today's air head scavenging system is the invention by Stephenson (1911), in which he fitted poppet valve which is fitted at the top of the transfer passage to fill the transfer passage with air during the upward stroke or the piston. Poppet valve closes when the piston approaches TDC and fuel and air mixture is admitted into the crankcase through piston controlled intake port. These are some or the improvements suggested and successfully implemented but only suitable for the small capacity engines *viz.*, 65cc to 75cc engines only.

An automatic Ball Valve admitting air from outside to the top of transfer passage for a 240cc engine is also attempted (Woolf, 1901; Stephenson, 1911) the number of patents issued or applied for indicates the research activities pertaining to the design and application of such engines. Engine manufacturers have commercialized some of the technological advancements as Air-Head scavenging particularly Komatsu-Zenoah of Japan.

Experimental Programme

The Table 27.1 shows the specifications of the single cylinder engine used for the experimentation.

The Figure 27.1 shows the modifications done to conventional spark ignition engine used for the experimentation. The two stroke engine is altered in the basic design as another port is created in the transfer port in the vicinity of the port entrance into the cylinder and is provided with a reed valve to have unidirectional flow into the cylinder along with the charge from the crankcase and a part of exhaust gas is recirculated into the cylinder. As the exhaust gas and the fresh air are also accommodated into the cylinder the quality of the charge that is supplied to the cylinder is to changed to have a proper and better combustion in the chamber.

Table 27.1: Specifications of the Experimental Engine

Cylinder capacity	152cc
Engine Power	1.1 Kw
Rated Speed	4500rpm
Port Size	12mm × 14.5mm
Bore	58.24mm
Stroke	58.40mm
Compression Ratio	8.5
Inlet Port Opening	146° (ATDC)
Inlet port Closing	1460 (ABDC)
Exhaust Port Opening	1280 ATDC
Type Cooling method	Air cooled

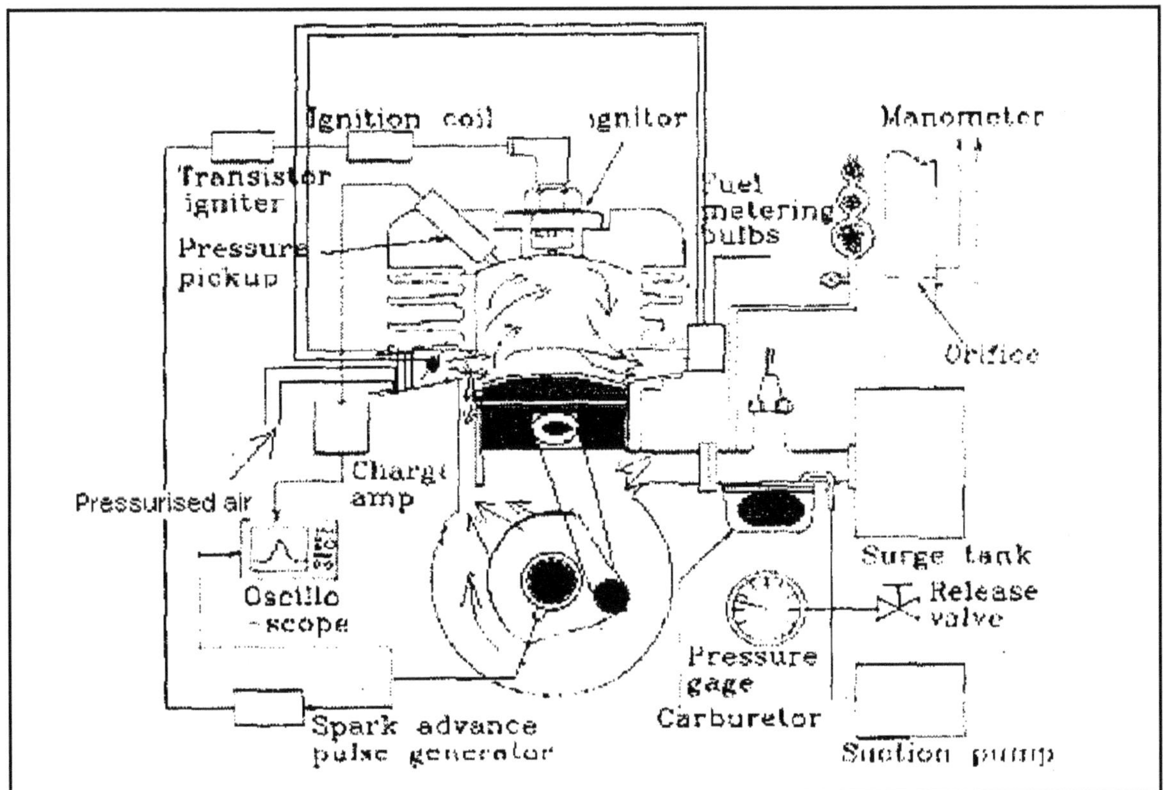

Figure 27.1: Stratified Scavenged Two Stroke Engine with EGR

As the fresh air and exhaust gases are being pumped into the cylinder the fresh charge that is entering into the cylinder through the crankcase may be diluted. The carburetor is provided with the suction pump to have a rich mixture when it is needed to overcome the above problem. A filter is provided to filter the particulate mater which may reenter the cylinder. The pollutants of CO and UBHC emissions are recorded with Netal Chromotograph CO/HC analyzer.

The fluid flow (air and charge) through the Engine under working conditions is shown in the Figure 27.2

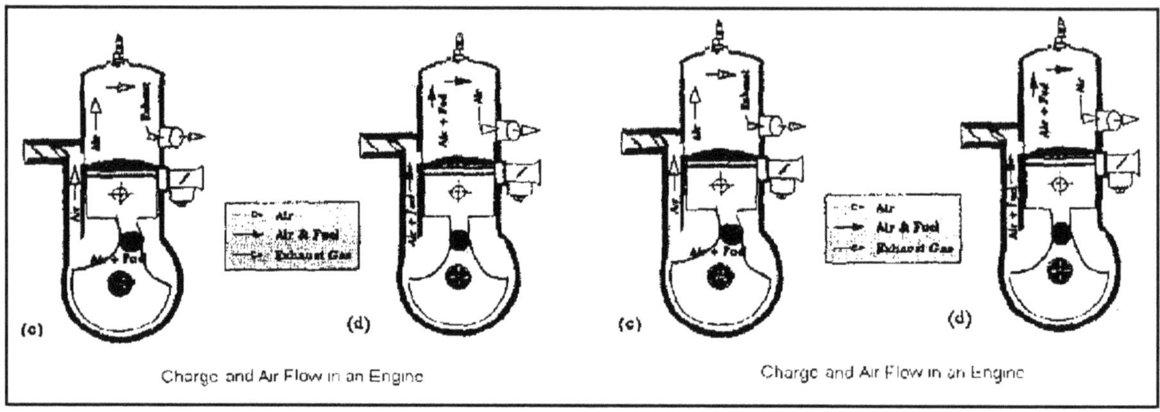

Figure 27.2: Charge and Air Flow in an Engine

Results and Discussions

Figures 27.3 and 27.4 shows the variation of UBHC and CO with fresh charge delivery ratio at different equivalence ratios respectively.

It is observed from the Figure 27.3 as fresh loss ratio increases Hydrocarbons decrease. The trends arc observed to be same for both residual and non residual cases. As the re-circulated gas which is sand witched between fresh air and fresh charge as the scavenging processes begins the fresh air pushes the burned gases out from the chamber and before the fresh charge enters the cylinder the re-circulated gas acts as an artificial piston to throwaway the Hue gases out completely leading to reduction in Hydrocarbon emissions.

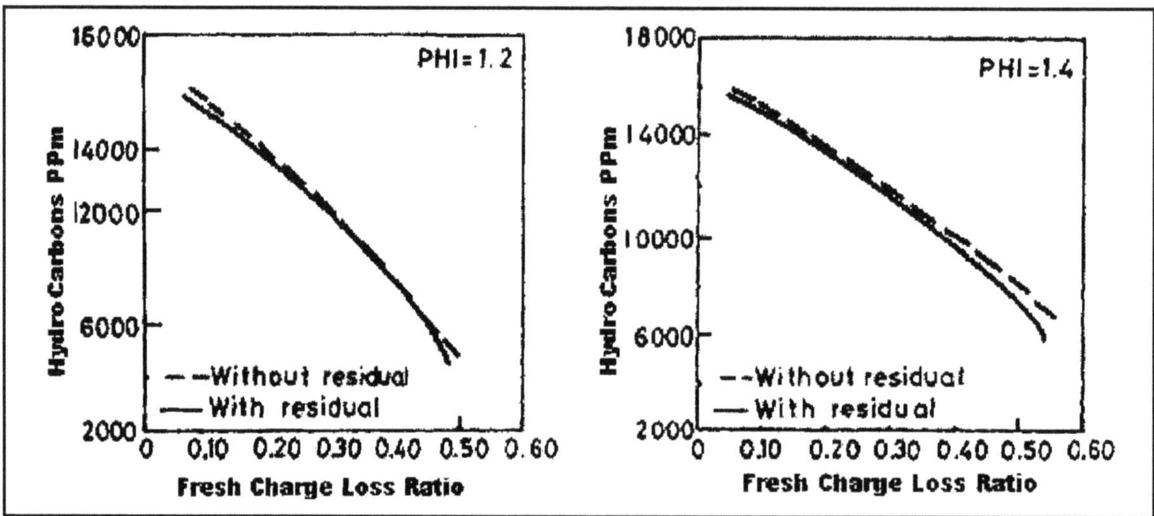

Figure 27.3: Effect of Fresh Charge Loss Ratio on HC Concentration at Different Equivalence Ratio

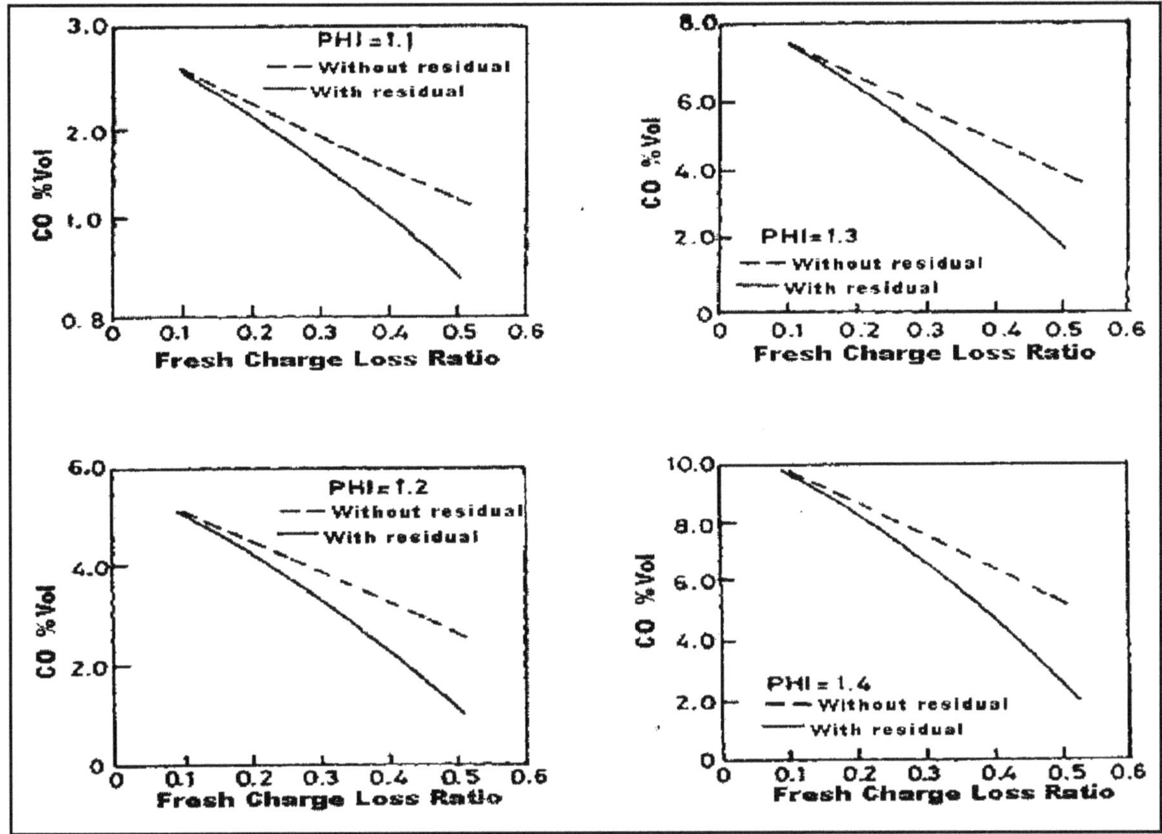

Figure 27.4: Effect of Fresh Charge Loss Ratio on Carbon Monoxide Concentration at Different Equivalence Ratio

From the Figure 27.4 it is observed that CO emissions decrease as the fresh charge loss ratio increase at all equivalence ratios and it is more pronounced at lower equivalence ratios. The CO emissions decrease due to the better oxidation and combustion reactions because of the admission of air into the combustion chamber. Similar trends are observed for both residual and non residual cases.

Conclusions

1. Based on the experimental investigations on a two stroke stratified scavenged engine with part-exhaust gas recirculation and fitted with extra port to provide the pressurized air through transfer port the pre an post combustion gases are clearly analyzed and found that the scavenging and trapping efficiencies are strong functions of delivery ratio.
2. The reduction of Hydrocarbon emissions is found to be 40 per cent when compared with conventional engine.
3. The CO emissions are decreased by 22 per cent to 25 per cent with stratified engine compared with non-stratified engine.

References

Blair, G.P. and Hill, B.W. Reduction of fuel consumption of a spark-ignition two-stroke cycle engine. *SAE*, 830093.

Blair, Gordon P., 1981. US Patent No. 4, 253, 433 March 3.

Fulekar, M.H., 1999. Chemical pollution: A threat to human life. *Indian J. of Environ. Prot.*, 1(3): 353–359.

Hill, B.W. and Blair, G.P., 1983. Further tests on reducing fuel consumption two stroke cycle engine. *SAE* Paper No. 831303.

Huber, Fritz and Lenz, Anton, 1943. US Patent No. 2. 317, 772, April 27.

Mavinahalli, Nagesh S. and Neelakantappa, B.P., 1991–92. Pedormance improvement or a two stroke engine using non-return valves in transfer passages. Student project report. Malnad College of Engineering, Hasan, India.

Rosskamp, Heiko and Hahnadorff, Markus, 2003. US Patent No. 6, 598, 568, July 29.

Sharma, B.K., 1996. *Engineering Chemistry*. Pragathi Prakashan (P) Ltd., Meerut.

Stephenson, William, 1911. US Patent No. 1, 012, 288, December 19.

Woolf, Ellis J., 1901. US Patent No. 683, 886, January 29.

www.Komatsu–Zenoah, Japan.

Yoshida, Y., 1999. Development of Stratified Tow Stroke Cycle Engines for Emission Reduction. *SAE*, 0103269.

Chapter 28

An Experimental Study of Enzymatic Activity of Artificial Soil in Controlled and Malathion Treated Pot

Kumari Swarnim

P.G. Department of Zoology, Ranchi University, Ranchi – 834 008

ABSTRACT

An experimental study on the enzyme activity of amylase and invertase has been carried in controlled and malathion treated pots. The results are significant in invertase activity among days and doses but in amylase activity it is significant among days but non-significant among doses.

Keywords: *Soil, Amylase, Invertase, Enzyme activity.*

Introduction

According to the definition of Kellogg, soils are considered natural bodies covering parts of the earth surface that support plant growth and that have properties due to the integrated effect of climate and organisms acting upon the parent material, as conditioned by relief over a period of time (Tan, 1994).

When a toxic chemical is applied to the soil as a pesticide, besides the target organism, many non-target organisms are also killed. The percentage of kill depends upon the dosage, method of application and the chemistry of the compound. Soil texture, soil moisture, soil temperature and organic matter contents of soil influence the mortality rate of the soil organism. A dispersed or waterlogged soil will retard the movement of the chemical. High clay or organic matter content increases the adsorption of the chemical and those by the mortality rate of the soil organisms is reduced (Adamson and Inch,

1973; Adhya *et al.*, 1987). After the treatment it was found that, the organisms not killed or from non treated areas quickly recolonize the soil (Martin, 1982).

Soil metabolism refers to the over all activities of all the soil organisms involving some biochemical processes which are the parts of their life of which microbes contribute a lot. It is computed by the measurement of carbon dioxide evolution, enzyme activities of the soil and essential nutrients like carbon and nitrogen (Wanger, 1975; Skujins, 1978). Seasonal variation in soil carbon dioxide evolution and soil enzyme activities are of ulmost, biological importance (Spiers *et al.*, 1980; Rashid and Schaefer, 1985; Tiwari *et al.*, 1986) since these reflect the changes in the soil processes. Degradation of naturally occurring organic matter (Spalding, 1980) and other xenobiotic compounds (Burns and Edwards, 1980) harmful metabolic wastes (Doelman and Hanstra, 1979) and pesticides (Lethbridge *et al.*, 1981) in the soil sub-system is possible by the metabolic activities of the soil organisms.

Soil enzymes are extracellular secretions by living soil organisms. Therefore, any alteration in the life and functions of these organisms bring about a change in the soil enzyme activity irrespective of their source of production such as, bacteria, fungi or even earthworms (Cervelli *et al.*, 1975).

When applied to soil pesticides are subjected to be either degraded and utilized by the soil organisms or plant root system. Conversely they also become cause of death of several soil organisms. This may lead to the cell lysis and temporary increase in the extracellular fractions of the soil enzymes. Moreover, dead cells are utilized by other soil organisms. If at all lethal effects of these agro-chemicals are not found, still they interfere in the metabolic activity of the soil organisms among which inhibition of enzyme activity, modification of biosynthetic mechanisms (Lichfield and Huben, 1973), regulation of protein biosynthesis (Ambroz, 1973) and plant growth regulation system (Rich, 1969) are important to note.

It has been reported by Petty *et al.* (1977) that, even fertilizers can have a secondary effect of being carrier of pesticides. During the formulation of agrochemicals the chemical aspects such as compound stability, physical behaviour, chemical nature and economical aspects are generally taken into account. But their secondary influence on soil metabolism and non-target organisms are completely ignored. Furthermore, their interference in the soil metabolism depends upon the quality and quantity. But the realistic effects of these substances are sometimes insufficiently considered based on laboratory tests wwhich warrants study on field application in different climatic conditions.

Materials and Methods

Estimation of Enzyme Activities in Soil

Soil Invertase and Amylase Activity in Soil

Chemicals

1. *Sorenson's Buffer*: pH 5.5, 0.06M 10.6896 gm Na_2 $HPO_4 2H_2O$ made to 100 ml and 8.1654g KH_2PO_4 made to 1000 ml. Then 5ml Na_2HPO_2. $2H_2O$ and 95 ml KH_2PO_4 are mixed and the pH is adjusted (if necessary) in pH meter using the above solutions.
2. Toluene.
3. Substrate: 1 per cent soluble starch for amylase and 5 per cent sucrose for invertase.
4. *Salicylic acid solution*: 1 g of 3, 5 dinitro salicylic acid in 20 ml of 2N NaOH and 50 ml H_2O, add 30 g of Rochelle salt (Na, K tartarate) and make up to 100 ml with distilled water. Protect the solution from CO_2.

Procedure

Fresh soil (or soil preserved at 4°C) was screened through a 2 mm sieve. 3 g soil was mixed with 0.2 ml toluene in a 150 ml flask and allowed to stand at room temperature for 15 minutes. 6 ml sorenson's buffer and 6 ml of substrate solution were added to the flask. After shaking they were placed in the incubator at 30°C. Control flasks with water instead of substrate was added to the flask. After 24 h the content of the flask were centrifuged. Colour was developed adding 3,5 dinitrosalycilic acid solution (2 ml) to 1 ml of supernatant. The colour reagent also stops enzyme activity. The reducing sugar forms a pink colour which is read at 540nm. The standard graph is prepared taking glucose as standard. The colour develops after keeping the solution for 5 min in boiling water bath.

Results

Enzyme Activity

Soil Amylase

Variation in soil amylase activity (mg glucose g^{-1} h^{-1}) as well as percentage change in the control pot during the period of investigation has been presented in Table 28.1 and Figure 28.1. The amylase activity (mg glucose g^{-1} h^{-1}) was maximum being 120.62±5.68 and minimum of 36.56±1.92 on 45th day and at the start of the experiment respectively.

Table 28.1: Variation in Soil Amylase Activity (mg glucose g^{-1} h^{-1}) and Percentage Change in Control Pot at an Interval of 15 Days

Month	*Week*	*Days of Inoculation*	*Soli Amylase Activity*	*Percentage Change*
February	4th	0	36.56±1.92	
March	2nd	15	56.82±2.72	55.415
March	4th	30	89.29±4.46	57.145
April	2nd	45	120.62±5.68	35.087
April	4th	60	111.75±4.46	–7.353
May	2nd	75	99.56±3.01	–10.908
May	4th	90	91.62±2.02	–7.975

At the start of the experiment, the soil amylase activity was 36.56±1.92. On 15th day of inoculation (2nd week of March, 2004) the soil amylase activity increased from 36.56±1.92 to 56.82±2.72 showing an increase of 55.415 per cent. The soil amylase activity increased from 56.82±2.72 to 89.29±4.46 after 30th day (4th week of March, 2004).

The increase in terms of percentage was 57.145. On 45th day (2nd week of April, 2004) the soil amylase activity increased from 89.29±4.46 to 120.62±5.68 (35.087 per cent) whereas the soil amylase activity decreased from 120.62±5.68 to 111.75±4.46 showing a decrease of 7.353 per cent after 60th day (4th week of April, 2004). On 75th day (2nd week of May, 2004) the soil amylase activity decreased from 111.75±4.46 to 99.56±3.01. The decrease in terms of percentage was 10.908. On 90th day (4th week of May, 2004) the soil amylase activity decreased from 99.56±3.01 to 91.62±2.02 showing a decrease of 7.975 per cent.

Variation in soil amylase activity (mg glucose g^{-1} h^{-1}) and percentage change in the malathion treated (2.2 mg kg^{-1} soil) pot during the period of investigation is shown in Table 28.2. At the start of the

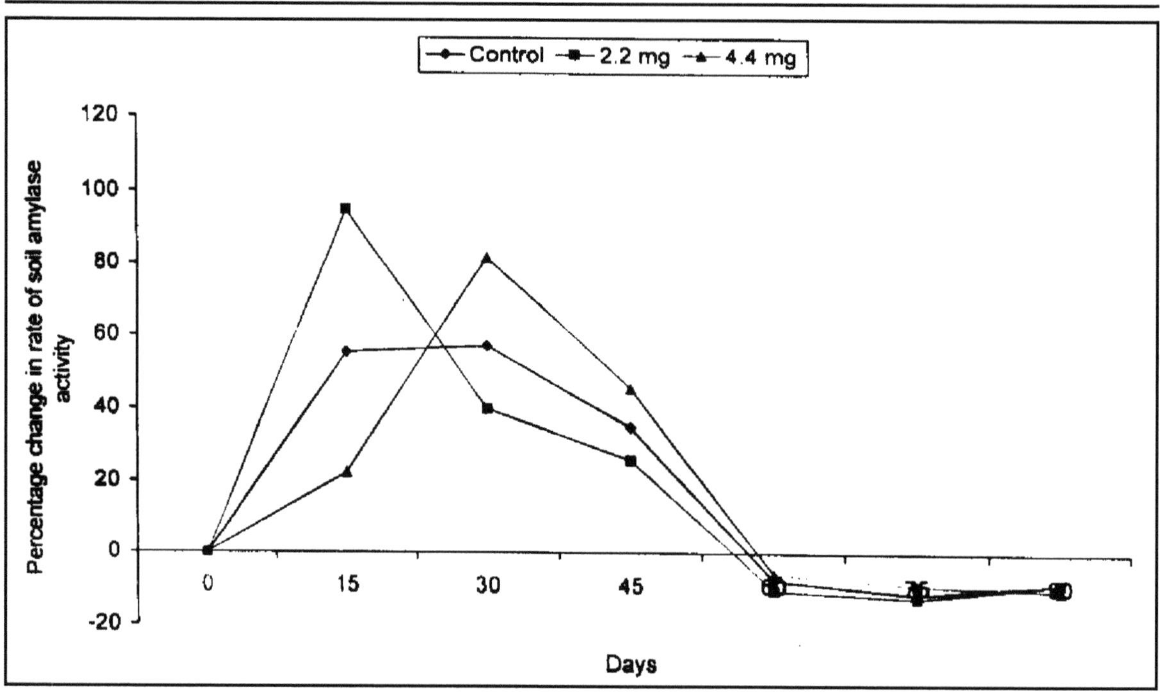

Figure 28.1: Percentage Change in Rate of Soil Amylase Activity in Control and Malathion Treated Pot

experiment, the soil amylase activity was 36.56±1.92. On 15th day (2nd week of March, 2004) the soil amylase activity increased from 36.56±1.92 to 71.23±3.36 (94.830 per cent). On 30 th day (4th week of March, 2004) the soil amylase activity increased from 71.23±3.36 to 99.76±4.26 showing an increase of 40.053 per cent. The soil amylase activity increased from 99.76 t 4.26 to 125.77 t 5.96 on 45 th day (2nd week of April, 2004). The increase in terms of percentage was 26.072. The soil amylase activity decreased from 125.77±5.96 to 113.23±5.23 showing a decrease of 9.970 per cent after 60 th day of experiment (4th week of April, 2004). The soil amylase activity decreased from 113.23±5.23 to 99.62±3.02 on 75 th day (2nd week of May, 2004). The decrease in terms of percentage was 12.019. On 90 th day (4th week of May, 2004) the soil amylase activity decreased from 99.62±3.02 to 91.65±1.82 showing a decrease of 8.00 per cent (Figure 28.1).

Table 28.2: Variation in Soil Amylase Activity (mg glucose g^{-1} h^{-1}) and Percentage Change in Malathion Treated Pot (2.2 mg kg^{-1} soil) at 15 Days Interval

Month	*Week*	*Days of Inoculation*	*Soli Amylase Activity*	*Percentage Change*
February	4th	0	36.56±1.92	
March	2nd	15	71.23±3.36	94.830
March	4th	30	99.76±4.26	40.053
April	2nd	45	125.77±5.96	26.072
April	4th	60	113.23±5.23	–9.970
May	2nd	75	99.62±3.02	–12.019
May	4th	90	91.65±1.82	–8.000

On 45th day of inoculation the amylase activity (mg glucose g^{-1} h^{-1}) was maximum (125.77±5.96) and minimum (36.56±1.92) at the start of the experiment respectively.

Variation in soil amylase activity (mg glucose g^{-1} h^{-1}) as well as percentage change in the malathion treated pot (4.4 mg kg^{-1} soil) during the study period has been shown in Table 28.3 and Figure 28.1. The soil amylase activity was 36.56±1.92 at the start of the experiment. On 15th day (2nd week of March, 2004) the soil amylase activity increased from 36.56±1.92 to 44.46±2.12. The increase in terms of percentage was 22.428.

Table 28.3: Variation in Soil Amylase Activity (mg glucose g^{-1} h^{-1}) and Percentage Change in Malathion Treated Pot (4.4 mg kg^{-1} Soil) at an Interval of 15 Days

Month	*Week*	*Days of Inoculation*	*Soli Amylase Activity*	*Percentage Change*
February	4th	0	36.56±1.92	
March	2nd	15	44.46±2.12	22.428
March	4th	30	80.75±3.16	81.623
April	2nd	45	117.72±3.82	45.783
April	4th	60	111.12±4.02	–5.606
May	2nd	75	101.71±3.82	–8.468
May	4th	90	91.84±1.89	–9.704

The soil amylase activity increased from 44.46±2.12 to 80.75±3.16 showing ah increase of 81.623 per cent on 30th day (4th week of March, 2004). On 45th day (2nd week of April, 2004) the soil amylase activity increased from 80.75±3.16 to 117.72±3.82 (45.783 per cent). On 60th day (4th week of April, 2004) the soil amylase activity decreased from 117.72±3.82 to 111.12±4.02 showing a decrease of 5.606 per cent. The soil amylase activity decreased from 111.12±4.02 to 101.71±3.82 on 75th day (2nd week of May, 2004). The decrease in terms of percentage was 8.468. On 90th day the soil amylase activity decreased from 101.71±3.82 to 91.84±1.89 showing a decrease of 9.704 per cent.

Amylase activity increased from a minimum (36.56±1.92 mg glucose g^{-1} h^{-1}) in 4th week of February, 2004 to a maximum (117.72±3.82 mg glucose g^{-1} h^{-1}) in 2nd week of April, 2004.

In 2.2 mg kg^{-1} soil treated soil the percentage increase was max (25.360) and minimum (0.0327) in the months of March, 2004 (2nd week, 15th day) and May 04 (4th week, 90th day) respectively, whereas the same in the double dose of malathion treated soil was maximum (2.159) and minimum (–21.752) in the months of May, 2004 (2nd week, 75th day) and March 04 (2nd week, 15th day) respectively.

Table 28.4a: Two Way ANOVA of Amylase Activity of Soil Among Different Doses and Duration of Malathion Treatment

Source of Variation	*Sum of Square*	*Degree of Freedom*	*Mean Square*	*Variance Ratio F*	*Significance*
Different duration	16289.15	6	2714.858	87.78664	$P < 0.001$
Different dose	207.8381	2	103.919	3.360287	NS
Residual	371.1077	12	30.92564		

A two way ANOVA relating to the amylase activity of the soil revealed significant differences in amylase activity among days whereas no significant difference was obtained between doses ($F = 87.78$; $df = 6, 12$; $p < 0.001$; $F = 3.36$; $df = 2, 12$) (Table 28.4a).

When the amylase activity of the soil were analyzed by 't' test a non significant difference was observed between control and single dose and double dose of malathion treatment ($t = -2.05$; $t = 1.57$; NS) (Table 28.4b).

Table 28.4b: t-test Between Amylase Activity of Soil in Control and Test Concentrations of Malathion

Treatment Level	*df*	*Calculated 't'*	*Tabulated*		*Significance*
			0.01	*0.05*	
Control–2.2 mg Kg^{-1} soil	6	-2.05	3.71	2.45	NS
Control–4.4 mg Kg^{-1} soil	6	1.57	3.71	2.45	NS

Soil Invertase

Variation in soil invertase activity (mg glucose g^{-1} h^{-1}) and percentage change in the control pot has been presented in Table 28.5 and Figure 28.2. The soil invertase activity was 43.66±2.36 at the start of the experiment. On 15th day of observation (2nd week of March, 2004) the soil invertase activity increased from 43.66±2.36 to 64.78±3.32 showing an increase of 48.373 per cent. The soil invertase

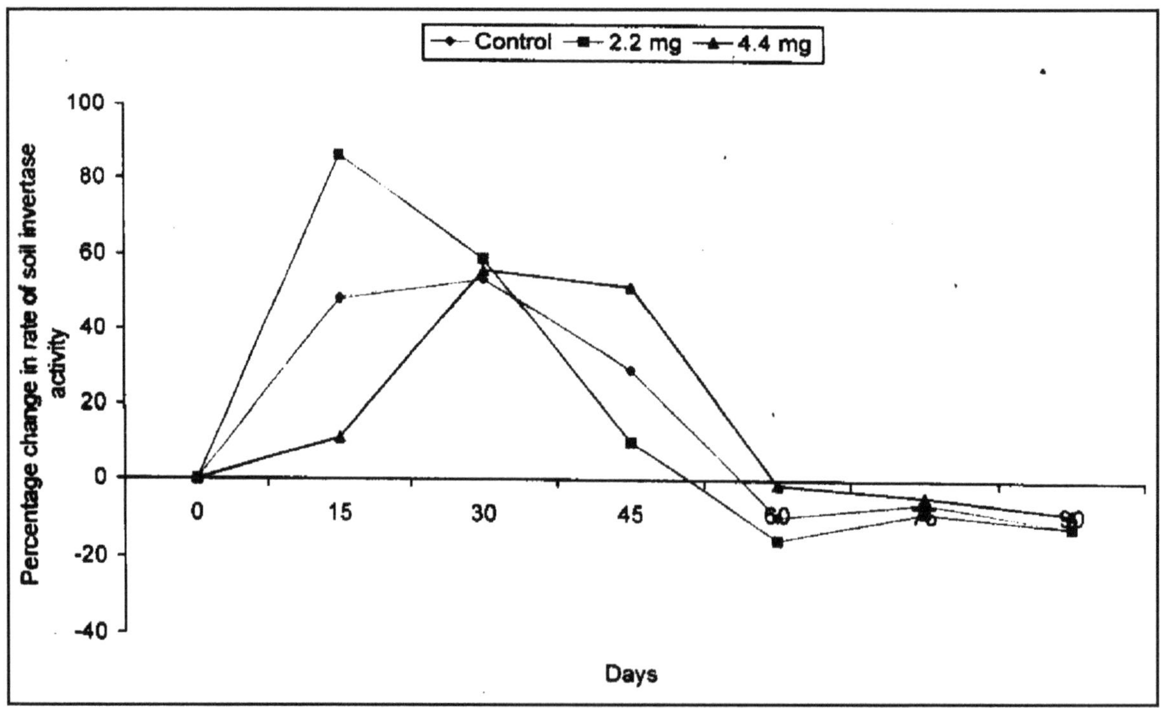

Figure 28.2: Percentage Change in Rate of Soil Invertase Activity in Control and Malathion Treated Pot

activity increased from 64.3.32±99.55±5.01 on 30th day (4th week of March, 2004). The increase in terms of percentage was 53.673. On 45th day (2nd week of April, 2004) the soil invertase activity increased from 99.55±5.01 to 128.7.03±7.03 (29.261 per cent). On 60th day of observation the soil invertase activity decreased from 128.68±7.03 to 116.44±5.23 showing a decrease of 9.511 per cent. The soil invertase activity decreased from 116.44±5.23 to 109.62±5.65 on 75th day (2nd week of May, 2004). The decrease in terms of percentage was 5.857. On 90th day of observation the soil invertase activity further decreased from 109.62±5.65 to 96.66±4.44 showing a decrease of 11.822 per cent (Figure 28.2).

The invertase activity (mg glucose $g^{-1} h^{-1}$) was maximum (128.68±7.03) and minimum (43.66±2.36) on 45th day and at the start of the experiment respectively (Table 28.5). Change of soil invertase activity in terms of percentage was maximum (53.673) and minimum –11.822 on 30th day (4th week of March, 2004) and on 90th day (4th week of May, 2004) respectively, (Table 28.5 and Figure 28.2).

Table 28.5: Change in Soil Invertase Activity (mg glucose $g^{-1} h^{-1}$) and Percentage Change in Control Pot at 15 Days Interval

Month	*Week*	*Days of Inoculation*	*Soli Amylase Activity*	*Percentage Change*
February	4th	0	43.66±2.36	
March	2nd	15	64.78±3.32	48.373
March	4th	30	99.55±5.01	53.673
April	2nd	45	128.68±7.03	29.261
April	4th	60	116.44±5.23	–9.511
May	2nd	75	109.62±5.65	–5.857
May	4th	90	96.66±4.44	–11.822

Variation in soil invertase activity (mg glucose $g^{-1} h^{-1}$) as well as percentage change in the malathion treated pot (2.2 mg kg^{-1} soil, the recommended agricultural dose) during the period of investigation has been presented in Table 28.6 and Figure 28.2 having a peak value of 142.33±5.89 mg glucose $g^{-1} h^{-1}$ and minimum of 43.66±2.36 mg glucose $g^{-1} h^{-1}$ was obtained on 45th day and at the start of the experiment respectively. The change of soil invertase activity in terms of percentage was maximum (86.303 per cent) and minimum (–15.421 per cent) on 15th day and 60th day of the experiment respectively (Figure 28.2).

The soil invertase activity was 43.66±2.36 at the start of the experiment. On 15th day (2nd week of March, 2004) the soil invertase activity increased from 43.66±2.36 to 81.36±4.22. The increase in terms of percentage was 86.303. The soil invertase activity increased from 81.34±4.22 to 129.29±4.66 showing an increase of 58.950 per cent on 30th day of the observation (4th week of March, 2004). On 45th day (2nd week of April, 2004) the soil invertase activity increased from 129.29±4.66 to 142.33±5.89 (10.085 per cent). The soil invertase activity decreased from 142.33±5.89 to 120.38±5.64 showing a decrease of 15.421 per cent on 60th day (4th week of April, 2004). On 75th day (2nd week of May, 2004) the soil invertase activity decreased from 120.38±5.64 to 10.36±5.55. The decrease in terms of percentage was –8.323. On 90th day (4th week of May, 2004) the soil invertase activity decreased from 110.36±5.55 to 97.20±4.36 showing a decrease of 11.924 per cent.

Table 28.6: Change in Soil Invertase Activity (mg glucose g^{-1} h^{-1}) and Percentage Change in Malathion Treated Pot (2.2 mg kg^{-1} Soil) at an Interval of 15 Days

Month	*Week*	*Days of Inoculation*	*Soli Amylase Activity*	*Percentage Change*
February	4th	0	43.66±2.36	
March	2nd	15	81.34±4.22	86.303
March	4th	30	129.29±4.66	58.950
April	2nd	45	142.33±5.89	10.085
April	4th	60	120.38±5.64	–15.421
May	2nd	75	110.36±5.55	–8.323
May	4th	90	99.20±4.36	–11.924

Table 28.7 and Figure 28.2 shows the variation in rate of soil invertase activity (mg glucose g^{-1} h^{-1}) as well as percentage change in the malathion treated pot (4.4 mg kg^{-1} soil, double the recommended agricultural dose) during the study period. At the start of the experiment the soil invertase activity was 43.66±2.36. On 15th day (2nd week of March, 2004) the soil invertase activity increased from 43.66±2.36 to 48.48±2.82 showing an increase of 11.039 per cent. The soil invertase activity increased from 48.48±2.82 to 75.65±4.01 (56.043 per cent) on 30 th day of the observation (4th week of March, 2004). On 45th day (2nd week of April, 2004) the soil invertase activity increased from 75.65±4.01 to 114.76±4.42. The increase in terms of percentage was 51.698. On 60 th day (4th week of April, 2004) the soil invertase activity decreased from 114.76±4.42 to 113.36±4.52 showing a decrease of 1.219 per cent. The soil invertase activity decreased from 113.36±4.52 to 108.62±4.82. The decrease in terms of percentage was 4.181 on 75th day (2nd week of May, 2004). On 90th day (4th week of May, 2004) the soil invertase activity decreased from 108.62±4.82 to 99.52±4.23 showing a decrease of 8.377 per cent.

Table 28.7: Variation in Soil Invertase Activity (mg glucose g^{-1} h^{-1}) and Percentage Change in Malathion Treated Pot (4.4 mg kg^{-1} Soil) at an Interval of 15 Days

Month	*Week*	*Days of Inoculation*	*Soli Amylase Activity*	*Percentage Change*
February	4th	0	43.66±2.36	
March	2nd	15	48.48±2.82	11.039
March	4th	30	75.65±4.01	56.043
April	2nd	45	114.76±4.42	51.698
April	4th	60	113.36±4.52	–1.219
May	2nd	75	108.62±4.82	–4.181
May	4th	90	99.52±4.23	–8.377

The soil invertase activity was (mg glucose g^{-1} h^{-1}) was maximum (114.76±4.42) and minimum (43.66±2.36) on 45th day (2nd week of April, 2004) and at the start of the experiment respectively. The percentage change of soil invertase activity was maximum (56.043 per cent) and minimum –8.377 per cent on 30th day and 90th day of experiment respectively.

In 2.2 mg kg^{-1} soil treated soil the percentage increase was maximum (29.874) and minimum (0.675) in the months of March 04 (4th week, 30th day) and May 04 (2nd week, 75th day) respectively, whereas the same in the double dose of malathion treated soil was maximum (2.959) and minimum (–25.162) in the months of May 04 (4th week, 90th day) and March 04 (2nd week, 15th day) respectively.

A two way ANOVA relating to the invertase activity of the soil revealed significant differences in invertase activity among days and among doses ($F = 24.58$; $df = 6, 12$; $p < 0.001$; $F = 4.88$; $df = 2, 12$; $p < 0.05$) (Table 28.8a).

Table 28.8a: Two Way ANOVA of Invertase Activity of Soil Among Different Doses and Duration of Malathion Treatment

Source of Variation	*Sum of Square*	*Degree of Freedom*	*Mean Square*	*Variance Ratio F*	*Significance*
Different duration	16227.45	6	2704.574	24.58532	P < 0.001
Different dose	1075.382	2	537.6911	4.887759	P < 0.05
Residual	1320.092	12	110.0077		

When the invertase activity of the soil were analyzed by 't' test a non significant difference was observed between control and single dose and double dose of malathion treatment ($t = -2.312$; $t = 2.071$; NS) (Table 28.8b).

Table 28.8b: t-test Between Invertase Activity of Soil in Control and Test Concentrations of Malathion

Treatment Level	*df*	*Calculated 't'*	*Tabulated*		*Significance*
			0.01	*0.05*	
Control–2.2 mg Kg^{-1} soil	6	-2.312	3.71	2.45	NS
Control–4.4 mg Kg^{-1} soil	6	2.071	3.71	2.45	NS

Conclusion

The amylase activity (mg glucose g^{-1} h^{-1}) was maximum being 120.62±5.68 and minimum of 36.56±1.92 on 45th day and at the start of the experiment respectively while in the malathion treated pot (2.2 mg per kg soil) the amylase activity (mg glucose g^{-1} h^{-1}) Was maximum (125.77±5.96) and minimum (36.56±1.92). Amylase activity increased from a minimum (36.56±1.92 mg glucose g^{-1} h^{-1}) in 2nd week of April, 2004 to a maximum (117.72±3.82 mg glucose g^{-1} h^{-1}) in 4th week of February, 2004. A two way ANOVA relating to the amylase activity of the soil revealed significant differences in amylase activity among days whereas no significant difference was obtained between doses ($F = 87.78$; $df = 6, 12$; $p<0.001$; $F = 3.36$; $df = 2,12$).

The invertase activity (mg glucose g^{-1} h^{-1}) was maximum (128.68±7.03) and minimum (43.66±2.36) on 45th day and at the start of the experiment respectively. In the malathion treated pot (2.2 mg kg^{-1} soil, the recommended agricultural dose) a peak value of 142.33±5.89 mg glucose g^{-1} h^{-1} and minimum of 43.66±2.36 mg glucose g^{-1} h^{-1} was obtained on 45th day and at the start of the experiment respectively. The soil invertase activity was (mg glucose g^{-1} h^{-1}) was maximum (114.76±4.42) and minimum (43.66±2.36) on 45th day (2nd week of April, 2004). In the malathion treated pot (4.4 mg kg^{-1} soil). A two way ANOVA relating to the invertase activity of the soil revealed significant differences in invertase activity among days and among doses ($F = 24.58$; $df = 6,12$; $p < 0.001$; $F = 4.88$; $df = 2,12$; $p < 0.05$).

Possibly the significant results as stated above is on account of synergistic activity of microbes and the enzymes.

Acknowledgement

Thanks are due to Dr. M. P. Sinha for his guidance.

References

Adamson, J. and Inch, T. D. 1973. Possible relationship between structure and mechanism of degradation of organophosphorous insecticide in the soil environment. In: *Proc. 7th Brit. Ins. Fungi Cont.*, p. 65–72.

Adhya, T.K., Wahid, P.A. and Sethunathan, N., 1987. Persistence and biodegradation of selected Organophosphorous insecticides in flooded versus non-flooded soil. *Biol. Fertil. Soils*, 4: 36–40.

Ambroz, Z., 1973. Study of cellulose complex in soil. *Rostl. Vyriba*, 19: 207–212.

Burns, R.G. and Edwards, J.A., 1980. Pesticides breakdown by soil enzymes. *Pesticides Sci.*, 11: 506–512.

Cervelli, S., Nannipieri, P., Giovannini, G. and Perna, A., 1975. Concerning the distribution of enzymes in soil organic matter: Studies about humus. *Trans. Int. Symp.* Humus et Planta VI. Prague, p. 291–296.

Doelman, P. and Haanstra, L., 1979. Effect of lead on soil respiration and dehydrogenase activity. *Soil Biol. Biochem.*, 11: 475–479.

Lichfield, C.D. and Huben, R.P., 1973. Effects of selected pollutants on the specific growth rate and extracellular enzyme synthesis of *Aeicomonas proteolytica. Bull. Ecol. Res. Comm. (Stockholm)*, 17: 464–466.

Lethbridge, G., Bull, A.T. and Burns, R.G., 1981. Effects of pesticides on 1,3-β-glucanose and urease activities in soil in the presence and absence of fertilizers, lime and organic materials. *Pesticides Science*, 12: 147–155.

Martin, N.A., 1982. The effects of herbicides used on asparagus on the growth rate of the earthworm *Allolobophora caliginosa.* In: *Proc. 35th N.Z. Weed and Pest Control Conf.*, pp. 328–331.

Petty, H.B., Burnside, O.C. and Bryant, S.P., 1977. Fertilizer combinations with herbicides or insecticides. In: *Fertilizer Technology and Use*, (Eds.) R.A. Olson, T.J. Army, J.J. Hanway and V.J. Kilmer, 2nd edn. Soil Sci. Soc., Am. Madison Wisconsis, p. 494–515.

Rashid, G.H. and Schaefer, I., 1985. The Seasonal pattern of carbon dioxide evolution from two temperate forests (Catera) Soils. *Rev. Ecol. Biol. Sol.*, 22: 419–431.

Rich, S., 1969. Quinones (Fungicides). In: *Fungicides: An Advanced Treatise*, (Ed.) D.C. Torgeson, Vol. 2, Academic Press, New York, p. 447–475.

Skujins, J., 1978. History of abiotic enzymes. In: *Soil Enzymes*, (Ed.) R.G. Burns. Academic Press, London, pp. 1–49.

Spiers, T.W., Lee, R., Paniser, E.A. and Cairns, A., 1980. A comparison of sulphatase, urease and protease activities in planted and fallow soils. *Soil Biol. Biochem.*, 12: 281–291.

Spalding, B.P., 1980. Enzymatic activities of coniferous leaf litter. *J. Soil Sci. Soc. of America*, 44: 760–764.

Tan, H.K., 1994. *Environmental Soil Science.* Marcel Dekker.

Tiwari, S.C., Tiwari, B.K. and Mishra, R.R. 1986. Temporal and depthwise variation in CO_2 evolution and microbial population in pineapple plantation soils. *J. of Soil Biol. Ecol.*, 6: 67–76.

Wanger, G.H., 1975. Microbial growth and carbon turn-over. In: *Soil Biochemistry*, Vol. 3, (Eds.) E.A. Paul and A.D. McLaven (Eds.). Dekker, New York, p. 269–305.

Chapter 29

Monitoring of Gramoxone Toxicity by Using Pollen of Apocynaceae*

S.A. Salgare

Salgare Research Foundation Pvt. Ltd., Prathamesh Society, Shivaji Chowk, Karjat – 410 201, M.S., India
E-mail: drsalgare@rediffmail.com and drsalgare@sancharnet.in

ABSTRACT

All the concentrations (10^{-17}–10^{-2}–10^{-3}, 1, 5, 10, 20–20–100 mg/ml) of gramoxone tried found to be toxic for the germination of pollen of F-72 series of pink-flowered cultivar of *Catharanthus roseus* (L.) G. Don. However, except for F-24 series of red-flowered cultivar of *Nerium odorum* Soland. all

----

* Monitoring of Herbicide (Gramoxone) Toxicity by Using Pollen as Indicators: Pollen of Five Cultivars of Apocynaceae: Further evidence of a criticisms of Banerjee and Gangulee (1937), Brewbaker and Kwack's (1963), Sudhakaran (1967, Ph.D. Thesis), Dharurkar (1971, Ph.D. Thesis), Berg (1973), Brandt (1974), Vick and Bevean (1976), Rasmussan (1977), Navara, Horvath and Kaleta (1978), Mhatre (1980, Ph.D. Thesis), Mhatre, Chaphekar, Ramani Rao, Patil, Haldar (1980), Shetye (1982, Ph.D. Thesis) and Giridhar (1984, Ph.D. Thesis): A Critical Review

* Paper was presented at the Proceedings of 4th Indian Paly. Conf., held on March 19–21, 1984 at Department of Environmental Sciences, Andhra University, Visakhapatnam – 530 003.

* Paper was presented at the Proceedings of 2nd Annual Conference of National Environmental Science Academy, held on May 25–27, 1985 at Awadh University, Faizabad.

* Paper was presented at the Proceedings of National Symposium Environmental Science and Warm Water Aquaculture–Finfish, held on November 6–9, 1985 at Department of Zoology, M.J.K. College, Bettiah – 845 438

* Paper was presented at the Proceedings of 1st National Symposium on Environmental Biology, held on December 30–31, 1986 at Department Zoology and Microbiology, Sir Krishnadevaraya University, Anantpur – 515 003

* Part of the Thesis submitted to University of Bombay for the partial fulfillment of Ph.D. in Botany.

the cultivars of all the series showed their germination even in the highest concentration (100 mg/ml) of gramoxone tried.

Keywords: *Palynology, Toxicology, Environmental sciences.*

Introduction

The use of vegetation as biological indicator of environmental quality has a long history dating back to the miners canary, to the recognition about 100 years ago. Recent studies have shown the feasibility of using natural vegetation for monitoring pollution (Berg, 1973; Brandt, 1974; Rasmussan, 1977; Navara, Horvath and Kaleta, 1978).

Materials and Methods

Pollen of successive flowers (*viz.* F, F-24, F-48, F-72 series *i.e.*, open flowers and the flower buds which require 24, 48, 72 hours to open respectively) of 5 cultivars of Apocynaceae *e.g.*, red-, pink- and white-flowered cultivars of *Nerium odorum* Soland. and pink- and white-flowered cultivars of *Catharanthlls rosells* (L.) G. Don. were collected at the stage of the dehiscence of anthers in the open flowers. Germination of pollen grains of successive flowers was studied by standing-drop technique in the optimum concentrations of sucrose supplemented with the wide range of concentrations (10^{-17}–10^{-2}–10^{-3}, 1, 5, 10, 20–20–100 mg/ml) of gramoxone (Table 29.1). The cultures then transferred to a moist filter chamber, stored at room temperature (19–31°C) having RH 51 per cent and in diffuse laboratory light. Observations were recorded 24 hours after incubation. For each experiment a random count of 200 grains was made to determine the percentage of pollen germination. For measurement of length of pollen tubes 50 tubes were selected randomly and measured at a magnification of 100x.

Results and Discussion

Pollen viability is a subject that has a great deal of practical as well as theoretical interest. In the present investigation even the different cultivars of the same species showed the variations in the percentage of their pollen viability (Table 29.1). Reduced pollen viability has been interpreted as an indication of suspected hybridity in wild populations. Nevertheless, variations in pollen viability may affect the breeding systems of the species concerned, and if the pollen viability can be altered by the environment, then the breeding system itself may be under some degree of environmental control.

As a rule the percentage of pollen germination is always less than the pollen viability. However, Banerji and Gangulee (1937) and Dharurkar (1971, Ph.D. Thesis) reported higher percentage of pollen germination than the pollen viability in *Eichhomia crassipes.* The claim of Banerjee and Gangulee (1937) and Dharurkar (1971) is challenged by Salgare (1986c, 95, 2000b, 06a, i, k, m, 0, 07) who stated that the observations of Banerjee and Gangulee (1937) and Dharurkar (1971) are exaggerating.

Potentiality of pollen germinability was recorded in F series of all the 5 cultivars of Apocynaceae studied. Pollen of F-24 series of red-flowered cultivar of *Nerillm odomm* and both the cultivars of *Catharanthus rose us* were found germinated in the optimum concentrations of sucrose. Pollen of F-48 and F-72 series of pink-flowered cultivar of C. *roseus* showed their germination in the optimum concentrations of sucrose. Thus, the potentiality of pollen germinability in Apocynaceae was observed in 10 out of 20 series investigated (Table 29.1).

Table 29.1: Inhibitory Effect of Gramoxone on Pollen Germination and Tube Growth of Successive Flowers of Five Cultivars of Apocynaccae

Cultivars	*Series*	*PV*	*IOCS*				*RCHI*		*PGTGSTCN*				*TC*
			SC	*PG*	*TG*	*V/O*	*RCPG*	*RCTG*	*HC*	*PG*	*TG*	*V/O*	
N. odorum pink-flowered	F	80	50	35	1485	12.38	10^{-17}–100	10^{-17}–100	NWO	NWO	NWO	NWO	NWO
N. odorum red- flowered	F	74	20	20	1250	11.36	60–100	10^{-9}–100	NWO	NWO	NWO	NWO	NWO
N. odorum white-flowered	F	62	50	20	0675	05.63	05–100	10^{-17}–100	NWO	NWO	NWO	NWO	NWO
C. roseus pink-flowered	F	90	20	60	1575	07.50	10^{-11}–100	10^{-11}–100	NWO	NWO	NWO	NWO	NWO
C. roseus white-flowered	F	88	20	40	1256	06.28	20–100	10^{-9}–100	NWO	NWO	NWO	NWO	NWO
N. odorum red-flowered	F–24	74	20	06	0485	04.41	Nil	10^{-9}–100	NWO	NWO	NWO	NWO	NWO
C. roseus pink-flowered	F–24	90	50	28	0240	01.14	01–100	10^{-17}–100	NWO	NWO	NWO	NWO	NWO
C. roseus white–flowered	F–24	88	50	16	0248	01.24	80–100	10–100	NWO	NWO	NWO	NWO	NWO
C. roseus pink-flowered	F–48	90	50	14	0095	00.45	80–100	60–100	NWO	NWO	NWO	NWO	NWO
C. roseus pink-flowered	F–72	90	80	10	0065	0031	Ng	Ng	Ng	Ng	Ng	Ng	Ng

HC: Concentrations of herbicide in mg/ml; IOCS: In optimum concentrations of sucrose germination of pollen and tube growth; Ng: No germination; PG: % inhibition caused by the herbicide in the germination of pollen; NOW: Not worked out; PGTGSTCH: Pollen germination a tube growth in sub-toxic concentrations of herbicide; PV: Pollen viability in per cent; RCHI: Range of concentrations of herbicide for inhibition of pollen germination and tube growth; RCPG: Range of concentrations of herbicide for inhibition of pollen germination; RCTG: Range of concentrations of herbicide for inhibition of pollen tube growth; SC: Optimum concentrations of sucrose in per cent; TG: % inhibition caused by t herbicide in tube growth; V/O: *in vitro* tube length in compare to *in vivo* in per cent.

Salgare (1983) observed the germination of pollen of F-72 series of pink-flowered cultivar of *Cathamnthus roseus in vitro* culture of sucrose. Trisa Palathingal (1990, M.Phil. Thesis) stated that the pollen of F-72 series of pink-flowered cultivar of *Catharanthus roseus* did not germinate in Brewbaker and Kwack's (1963) culture medium. This proves that the culture medium is also having the bearing on the germination of pollen. This points out that Brewbaker and Kwack's (1963).culture medium is not ideal for the pollen culture. This was also pointed out earlier by the author (2006a, i, m, 0, q, t, 07).

Even the lowest concentration (10–17 mg/ml) of gramoxone tried found to be toxic for the germination of pollen of F-72 series of pink-flowered cultivar of *C. roseus* (Table 29.1). Sharma (1984) reported the failure of the germination of pollen of F and F-24 series of duet and sonata and F-48 series of pink and red cascades by the lowest concentration (10–17 mg/ml) of gram ox one tried. All of them are the cultivars of *Petunia grandiflora*. Germination of pollen of F and F-24 series of light-violet-flowered and F-24 and F-48 series of white-flowered cultivars of *Petunia axillaris* was suppressed even by the lowest concentration (10-17 mg/ml) of gramoxone tried (Salgare, 1986a). This proves that the pollen of the said series are highly sensitive and acts as an ideal indicators of pollution. Thus it is confirmed that the pollen development and activity are more sensitive indicators of adverse factors in the botanical environment and the use of an entire vascular plant (Berg, 1973; Brandt, 1974; Vick and Bevan, 1976; Rasmussan, 1977; Navara, Horvath and Kaleta, 1978; Mhatre, 1980; Mhatre, Chaphekar, Ramani Rao, Patil, Haldar, 1980; Shetye, 1982 and Giridhar, 1984) as an indicator of pollution is a very crud method and rather a wrong choice. There is no evidence of any entire vascular plant exhibiting this much degree of sensitivity. This is very clearly confirmed in the present critical review (Table 1). This was already proved earlier by Salgare (1983, 84, 85a–c, 86a, d–e, 2000a, 01a–b, 05a–c, 06a, e–j, l–q, t, 07), Salgare and Theresa Sebastian (1986, 2006), Salgare and Phunguskar (2002), Salgare and Sanju Singh (2002, 06a–b), Salgare and Sancmta Pathak (2005), Sharma (1984) and Trisa Palathingal (1990) also supports the present findings.

Gramoxone inhibited the pollen germination in 8, 7, 7 series of Apocynaceae (Table 29.1, Salgare, 1983), *Petunia grandiflora* (Sharma, 1984) and *Petunia axillaris* (Salgare, 1986a) respectively.

The widest range of concentrations of gramoxone found to be 10^{-17}–100, 10^{-9}–100, 10^{-17}–10^{-11} mg/ml which inhibited the germination of pollen of Apocynaceae (in F series of pink-flowered cultivar of *Nerium odomm*) (Table 29.1, Salgare, 1983), *Petunia grandiflora* (in F series of red cascade) (Sharma, 1984) and *Petunia axillaris* (in F and F-24 series of series of violet-flowered cultivar) (Salgare, 1986a) respectively.

Sub-toxic concentration of gramoxone caused as high as 96.77 per cent inhibition in the pollen germination of *P. grandiflora* (in F-24 series of pink cascade) (Sharma, 1984).

Ratio between the series and inhibition caused by gramoxone (in sub-toxic concentration) in the germination of pollen is as (Inhibition is represented in the form of percentage):

F : F-24 : F-48 : F-72 = 00.00 : 00.00 : 00.00 : 00.00 in Apocynaceae (Table 29.1, Salgare, 1983)

F : F-24 : F-48 : F-72 = 96.28 : 91.40 : 94.74 : 00 : 00 in *Petunia grandiflora* (Sharma, 1984)

F : F-24 : F-48 : F-72 = 00.00 : 00.00 : 00.00 : 00.00 in *Petunia axillaris* (Table 29.1, Salgare, 1986a)

This shows that gramoxone caused maximum inhibition in the germination of pollen of F series of *Petunia grandiflora* (Sharma, 1984).

In many instances due to hyper- or hypo-nutrition the percentage of germination and length of the tube are considerably reduced. Bursting of pollen also increases and occasionally the pollen tubes were observed to eject their content. In addition to this various pollen tube deformities *viz.*, 'bloating' or 'bulla' formation resulting in the swelling of the tip of the pollen tube were also observed. In the pollen tubes that grew in the coiled or zig-zag manner the wall was not straight. *Catharanthus roseus* though characterized by the presence of monosiphonous condition at a low frequency bisiphonous and trisiphonous condition was also recorded in the present investigation along with the branched pollen tubes. In this connection it should be pointed out that Sudhakaran (1967) stated that in *Vinca rosea* L. [*Catharanthus roseus* (L.) G. Don.] besides pollen grains which produced single pollen tube, it has also been noticed that tetraploid grains frequently produce more than one pollen tube. Pollen tubes are branched quite frequently. Aberrations of this type in the pollen tube development are not observed in diploid pollen tubes, but quite frequently met with the pollen grains of irradiated plants. Salgare (1983) made it very clear that Sudhakaran (1967) had failed to trace out the branched pollen tubes and polysiphonous condition which is fairly common even in diploid pollen grains. Apart from this Sudhakaran (1967) was not able to report the various types of pollen tube deformities either with diploid or tetraploid gldins. Present investigations well as the extensive work of Salgare (1983, 86b, 2006b–d, f–g, i, m, o–p, r–t, 07) also proved that Sudhakaran's (1967) observations are superficial and misleading.

References

Berg, H., 1973. Plants as indicators of air pollution. *Toxicol.*, 1: 79–89.

Brandt, C.C., 1984. Indicators of environmental quality. In: (Ed.) W.A. Thomas. Plenum Press, New York, pp. 101.

Brewbaker, J.L. and Kwack, B.H., 1963. The essential role of Ca ion in pollen germination and pollen tube growth. *Amer. J. Bot.*, 50: 859–865.

Giridhar, B.A., 1984. Study of interactions between industrial air pollutants and plants. *Ph.D. Thesis,* University Bombay.

Mhatre, G.N., 1980. Studies in responses to heavy metals in industrial Environment. *Ph.D. Thesis,* University Bombay.

Mhatre, G.N., Chaphekar, S.B., Ramani Rao, I.V., Patil, M.R. and Haldar, B.C., 1980. Effect of industrial pollution on the Kalu river ecosystem. *Environ. Pollut.,* Series A 23: 67–78.

Navara, J., Horvath, I. and Kaleta, M., 1978. Contribution to the determination of limiting of values of SO_2 for vegetation in the region of Bratislava. *Environ. Pollut.,* Series A 16: 249–262.

Ram Indar, 1981. Effect of herbicides on plants of the Solanaceae–I. *M.Sc. Thesis,* University Bombay.

Rasmussan, L., 1977. Epiphytic bryophytes as indicators of changes in the background levels of air borne metals from 1951 to 1975. *Environ. Pollut.,* Series A 14: 34–45.

Salgare, S.A., 1983. Pollen physiology of Angiosperms. *Ph.D. Thesis,* University Bombay.

Salgare, S.A., 1984. Further evidence of a criticism of the hypothesis of Berg (1973), Brandt (1974), Vick and Bevan (1976), Rasmussan (1977), Navara, Horvath and Kaleta (1978), Mhatre (1980, Ph.D. Thesis), Mhatre, Chaphekar, Ramani Rao, Patil, Haldar (1980), Shetye (1982, Ph.D. Thesis) and Giridhar (1984, Ph.D. Thesis). In: *Proceedings* of 4th *Indian Palyno. Conference*, held on March 19–

21, 1984 at the Department of Environmental Science, Andhra University, Visakhapatnam – 530 003, Abstract No. SIII–06.

Salgare, S.A., 1985a. A criticism on the findings of Mhatre (1980, Ph.D. Thesis), Mhatre, Chaphekar, Ramani Rao, Patil, Haldar (1980), Shetye (1982, Ph.D. Thesis) and Giridhar (1984, Ph.D. Thesis). In: *Proceedings* of 2nd *Annual Conference of the National Environmental Science Academy*, held on May 25–27, 1985 at the Awadh University, Faizabad. Abstract No. 2.

Salgare, S.A., 1985b. Further evidence of the findings of Mhatre (1980, Ph.D. Thesis), Mhatre, Chaphekar, Ramani Rao, Patil, Haldar (1980), Shetye (1982, Ph.D. Thesis) and Giridhar (1984, Ph.D. Thesis). In: *Proceedings* of *National Symposium on Environmental Science and Warm Water Aquaculture–Finfish*, held on November 6–9, 1985 at the Department of 2001, MJ.K. College of Bihar University, Bettiah–845 438, Abstract No. 2.

Salgare, S.A., 1985c. A criticism on the findings of Mhatre (1980, Ph.D. Thesis), Mhatre, Chaphekar, Ramani Rao, Patil, Haldar (1980), Shetye (1982, Ph.D. Thesis) and Giridhar (1984, Ph.D. Thesis). In: *Proceedings* of *National Symposium on Environmental Science and Warm Water Aquaculture–Finfish*, held on November 6–9, 1985 at the Department of Zoology, M.J.K. College of Bihar University, Bettiah – 845 438, Abstract No. 73.

Salgare, S.A., 1986a. Effect of herbicides on pollen physiology of *Petunia axillaris* BSP. *Ph.D. Thesis,* World University.

Salgare, S.A., 1986b. A Criticism on *Ph.D. Thesis* of Sudhakaran (1967) entitled, '*Cytogenetic Studies in Vinca rosea Linn.*'. *Proceedings 1st National Symposium Environmental Biology*, held on December 30–31, 1986 at Department of Zoology and Microbiology, S.K. University, Anantapur – 515 003, Abstract No. 5.

Salgare, S.A., 1986c. A criticism on *Ph.D. Thesis* of Dharurkar (1971) entitled, 'Effect of herbicides on the cytomorphology of *Eichhornia crassipes* Solms (Mart.). *Ph.D. Thesis,* University of Bombay. In: *Proceedings of 1st National Symposium on Environmental Biology,* held on December 30–31, 1986 at Department of Zoology and Microbiology, S.K. University, Anantapur – 515 003, Abstract No. 6.

Salgare, S.A., 1986d. A criticism on *Ph.D. Thesis* of Giridhar (1984) entitled, 'Study of interactions between industrial air pollutants and plants'. *Ph.D. Thesis,* University Bombay. In: *Proceedings on 1st National Symposium on Environmental Biology*, held on December 30–31, 1986 at Department of Zoology and Microbiology, S.K. University, Anantapur – 515 003, Abstract No. 14.

Salgare, S.A., 1986e. A criticism on *Ph.D. Thesis* of Shetye (1982) entitled, 'Effect of heavy metals on plants'. *Ph.D. Thesis*, University Bombay. In: *Proceedings of 1st National Symposium on Environmental Biology,* held on December 30–31, 1986 at Department of Zoology and Microbiology, S.K. University, Anantapur – 515 003, Abstract No. 16.

Salgare, S.A. 1995. Interesting observations on the physiology of pollen of *Eichhornia crassipes.* Solam. (Mart). In: *Proceedings of 13th National Symposium on Life Science*, Proceeded with Annual Session of Indian Society of Life Science, held on December 30–31, 1995 and January 1, 1996 at Ch. Charan Singh University, Meerut – 250 004, Abstract No. 83.

Salgare, S.A., 2000a. Additional evidence of a criticism on the hypothesis of Berg (1973), Brandt (1974), Vick and Bevan (1976), Rasmussan (1977), Navara, Horvath and Kaleta (1978), Mhatre (1980, Ph.D. Thesis), Mhatre, Chaphekar, Ramani Rao, Patil, Haldar (1980), Shetye (1982, Ph.D. Thesis) and Giridhar (1984, Ph.D. Thesis). In: *Proceedings of 1st National Conference on Recent Trends in Lift Management*, held on October 22–23, 2000 at Bipin Bihari P.G. College, Jhansi – 284 001 (U.P.), Abstract No. 92, pp. 64–65.

Salgare, S.A., 2000b. A Criticism on the findings of Banerjee and Gangulee (1937) and Dharurkar (1971, Ph.D. Thesis). *Him. J. Environ. Zool.*, 14: 159–160.

Salgare, S.A., 2001a. Monitoring of herbicide (nitrogen) toxicity by using pollen as indicators: A critical review–I. Website www.microbiologyou.com. US.A's Publications (35 pages).

Salgare, S.A., 2001b. Monitors of pollution: A Criticism on the Hypothesis of Berg (1973), Brandt (1974), Vick and Bevan (1976), Rasmussan (1977), Navara, Horvath and Kaleta (1978), Mhatre (1980, Ph.D. Thesis), Mhatre, Chaphekar, Ramani Rao, Patil, Haldara (1980), Shetye (1982, Ph.D. Thesis) and Giridhar (1984, Ph.D. Thesis): A Critical Review–V. *Biojournal*, 13: 39–44.

Salgare, S.A., 2005a. Monitoring of herbicide (simazine) toxicity–A Criticism on the Hypothesis of Berg (1973), Brandt (1974), Vick and Bevan (1976), Rasmussan (1977), Navara, Horvath and Kaleta (1978), Mhatre (1980, Ph.D. Thesis), Mhatre, Chaphekar, Ramani Rao, Patil, Haldar (1980), Shetye (1982, Ph.D. Thesis) and Giridhar (1984, Ph.D. Thesis)–A Critical Review. *Him. J. Environ. Zool.*, 19: 69–71.

Salgare, S.A., 2005b. Monitoring of herbicide (dalapon) toxicity by using pollen as indicators Pollen of some cultivars of Apocynaceae and Further Evidence of a Criticism on the Hypothesis of Berg (1973), Brandt (1974), Vick and Bevan (1976), Rasmussan (1977), Navara, Horvath and Kaleta (1978), Mhatre (1980, Ph.D. Thesis), Mhatre, Chaphekar, Ramani Rao, Patil, Haldar (1980), Shetye (1982, Ph.D. Thesis) and Giridhar (1984, Ph.D. Thesis)–A Critical Review. *Flora and Fauna*, 11: 49–50

Salgare, S.A., 2005c. Monitoring of herbicide (basalin EC) toxicity–A Criticism on the Hypothesis of Berg (1973), Brandt (1974), Vick and Bevan (1976), Rasmussan (1977), Navara, Horvath and Kaleta (1978), Mhatre (1980, Ph.D. Thesis), Mhatre, Chaphekar, Ramani Rao, Patil, Haldar (1980), Shetye (1982, Ph.D. Thesis) and Giridhar (1984, Ph.D. Thesis)–A Critical Review. *Internat. J. Biosci. Reporter*, 3: 298–300.

Salgare, S.A., 2006a. Monitoring of herbicide (femoxone) toxicity by using pollen as indicators pollen of five cultivars of *Petunia axillaris:* Further Evidence of a Criticism of Banerjee and Gangulee (1937), Brewbaker and Kwack (1963), Dharurkar (1971, Ph.D. Thesis), Berg (1973), Brandt (1974), Vick and Bevan (1976), Rasmussan (1977), Navara, Horvath and Kaleta (1978), Mhatre (1980, Ph.D. Thesis), Mhatre, Chaphekar, Ramani Rao, Patil, Haldar (1980), Shetye (1982, Ph.D. Thesis) and Giridhar (1984, Ph.D. Thesis)–A Critical Review. *Environ. Conservation J.*, 7: 1–7.

Salgare, S.A., 2006b. Further evidence of a criticism of the findings of Sudhakaran (1967, Ph.D. Thesis) and Katre and Ghatnekar (1978). *Internal J. Biosci. Reporter*, 4: 1920.

Salgare, S.A., 2006c. Effect of acrolein on pollen germination and tube growth of stored pollen of Apocynaceae and further evidence of a criticism of the hypothesis of Sudhakaran (1967)–A Critical Review. *Him. J. Environ. Zool.*, 20: 137–139.

Salgare, S.A., 2006d. Effect of femoxone on pollen germination and tube growth of twelve hours stored pollen of Apocynaceae and further evidence of a criticism of the hypothesis of Sudhakaran (1967). *Asian J. Bio. Sci.*, 1: 160.

Salgare, S.A., 2006e. Monitoring of herbicide (basalin EC) toxicity by using pollen as indicators–Pollen of *Petunlia axillaries* and further evidence of a criticism of the ypothesis of Berg (1973), Brandt (1974), Vick and Bevan (1976), Rasmussan (1977), Navara, Horvath and Kaleta (1978), Mhatre (1980, Ph.D. Thesis), Mhatre, Chaphekar, Ramani Rao, Patil, Haldar (1980), Shetye (1982, Ph.D. Thesis) and Giridhar (1984, Ph.D. Thesis)–A critical review. *Him. J. Environ. Zool.*, 20: 269–271.

Salgare, S.A, 2006f. Alteration of resting period of pollen of Apocynaceae by herbicide (nitrofen): Further evidence of a criticism of Sudhakaran (1967–Ph.D. Thesis), Saoji and Chitaley (1972), Berg (1973), Brandt (1974), Vick and Bevan (1976), Rasmussan (1977), Navara, Horvath and Kaleta (1978), Mhatre (1980, Ph.D. Thesis), Mhatre, Chaphekar, Ramani Rao, Patil, Haldar (1980), Shetye (1982, Ph.D. Thesis) and Giridhar (1984, Ph.D. Thesis). *Internat. J Agri. Sci.*, 3: 239–243.

Salgare, S.A. 2006g. Effect of herbicide (acrolein) on pollen germination and tube growth of twelve hours stored pollen of five cultivars of Apocynaceae: Further Evidence of a Criticism of Sudhakaran (1967, Ph.D. Thesis), Berg (1973), Brandt (1974), Vick and Bevan (1976), Rasmussan (1977), Navara, Horvath and Kaleta (1978), Mhatre (1980, Ph.D. Thesis), Mhatre, Chaphekar, Ramani Rao, Patil, Haldar (1980), Shetye (1982, Ph.D. Thesis) and Giridhar (1984, Ph.D. Thesis)–A Critical Review. *J. Natcon.*, 8: 283–290.

Salgare, S.A, 2006h. Monitoring of Herbicide (2,4–D) toxicity by using pollen as indicators–Pollen of apocynaceae and further evidence of a criticism of the hypothesis of Berg (1973), Brandt (1974), Vick and Bevan (1976), Rasmussan (1977), Navara, Horvath and Kaleta (1978), Mhatre (1980, Ph.D. Thesis), Mhatre, Chaphekar, Ramani Rao, Patil, Haldar (1980), Shetye (1982, Ph.D. Thesis) and Giridhar (1984, Ph.D. Thesis)–A Critical Review. *Internat. J. Plant Sci.*, 1: 134–136.

Salgare, S.A., 2006i. Whether optimum pollen germination and tube length attained in the same growth medium (sucrose + 2,4–D) by five cultivars of the Apocynaceae: Further evidence of a criticism of Banerjee and Gangulee (1937), Brewbaker and Kwack's (1963), Sudhakaran (1967, Ph.D. Thesis), Dharurkar (1971, Ph.D. Thesis), Nair, Nambudiri and Thomas (1973), Berg (1973), Brandt (1974), Vick and Bevan (1976), Rasmussan (1977), Navara, Horvath and Kaleta (1978), Mhatre (1980, Ph.D. Thesis), Mhatre, Chaphekar, Ramani Rao, Patil, Haldar (1980), Shetye (1982, Ph.D. Thesis) and Giridhar (1984, Ph.D. Thesis)–A Critical Review. *Environ. Conservation J.*, 7: 21–29.

Salgare, S.A., 2006j. Evaluation of MH as male gametocide on *Phaseolus aureus* and a new method of plant breeding and further evidence of a criticism of the hypothesis of Nair, Nambudiri, Thomas (1973), Berg (1973), Brandt (1974), Vick and Bevan (1976), Rasmussan (1977), Navara, Horvath and Kaleta (1978), Mhatre (1980, Ph.D. Thesis), Mhatre, Chaphekar, Ramani Rao, Patil, Haldar (1980), Shetye (1982, Ph.D. Thesis) and Giridhar (1984, Ph.D. Thesis). *Internat. J. Plant Sci.*, 1: 352–356.

Salgare, S.A., 2006k. Further evidence of a criticism of the findings of Banerjee and Gangulee (1937) and Dharurkar (1971). *Internat. J. Biosci. Reporter*, 4: 69–170.

Salgre, S.A., 20061. Alteration of resting period of pollen of five cultivars of *Petunia axillaris* BSP by atrataf 50W: Further evidence of a criticism of Saoji and Chitaley (1972), Berg (1973), Brandt (1974), Vick and Bevan (1976), Rasmussan (1977), Navara, Horvath and Kaleta (1978), Mhatre (1980, Ph.D. Thesis), Mhatre, Chaphekar, Ramani Rao, Patil, Haldar (1980), Shetye (1982, Ph.D. Thesis) and Giridhar (1984, Ph.D. Thesis)–A Critical Review. *J. Natcon.*, 18: 353–360.

Salgare, S.A., 2006m. Monitoring of herbicide (2,4-dinitrophenol) toxicity by using pollen as indicators–Pollen of five cultivars of Apocynaceae: Further evidence of a criticism of Banerjee and Gangulee (1937), Brewbaker and Kwack (1963), Sudhakaran (1967, Ph.D. Thesis), Dharurkar (1971, Ph.D. Thesis), Berg (1973), Brandt (1974), Vick and Bevan (1976), Rasmussan (1977), Navara, Horvath and Kaleta (1978), Mhatre (1980, Ph.D. Thesis), Mhatre, Chaphekar, Ramani Rao, Patil, Haldar (1980), Shetye (1982, Ph.D. Thesis) and Giridhar (1984, Ph.D. Thesis)–A Critical Review. *Research Hunt*, 1: 1–6.

Salgre, S.A., 2006n. Alteration of resting period of pollen of five cultivars of *Petunia axillaris* BSP by atrataf SOW: Further evidence of a criticism of Saoji and Chitaley (1972), Berg (1973), Brandt (1974), Vick and Bevan (1976), Rasmussan (1977), Navara, Horvath and Kaleta (1978), Mhatre (1980, Ph.D. Thesis), Mhatre, Chaphekar, Ramani Rao, Patil, Haldar (1980), Shetye (1982, Ph.D. Thesis) and Giridhar (1984, Ph.D. Thesis)–A Critical Review. *J Natcon.*, 18: 357–364.

Salgare, S.A., 2006o. Whether optimum pollen germination and tube length attained in the same growth medium (sucrose + simazine) by five cultivars of Apocynaceae: Further evidence of a criticism of Banerjee and Gangulee (1937), Brewbaker and Kwack's (1963), Sudhakaran (1967, Ph.D. Thesis), Dharurkar (1971, Ph.D. Thesis), Nair, Nambudiri and Thomas (1973), Berg (1973), Brandt (1974), Vick and Bevan (1976), Rasmussan (1977), Navara, Horvath and Kaleta (1978), Mhatre (1980, Ph.D. Thesis), Mhatre, Chaphekar, Ramani Rao, Patil, Haldar (1980), Shetye (1982, Ph.D. Thesis) and Giridhar (1984, Ph.D. Thesis)–A Critical Review. *Res. Hunt*, 1: 146–155.

Salgare, S.A., 2006p. Effect of herbicide (acrolein) on pollen germination and tube growth of twelve hours stored pollen of five cultivars of Apocynaceae: Further evidence of a criticism of Sudhakaran (1967, Ph.D. Thesis), Berg (1973), Brandt (1974), Vick and Bevan (1976), Rasmussan (1977), Navara, Horvath and Kaleta (1978), Mhatre (1980, Ph.D. Thesis), Mhatre, Chaphekar, Ramani Rao, Patil, Haldar (1980), Shetye (1982, Ph.D. Thesis) and Giridhar (1984, Ph.D. Thesis)–A Critical Review. *J Natcon.*, 18: 361–368.

Salgare, S.A., 2006q. Alteration of resting period of pollen of five cultivars of *Petunia axillaris* BSP by Gramoxone: Further evidence of a criticism of Brewbaker and Kwack's (1963), Saoji and Chitaley (1972), Berg (1973), Brandt (1974), Vick and Bevan (1976), Rasmussan (1977), Navara, Horvath and Kaleta (1978), Mhatre (1980, Ph.D. Thesis), Mhatre, Chaphekar, Ramani Rao, Patil, Haldar (1980), Shetye (1982, Ph.D. Thesis) and Giridhar (1984, Ph.D. Thesis)–A Critical Review. *Environ. Conservation J.*, 7: 51–57.

Salgare, S.A., 2006r. Effect of gramoxone on pollen germination and tube growth of twelve hours stored pollen of Apocynaceae and further evidence of a criticism of the hypothesis of Sudhakaran (1967)–A Critical Review–II. *Plant Archives*, 6: 389–390.

Salgare, S.A., 2006s. Effect of atrataf 50W on pollen germination and tube growth of twelve hours stored pollen of Apocynaceae and further evidence of a criticism of Sudhakaran (1967)–A critical review–I. *Internat. J. Biosci. Reporter*, 4: 204–206.

Salgare, S.A., 2006t. Whether optimum pollen germination and tube length attained in the same growth medium (sucrose + simazine) by five cultivars of Apocynaceae: Further evidence of a criticism of Banerjee and Gangulee (1937), Brewbaker and Kwack's (1963), Sudhakaran (1967, Ph.D. Thesis), Dharurkar (1971, Ph.D. Thesis), Nair, Nambudiri and Thomas (1973), Berg (1973), Brandt (1974), Vick and Bevan (1976), Rasmussan (1977), Navara, Horvath and Kaleta (1978), Mhatre (1980, Ph.D. Thesis), Mhatre, Chaphekar, Ramani Rao, Patil, Haldar (1980), Shetye (1982, Ph.D. Thesis) and Giridhar (1984, Ph.D. Thesis)–A Critical Review. *Res. Hunt*, 1: 146–155.

Salgare, S.A., 2007. Effect of herbicide (MH) on pollen germination and tube growth of twelve hours stored pollen of Apocynaceae: Further evidence of a criticism of Banerjee and Gangulee (1937), Brewbaker and Kwack (1963), Sudhakaran (1967, Ph.D. Thesis), Dharurkar (1971, Ph.D. Thesis), Berg (1973), Brandt (1974), Vick and Bevan (1976), Rasmussan (1977), Navara, Horvath and Kaleta (1978), Mhatre (1980, Ph.D. Thesis), Mhatre, Chaphekar, Ramani Rao, Patil, Haldar (1980), Shetye (1982, Ph.D. Thesis) and Giridhar (1984, Ph.D. Thesis)–A critical review. *Internat. J. Plant Sci.,* 2: 215–219.

Salgare, S.A. and Phunguskar, K.P., 2002. Monitors of pesticide (alphamethrin) toxicity by using pollen as indicators–Pollen of pink-flowered *Catharanthus roseus*–A criticism on the hypothesis of Berg (1973), Brwdt (1974), Vick and Bevan (1976), Rasmussan (1977), Navara, Horvath and Kaleta (1978), Mhatre (1980, Ph.D. Thesis), Mhatre, Chaphekar, Ramani Rao, Patil, Haldara (1980), Shetye (1982, Ph.D. Thesis) and Giridhar (1984, Ph.D. Thesis)–A Critical Review–III. *Biojournal,* 14: 1–4.

Salgare, S.A. and Pathak, Sanchita, 2005. Monitoring of heavy metal (bismuth nitrate) toxicity by using pollen as indicators–Pollen of white-flowered *Catharanthus roseus* and further evidence of a criticism on the hypothesis of Berg (1973), Brandt (1974). Vick and Bevan (1976), Rasmussan (977), Navara, Horvath and Kaleta (1978), Mhatre (1980, Ph.D. Thesis), Mhatre, Chaphekar, Ramani Rao, Patil, Haldar (1980), Shetye (1982, Ph.D. Thesis), and Giridhar (1984, Ph.D. Thesis)–A Critical Review. *Flora and Fauna,* 11: 69–70.

Salgare, S.A. and Singh, Sanju, 2002. Monitors of heavy metal (ferrous sulphate) toxicity by using pollen as indicators–Pollen of white-flowered *Catharanthus roseus*–A criticism on the hypothesis of Berg (1973), Brandt (1974), Vick and Bevan (1976), Rasmussan (1977), Navara, Horvath and Kaleta (1978), Mhatre (1980, Ph.D. Thesis), Mhatre, Chaphekar, Ramani Rau, Patil, Haldara (1980), Shetye (1982, Ph.D. Thesis) and Giridhar (1984, Ph.D. Thesis)–A Critical Review–II. *Biojournal,* 14: 5–7.

Salgare, S.A. and Singh, Sanju, 2006a. Monitoring of heavy metal (copper sulphate) toxicity by using pollen as indicators–Pollen of white-flowered *Catharanthlls roseus* and further evidence of a criticism on the hypothesis of Berg (1973), Brandt (1974), Vick and Bevan (1976), Rasmussan (1977), Navara, Horvath and Kaleta (1978), Mhatre (1980, Ph.D. Thesis), Mhatre, Chaphekar, Ramani Rao, Patil, Haldar (1980), Shetye (1982, Ph.D. Thesis), and Giridhar (1984, Ph.D. Thesis)–A critical review–I. *Him. J. Environ. Zool.,* 23: 276–278.

Salgare, S.A. and Singh, Sanju, 2006b. Monitoring of heavy metal (cobalt nitrate) toxicity by using pollen as indicators–Pollen of pink-flowered cultivar of *Catharanthus roseus* and further evidence of a criticism of the hypothesis of Berg (1973), Brandt (1974), Vick and Bevan (1976), Rasmussan (1977), Navara, Horvath and Kaleta (1978), Mhatre (1980, Ph.D. Thesis), Mhatre, Chaphekar, Ramani Rao, Patil, Haldar (1980), Shetye (1982, Ph.D. Thesis) and Giridhar (1984, Ph.D. Thesis)–A critical review– II. *Him. J. Environ. Zool.,* 20: 283–284.

Salgare, S.A. and Sebastian, Theresa, 1986. Further evidence of a criticism of the hypothesis of Berg (1973), Brandt (1974), Vick and Bevan (1976), Rasmussan (1977), Navara, Horvath and Kaleta (1978), Mhatre (1980, Ph.D. Thesis), Mhatre, Chaphekar, Ramani Rao, Patil, Haldar (1980), Shetye (1982, Ph.D. Thesis) and Giridhar (1984, Ph.D. Thesis). In: *1st National Symposium on Environmental Biology,* held on December 30–31, 1986 at Department of Zoology and Microbiology, S.K. University, Anantapur – 515 003. Abstract No. 15.

Salgare, S.A. and Sebastian, Theresa, 2006. Monitoring of herbicide (dalapon) toxicity by using pollen as indicators–Pollen of urid and further evidence of a criticism of Berg (1973), Brandt (1974), Vick and Bevan (1976), Rasmussan (1977), Navara, Horvath and Kaleta (1978), Mhatre (1980, Ph.D. Thesis), Mhatre, Chaphekar, Ramani Rao, Patil, Haldar (1980), Shetye (1982, Ph.D. Thesis) and Giridhar (1984, Ph.D. Thesis)–A critical review–I. *Internat. J. Biosci. Reporter*, 4: 323–329.

Sharma, R.I., 1984. Effect of herbicides on plants of the Solanaceae–II. *Ph.D. Thesis,* University Bombay.

Shetye, R.P., 1982. Effect of heavy metals on plants. *Ph.D. Thesis,* University Bombay.

Trisa, Palathingal, 1990. Evaluation of industrial pollution of Bombay by pollen–I. *M.Phil. Thesis,* University Mumbai.

Chapter 30

Alteration of Resting Period of Pollen of Apocynaceae by MH*

S.A. Salgare

Salgare Research Foundation Pvt. Ltd., Prathamesh Society, Shivaji Chowk, Karjat – 410 201, M.S., India
E-mail: drsalgare@rediffmail.com and drsalgare@sancharnet.in

ABSTRACT

Maleic Hydrazide (MH) (1,2-dihydropyridazine, 3-6-dione) altered the resting period of pollen of all the series except for F series of pink-flowered cultivar of *Catharanthus roseus* (L.) G. Don. Pollen of F-24, F-48 and F-72 series of pink-flowered cultivar of *Catharanthus roseus* (L.) G.Don. failed to germinate even 24 hours of their sowing indicating the extension of their resting period. However, MH reduced the resting period of pollen in rest of the series.

Keywords: *Palynology, Toxicology, Environmental sciences.*

---

* Alteration of Resting Period of pollen of Pollen of Five Cultivars of Apocynaceae by MH: Further evidence of a criticisms of Brewbaker and Kwack's (1963), Sudhakaran (1967, Ph.D. Thesis), Saoji and Chitaley (1972), Berg (1973), Brandt (1974), Vick and Bevan (1976), Rasmussan (1977), Navara, Horvath and Kaleta (1978), Mhatre (1980, Ph.D. Thesis), Mhatre, Chaphekar, Ramani Rao, Patil, Haldar (1980), Shetye (1982, Ph.D. Thesis) and Giridhar (1984, Ph.D. Thesis): A Critical Review

* Paper was presented at the Proceedings of 4th Indian Paly. Conf., held on March 19–21, 1984 at Department of Environmental Sciences, Andhra University, Visakhapatnam – 530 003

* Paper was presented at the Proceedings of 2nd Annual Conference of National Environmental Science Academy, held on May 25–27, 1985 at Awadh University, Faizabad

* Paper was presented at the Proceedings of 1st National Symposium on Environmental Biology, held on December 30–31, 1986 at Department Zoology and Microbiology, Sir Krishnadevaraya University, Anantpur – 515 003

* Part of the Thesis submitted to University of Bombay for the partial fulfillment of Ph.D. in Botany.

Introduction

Palynology, in recent years has attracted the attention of workers of different disciplines on account of its numerous applications to problems of plant taxonomy, genetics, geology, medical and agricultural sciences. Pollen physiology furnishes the information required for effecting hybridization of plants growing in different geographical and climatic regions with blooms in different seasons.

Materials and Methods

Pollen of successive flowers (*viz.* F, F-24, F-48, F-72 series *i.e.*, open flowers and the flower buds which require 24, 48, 72 hours to open respectively) of 5 cultivars of Apocynaceae *e.g.*, red-, pink- and white-flowered cultivars of *Nerium odorum* Soland. and pink- and white-flowered cultivars of *Catharanthus roseus* (L.) G. Don. were collected at the stage of the dehiscence of anthers in the open flowers. Germination of pollen grains of successive flowers was studied by standing drop technique in the optimum concentrations of sucrose as well as in the optimum concentrations of sucrose supplemented with the optimum concentrations of Maleic Hydrazide (MH) (1,2-dihydropyridazine, 3-6-dione) (Table 30.1). The rate of pollen germination of successive flowers was determined by fixing the cultures at one hour intervals. Such preparations were continued for 10 hours. Observations on the germination of pollen were recorded 24 hours after incubation.

Table 31.1: Effect of MH on the Rate of Pollen Germination of Successive Flowers of Apocynaceae

Cultivars	Series	PV	Conc.		TRFPG	
			SC	HC	C	T
N. odorum pink–flowered	F	80	50	10^{-17}	Ng_1	3
N. odorum red–flowered	F	74	20	10^{-17}	Ng_1	8
N. odorum white–flowered	F	62	50	10^{-17}	Ng_1	10
C. roseus pink–flowered	F	90	20	10^{-17}	1	1
C. roseus white–flowered	F	88	20	10^{-17}	2	1
N. odorum red–flowered	F–24	74	20	10^{-17}	Ng_1	9
C. roseus pink–flowered	F–24	90	50	Ng_2	Ng_2	Ng_2
C. roseus white–flowered	F–24	88	50	10^{-17}	10	2
C. roseus pink–flowered	F–48	90	50	Ng_2	Ng_2	Ng_2
C. roseus pink–flowered	F–72	90	80	Ng_2	Ng_2	Ng_2

C: In control sets time required for germination of pollen in optimum concentrations of sucrose; CH: Optimum concentrations of herbicide in mg/ml; Conc.: Optimum concentrations of sucrose and herbicide; SC: Optimum concentrations of sucrose in per cent; Ng_1, and Ng_2: No germination of pollen even after 10 and 24 hours of sowing respectively; T: Time required for germination of pollen in optimum concentrations of sucrose + herbicide (in treated sets); TRFPG: Time required for the germination of pollen in control sets and treated sets.

Results and Discussion

Potentiality of pollen germinability was recorded in F series of all the 5 cultivars of the Apocynaceae studied. It was the pollen of F-24 series of red-flowered cultivar of *Nerium odorum* and both the cultivars of *Catharanthus roseus* found germinated in the optimum concentrations of sucrose. It should be pointed out that the pollen of F-48 and F-72 series of pink-flowered cultivar of *C. roseus* showed their germination in the optimum concentrations of sucrose. Thus, the potentiality of pollen germinability in Apocynaceae was observed in 10 out of 20 series investigated (Table 30.1).

Germination of pollen of F-72 series of pink-flowered cultivar of *Catharanthus roseus in vitro* culture of sucrose was noted in the present investigation. However, Trisa Palathingal (1990, M.Phil. Thesis) failed to germinate the pollen of F-72 series of pink-flowered cultivar of *C. roseus* in Brewbaker and Kwack's (1963) culture medium. This proves that the culture medium is also having the bearing on the germination of pollen. This also confirms that Brewbaker and Kwack's (1963) culture medium is not ideal for pollen cultures. This was also pointed out earlier by the author (2006a, k, q, s, u, 07a).

Even the lowest concentration (10–17 mg/ml) of MH tried suppressed the germinability of pollen of F-24, F-48 and F-72 series of pink-flowered cultivar of *Catharanthus roseus* (Table 30.1). Sharma (1984) stated that the germinability of pollen of *F* series of duet and sonata, F-24 series of red and white cascades, duet and sonata and F-48 series of all the 3 cascades was prevented even by the lowest concentration (10–17 mg/ml) of MH tried. All of them are the cultivars of *Petunia grandiflora.* Singh (1985) reported the failure of germination of pollen of F series of brinjal round and F-24 of brinjallong, muktakeshi and round even by the lowest concentration (10–17 mg/ml) of MH tried. All of them are the cultivars of *Solanum melongena.* This proves that the pollen of the said series are highly sensitive and acts as an ideal indicators of pollution. Thus, it is confirmed that the pollen development and activity are more sensitive indicators of adverse factors in the botanical environment and the use of an entire vascular plant (Berg, 1973; Brandt, 1974; Vick and Bevan, 1976; Rasmussan, 1977; Navara, Horvath and Kaleta, 1978; Mhatre, 1980, Ph.D.Thesis; Mhatre, Chaphekar, Ramani Rao, Patil, Haldar, 1980; Shetye, 1982, Ph.D. Thesis and Giridhar, 1984, Ph.D.Thesis) as an indicator of pollution is a very crud method and rather a wrong choice. There is no evidence of any entire vascular plant exhibiting this much degree of sensitivity. This is confirmed in the present critical review (Table 30.1) as well as by the previous extensive work of Salgare (1983, 84b, 85a, c–d, 86a, c–d, 2000, 01a–b, 05a, c, e, 06a, f–g, i–k, m, o, q–u, 07a), Salgare and Theresa Sebastian (1986a), Salgare and Phunguskar (2002), Salgare and Sanju Singh (2002), Salgare and Sanchita Pathak (2005).

The delay in pollen germination was interpreted by Saoji and Chitaley (1972) as being due to the grains not being mature enough to effect pollination, immediately after being shed from the anther. Further they stated that 4-5 hours are required for the complete maturation of pollen grains. It was Salgare (1983) who pointed out of the first time that the pollen require resting period before germination and it was the failure of Saoji and Chitaley (1972) who misinterpreted the resting period for pollen maturity. Further he (1983) stated that this resting period differs species to species which is also noted in the present investigation (Table 30.1). This resting period is altered by different chemicals. Present work as well as the extensive work of Salgare (1983, 84, 85, 2001, 04, 05a, b), Salgare and Theresa Sebastian (1986b), Sharma (1984) and Singh (1985), Salgare and Shashi Yadav (2002, *05*). Recently Salgare and Sanchita Pathak (2002, 05b) and Salgare and Sanju Singh (2006a) made it very clear that Saoji and Chitaley's (1972) arguments are superficial and misleading.

MH altered the resting period of pollen of 6 and failed only in one series of the Apocynaceae (Table 30.1). The herbicide reduced the resting period of pollen of all the 6 series. MH caused maximum reduction in the resting period of the pollen of F series of pink-flowered cultivar of *Nerium odorum.* Alteration of the resting period of pollen by the herbicide was also noted by the author (1983, 84a, 85b, 86a, e, 2001c, 04, 05b, d, 06e, g–h, i, o, r, u, 07b), Salgare and Theresa Sebastian (1986b), Sharma (1984) and Singh (1985). Alteration of resting period of pollen of successive flowers by the minerals was noted by Salgare and Shashi Yadav (2002, 05). Recently Salgare and Sanchita Pathak (2002, 05b) and Salgare and Sanju Singh (2006a) noted the alteration of resting period of pollen by the heavy metal.

Sudhakaran (1967) stated that in *Vinca rosea* L. [*Catharanthus roseus* (L.) G. Don.] besides pollen grains which produced single pollen tube, it has also been noticed that tetraploid grains trequently produce more than one pollen tube. Pollen tubes are branched quite frequently. Aberrations of this type in the pollen tube development are not observed in diploid pollen tubes, but quite trequently met with the pollen grains of irradiated plants. Salgare (1983, 86b, 2006a, c, f) made it very clear that Sudhakaran (1967) had failed to trace out the branched pollen tubes and polysiphonous condition which is fairly common even in diploid pollen grains. Apart from this Sudhakaran (1967) was not able to report the various types of pollen tube deformities either with diploid or tetraploid grains. Present findings as well as the previous work of Salgare (1983, 86b, 2006b–d, g, i, k, n, p–q, s–t, 07a) also proved that Sudhakaran's (1967) observations are superficial and misleading.

References

Berg, H., 1973. Plants as indicators of air pollution. *Toxicol.,* 1: 79–89.

Brandt, C.C., 1984. Indicators of environmental quality. In: (Ed.) W.A. Thomas. Plenum Press, New York, pp. 101.

Brewbaker, J.L. and Kwack, B.H., 1963. The essential role of Ca ion in pollen germination and pollen tube growth. *Amer. J. Bot.,* 50: 859–865.

Giridhar, B.A., 1984. Study of interactions between industrial air pollutants and plants. *Ph.D. Thesis,* University Bombay.

Mhatre, G.N., 1980. Studies in responses to heavy metals in industrial Environment. *Ph.D. Thesis,* University Bombay.

Mhatre, G.N., Chaphekar, S.B., Ramani Rao, I.V., Patil, M.R. and Haldar, B.C., 1980. Effect of industrial pollution on the Kalu river ecosystem. *Environ. Pollut.,* Series A 23: 67–78.

Navara, J., Horvath, I. and Kaleta, M., 1978. Contribution to the determination of limiting of values of SO_2 for vegetation in the region of Bratislava. *Environ. Pollut.,* Series A 16: 249–262.

Rasmussan, L., 1977. Epiphytic bryophytes as indicators of changes in the background levels of air borne metals from 1951 to 1975. *Environ. Pollut.,* Series A 14: 34–45.

Salgare, S.A., 1983. Pollen physiology of Angiosperms. *Ph.D. Thesis,* University Bombay.

Salgare, S. A. 1984a. A criticism on the paper of Saoji and Chitaley (1972) entitled, 'Palynological studies in *Bauhinia variegata* Linn.'. In: *Proceedings of 4th Indian Palyno. Conference,* held on March 19–21, 1984 at Department of Environmental Science, Andhra University, Visakhpatnam – 530 003, Abstract No. SIII–05.

Salgare, S.A., 1984. Further evidence of a criticism of the hypothesis of Berg (1973), Brandt (1974), Vick and Bevan (1976), Rasmussan (1977), Navara, Horvath and Kaleta (1978), Mhatre (1980, Ph.D. Thesis), Mhatre, Chaphekar, Ramani Rao, Patil, Haldar (1980), Shetye (1982, Ph.D. Thesis) and Giridhar (1984, Ph.D. Thesis). In: *Proceedings of 4th Indian Palyno. Conference,* held on March 19–21, 1984 at the Department of Environmental Science, Andhra University, Visakhapatnam – 530 003, Abstract No. SIII–06.

Salgare, S.A., 1985a. A criticism on the findings of Mhatre (1980, Ph.D. Thesis), Mhatre, Chaphekar, Ramani Rao, Patil, Haldar (1980), Shetye (1982, Ph.D. Thesis) and Giridhar (1984, Ph.D. Thesis). In: *Proceedings* of 2nd *Annual Conference of the National Environmental Science Academy,* held on May 25–27, 1985 at the Awadh University, Faizabad. Abstract No. 2.

Salgare, S.A., 1985b. A criticism on the hypothesis of Saoji and Chitaley (1972). In: *Proceeding of 2nd Annual Conference of National Environmental Science Academy*, held on May 25–27, 1985 at Awadh University, Faizabad, Abstract No. 72.

Salgare, S.A., 1985c. Further evidence of the findings of Mhatre (1980, Ph.D. Thesis), Mhatre, Chaphekar, Ramani Rao, Patil, Haldar (1980), Shetye (1982, Ph.D. Thesis) and Giridhar (1984, Ph.D. Thesis). In: *Proceedings* of *National Symposium on Environmental Science and Warm Water Aquaculture–Finfish*, held on November 6–9, 1985 at the Department of 2001, MJ.K. College of Bihar University, Bettiah–845 438, Abstract No. 2.

Salgare, S.A., 1985d. A criticism on the findings of Mhatre (1980, Ph.D. Thesis), Mhatre, Chaphekar, Ramani Rao, Patil, Haldar (1980), Shetye (1982, Ph.D. Thesis) and Giridhar (1984, Ph.D. Thesis). In: *Proceedings* of *National Symposium on Environmental Science and Warm Water Aquaculture–Finfish*, held on November 6–9, 1985 at the Department of Zoology, M.J.K. College of Bihar University, Bettiah – 845 438, Abstract No. 73.

Salgare, S.A., 1986a. Effect of herbicides on pollen physiology of *Petunia axillaris* BSP. *Ph.D. Thesis,* World University.

Salgare, S.A., 1986b. A Criticism on *Ph.D. Thesis* of Sudhakaran (1967) entitled, '*Cytogenetic Studies in Vinca rosea Linn.*'. *Proceedings of 1st National Symposium Environmental Biology*, held on December 30–31, 1986 at Department of Zoology and Microbiology, S.K. University, Anantapur – 515 003, Abstract No. 5.

Salgare, S.A., 1986c. A criticism on *Ph.D. Thesis* of Giridhar (1984) entitled, 'Study of interactions between industrial air pollutants and plants'. *Ph.D. Thesis,* University Bombay. In: *Proceedings on 1st National Symposium on Environmental Biology*, held on December 30–31, 1986 at Department of Zoology and Microbiology, S.K. University, Anantapur – 515 003, Abstract No. 14.

Salgare, S.A., 1986d. A criticism on *Ph.D. Thesis* of Shetye (1982) entitled, 'Effect of heavy metals on plants'. *Ph.D. Thesis*, University Bombay. In: *Proceedings of 1st National Symposium on Environmental Biology,* held on December 30–31, 1986 at Department of Zoology and Microbiology, S.K. University, Anantapur – 515 003, Abstract No. 16.

Salgare, S.A., 1986e. Further evidence of a criticism on the findings of Saoji and Chitaley (1972). In: *Proceeding of 1st National Symposium on Environmental Biology*, held on December 30–31, 1986 at S.K. University, Anantapur – 515 003, Abstract No. 17.

Salgare, S.A., 2000. Additional evidence of a criticism on the hypothesis of Berg (1973), Brandt (1974), Vick and Bevan (1976), Rasmussan (1977), Navara, Horvath and Kaleta (1978), Mhatre (1980, Ph.D. Thesis), Mhatre, Chaphekar, Ramani Rao, Patil, Haldar (1980), Shetye (1982, Ph.D. Thesis) and Giridhar (1984, Ph.D. Thesis). In: *Proceedings of 1st National Conference on Recent Trends in Lift Management*, held on October 22–23, 2000 at Bipin Bihari P.G. College, Jhansi – 284 001 (U.P.), Abstract No. 92, pp. 64–65.

Salgare, S.A., 2001a. Monitoring of herbicide (nitrogen) toxicity by using pollen as indicators: A critical review–I. Website www.microbiologyou.com. US.A's Publications (35 pages).

Salgare, S.A., 2001b. Monitors of pollution: A Criticism on the Hypothesis of Berg (1973), Brandt (1974), Vick and Bevan (1976), Rasmussan (1977), Navara, Horvath and Kaleta (1978), Mhatre (1980, Ph.D. Thesis), Mhatre, Chaphekar, Ramani Rao, Patil, Haldara (1980), Shetye (1982, Ph.D. Thesis) and Giridhar (1984, Ph.D. Thesis): A Critical Review–V. *Biojournal,* 13: 39–44.

Salgare, S.A., 2001c. Resting period of pollen–A criticism on the hypothesis of Saoji and Chitaley (1972)–A Critical Review–V. *Biojournal*, 13: 45–48.

Salgare, S.A., 2004. Resting period of pollen (sucrose + 2,4–Dinitrophenol)–A criticism of the hypothesis of Saoji and Chitaley (1972)–A Critical Review. *Him. J. Environ. Zool.*, 18: 69–71.

Salgare, S.A., 2005a. Monitoring of herbicide (simazine) toxicity–A Criticism on the Hypothesis of Berg (1973), Brandt (1974), Vick and Bevan (1976), Rasmussan (1977), Navara, Horvath and Kaleta (1978), Mhatre (1980, Ph.D. Thesis), Mhatre, Chaphekar, Ramani Rao, Patil, Haldar (1980), Shetye (1982, Ph.D. Thesis) and Giridhar (1984, Ph.D. Thesis)–A Critical Review. *Him. J. Environ. Zool.*, 19: 69–71.

Salgare, S.A., 2005b. Alteration of resting period of pollen of Apocynaceae by herbicide (femoxone) and a criticism on the hypothesis of Saoji and Chitaley (1972)–A Critical Review. *Him. J. Environ. Zool.*, 19:73–75.

Salgare, S.A., 2005c. Monitoring of herbicide (dalapon) toxicity by using pollen as indicators Pollen of some cultivars of Apocynaceae and Further Evidence of a Criticism on the Hypothesis of Berg (1973), Brandt (1974), Vick and Bevan (1976), Rasmussan (1977), Navara, Horvath and Kaleta (1978), Mhatre (1980, Ph.D. Thesis), Mhatre, Chaphekar, Ramani Rao, Patil, Haldar (1980), Shetye (1982, Ph.D. Thesis) and Giridhar (1984, Ph.D. Thesis)–A Critical Review. *Flora and Fauna*, 11: 49–50.

Salgare, S.A., 2005d. Alteration of resting period of pollen of some cultivars of Apocynaceae by herbicide (gramoxone) and further evidence of a criticism on the hypothesis of Saoji and Chitaley (1972)–A Critical Review. *Flora and Fauna*, 11: 161–162.

Salgare, S.A., 2005e. Monitoring of herbicide (basalin EC) toxicity–A Criticism on the Hypothesis of Berg (1973), Brandt (1974), Vick and Bevan (1976), Rasmussan (1977), Navara, Horvath and Kaleta (1978), Mhatre (1980, Ph.D. Thesis), Mhatre, Chaphekar, Ramani Rao, Patil, Haldar (1980), Shetye (1982, Ph.D. Thesis) and Giridhar (1984, Ph.D. Thesis)–A Critical Review. *Internat. J. Biosci. Reporter*, 3: 298–300.

Salgare, S.A., 2006a. Monitoring of herbicide (femoxone) toxicity by using pollen as indicators pollen of five cultivars of *Petunia axillaris:* Further Evidence of a Criticism of Banerjee and Gangulee (1937), Brewbaker and Kwack (1963), Dharurkar (1971, Ph.D. Thesis), Berg (1973), Brandt (1974), Vick and Bevan (1976), Rasmussan (1977), Navara, Horvath and Kaleta (1978), Mhatre (1980, Ph.D. Thesis), Mhatre, Chaphekar, Ramani Rao, Patil, Haldar (1980), Shetye (1982, Ph.D. Thesis) and Giridhar (1984, Ph.D. Thesis)–A Critical Review. *Environ. Conservation J.*, 7: 1–7.

Salgare, S.A., 2006b. Further evidence of a criticism of the findings of Sudhakaran (1967, Ph.D. Thesis) and Katre and Ghatnekar (1978). *Internal J. Biosci. Reporter*, 4: 1920.

Salgare, S.A., 2006c. Effect of acrolein on pollen germination and tube growth of stored pollen of Apocynaceae and further evidence of a criticism of the hypothesis of Sudhakaran (1967)–A Critical Review. *Him. J. Environ. Zool.*, 20: 137–139.

Salgare, S.A., 2006d. Effect of femoxone on pollen germination and tube growth of twelve hours stored pollen of Apocynaceae and further evidence of a criticism of the hypothesis of Sudhakaran (1967). *Asian J. Bio. Sci.*, 1: 160.

Salgare, S.A., 2006e. Alteration of resting period of pollen of Apocynaceae by herbicide (acrolein) and a criticism on the hypothesis of Saoji and Chitaley (1972)–A Critical Review. *Internat. J. Biosci. Reporter*, 4: 33–35.

Salgare, S.A., 2006f. Monitoring of herbicide (basalin EC) toxicity by using pollen as indicators–Pollen of *Petunlia axillaries* and further evidence of a criticism of the ypothesis of Berg (1973), Brandt (1974), Vick and Bevan (1976), Rasmussan (1977), Navara, Horvath and Kaleta (1978), Mhatre (1980, Ph.D. Thesis), Mhatre, Chaphekar, Ramani Rao, Patil, Haldar (1980), Shetye (1982, Ph.D. Thesis) and Giridhar (1984, Ph.D. Thesis)–A critical review. *Him. J. Environ. Zool.*, 20: 269–271.

Salgare, S.A., 2006g. Alteration of resting period of pollen of Apocynaceae by herbicide (nitrofen): Further evidence of a criticism of Sudhakaran (1967–Ph.D. Thesis), Saoji and Chitaley (1972), Berg (1973), Brandt (1974), Vick and Bevan (1976), Rasmussan (1977), Navara, Horvath and Kaleta (1978), Mhatre (1980, Ph.D. Thesis), Mhatre, Chaphekar, Ramani Rao, Patil, Haldar (1980), Shetye (1982, Ph.D. Thesis) and Giridhar (1984, Ph.D. Thesis). *Internat. J Agri. Sci.*, 3: 239–243.

Salgare, S.A., 2006h. Alteration of resting period of pollen of Apocynaceae by herbicide (sodium penta chloro phenate) and further evidence of a criticism of the hypothesis of Saoji and Chitaley (1972). *Asian J. Bio Sci.*, 1: 161–162.

Salgare, S.A. 2006i. Effect of herbicide (acrolein) on pollen germination and tube growth of twelve hours stored pollen of five cultivars of Apocynaceae: Further Evidence of a Criticism of Sudhakaran (1967, Ph.D. Thesis), Berg (1973), Brandt (1974), Vick and Bevan (1976), Rasmussan (1977), Navara, Horvath and Kaleta (1978), Mhatre (1980, Ph.D. Thesis), Mhatre, Chaphekar, Ramani Rao, Patil, Haldar (1980), Shetye (1982, Ph.D. Thesis) and Giridhar (1984, Ph.D. Thesis)–A Critical Review. *J. Natcon.*, 8: 283–290.

Salgare, S.A, 2006j. Monitoring of Herbicide (2,4–D) toxicity by using pollen as indicators–Pollen of apocynaceae and further evidence of a criticism of the hypothesis of Berg (1973), Brandt (1974), Vick and Bevan (1976), Rasmussan (1977), Navara, Horvath and Kaleta (1978), Mhatre (1980, Ph.D. Thesis), Mhatre, Chaphekar, Ramani Rao, Patil, Haldar (1980), Shetye (1982, Ph.D. Thesis) and Giridhar (1984, Ph.D. Thesis)–A Critical Review. *Internat. J. Plant Sci.*, 1: 134–136.

Salgare, S.A., 2006k. Whether optimum pollen germination and tube length attained in the same growth medium (sucrose + 2,4–D) by five cultivars of the Apocynaceae: Further evidence of a criticism of Banerjee and Gangulee (1937), Brewbaker and Kwack's (1963), Sudhakaran (1967, Ph.D. Thesis), Dharurkar (1971, Ph.D. Thesis), Nair, Nambudiri and Thomas (1973), Berg (1973), Brandt (1974), Vick and Bevan (1976), Rasmussan (1977), Navara, Horvath and Kaleta (1978), Mhatre (1980, Ph.D. Thesis), Mhatre, Chaphekar, Ramani Rao, Patil, Haldar (1980), Shetye (1982, Ph.D. Thesis) and Giridhar (1984, Ph.D. Thesis)–A Critical Review. *Environ. Conservation J.*, 7: 21–29.

Salgare, S.A., 20061. Alteration of resting period of pollen of Apocynaceae by herbicide (dalapon) and further evidence of a criticism of the hypothesis of Saoji and Chitaley (1972)–A Critical Review. *Plant Archives*, 6: 385–386.

Salgare, S.A., 2006m. Evaluation of MH as male gametocide on *Phaseolus aureus* and a new method of plant breeding and further evidence of a criticism of the hypothesis of Nair, Nambudiri, Thomas (1973), Berg (1973), Brandt (1974), Vick and Bevan (1976), Rasmussan (1977), Navara, Horvath and Kaleta (1978), Mhatre (1980, Ph.D. Thesis), Mhatre, Chaphekar, Ramani Rao, Patil, Haldar (1980), Shetye (1982, Ph.D. Thesis) and Giridhar (1984, Ph.D. Thesis). *Internat. J. Plant Sci.*, 1: 352–356.

Salgare, S.A., 2006n. Effect of gramoxone on pollen germination and tube growth of twelve hours stored pollen of Apocynaceae and further evidence of a criticism of the hypothesis of Sudhakaran (1967)–A Critical Review–II. *Plant Archives*, 6: 389–390.

Salgre, S.A., 2006o. Alteration of resting period of pollen of five cultivars of *Petunia axillaris* BSP by atrataf 50W: Further evidence of a criticism of Saoji and Chitaley (1972), Berg (1973), Brandt (1974), Vick and Bevan (1976), Rasmussan (1977), Navara, Horvath and Kaleta (1978), Mhatre (1980, Ph.D. Thesis), Mhatre, Chaphekar, Ramani Rao, Patil, Haldar (1980), Shetye (1982, Ph.D. Thesis) and Giridhar (1984, Ph.D. Thesis)–A Critical Review. *J. Natcon.*, 18: 353–360.

Salgare, S.A., 2006p. Effect of atrataf 50W on pollen germination and tube growth of twelve hours stored pollen of Apocynaceae and further evidence of a criticism of Sudhakaran (1967)–A critical review–I. *Internat. J. Biosci. Reporter*, 4: 204–206.

Salgare, S.A., 2006q. Monitoring of herbicide (2,4-dinitrophenol) toxicity by using pollen as indicators–Pollen of five cultivars of Apocynaceae: Further evidence of a criticism of Banerjee and Gangulee (1937), Brewbaker and Kwack (1963), Sudhakaran (1967, Ph.D. Thesis), Dharurkar (1971, Ph.D. Thesis), Berg (1973), Brandt (1974), Vick and Bevan (1976), Rasmussan (1977), Navara, Horvath and Kaleta (1978), Mhatre (1980, Ph.D. Thesis), Mhatre, Chaphekar, Ramani Rao, Patil, Haldar (1980), Shetye (1982, Ph.D. Thesis) and Giridhar (1984, Ph.D. Thesis)–A Critical Review. *Research Hunt*, 1: 1–6.

Salgre, S.A., 2006r. Alteration of resting period of pollen of five cultivars of *Petunia axillaris* BSP by atrataf SOW: Further evidence of a criticism of Saoji and Chitaley (1972), Berg (1973), Brandt (1974), Vick and Bevan (1976), Rasmussan (1977), Navara, Horvath and Kaleta (1978), Mhatre (1980, Ph.D. Thesis), Mhatre, Chaphekar, Ramani Rao, Patil, Haldar (1980), Shetye (1982, Ph.D. Thesis) and Giridhar (1984, Ph.D. Thesis)–A Critical Review. *J Natcon.*, 18: 357–364.

Salgare, S.A., 2006s. Whether optimum pollen germination and tube length attained in the same growth medium (sucrose + simazine) by five cultivars of Apocynaceae: Further evidence of a criticism of Banerjee and Gangulee (1937), Brewbaker and Kwack's (1963), Sudhakaran (1967, Ph.D. Thesis), Dharurkar (1971, Ph.D. Thesis), Nair, Nambudiri and Thomas (1973), Berg (1973), Brandt (1974), Vick and Bevan (1976), Rasmussan (1977), Navara, Horvath and Kaleta (1978), Mhatre (1980, Ph.D. Thesis), Mhatre, Chaphekar, Ramani Rao, Patil, Haldar (1980), Shetye (1982, Ph.D. Thesis) and Giridhar (1984, Ph.D. Thesis)–A Critical Review. *Res. Hunt*, 1: 146–155.

Salgare, S.A., 2006t. Effect of herbicide (acrolein) on pollen germination and tube growth of twelve hours stored pollen of five cultivars of Apocynaceae: Further evidence of a criticism of Sudhakaran (1967, Ph.D. Thesis), Berg (1973), Brandt (1974), Vick and Bevan (1976), Rasmussan (1977), Navara, Horvath and Kaleta (1978), Mhatre (1980, Ph.D. Thesis), Mhatre, Cbaphekar, Ramani Rao, Patil, Haldar (1980), Shetye (1982, Ph.D. Thesis) and Giridhar (1984, Ph.D. Thesis)–A Critical Review. *J. Natcon.*, 18: 361–368.

Salgare, S.A., 2006u. Alteration of resting period of pollen of five cultivars of *Petunia axillaris* BSP by Gramoxone: Further evidence of a criticism of Brewbaker and Kwack's (1963), Saoji and Chitaley (1972), Berg (1973), Brandt (1974), Vick and Bevan (1976), Rasmussan (1977), Navara, Horvath and Kaleta (1978), Mhatre (1980, Ph.D. Thesis), Mhatre, Chaphekar, Ramani Rao, Patil, Haldar (1980), Shetye (1982, Ph.D. Thesis) and Giridhar (1984, Ph.D. Thesis)–A Critical Review. *Environ. Conservation J.*, 7: 51–57.

Salgare, S.A., 2007a. Effect of herbicide (MH) on pollen germination and tube growth of twelve hours stored pollen of Apocynaceae: Further evidence of a criticism of Banerjee and Gangulee (1937), Brewbaker and Kwack (1963), Sudhakaran (1967, Ph.D. Thesis), Dharurkar (1971, Ph.D. Thesis), Berg (1973), Brandt (1974), Vick and Bevan (1976), Rasmussan (1977), Navara, Horvath and

Kaleta (1978), Mhatre (1980, Ph.D. Thesis), Mhatre, Chaphekar, Ramani Rao, Patil, Haldar (1980), Shetye (1982, Ph.D. Thesis) and Giridhar (1984, Ph.D. Thesis)–A critical review. *Internat. J. Plant Sci.*, 2: 215–219.

Salgare, S.A 2007b. Variation in the resting period of pollen of successive flowers of five forms of *Petunia axillaris* BSP *in vitro* culture of sugars (D-glucose and sucrose) and further evidence of a criticism of Saoji and Chitaley (1972)–A critical review. *Internat. J. Plant Sci.*, 2: 231–233.

Salgare, S.A. and Phunguskar, K.P., 2002. Monitors of pesticide (alphamethrin) toxicity by using pollen as indicators–Pollen of pink-flowered *Catharanthus roseus*–A criticism on the hypothesis of Berg (1973), Brwdt (1974), Vick and Bevan (1976), Rasmussan (1977), Navara, Horvath and Kaleta (1978), Mhatre (1980, Ph.D. Thesis), Mhatre, Chaphekar, Ramani Rao, Patil, Haldara (1980), Shetye (1982, Ph.D. Thesis) and Giridhar (1984, Ph.D. Thesis)–A Critical Review–III. *Biojournal*, 14: 1–4.

Salgare, S.A. and Pathak, Sanchita, 2002. Effect of heavy metal (manganous sulphate) on the resting period of pollen of pink-flowered *Catharanthus roseus* and a criticism on the hypothesis of Saoji and Chitaley (1972)–A Critical Review–I. *Biojournal*, 14: 9–12.

Salgare, S.A. and Pathak, Sanchita, 2005a. Monitoring of heavy metal (bismuth nitrate) toxicity by using pollen as indicators–Pollen of white-flowered *Catharanthus roseus* and further evidence of a criticism on the hypothesis of Berg (1973), Brandt (1974). Vick and Bevan (1976), Rasmussan (977), Navara, Horvath and Kaleta (1978), Mhatre (1980, Ph.D. Thesis), Mhatre, Chaphekar, Ramani Rao, Patil, Haldar (1980), Shetye (1982, Ph.D. Thesis), and Giridhar (1984, Ph.D. Thesis)–A Critical Review. *Flora and Fauna*, 11: 69–70.

Salgare, S.A. and Pathak, Sanchita, 2005b. Effect of heavy metal (manganous sulphate) on the resting period of pollen of pink-flowered *Catharanthus roseus* and further evidence of a criticism to the hypothesis of Saoji and Chitaley (1972)–A Critical Review. *Flora and Fauna*, 11: 197–198.

Salgare, S.A. and Singh, Sanju, 2002. Monitors of heavy metal (ferrous sulphate) toxicity by using pollen as indicators–Pollen of white-flowered *Catharanthus roseus*–A criticism on the hypothesis of Berg (1973), Brandt (1974), Vick and Bevan (1976), Rasmussan (1977), Navara, Horvath and Kaleta (1978), Mhatre (1980, Ph.D. Thesis), Mhatre, Chaphekar, Ramani Rau, Patil, Haldara (1980), Shetye (1982, Ph.D. Thesis) and Giridhar (1984, Ph.D. Thesis)–A Critical Review–II. *Biojournal*, 14: 5–7.

Salgare, S.A. and Singh, Sanju, 2006a. Effect of heavy metal (cobalt nitrate) on resting period of pollen of pink-flowered *Catharanthus roseus* and further evidence of a criticism on the hypothesis of Saoji and Chitaley (1972)–A critical review. *Him. J. Environ. Zool.*, 20: 135–136.

Salgare, S.A. and Singh, Sanju, 2006b. Monitoring of heavy metal (copper sulphate) toxicity by using pollen as indicators–Pollen of white-flowered *Catharanthlls roseus* and further evidence of a criticism on the hypothesis of Berg (1973), Brandt (1974), Vick and Bevan (1976), Rasmussan (1977), Navara, Horvath and Kaleta (1978), Mhatre (1980, Ph.D. Thesis), Mhatre, Chaphekar, Ramani Rao, Patil, Haldar (1980), Shetye (1982, Ph.D. Thesis), and Giridhar (1984, Ph.D. Thesis)–A critical review–I. *Him. J. Environ. Zool.*, 23: 276–278.

Salgare, S.A. and Singh, Sanju, 2006c. Monitoring of heavy metal (cobalt nitrate) toxicity by using pollen as indicators–Pollen of pink-flowered cultivar of *Catharanthus roseus* and further evidence of a criticism of the hypothesis of Berg (1973), Brandt (1974), Vick and Bevan (1976), Rasmussan (1977), Navara, Horvath and Kaleta (1978), Mhatre (1980, Ph.D. Thesis), Mhatre, Chaphekar,

Ramani Rao, Patil, Haldar (1980), Shetye (1982, Ph.D. Thesis) and Giridhar (1984, Ph.D. Thesis)–A critical review– II. *Him. J. Environ. Zool.*, 20: 283–284.

Salgare, S.A. and Yadav, Shashi, 2002. Alteration of resting period of pollen of white-flowered *Catharanthus roseus* by mineral (calcium nitrate) and a criticism on the hypothesis of Saoji and Chitaley (1972)–A Critical Review–I. *Biojournal*, 14: 17–20.

Salgare, S.A. and Yadav, Shashi, 2005. Effect of mineral (potassium sulphate) on the resting period of pollen of white-flowered *catharanthus roseus* and further evidence of a criticism on the hypothesis of Saoji and Chitaley (1972)–A Critical Review. *Flora and Fauna*, 11: 171–172.

Salgare, S.A. and Sebastian, Theresa, 1986a. Further evidence of a criticism of the hypothesis of Berg (1973), Brandt (1974), Vick and Bevan (1976), Rasmussan (1977), Navara, Horvath and Kaleta (1978), Mhatre (1980, Ph.D. Thesis), Mhatre, Chaphekar, Ramani Rao, Patil, Haldar (1980), Shetye (1982, Ph.D. Thesis) and Giridhar (1984, Ph.D. Thesis). In: *1st National Symposium on Environmental Biology*, held on December 30–31, 1986 at Department of Zoology and Microbiology, S.K. University, Anantapur – 515 003. Abstract No. 15.

Salgare, S.A. and Sebastian, Theresa, 1986b. Further evidence of a criticism on the findings of Saoji and Chitaley (1972). In: *Proceeding of 1st National Symposium Environmental Biology*, held on December 30–31, 1986 at Department of Zoology and Microbiology, S.K. University, Anantapur – 515 003. Abstract No. 17.

Salgare, S.A. and Sebastian, Theresa, 2006. Monitoring of herbicide (dalapon) toxicity by using pollen as indicators–Pollen of urid and further evidence of a criticism of Berg (1973), Brandt (1974), Vick and Bevan (1976), Rasmussan (1977), Navara, Horvath and Kaleta (1978), Mhatre (1980, Ph.D. Thesis), Mhatre, Chaphekar, Ramani Rao, Patil, Haldar (1980), Shetye (1982, Ph.D. Thesis) and Giridhar (1984, Ph.D. Thesis)–A critical review–I. *Internat. J. Biosci. Reporter*, 4: 323–329.

Saoji, A.A. and Chitaley, S.D., 1972. Palynological studies in *Bauhinia variegata* Linn. *Botanique*, 3: 7–12.

Sharma, R.I., 1984. Effect of herbicides on plants of the Solanaceae–II. *Ph.D. Thesis*, University Bombay.

Shetye, R.P., 1982. Effect of heavy metals on plants. *Ph.D. Thesis*, University Bombay.

Singh, S.R., 1985. Effect of herbicides on pollen physiology of Brinjal. *M.Sc. Thesis*, University Bombay.

Trisa, Palathingal, 1990. Evaluation of industrial pollution of Bombay by pollen–I. *M.Phil. Thesis*, University Mumbai.

Chapter 31

Stimulatory Effect of Atrataf 50W on Pollen Germination and Tube Growth of *Petunia axillaries**

S.A. Salgare

Salgare Research Foundation Pvt. Ltd., Prathamesh Society, Shivaji Chowk, Karjat – 410 201, M.S., India
E-mail: drsalgare@rediffmail.com and drsalgare@sancharnet.in

ABSTRACT

All the concentrations (100^{-17}–10^{-2}–10^{-3}, 1, 5, 10, 20–20–100 mg/ml) of atrataf 50W tried failed to stimulate the germination as well as tube growth of all the 5 cultivars of *Petunia axillaries*.

Keywords: *Physiology of pollen, Palylnology, Toxicology, Environmental sciences,*

Introduction

The residual matter of the herbicides left over in the soil may, further, proves to be an important factor in the growth of the subsequent crops.

---

* Whether Optimum Pollen Germination and Tube Length Attained in the same Growth Medium (Sucrose + Atrataf 50W) by Five Cultivars of *Petunia Axillaris* BSP: Further evidence of a criticisms of Banerjee and Ganguli (1937), Brewbaker and Kwack's (1963), Dharukar (1971, Ph.D. Thesis), Nair, Nambudiri and Thomas (1973), Berg (1973), Brandt (1974), Vick and Bevan (1976), Rasmussan (1977), Navara, Horvath and Kaleta (1978), Mhatre (1980, Ph.D. Thesis), Mhatre, Chaphekar, Ramani Rao, Patil, Haldar (1980), Shetye (1982, Ph.D. Thesis) and Giridhar (1984, Ph.D. Thesis): A Critical Review.

Materials and Methods

Pollen of successive flowers (*viz.*, F, F–24, F–48, F–72 series *i.e.*, open flowers and the flower buds which require 24, 48, 72 hours to open respectively) of 5 cultivars (light-violet-, pink-, violet-, white- and white-violet-flowered) of *Petunia axillaris* BSP were collected soon after the dehiscence of anthers in the open flowers. Germination of pollen grains was studied by standing drop technique in the optimum concentrations of sucrose which acts as control as well as in the optimum concentrations of sucrose supplemented with the wide range of concentrations (10^{-17}–10^{-2}–10^{-3}, 1, 5, 10, 20–20–100 mg/ml) of atrataf 50W. Pollen grains were incubated soon after the dehiscence of anthers. The cultures then transferred to a moist filter chamber, stored at room temperature (29.3–32.5°C) having RH 64 per cent and in diffuse laboratory light. The experiments were run in triplicate and average results were recorded. Observations on the germination of pollen and tube growth were recorded 24 hours after incubation. For each experiment a random count of 200 grains was made to determine the percentage of pollen germination. For measurement of length of pollen tubes, 50 tubes were selected randomly and measured at a magnification of 100X.

Results and Discussion

Pollen viability is a subject that has a great deal of practical as well as theoretical interest. In the present investigation even the different cultivars of the same species showed the variations in the percentage of their pollen viability (Table 31.1). Reduced pollen viability has been interpreted as an indication of suspected hybridity in wild populations. Nevertheless, variations in pollen viability may affect the breeding systems of the species concerned, and if the pollen viability can be altered by the environment, then the breeding system itself may be under some degree of environmental control.

As a rule the percentage of pollen germination is always less than the pollen viability. However, Banerjee and Gangulee (1937) and Dharurkar (1971, Ph.D. Thesis) reported higher percentage of pollen germination than the pollen viability in *Eichhomia crassipes.* The claim of Banerjee and Gangulee (1937) and Dharurkar (1971) is challenged by Salgare (1986b, 95, 2000b, 06a, g, j, l, n, 07) who stated that the observations of Banerjee and Gangulee (1937) and Dharurkar (1971) are exaggerating.

Potentiality of pollen germinability was noted in F and F-24 series of all the 5 cultivars of *Ptunia axillaris* and in F-48 series of white-flowered cultivar of *P. axillaris.* Thus the potentiality of pollen germinability in *P. axillaris* was recorded in 11 out of 20 series investigated (Table 31.1).

Salgare (1983) observed the germination of pollen of F-72 series of pink-flowered cultivar of *Catharanthus roseus in vitro* culture of sucrose. Trisa Palathingal (1990, M.Phil. Thesis) stated that the pollen of F-72 series of pink-flowered cultivar of *C. roseus* did not germinate in Brewbaker and Kwack's (1963) culture medium. This confirms that Brewbaker and Kwack's (1963) culture medium is not perfect. This also proves that the culture medium is also having the bearing on the germination of pollen. This pointed out that Brewbaker and K wack's (1963) culture medium is not ideal for pollen culture. This was also pointed out earlier by the author (2006a, g, l, n, p, 07).

Even the lowest concentration (10^{-17} mg/ml) of atrataf 50W tried found to be toxic for the germination of F and F-48 series of white-flowered cultivar and F-24 series of light-violet-, pink, violet-, white-flowered cultivars of *Petunia axillaris* (Table 31.1, Salgare, 1986a). Pollen of F–24 series of red-flowered cultivar of *Nerium odorum* and F–72 series of pink-flowered cultivar of *Catharanthus roseus* failed to germinate in the sucrose medium supplemented with the 10^{-17} mg/ml of atrataf 50W (Salgare, 1983). Even the lowest concentration (10^{-17} mg/ml) of atrataf 50W tried suppressed the germination of pollen of F series of duet and sonata, F-24 series of pink and white cascades, duet and sonata and F-48 series

Table 31.1: Effect of Atrataf 50W on Pollen Germination and Tube Growth of Successive Flowers of Five Cultivars of *Petunia axillaris* BSP

Cultivars	*Series*	*PV*	*IOCS*				*RCHS*		*PGTGSTCH*				*V/O*
			SC	*PG*	*TG*	*V/O*	*RCPG*	*RCTG*	*OCH*	*SPG*	*OCH*	*STG*	
Light-violet-	F	76	50	32	030	0.09	Nil	Nil	10^{-15}	Nil	10^{-17}	Nil	0.09
Pink-	F	93	50	28	035	0.11	Nil	Nil	10^{-15}	Nil	10^{-17}	Nil	0.11
Violet-	F	80	50	25	038	0.11	Nil	Nil	10^{-15}	Nil	10^{-17}	Nil	0.09
White-	F	95	30	34	080	0.24	No	Ng	Ng	Ng	Ng	Ng	Ng
White-violet-	F	90	30	30	325	0.88	Nil	Nil	10^{-15}	Nil	10^{-17}	Nil	0.07
Light-violet-	F-24	76	30	25	045	0.14	No	Ng	Ng	Ng	Ng	Ng	Ng
Pink-	F-24	93	10	16	030	0.09	Ng	Ng	Ng	Ng	Ng	Ng	Ng
Violet-	F-24	80	60	25	030	0.09	No	Ng	Ng	Ng	Ng	Ng	Ng
White–	F-24	95	10	26	030	0.09	Ng	No	Ng	Ng	Ng	Ng	Ng
White-violet-	F-24	90	30	30	210	0.57	Nil	Nil	10^{-15}	Nil	10^{-17}	Nil	0.08
White-	F-48	95	10	13	40	0.12	Ng	Na	Ng	No	Ng	Ng	Ng

IOCS: In optimum concentrations of sucrose germination of pollen and tube growth; OCH: Optimum concentrations of herbicide in mg/ml for germination of pollen and tube growth; PGTGSTCH: Pollen germination and tube growth in optimum concentrations of the herbicide; PV: Pollen viability in per cent; RCHS: range of concentrations of herbicide for stimulation of pollen germination and tube growth; RCPG: Range of concentrations of herbicide for stimulation of pollen germination; RCTG: Range of concentrations, herbicide for stimulation of pollen tube growth; SC: Optimum concentrations of sucrose in per cent; SPG: Stimulation in pollen germination in per cent; STG: Stimulation in pollen tube growth (in μm) in per cent; V/O: *in vitro* tube length in compare to *in vivo* in per cent.

of all the 3 cascades. All of them are the cultivars of *Petunia grandiflora* (Sharma, 1984). This proves that the pollen of the said series are highly sensitive and acts as an ideal indicator of pollution. Thus, it is confirmed that the pollen development and activity are more sensitive indicators of adverse factors in the botanical environment and the use of an entire vascular plant (Berg, 1973; Brandt, 1974; Vick and Bevan 1976; Rasmussan, 1977; Navara, Horvath and Kaleta, 1978; Mhatre, 1980, Ph.D. Thesis; Mhatre, Chaphekar, Ramani Rao, Patil, Haldar, 1980; Shetye, 1982, Ph.D. Thesis and Giridhar, 1984, Ph.D. Thesis) as an indicator of pollution is a very crud method and rather a wrong choice. There is no evidence of any entire vascular plant exhibiting this much degree of sensitivity. This is also confirmed in the present critical review (Table 31.1). This was already proved earlier by the extensive work of Salgare (1983, 84, 85a–c, 86a, d–e, 2000a, 01a–b, 04, 05a, c, e, 06a, c–g, i , k–p, 07), Salgare and Theresa Sebastian (1986), Salgare and Phunguskar (2002), Salgare and Sanju Singh (2002) and Salgare and Sanchita Pathak (2005), Sharma (1984) and Trisa Palathingal (1990) also supports the present findings.

All the concentrations (10^{-17}–10^{-2}–10^{-3}, 1, 5, 10, 20–20–100 mg/ml) of atrataf 50W tried failed to stimulate the germination as well as tube growth of all the 5 cultivars of *Petllnia axillaris* (Table 31.1, Salgare, 1986a). Salgare (1983) confirmed 10^{-17}–100 mg/ml atrataf 50W as the widest range of concentrations for the Apocynaceae (in F-24 series of white-flowered cultivar of *Catharanthus roseus*) which stimulated the germination of pollen. It was Sharma (1984) who confirmed 10^{-17}– 10^{-15} mg/ml atrataf 50W as the widest range of concentrations which stimulated the germination of pollen of *Petunia grandiflora* (in F and F-24 series of red cascade).

Atrataf 50W stimulated the germination of pollen of 7 and 2 series of the Apocynaceae (Salgare, 1983) and *Petunia grandiflora* (Sharma, 1984) respectively.

Ratio between the series and stimulation produced by atrataf 50W (in an optimum concentration) in the germination of pollen in the number of series is as

F : F–24 : F–48 : F–72 = 0 : 0 : 0 : 0 in *Petunia axillaris* (Table 31.1–Salgare, 1986a)

F : F–24 : F–48 : F–72 = 4 : 2 : 1 : 0 in Apocynaceae (Salgare, 1983)

F : F–24 : F–48 : F–72 = 1 : 1 : 0 : 0 in *Petunia grandiflora* (Sharma, 1984)

This shows that atrataf 50W stimulated the germination of pollen in maximum number of series of F series in the Apocynaceae and in an equal number of series in F and F-24 series in *Petunia grandiflora*.

All the concentrations (10^{-17}–10^{-2}–10^{-3}, 1, 5, 10, 20–20–100 mg/ml) of atrataf 50W tried failed to stimulate the pollen tube growth of *Petunia axillaris* (Table 31.1, Salgare, 1986a) and *Petunia grandiflora* (Sharma, 1984). However, atrataf 50W stimulated the pollen tube growth of 5 series of the Apocynaceae (Salgare, 1983).

Tube length *in vitro* culture (sucrose + atrataf 50W) of atrataf 50W (in sub-toxic concentration) is 0.03 per cent in *Fetllnia axillaris* (in F series of light-violet, pink-, violet- and white-violet-flowered and F-24 series of white-violet-flowered cultivars) (Table 31.1, (Salgare, 1986a) and 0.71 per cent in the Apocynaceae (in F series of white-flowered cultivar of *Nerillm odorum*) (Salgare, 1983) of the tube length found *in vivo* is the longest of all the cultivars investigated.

It should be pointed out that in a few cases the length of the tubes in cultures does equal that in nature (Knight, 1917; Schoch-Bodmer, 1921; Brink, 1924; Branscheidt, 1929, 30; Ehlers, 1951; Vasil, 1960).

Pollen germination and tube elongation are two distinct processes differing in their sensitivity to different concentrations of the herbicide was also confirmed with the present work (Table 31.1, Salgare, 1986a). However, Nair, Nambudiri and Thomas (1973) stated that it has been significant that the optimum percentage of germination and tube length were attained in the same growth medium. However, with the present work (Table 31.1) as well as previous extensive work of Salgare (1979, 83, 86a, c, 2004, 05b, d, 06b, g–i , n), Sharma (1984), Salgare and Bindu (2002, 05) and Salgare and Tessy Mol Antony (2005a, b) it could be concluded that the observations of Nair, Nambudiri and Thomas (1973) are superficial and misleading.

References

Berg, H., 1973. Plants as indicators of air pollution. *Toxicol.,* 1: 79–89.

Brandt, C.C., 1984. Indicators of environmental quality. In: (Ed.) W.A. Thomas. Plenum Press, New York, pp. 101.

Branscheidt, P., 1929. Die Befruchtungs verhaltnisse beim Obst und der Rebe. *Gartenbauwiss,* 2: 158–270.

Branscheidt, P., 1930. Zur Physiologie der Pollenkeimung und ihere experimentellen Beeinflussung. *Planta,* 11: 368–453.

Brewbaker, J.L. and Kwack, B.H., 1963. The essential role of Ca ion in pollen germination and pollen tube growth. *Amer. J. Bot.,* 50: 859–865.

Brink, R.A., 1921. The physiology of pollen. III. Growth *in vitro* and *in vivo. Amer. J. Bot.,* 11: 351–364.

Ehlers, H., 1951. Untersuchungen Zur Emahrungsphysiologie der Pollenschlauche. *Biol. Zentralbl.,* 70:432–451.

Giridhar, B.A., 1984. Study of interactions between industrial air pollutants and plants. *Ph.D. Thesis,* University Bombay.

Knight, L.I., 1917. Physiological aspects of the self-sterility of the apple. In: *Proc. Amer. Soc. Hort. Sci.,* 14: 101–105.

Mhatre, G.N., 1980. Studies in responses to heavy metals in industrial Environment. *Ph.D. Thesis,* University Bombay.

Mhatre, G.N., Chaphekar, S.B., Ramani Rao, I.V., Patil, M.R. and Haldar, B.C., 1980. Effect of industrial pollution on the Kalu river ecosystem. *Environ. Pollut.,* Series A 23: 67–78.

Nair, P.K.K., Nambudiri, E.M.V. and Thomas, M.K., 1973. A note on pollen germination at various stages of development of flower buds of Balsam (*Impatians balsam*). *J. Palyno.,* 4:29–33.

Navara, J., Horvath, I. and Kaleta, M., 1978. Contribution to the determination of limiting of values of SO_2 for vegetation in the region of Bratislava. *Environ. Pollut.,* Series A 16: 249–262.

Rasmussan, L., 1977. Epiphytic bryophytes as indicators of changes in the background levels of air borne metals from 1951 to 1975. *Environ. Pollut.,* Series A 14: 34–45.

Salgare, S.A., 1979. A criticism of Nair, Nambudiri and Thomas' paper entitled, 'A note on pollen germination at various stages of development of flower buds of balsam (*Impatiens balsam*)'. In: *Proceedings of 66th Session Indian Science Congress,* held on January 3–7, 1979 at Hyderabad, Section of Botany, 3: 61–62, Abstract No. 131.

Salgare, S.A., 1983. Pollen physiology of Angiosperms. *Ph.D. Thesis,* University Bombay.

Salgare, S.A., 1984. Further evidence of a criticism of the hypothesis of Berg (1973), Brandt (1974), Vick and Bevan (1976), Rasmussan (1977), Navara, Horvath and Kaleta (1978), Mhatre (1980, Ph.D. Thesis), Mhatre, Chaphekar, Ramani Rao, Patil, Haldar (1980), Shetye (1982, Ph.D. Thesis) and Giridhar (1984, Ph.D. Thesis). In: *Proceedings of 4th Indian Palyno. Conference*, held on March 19–21, 1984 at the Department of Environmental Science, Andhra University, Visakhapatnam – 530 003, Abstract No. SIII–06.

Salgare, S.A., 1985a. A criticism on the findings of Mhatre (1980, Ph.D. Thesis), Mhatre, Chaphekar, Ramani Rao, Patil, Haldar (1980), Shetye (1982, Ph.D. Thesis) and Giridhar (1984, Ph.D. Thesis). In: *Proceedings* of *2nd Annual Conference of the National Environmental Science Academy*, held on May 25–27, 1985 at the Awadh University, Faizabad. Abstract No. 2.

Salgare, S.A., 1985b. Further evidence of the findings of Mhatre (1980, Ph.D. Thesis), Mhatre, Chaphekar, Ramani Rao, Patil, Haldar (1980), Shetye (1982, Ph.D. Thesis) and Giridhar (1984, Ph.D. Thesis). In: *Proceedings* of *National Symposium on Environmental Science and Warm Water Aquaculture–Finfish*, held on November 6–9, 1985 at the Department of 2001, MJ.K. College of Bihar University, Bettiah–845 438, Abstract No. 2.

Salgare, S.A., 1985c. A criticism on the findings of Mhatre (1980, Ph.D. Thesis), Mhatre, Chaphekar, Ramani Rao, Patil, Haldar (1980), Shetye (1982, Ph.D. Thesis) and Giridhar (1984, Ph.D. Thesis). In: *Proceedings* of *National Symposium on Environmental Science and Warm Water Aquaculture–Finfish*, held on November 6–9, 1985 at the Department of Zoology, M.J.K. College of Bihar University, Bettiah – 845 438, Abstract No. 73.

Salgare, S.A., 1986a. Effect of herbicides on pollen physiology of *Petunia axillaris* BSP. *Ph.D. Thesis*, World University.

Salgare, S.A., 1986b. A criticism on Ph.D. Thesis of Dharurkar (1971) entitled, 'Effect of herbicides on the cytomorphology of *Eichhomia crassipes* Solms (Mart.). *Ph.D. Thesis*, University of Bombay. In: *Proceedings of 1st National. Symposium on Environmental Biology*, held on December 30–31, 1986 at Department of Zoology and Microbiology, S.K. University, Anantapur – 515 003, Abstract No. 6.

Salgare, S.A., 1986c. A criticism on the findings of Nair, Nambudiri and Thomas (1973). In: *Proceedings of 1st National. Symposium on Environmental Biology*, held on December 30–31, 1986 at Department of Zoology and Microbiology, S.K. University, Anantapur – 515 003, Abstract No. 13.

Salgare, S.A., 198cc. A criticism on *Ph.D. Thesis* of Giridhar (1984) entitled, 'Study of interactions between industrial air pollutants and plants'. *Ph.D. Thesis*, University Bombay. In: *Proceedings on 1st National Symposium on Environmental Biology*, held on December 30–31, 1986 at Department of Zoology and Microbiology, S.K. University, Anantapur – 515 003, Abstract No. 14.

Salgare, S.A., 1986e. A criticism on *Ph.D. Thesis* of Shetye (1982) entitled, 'Effect of heavy metals on plants'. *Ph.D. Thesis*, University Bombay. In: *Proceedings of 1st National Symposium on Environmental Biology*, held on December 30–31, 1986 at Department of Zoology and Microbiology, S.K. University, Anantapur – 515 003, Abstract No. 16.

Salgare, S.A., 1995. Interesting observations on the physiology of pollen of *Eichhomia crassipes* Solam. (Mart). In: *Proceedings of 13th National Symposium on Life Science*, Proceeded with Annual Session of Indian Society of Life Science, held on December 30–31, 1995 and January 1, 1996 at Ch. Charan Singh University, Meerut – 250 004, Abstract No. 83.

Salgare, S.A., 2000a. Additional evidence of a criticism on the hypothesis of Berg (1973), Brandt (1974), Vick and Bevan (1976), Rasmussan (1977), Navara, Horvath and Kaleta (1978), Mhatre (1980, Ph.D. Thesis), Mhatre, Chaphekar, Ramani Rao, Patil, Haldar (1980), Shetye (1982, Ph.D. Thesis) and Giridhar (1984, Ph.D. Thesis). In: *Proceedings of 1st National Conference on Recent Trends in Lift Management*, held on October 22–23, 2000 at Bipin Bihari P.G. College, Jhansi – 284 001 (U.P.), Abstract No. 92, pp. 64–65.

Salgare, S.A 2000b. A criticism on the findings of Banerjee and Gangulee (1937) and Dharurkar (1971, Ph.D. Thesis). *Him. J. Environ. Zool.*, 14: 159–160.

Salgare, S.A., 2001a. Monitoring of herbicide (nitrogen) toxicity by using pollen as indicators: A critical review–I. Website www.microbiologyou.com. US.A's Publications (35 pages).

Salgare, S.A., 2001b. Monitors of pollution: A Criticism on the Hypothesis of Berg (1973), Brandt (1974), Vick and Bevan (1976), Rasmussan (1977), Navara, Horvath and Kaleta (1978), Mhatre (1980, Ph.D. Thesis), Mhatre, Chaphekar, Ramani Rao, Patil, Haldara (1980), Shetye (1982, Ph.D. Thesis) and Giridhar (1984, Ph.D. Thesis): A Critical Review–V. *Biojournal*, 13: 39–44.

Salgare, S.A., 2004. Whether optimum pollen germination and tube length attained in the same growth medium (sucrose + 2,4-D) by some species of Apocynaceae–A criticism to the hypothesis of Nair, Nambudiri and Thomas (1973)–A Critical Review–IV. Him. J. Environ. *Zool.*, 18: 53–56.

Salgare, S.A., 2005a. Monitoring of herbicide (simazine) toxicity–A Criticism on the Hypothesis of Berg (1973), Brandt (1974), Vick and Bevan (1976), Rasmussan (1977), Navara, Horvath and Kaleta (1978), Mhatre (1980, Ph.D. Thesis), Mhatre, Chaphekar, Ramani Rao, Patil, Haldar (1980), Shetye (1982, Ph.D. Thesis) and Giridhar (1984, Ph.D. Thesis)–A Critical Review. *Him. J. Environ. Zool.*, 19: 69–71.

Salgare, S.A., 2005b. Whether optimum pollen germination and tube length attained in the same growth medium (sucrose + 2,4,5-T) by some cultivars of Apocynaceae and further evidence of a criticism to the hypothesis of Nair, Nambudiri and Thomas (1973)–A Critical Review. *Flora and Fauna*, 11: 42–44.

Salgare, S.A., 2005c. Monitoring of herbicide (dalapon) toxicity by using pollen as indicators Pollen of some cultivars of Apocynaceae and Further Evidence of a Criticism on the Hypothesis of Berg (1973), Brandt (1974), Vick and Bevan (1976), Rasmussan (1977), Navara, Horvath and Kaleta (1978), Mhatre (1980, Ph.D. Thesis), Mhatre, Chaphekar, Ramani Rao, Patil, Haldar (1980), Shetye (1982, Ph.D. Thesis) and Giridhar (1984, Ph.D. Thesis)–A Critical Review. *Flora and Fauna*, 11: 49–50.

Salgare, S.A., 2005d. Whether optimum pollen germination and tube length attained in the same growth medium (sucrose + nitrogen)–A criticism to the hypothesis of Nair, Nambudiri and Thomas (1973)–A critical review. *Him. J. Environ. Zool.*, 19: 93–95.

Salgare, S.A., 2005e. Monitoring of herbicide (basalin EC) toxicity–A Criticism on the Hypothesis of Berg (1973), Brandt (1974), Vick and Bevan (1976), Rasmussan (1977), Navara, Horvath and Kaleta (1978), Mhatre (1980, Ph.D. Thesis), Mhatre, Chaphekar, Ramani Rao, Patil, Haldar (1980), Shetye (1982, Ph.D. Thesis) and Giridhar (1984, Ph.D. Thesis)–A Critical Review. *Internat. J. Biosci. Reporter*, 3: 298–300.

Salgare, S.A., 2006a. Monitoring of herbicide (femoxone) toxicity by using pollen as indicators pollen of five cultivars of *Petunia axillaris:* Further Evidence of a Criticism of Banerjee and Gangulee

(1937), Brewbaker and Kwack (1963), Dharurkar (1971, Ph.D. Thesis), Berg (1973), Brandt (1974), Vick and Bevan (1976), Rasmussan (1977), Navara, Horvath and Kaleta (1978), Mhatre (1980, Ph.D. Thesis), Mhatre, Chaphekar, Ramani Rao, Patil, Haldar (1980), Shetye (1982, Ph.D. Thesis) and Giridhar (1984, Ph.D. Thesis)–A Critical Review. *Environ. Conservation J.*, 7: 1–7.

Salgare, S.A., 2006b. Whether optimum pollen germination and tube length attained in the same growth medium (sucrose + sodium arsenite) and further evidence of a criticism of the hypothesis of Nair, Nambudiri and Thomas (1973)–A critical review. *Internat. J. Plant Sci.*, 1: 132–133.

Salgare, S.A., 2006c. Monitoring of herbicide (basalin EC) toxicity by using pollen as indicators–Pollen of *Petunlia axillaries* and further evidence of a criticism of the ypothesis of Berg (1973), Brandt (1974), Vick and Bevan (1976), Rasmussan (1977), Navara, Horvath and Kaleta (1978), Mhatre (1980, Ph.D. Thesis), Mhatre, Chaphekar, Ramani Rao, Patil, Haldar (1980), Shetye (1982, Ph.D. Thesis) and Giridhar (1984, Ph.D. Thesis)–A critical review. *Him. J. Environ. Zool.*, 20: 269–271.

Salgare, S.A., 2006d. Alteration of resting period of pollen of Apocynaceae by herbicide (nitrofen): Further evidence of a criticism of Sudhakaran (1967–Ph.D. Thesis), Saoji and Chitaley (1972), Berg (1973), Brandt (1974), Vick and Bevan (1976), Rasmussan (1977), Navara, Horvath and Kaleta (1978), Mhatre (1980, Ph.D. Thesis), Mhatre, Chaphekar, Ramani Rao, Patil, Haldar (1980), Shetye (1982, Ph.D. Thesis) and Giridhar (1984, Ph.D. Thesis). *Internat. J Agri. Sci.*, 3: 239–243.

Salgare, S.A. 2006e. Effect of herbicide (acrolein) on pollen germination and tube growth of twelve hours stored pollen of five cultivars of Apocynaceae: Further Evidence of a Criticism of Sudhakaran (1967, Ph.D. Thesis), Berg (1973), Brandt (1974), Vick and Bevan (1976), Rasmussan (1977), Navara, Horvath and Kaleta (1978), Mhatre (1980, Ph.D. Thesis), Mhatre, Chaphekar, Ramani Rao, Patil, Haldar (1980), Shetye (1982, Ph.D. Thesis) and Giridhar (1984, Ph.D. Thesis)–A Critical Review. *J. Natcon.*, 8: 283–290.

Salgare, S.A, 2006f. Monitoring of Herbicide (2,4–D) toxicity by using pollen as indicators–Pollen of apocynaceae and further evidence of a criticism of the hypothesis of Berg (1973), Brandt (1974), Vick and Bevan (1976), Rasmussan (1977), Navara, Horvath and Kaleta (1978), Mhatre (1980, Ph.D. Thesis), Mhatre, Chaphekar, Ramani Rao, Patil, Haldar (1980), Shetye (1982, Ph.D. Thesis) and Giridhar (1984, Ph.D. Thesis)–A Critical Review. *Internat. J. Plant Sci.*, 1: 134–136.

Salgare, S.A., 2006g. Whether optimum pollen germination and tube length attained in the same growth medium (sucrose + 2,4–D) by five cultivars of the Apocynaceae: Further evidence of a criticism of Banerjee and Gangulee (1937), Brewbaker and Kwack's (1963), Sudhakaran (1967, Ph.D. Thesis), Dharurkar (1971, Ph.D. Thesis), Nair, Nambudiri and Thomas (1973), Berg (1973), Brandt (1974), Vick and Bevan (1976), Rasmussan (1977), Navara, Horvath and Kaleta (1978), Mhatre (1980, Ph.D. Thesis), Mhatre, Chaphekar, Ramani Rao, Patil, Haldar (1980), Shetye (1982, Ph.D. Thesis) and Giridhar (1984, Ph.D. Thesis)–A Critical Review. *Environ. Conservation J.*, 7: 21–29.

Salgare, S.A., 2006h. Whether optimum pollen germination and tube length attained in the same growth medium (sucrose + atrataf 50w)–A criticism to the hypothesis of Nair, Nambudiri and Thomas (1973)–A critical review. *Internat. J. Biosci. Reporter*, 4: 145–146.

Salgare, S.A., 2006i. Evaluation of MH as male gametocide on *Phaseolus aureus* and a new method of plant breeding and further evidence of a criticism of the hypothesis of Nair, Nambudiri, Thomas (1973), Berg (1973), Brandt (1974), Vick and Bevan (1976), Rasmussan (1977), Navara, Horvath and Kaleta (1978), Mhatre (1980, Ph.D. Thesis), Mhatre, Chaphekar, Ramani Rao, Patil, Haldar (1980), Shetye (1982, Ph.D. Thesis) and Giridhar (1984, Ph.D. Thesis). *Internat. J. Plant Sci.,* 1: 352–356.

Salgare, S.A., 2006j. Further evidence of a criticism of the findings of Banerjee and Gangulee (1937) and Dharurkar (1971). *Internat. J. Biosci. Reporter,* 4: 169–170.

Salgre, S.A., 2006k. Alteration of resting period of pollen of five cultivars of *Petunia axillaris* BSP by atrataf 50W: Further evidence of a criticism of Saoji and Chitaley (1972), Berg (1973), Brandt (1974), Vick and Bevan (1976), Rasmussan (1977), Navara, Horvath and Kaleta (1978), Mhatre (1980, Ph.D. Thesis), Mhatre, Chaphekar, Ramani Rao, Patil, Haldar (1980), Shetye (1982, Ph.D. Thesis) and Giridhar (1984, Ph.D. Thesis)–A Critical Review. *J. Natcon.,* 18: 353–360.

Salgare, S.A., 2006l. Monitoring of herbicide (2,4-dinitrophenol) toxicity by using pollen as indicators–Pollen of five cultivars of Apocynaceae: Further evidence of a criticism of Banerjee and Gangulee (1937), Brewbaker and Kwack (1963), Sudhakaran (1967, Ph.D. Thesis), Dharurkar (1971, Ph.D. Thesis), Berg (1973), Brandt (1974), Vick and Bevan (1976), Rasmussan (1977), Navara, Horvath and Kaleta (1978), Mhatre (1980, Ph.D. Thesis), Mhatre, Chaphekar, Ramani Rao, Patil, Haldar (1980), Shetye (1982, Ph.D. Thesis) and Giridhar (1984, Ph.D. Thesis)–A Critical Review. *Research Hunt,* 1: 1–6.

Salgre, S.A., 2006m. Alteration of resting period of pollen of five cultivars of *Petunia axillaris* BSP by atrataf SOW: Further evidence of a criticism of Saoji and Chitaley (1972), Berg (1973), Brandt (1974), Vick and Bevan (1976), Rasmussan (1977), Navara, Horvath and Kaleta (1978), Mhatre (1980, Ph.D. Thesis), Mhatre, Chaphekar, Ramani Rao, Patil, Haldar (1980), Shetye (1982, Ph.D. Thesis) and Giridhar (1984, Ph.D. Thesis)–A Critical Review. *J Natcon.,* 18: 357–364.

Salgare, S.A., 2006n. Whether optimum pollen germination and tube length attained in the same growth medium (sucrose + simazine) by five cultivars of Apocynaceae: Further evidence of a criticism of Banerjee and Gangulee (1937), Brewbaker and Kwack's (1963), Sudhakaran (1967, Ph.D. Thesis), Dharurkar (1971, Ph.D. Thesis), Nair, Nambudiri and Thomas (1973), Berg (1973), Brandt (1974), Vick and Bevan (1976), Rasmussan (1977), Navara, Horvath and Kaleta (1978), Mhatre (1980, Ph.D. Thesis), Mhatre, Chaphekar, Ramani Rao, Patil, Haldar (1980), Shetye (1982, Ph.D. Thesis) and Giridhar (1984, Ph.D. Thesis)–A Critical Review. *Res. Hunt,* 1: 146–155.

Salgare, S.A., 2006o. Effect of herbicide (acrolein) on pollen germination and tube growth of twelve hours stored pollen of five cultivars of Apocynaceae: Further evidence of a criticism of Sudhakaran (1967, Ph.D. Thesis), Berg (1973), Brandt (1974), Vick and Bevan (1976), Rasmussan (1977), Navara, Horvath and Kaleta (1978), Mhatre (1980, Ph.D. Thesis), Mhatre, Chaphekar, Ramani Rao, Patil, Haldar (1980), Shetye (1982, Ph.D. Thesis) and Giridhar (1984, Ph.D. Thesis)–A Critical Review. *J. Natcon.,* 18: 361–368.

Salgare, S.A., 2006p. Alteration of resting period of pollen of five cultivars of *Petunia axillaris* BSP by Gramoxone: Further evidence of a criticism of Brewbaker and Kwack's (1963), Saoji and Chitaley (1972), Berg (1973), Brandt (1974), Vick and Bevan (1976), Rasmussan (1977), Navara, Horvath and Kaleta (1978), Mhatre (1980, Ph.D. Thesis), Mhatre, Chaphekar, Ramani Rao, Patil, Haldar (1980), Shetye (1982, Ph.D. Thesis) and Giridhar (1984, Ph.D. Thesis)–A Critical Review. *Environ. Conservation J.,* 7: 51–57.

Salgare, S.A., 2007. Effect of herbicide (MH) on pollen germination and tube growth of twelve hours stored pollen of Apocynaceae: Further evidence of a criticism of Banerjee and Gangulee (1937),

Brewbaker and Kwack (1963), Sudhakaran (1967, Ph.D. Thesis), Dharurkar (1971, Ph.D. Thesis), Berg (1973), Brandt (1974), Vick and Bevan (1976), Rasmussan (1977), Navara, Horvath and Kaleta (1978), Mhatre (1980, Ph.D. Thesis), Mhatre, Chaphekar, Ramani Rao, Patil, Haldar (1980), Shetye (1982, Ph.D. Thesis) and Giridhar (1984, Ph.D. Thesis)–A critical review. *Internat. J. Plant Sci.,* 2: 215–219.

Salgare, S.A. and Bindu, 2002. Whether optimum pollen germination and tube length attained in the same growth medium (sucrose + calcium chloride) by pink-flowered *Catharanthus roseus*–A criticism to the hypothesis of Nair, Narnbudiri and Thomas (1973)–A critical review–I. *Biojournal,* 14: 21–23.

Salgare, S.A. and Bindu, 2005. Whether optimum pollen germination and tube length attained in the same growth medium (sucrose + calcium chloride) by white-flowered *Catharanthus roseus* and further evidence of a criticism on the hypothesis of Nair, Nambudiri and Thomas (1973)–A critical review. *Flora and Fauna,* 11: 183–184.

Salgare, S.A. and Phunguskar, K.P., 2002. Monitors of pesticide (alphamethrin) toxicity by using pollen as indicators–Pollen of pink-flowered *Catharanthus roseus*–A criticism on the hypothesis of Berg (1973), Brwdt (1974), Vick and Bevan (1976), Rasmussan (1977), Navara, Horvath and Kaleta (1978), Mhatre (1980, Ph.D. Thesis), Mhatre, Chaphekar, Ramani Rao, Patil, Haldara (1980), Shetye (1982, Ph.D. Thesis) and Giridhar (1984, Ph.D. Thesis)–A Critical Review–III. *Biojournal,* 14: 1–4.

Salgare, S.A. and Pathak, Sanchita, 2005. Monitoring of heavy metal (bismuth nitrate) toxicity by using pollen as indicators–Pollen of white-flowered *Catharanthus roseus* and further evidence of a criticism on the hypothesis of Berg (1973), Brandt (1974). Vick and Bevan (1976), Rasmussan (977), Navara, Horvath and Kaleta (1978), Mhatre (1980, Ph.D. Thesis), Mhatre, Chaphekar, Ramani Rao, Patil, Haldar (1980), Shetye (1982, Ph.D. Thesis), and Giridhar (1984, Ph.D. Thesis)–A Critical Review. *Flora and Fauna,* 11: 69–70.

Salgare, S.A. and Singh, Sanju, 2002. Monitors of heavy metal (ferrous sulphate) toxicity by using pollen as indicators–Pollen of white-flowered *Catharanthus roseus*–A criticism on the hypothesis of Berg (1973), Brandt (1974), Vick and Bevan (1976), Rasmussan (1977), Navara, Horvath and Kaleta (1978), Mhatre (1980, Ph.D. Thesis), Mhatre, Chaphekar, Ramani Rau, Patil, Haldara (1980), Shetye (1982, Ph.D. Thesis) and Giridhar (1984, Ph.D. Thesis)–A Critical Review–II. *Biojournal,* 14: 5–7.

Salgare, S.A. and Singh, Sanju, 2006a. Monitoring of heavy metal (copper sulphate) toxicity by using pollen as indicators–Pollen of white-flowered *Catharanthlls roseus* and further evidence of a criticism on the hypothesis of Berg (1973), Brandt (1974), Vick and Bevan (1976), Rasmussan (1977), Navara, Horvath and Kaleta (1978), Mhatre (1980, Ph.D. Thesis), Mhatre, Chaphekar, Ramani Rao, Patil, Haldar (1980), Shetye (1982, Ph.D. Thesis), and Giridhar (1984, Ph.D. Thesis)–A critical review–I. *Him. J. Environ. Zool.,* 23: 276–278.

Salgare, S.A. and Singh, Sanju, 2006b. Monitoring of heavy metal (cobalt nitrate) toxicity by using pollen as indicators–Pollen of pink-flowered cultivar of *Catharanthus roseus* and further evidence of a criticism of the hypothesis of Berg (1973), Brandt (1974), Vick and Bevan (1976), Rasmussan (1977), Navara, Horvath and Kaleta (1978), Mhatre (1980, Ph.D. Thesis), Mhatre, Chaphekar, Ramani Rao, Patil, Haldar (1980), Shetye (1982, Ph.D. Thesis) and Giridhar (1984, Ph.D. Thesis)–A critical review– II. *Him. J. Environ. Zool.,* 20: 283–284.

Salgare, S.A. and Tessy Mol Antony, 2005a. Whether optimum pollen germination and tube length attained in the same growth medium (sucrose + potassium chloride) by pink-flowered *Catharanthus roseus* and further evidence of a criticism to the hypothesis of Nair, Nambudiri and Thomas (1973)–A critical review. *Flora and Fauna,* 11: 61–62.

Salgare, S.A. and Tessy Mol Antony, 2005b. Whether optimum pollen germination and tube length attained in the same growth medium (sucrose+sodium oxalate) by white-flowered *Catharanthus roseus* and further evidence of a criticism to the hypothesis of Nair, Nambudiri and Thomas (1973)–A critical review. *Flora and Fauna,* 11: 75–76.

Salgare, S.A. and Sebastian, Theresa, 1986a. Further evidence of a criticism of the hypothesis of Berg (1973), Brandt (1974), Vick and Bevan (1976), Rasmussan (1977), Navara, Horvath and Kaleta (1978), Mhatre (1980, Ph.D. Thesis), Mhatre, Chaphekar, Ramani Rao, Patil, Haldar (1980), Shetye (1982, Ph.D. Thesis) and Giridhar (1984, Ph.D. Thesis). In: *1st National Symposium on Environmental Biology,* held on December 30–31, 1986 at Department of Zoology and Microbiology, S.K. University, Anantapur – 515 003. Abstract No. 15.

Salgare, S.A. and Sebastian, Theresa, 2006. Monitoring of herbicide (dalapon) toxicity by using pollen as indicators–Pollen of urid and further evidence of a criticism of Berg (1973), Brandt (1974), Vick and Bevan (1976), Rasmussan (1977), Navara, Horvath and Kaleta (1978), Mhatre (1980, Ph.D. Thesis), Mhatre, Chaphekar, Ramani Rao, Patil, Haldar (1980), Shetye (1982, Ph.D. Thesis) and Giridhar (1984, Ph.D. Thesis)–A critical review–I. *Internat. J. Biosci. Reporter,* 4: 323–329.

Schoch-Bodmer, H., 1921. Reservestoffe bei einigen anemophilen Pollenarten. *Vjschr. Naturf Gas. Zmich,* 66: 339–346.

Sharma, R.I., 1984. Effect of herbicides on plants of the Solanaceae–II. *Ph.D. Thesis,* University Bombay.

Shetye, R.P., 1982. Effect of heavy metals on plants. *Ph.D. Thesis,* University Bombay.

Trisa, Palathingal, 1990. Evaluation of industrial pollution of Bombay by pollen–I. *M.Phil. Thesis,* University Mumbai.

Vasil, I.K., 1960. Pollen germination in some Gramineae: *Penllisetum typhoideum. Nature,* 187: 1134–1135.

Chapter 32

Assessment of Primary Productivity of Phytoplankton of Jagatpur Wetland, Bhagalpur, Bihar

Braj Nandan Kumar and S.K. Choudhary

Environmental Biology Research Laboratory, University Department of Botany, Tilka Manjhi Bhagalpur University, Bhagalpur – 812 007, Bihar

ABSTRACT

The Primary Productivity of Phytoplankton of Jagatpur Wetland, Bhagalpur has been assessed for two years from August, 2003 to July, 2005. The net and gross Primary Productivity was recorded with maximum value of 2.02 gc/m^3/d and 3.7 gc/m^3/d, while the minimum were 0.16 gc/m^3/d and 0.05 gc/m^3/d respectively. The community respiration ranged between 0.33 gc/m^3/d to 3.03 gc/m^3/d. The ratio of NPP and GPP fluctuated from 0.99 to 0.85. The PIR ratio was found to vary between 0.09 to 6.0. Net Primary productivity value and P/R ratio indicated eutrophic nature of Jagatpur Wetland.

Keywords: *Jagatpur wetland, Primary productivity, Phytoplankton.*

Introduction

The photosynthetic fixation of carbon dioxide and its quantitative measurement is considered a vital index of the productive potential of any aquatic system. Productivity in a broad sense refers to a concept of organic matter synthesis potential which measures ability of an area to support a biological population and sustain a level of growth and respiration (Raymount, 1966). Primary Productivity of different lentic water bodies in and around Bhagalpur, Bihar had been studied by Nasar and Dutta

Munshi (1975), Saha *et al.* (1985), Pandey *et al.* (1995), Mandal (2002) and, Kumar and Singh (2006). The present study to evaluate phytoplanktonik primary productivity of the Jagatpur wetland for the first time is part of the SAP (DRS) programme of UGC being conducted by the Botany Department of T.M. Bhagalpur University. Jagatpur wetland is perennial one and quite rich in biodiversity components (Choudhary *et al.*, 2006). We have noticed extensive agricultural practices around the wetland. The agro-chemicals from the crop fields are channelized to the wetland through run off. Fishing is common in the wetland and a large number of people from neighbouring villages depend for their livelihood on the wetland. The wetland is home to a variety of resident and winter migratory birds including rare and threatened species like Greater adjutant storks *Letoptilos dubius* and Indian Skimmers *Rynchops albicosis*. Seepage water from the river Ganga and rain water are main sources of water supplies for the wetland. The present study on productivity of phytoplankton will help in formulating the management plan in future for the wetland.

Materials and Methods

Jagatpur wetland is located in the middle Ganga flood plain near Bhagalpur in between 25°22′ North latitude and 87°02′ East longitudes. It consists of a total area of 0.40 km^2.

Sampling and Analysis

The primary productivity was determined at monthly intervals for the period of two years (August, 2003–July, 2005), by 'light and dark bottle' method of Garder and Gran (1927). The dissolved oxygen content of the initial bottles was determined immediately after collecting the water samples, following the modified Winkler's method (APHA, 1998). Two sets of light and dark bottles (250 ml) filled with water were suspended at the depth of 10 cm below the water surface for 4 hours (9.00–13.00 h). After the incubation period of 4 hours, the suspended bottles were removed and dissolved oxygen content in each bottle was determined. Oxygen values produced due to photosynthesis were converted into carbon by Wood's (1975).

Results and Discussion

Monthly values of productivity (August, 2003 to July, 2005) in the Jagatpur wetland are given in Table 32.1. The values of Gross Primary Production (GPP) of phytoplankton ranged in between 0.5 gc/m^3/d and 3.7 gc/m^3/d during the first year and in between 1.01 gc/m^3/d and 3.03 gc/m^3/d during the second year of investigations. High gross production was recorded during winter and summer months and the value was low in monsoon months (July–August). Similar to the observations of Singh (1993), Harikrishnan and Azis Abdul (2000), and Kumar and Singh (2006), the present study revealed distinct seasonal and bimodal pattern of variation in the GPP value, having winter and summer peaks. Solar radiation was perhaps the prime factor contributing to the higher values.

The net primary production of phytoplankton varied from 0.16 gc/m^3/d to 1.68 gc/m^3/d and from 0.16 gc/m^3/d to 2.02 gc/m^3/d during the first and second year of investigations respectively. NPP values showed the same seasonal and bimodal pattern of variations as shown by GPP values, *i.e.*, maxima attained during summer and early winter months and minima in rainy months. The net primary productivity had direct relationship with the temperature in the present study. The decreasing trend of NPP during monsoon could be due to the reduction in the size of productive zones owing to rains, increase in water level, dilution of nutrient concentration and cloudy weather (Saran *et al.*, 1985).

Table 32.1: Phytoplankton Productivity of Jagatpur Wetland, Bhagalpur (August, 2003–July, 2005) Production and Respiration Values expressed in gc/m³/d

Parameters	Gross Primary Productivity (G.P.P)	Net Primary Productivity (N.P.P)	NP/GP	Community Respiration (C.R)	Respiration (R) as % of G.P.P.	P/R Ratio
			Months			
August 2003	0.50	0.16	0.32	0.33	66.0	0.48
September	2.61	1.51	0.57	1.09	41.76	1.38
October	3.03	1.68	0.55	1.35	44.55	1.24
November	1.77	0.84	0.47	0.92	51.97	0.91
December	0.67	0.33	0.49	0.33	49.25	1.0
January 2004	2.7	1.01	0.37	1.68	62.48	0.59
February	1.35	1.01	0.74	0.33	24.96	3.0
March	1.01	0.33	0.33	0.67	66.69	0.49
April	1.35	0.67	0.50	0.67	50.0	1.0
May	3.7	0.67	0.18	3.03	81.79	0.22
June	0.67	0.33	0.49	0.33	49.22	1.0
July	0.59	0.22	0.38	0.33	55.93	0.66
August	1.18	0.16	0.14	1.1	85.69	0.16
September	3.03	1.51	0.49	1.51	49.98	1.0
October	3.03	1.68	0.55	1.35	44.55	1.24
November	2.02	1.35	0.66	0.67	3.11	2.01
December	1.35	0.33	0.24	1.01	74.96	0.33
January 2005	1.01	0.67	0.66	0.33	33.3	2.0
February	2.7	2.02	0.75	0.67	25.0	3.0
March	1.35	0.33	0.24	1.01	74.96	0.33
April	2.19	0.67	0.30	2.02	92.33	0.33
May	1.85	0.16	0.09	1.68	90.89	0.09
June	2.36	2.02	0.85	0.33	14.26	6.0
July	1.68	0.33	0.19	1.35	80.02	0.24

In the first year of investigation the community respiration of phytoplankton ranged in between 0.33 gc/m³/d and 3.03 gc/m³/d while in the second year it was from 0.33 gc/m³/d to 2.02 gc/m³/d. The high rate of oxygen consumption (May, 2003–3.03 gc/m³/d; and April, 2005–2.02 gc/m³/d) during warmer months was about 81.79 per cent and 92.33 per cent of gross production. It was probably due to increase in the mineralization rate caused by quick warming of water up to 34.7°C and also due to enhanced O_2 consumption by the biota of the wetland.

The net and gross ratio ranged from 0.18 to 0.74 during the first year while in the second year, it varied from 0.09 to 0.85. Lower values of net gross ratio during monsoon may be attributed to the fact that when the wetland receives organic matter along with run off from the catchment area, the decomposition of these wastes demands more oxygen resulting in enhanced respiratory values which in turn gave a low value of net: gross ratio. Higher net: gross ratio during summers and early winters were perhaps due to higher net primary productivity during these periods. Similar findings were made by Qasim *et al.* (1969), Harikrishnan and Azis Abdul (2000) and, Kumar and Singh (2006). The value of net: gross ratio was less than unity in all the months of investigation and that revealed that respiration exceeded production. This suggested that the wetland under study was not healthy.

The ratio of net primary production to respiration (P/R ratio) was found to vary from 0.22 to 3.0 during the first year of investigation while in the second year; it varied from 0.09 to 6.0. It is an important functional index of relative maturity of the system. The P/R ratio shows the wetland may be autotrophic in terms of Odum (1969).

From the net primary productivity value, net: gross ratio and P/R ratio obtained in the present study, it may be suggested that the Jagatpur wetland is progressing towards eutrophication, and it needs immediate remedial measures as the wetland is the source of livelihood for thousands of local people, and is important for floral and faunal diversity.

Acknowledgement

The authors are thankful to the Head, University Department of Botany, T.M. Bhagalpur University for providing laboratory facilities. Financial assistance to Sri Braj Nandan Kumar as Research Fellow under SAP (DRS-III) of UGC is duly acknowledged.

References

APHA, 1998. *Standard Methods for the Examination of Water and Wastewater*, 20th Edn. American Public Health Association, Washington D.C.

Choudhary, S.K., Smith, Brian D., Dey, S., Dey, S. and Prakash Satya, 2006. Conservation and biomonitoring in the Vikramshila Gangetic Dolphin Sanctuary, Bihar, India. *Oryza*, 40(2): 189–197.

Gaarder, T. and Gran, H.H., 1927. Investigations of the production of plankton in the Oslo Fjord, Rapp. et. Proc. verb. Cons. Internat. Explor. Ma2., 42: 1 – 48.

Harikrishnan, K. and Aziz Abdul, P.K., 2000. Primary production studies in a freshwater temple tank in Kerala. *Indian J. Environ and Eco-plan.*, 3(1): 127–130.

Kumar, Ashok and Singh, N.K., 2006. Phytoplankonology of pond at Deoghar, India. *J. Haematol and Ecotoxicol.*, 1(2): 61–66.

Mandal, Om Prakash, 2002. Primary productivity in relation to nutrient status of Kawar wetland, North Bihar. *Ph.D. Thesis*, B.N. Mandal University, Madhepura, (Bihar).

Nasar, S.A.K. and Dutta Munshi, J.S., 1975. Studies on primary production in a fresh water pond. *Jap. J. Ecol.*, 25: 21–23.

Odum, E.P., 1969. The strategy of ecosystem development. *Science*, 164: 262–270.

Pandey, K.N., Prakasan, V. and Sharma, U.P., 1995. Assessment of primary productivity of phytoplankton and macrophytes of Kawar Lake wetland (Begusarai) Bihar. *J. Freshwater Biol.*, 7(4): 237–239.

Qasim, S.Z., Wellers Show, S., Bhattathiri, P.M.A and Abidi, S.M.A., 1969. Organic production in a tropical estuary. In: *Proc. Ind. Acad. Sci.*, 69, 13(2): 51–54.

Raymount, J.E.G., 1966. The production of marine plankton. In: *Advances in Ecological Resources*, (Ed.) J.B. Crag. Academic Press, New York, 3: 51–94.

Saha, L.C., Choudhary, S.K. and Singh, N.K., 1985. Factors affecting phytoplankton productivity in the river Ganges at Bhagalpur. *Geobios*, 12(2): 63–65.

Saran, H.M. and Adoni, A.D., 1985. Limnological studies on seasonal variation in phytoplanktonic primary productivity in Sagar lake. In: *Proc. Nat. Symp. Pure and Appl.. Limnology,* (Ed.) A.D. Adoni. Bull. Bot. Soc., Sagar, 32: 71–76.

Singh, N.K., 1963. Studies on density, productivity and species composition of phytoplankton in relation to abiotic spectrum of Ganges at Sahibganj. *J. Freshwater Biol.*, 5(1): 1–8.

Singh, Meena and Sinha, R.K., 1994. Primary productivity and limnological profile of two ponds at Patna, Bihar. *J. Freshwater Biol.*, 6(2): 127–133.

Wood, R.D., 1975. *Hydrobiological Methods*. University Park Press, Baltimore, London, Tokyo, pp. 1–165.

Chapter 33

Effect of Paper Mill Effluent on Soil Microflora and Soil Enzymes of Soyabean

P. Dhevagi and G. Oblisami*

Department of Environmental Sciences,
Tamil Nadu Agricultural University, Coimbatore – 641 003

ABSTRACT

An experiment was conducted to study the effect of paper mill effluent on soil micro flora and soil enzymes of soyabean. Irrigation with treated effluent increased the beneficial microflora like *Azospirillum, Azotobacter* and *Rhizobium,* soil enzyme activities and available nutrients. The soil amylase activity increased in effluent irrigated plots and it varied from 0.52 to 0.69 mg and 0.33 to 0.56 mg in the effluent and well water irrigated plots respectively. The activity of invertase was higher in amendment received plots than the unamended plots. The cellulase activity of the effluent irrigated plots varied from 10.91 to 14.37 µg but it varied from 10.60 to 12.49 µg in well water irrigated plots. The CO_2 evolved showed a variation of 4.01 to 5.70 µg in effluent irrigated plots and 3.45 to 4.31 mg in well water irrigated plots.

Keywords: *Paper mill, Treated wastewater, Soyabean, Spermosphere, Soil microflora, Soil enzymes.*

Introduction

The annual production of paper and board was about 310 mt. It has been estimated that about 330 m^3 of effluent is produced per ton of paper manufactured in a small mill with a capacity of

* Assistant Professor; E-mail: devagihfr@yahoo.com.

20 td^{-1}; whereas 220 m^3 t^{-1} is generated in a large mill with a manufacturing capacity of 2000 t d^{-1}. Nearly 80 per cent of fresh water used in the paper and pulp mill is discharged as effluent containing organic and inorganic pollutants requiring treatment and disposal (Kallas and Munter, 1994). Effluents are used for irrigation in dry land areas after treatment. These effluents not only contain nutrients that enhance the growth of crop plants but also have toxic materials that interfere with the soil ecosystem. Hence it is essential to investigate the effect of paper mill effluent under field situation. This paper reveals the effect of paper mill effluent on soil microflora and its activities in a soybean grown soil.

Materials and Methods

Characterization of the Effluent

The treated effluent samples were collected from a bagasse based paper mill situated in Tamil Nadu. The characteristics of the effluent samples were analyzed as per the methods detailed in the standard methods for the examination of water and waste water (Anon., 1965). The populations of the different groups of microorganisms in the effluent and soil samples were enumerated using the standard serial dilution plating technique (Waksman and Fred, 1922). The raw and treated effluents from the Bagasse based paper mill was studied for their effect on seed germination and vigour index (Abdul-Baki and Anderson, 1973). The spermosphere microflora enumerated using the standard serial dilution plating technique. was

Field Trial

A field trial was conducted to assess the effect of bagasse based paper mill effluent on soil, crop growth and microbial load. Treatment details used for field experiment includes:

Main plot: Irrigation

- I_1: Well water irrigation
- I_2: Effluent water irrigation

Sub plots: Amendments

- A_1: Control (100 per cent NPK)
- A_2: Farm yard manure @ 12.5 t ha^{-1}
- A_3: Pressmud @ 5 t ha^{-1}
- A_4: Fortified Pressmud @ 2.5 t ha^{-1}
- A_5: TNPL sludge @ 2.5 t ha^{-1}

The physico-chemical properties like Bulk density (Chopra and Kanwar, 1982), pH and EC (Jackson, 1973), Organic carbon (Piper, 1966), Available N (Subbiah and Asija, 1956), Available P (Olsen *et al.*, 1954), Available K (Stanford and English, 1948) and biological properties like Bacteria, Fungi, actinomycetes, *Azotobacter*, *Rhizobium* (Subba Rao, 1988), *Azospirillum* (Dobereiner, 1980), Carbon dioxide (Pramer and Schmidt, 1965), Amylase, Invertase (Galystyan, 1965) and Cellulase (Benefield, 1971) were analyzed.

The experimental data were statistically analyzed as suggested by Panse and Sukhatme (1985). The critical difference was worked out at 5 per cent (0.05) probability level.

Results and Discussion

The effect of treated paper factory effluent from Tamil Nadu Newsprint and Paper Limited on soil characteristics, crop growth and microbial activities were studied and the results obtained are presented in Tables 33.2–33.8.

Characteristics of the Effluent

The physico-chemical and biological properties of the effluent were given in Table 33.1. The characteristic dark brown colour of the effluents is due to formation of lignin degradation products during the processing of lignocellulosics. The dissolved oxygen content of the effluent ranged from 0.87 to 4.7 mg l^{-1} with BOD from 22 to 37 mg l^{-1}. Though this was in accordance with the report of Verma *et al.*, 1988, is contradictory to the findings of Gomathi and Oblisami (1992). Few studies explored very low concentration of N, P and K in the treated effluent (Dhevagi and Oblisami, 2000), The treated effluent recorded appreciable numbers of microbial population. This was in accordance with the report of Niemelae and Vaeaetaenen (1982), Kannapiran *et al.* (1997a) and Dhevagi *et al.* (2000, 2004 and 2006).

Table 33.1: Physico-Chemical and Biological Characteristics of the Paper Mill Effluent

Characteristics	*Concentration*
Colour	Straw yellow
pH	7.3
EC (dSm^{-1})	0.45
Total solids (mg l^{-1})	296.0
Suspended solids (mg l^{-1})	162.0
Dissolved oxygen (mg l^{-1})	4.7
BOD (mg l^{-1})	37.0
COD (mg l^{-1})	565.0
Nitrogen (mg l^{-1})	24.0
Phosphorus (mg l^{-1})	1.13
Potassium (mg l^{-1})	14.8
Phenols (mg l^{-1})	26.9
Actinomycetes × 10^2 ml^{-1}	4.5
Fungi × 10^3 ml^{-1}	14.7
Azotobacter × 10^2 ml^{-1}	40.0
Yeast × 10^2 ml^{-1}	52.0

Effect on Germination

The effect of raw and treated effluent on germination of seed and vigour index of seedlings were studied at different concentrations (25, 50, 75 and 100 per cent) under laboratory condition. Tap water was used as control (0 per cent effluent). The results were presented in Table 33.2.

The vigour index of the seedlings in control (T_1) was 440, while in other treatments it ranged from 313.7 to 427.0. Subrahmanyam *et al.* (1984) had revealed that crops like paddy, wheat, barley and sugarcane would be successfully grown with practically no reduction in yield. Sandana and Oblisami (1996) and Dhevagi *et al.* (2006, 2007) also observed the same results. The effluent treated by microbial community [*Klebsiella* sp. (one isoate) and *Pseudomonas* sp.] is applied for the germination of seeds of paddy and wheat and growth of seedling exhibits significant reduction in toxicity of contaminant present in effluent (Singh *et al.*, 2002)

Table 33.2: Effect of TNPL Effluent on Spermophere Growth Microflora and Seedling

Treatment	Total Microbial Population	Soybean (CO1)			
		G (%)	R (cm)	S (cm)	VI
Control	44.3	66.7	2.9	3.7	440.0
25 per cent	46.7	66.7	2.8	3.6	427.0
50 per cent	49.1	64.2	2.6	3.4	385.2
75 per cent	50.7	60.0	2.4	3.3	343.0
100 per cent	47.2	58.0	2.4	3.0	313.7
SEd	2.48	2.15	1.23	0.154	21.20
CD (0.05)	4.39	6.33	3.63	0.460	62.52

TMP: Total Microbial Population; R: Root length (cm); S: Shoot length (cm); G: Germination percentage; VI: Vigour Index

Effect on Spermosphere Micronora

Spermosphere organism plays an important role in better establishment of the plants in soil. In soybean treatment T_3 recorded the highest bacterial, actinomycetes and fungal population, which was on par with T_1 and T_2. Killham (1994) found that plant growth and nutrient uptake through the seed in initial stages of growth was influenced by microbial mediated changes in seed morphology, seed physiology.

As the effluent concentration increases the population loads as well as the germination percentage of the seeds also increased till 50 to 75 per cent effluent concentration (Dhevagi *et al.*, 2000 and 2006).

Field Trial

Field experiment was conducted to assess the impact of bagasse based paper mill effluent along with amendments. The field soil employed for this study was sandyloam in texture with a clay content of 6.05 per cent. The total bacterial, actinomycetes and fungal populations were 7.2×10^6, 1.2×10^4 and 3.1×10^4 g^{-1} of dry soil respectively. The *Azospirillum, Azotobacter* and *Rhizobium* population were 0.7×10^4, 6.9×10^2 and 0.3×10^2 g^{-1} of soil respectively. The water was colourless and odourless with pH of 7.60 and EC of 0.68 dSm^{-1}. The dissolved oxygen content of the water was 6.11 mg l^{-1}. TNPL sludge showed the pH of 9.1 with an EC of 1.2 dSm^{-1}. The available nutrient contents were low in TNPL sludge.

Effect of Effluent and Amendments on Physico-chemical Properties of Soil

The results on the physico-chemical characteristics of the soil as influenced by treated effluent irrigation and amendments are presented in Table 33.3. Among the amendments, A_3 (pressmud) recorded higher bulk density (1.31 g CC^{-1}) which was significantly different from rest of the treatments. As far as interaction effect was concerned, the treatments A_2 (FYM), A_3 (PM) and A_4 (FPM) were on par with each other and significantly different from other treatments. Effluent irrigation along with amendments caused changes in bulk density of the soil but the change was below the critical level. Yingming and Corey (1993) and Foley and Cooperband (2002) observed that sludge application can alter the bulk density of soil.

Effluent irrigated plots showed higher EC and organic matter than the well water irrigated plots. In the present study the progressive increase in pH of the soil in the treated effluent received plots might be due to continuous irrigation with the effluent which was alkaline in nature and increased the salt accumulation in the soil.

Table 33.3: Effect of Treated Paper Mill Effluent on Soil Physico-chemical Properties

Treatments	*Bulk Density (g CC^{-1})*		*pH*		*EC (dSm^{-1})*		*Organic Matter (per cent)*	
	I_1	I_2	I_1	I_2	I_1	I_2	I_1	I_2
A_1 (Control)	1.08	1.06	8.29	7.18	0.49	0.26	0.76	0.61
A_2 (FYM)	1.27	1.25	8.33	7.29	0.53	0.27	0.86	0.66
A_3 (PM)	1.31	1.29	8.34	7.39	0.54	0.30	0.81	0.67
A_4 (FPM)	1.30	1.28	8.15	7.46	0.52	0.29	0.84	0.69
A_5 (TNPLS)	1.21	1.18	8.39	7.48	0.53	0.32	0.75	0.64
	SEd	*CD*	*SEd*	*CD*	*SEd*	*CD*	*SEd*	*CD*
Amendment (A)	0.002	0.0032	0.009	0.018	0.015	0.029	0.015	0.029
Irrigation (I)	0.001	0.002	0.006	0.012	0.009	0.018	0.009	0.018
A × I	0.001	0.002	0.008	0.016	0.012	0.024	0.015	0.029

This was in line with the findings of Reddy *et al.* (1991); Narashimha Rao and Narashimha Rao (1992) and Udayasoorian *et al.* (1999a and 1999b). The available nitrogen, phosphorus and potassium status was higher in the effluent irrigated plots than the well water irrigated plots (Figure 33.1). Moura (1987), Palanisami and Sree Ramulu (1994) and Dhevagi *et al.* (2006) also had the same opinion.

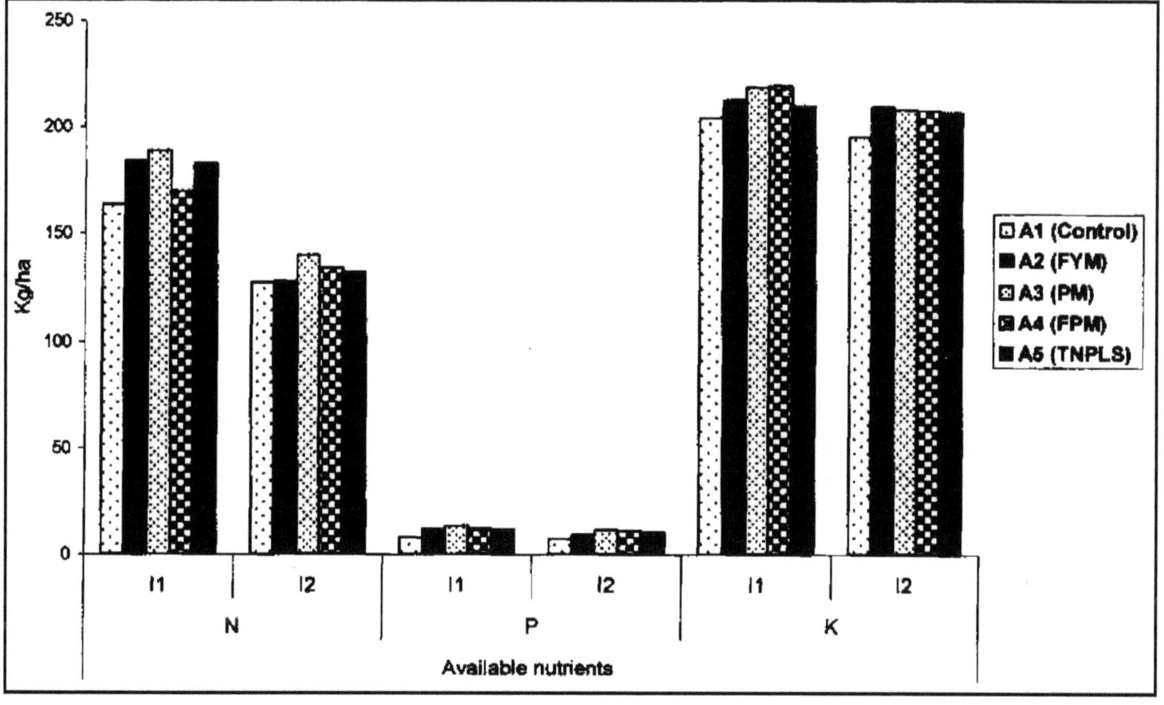

Figure 33.1: Effect of Paper Mill Effluent on Soil Available Nutrients

Effect of Effluent and Amendments on Soil Microbial Properties

The results obtained on the enumeration of microbial load from the field trial were indicated in Table 33.4. Effluent irrigated plots recorded higher bacterial population than the well water irrigated plots. Effluent combined with amendments recorded higher microbial load than the effluent alone irrigated treatments. Treatment A_4 (FPM) recorded the highest rhizosphere bacterial population followed by the TNPL sludge.

Table 33.4: Effect of Treated Paper Mill Effluent on Biological Properties of the Soil Rhizosphere (Population/g of Rhizosphere Soil)

Treatments	*Bacteria × 10^7*		*Actinomycetes × 10^3*		*Fungi × 10^4*	
	I_1	I_2	I_1	I_2	I_1	I_2
A_1 (Control)	153.83	113.73	34.33	23.20	36.37	24.03
A_2 (FYM)	166.50	118.57	60.20	33.67	46.03	25.07
A_3 (PM)	171. 70	127.43	46.97	31.53	49.16	27.37
A_4 (FPM)	190.90	141.47	44.27	27.77	52.07	29.10
A_5 (TNPLS)	179.00	139.60	46.29	30.70	45.50	26.17
	SEd	*CD*	*SEd*	*CD*	*SEd*	*CD*
Amendment (A)	1.708	3.369	0.283	0.558	0.247	0.487
Irrigation (I)	1.081	2.133	0.179	0.353	0.157	0.309
A × I	1.324	2.612	0.219	0.432	0.037	0.073

The *Azotobacter* population varied from 9.40 to 14.90 × 102 g^{-1} of dry soil for the effluent irrigated plots and 7.31 to 9.94 × 102 g^{-1} of dry soil for the well water irrigated plots. A_3 (PM) recorded the highest population. In both the cases, pressmud (A_3) received plots recorded higher population than the other treatments (Figure 33.2). The same trend was observed in *Azospirillum* also. Chauhan and Kaur (1991), Kannan and Oblisami (1992) and Dhevagi *et al.* (2003) also had the same observation.

Epstein *et al.* (1976), Brendecke *et al.* (1994), Bolton *et al.* (1985) and Hasbe *et al.* (1985) observed that organic fertilizer caused a greater increase in soil microbial biomass than inorganic fertilizer. Ilyaltdinov *et al.* (1990) reported that the application of decomposed paper mill waste at the rate of 10–20 t ha^{-1} improved the biological activity.

Rhizosphere microorganisms would stimulate the growth of plants by mobilization of nutrients and production of phyto-effective metabolites, protecting the plants from pathogens, degrading toxic substances, increasing the level of tolerance and stabilizing the soil structure. The rhizosphere effect was more pronounced in bacteria, fungi followed by actinomycetes, *Azotobacter, Azospirillum* and *Rhizobium.* The presence of more rhizosphere population might be due to the presence of root exudates (Kannan and Oblisami, 1990a and b; Datta Roy and Malty, 1984; Manikya Reddy and Venkateswarlu, 1985; Krueger and Sheikh, 1987; Bharathi and Krishnamoorthy, 1988 and Oblisami and Palanisami, 1991).

Sandhya *et al.* (1990), Sun and Lie (1992), Beijarano and Madrid (1992) reported that paper factory effluent irrigation resulted in an increase in soil amylase, invertase, cellulase, dehydrogenase and phosphatase activity. Kannapiran (1995) observed that more microbial activity in the rhizosphere with 50 and 75 per cent effluent concentration. Sandana and Oblisami 1996 reported that the population

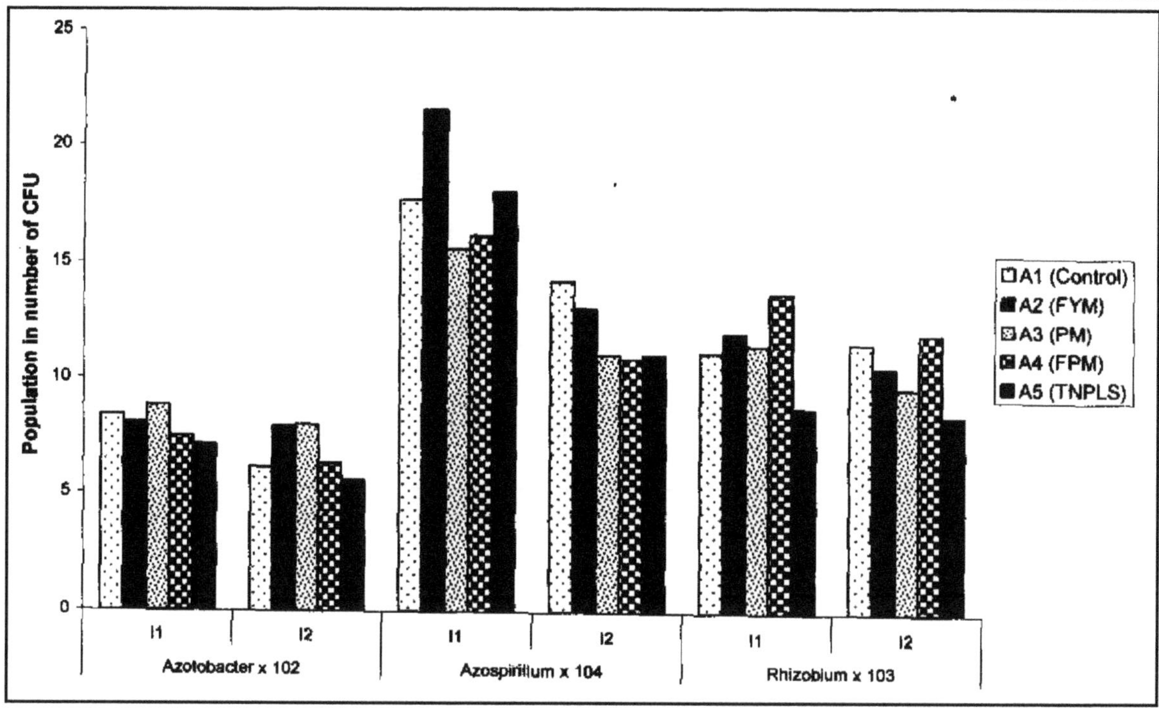

Figure 33.2: Effect of Paper Mill Effluent on Soil Beneficial Microbes

of spermosphere microbial load was lower at higher concentration of effluent whereas, higher counts were recorded at lower effluents concentration. She also observed that marked differences were exhibited in non rhizosphere and rhizosphere microflora of tomato by the effluent irrigation along with amendment.

Effect of Effluent and Amendments on Soil Enzyme Activity

The amylase activity varied from 0.35 to 0.61 and 0.29 to 0.46 mg in the effluent and well water irrigated plots respectively. The activity of invertase was higher in amendment received plots than the unamended plots. The cellulase activity of the effluent irrigated plots varied from 12.04 to 17.25 µg but it varied from 10.96 to 14.46 µg in well water irrigated plots. Mahmood *et al.* (1985), Juwarker *et al.* (1988) and Lahdesmaki and Piispanen (1988) also observed the same result (Table 33.5).

All the treatments showed significant variations in the main as well as interaction effects. In the present investigation there was an increase in amylase, invertase and celluase activity in the soils with effluent irrigation and this might be due to increased microbial population which helped in mineralization and degradation of organic matter.

Effect of Effluent on CO_2 Evolution

The CO_2 evolved showed an increasing trend upto 80 days of effluent irrigation soybean were concerned, the CO_2 evolved was slightly higher than maize and sunflower. In soybean, the CO_2 evolved at vegetative stage in the effluent irrigated plots showed a variation of 3.53 to 5.02 mg and in the well water irrigated plots but it varied from 3.15 to 3.84 mg (Table 33.5). This might be due to increase in organic matter content of the soil due to effluent irrigation and this might have accelerated the microbial activity which in turn accelerated the substrate decomposition with the release of carbon dioxide.

Table 33.5: Effect of Treated Paper Mill Effluent on Soil Enzyme Activities

Treatments	Amylase Activities (mg of reducing sugars/g of soil/24 hours)		Soil Invertase (mg of reducing sugars/g of soil/24 hrs.)		Cellulase (μg of reducing sugars/g of soil/24 hours		CO_2 Evolution (mg 100 g^{-1} of soil)	
	I_1	I_2	I_1	I_2	I_1	I_2	I_1	I_2
A_1 (Control)	0.52	0.33	0.29	0.24	10.91	10.60	4.01	3.45
A_1 (FYM)	0.64	0.51	0.33	0.27	12.67	11.66	5.54	4.31
A_3 (PM)	0.65	0.47	0.38	0.29	12.78	11.49	5.14	3.88
A_4 (FPM)	0.69	0.56	0.45	0.31	14.37	12.49	5.70	4.26
A_5 (TNPLS)	0.62	0.48	0.37	0.30	12.90	11.31	5.00	3.88
	SEd	CD	SEd	CD	SEd	CD	SEd	CD
Amendment (A)	0.007	0.014	0.004	0.008	0.070	0.138	0.016	0.031
Irrigation (I)	0.004	0.008	0.003	0.006	0.045	0.089	0.009	0.019
A × I	0.006	0.012	0.003	0.006	0.056	0.111	0.012	0.024

Effect of Effluent and Amendments on Nodulation

The number of nodules was more in effluent irrigated treatment along with amendment. This might be due to increased rhizobial population in the rhizosphere. The NPK uptake was higher in treatments received pressmud and fortified pressmud (Kannapiran, 1997b). The K uptake showed an increasing trend towards the stages of the crop. Indira Raja and Raj (1981) and Mariappan *et al.* (1983) had the same opinion.

Table 33.6: Effect of Paper Mill Effluent on Yield Parameters

Amendment	Nodule Numbers		No of Pods/Plant		Yield (kg of pods/ha)	
	I_1	I_2	I_1	I_2	I_1	I_2
Al	6.30	5.77	25.00	24.00	1175.70	1025.40
A2	11.47	9.47	26.50	25.00	1325.80	1076.35
A3	9.60	8.49	29.50	26.00	1750.40	1395.30
A4	8.57	6.97	27.00	25.50	1455.40	1209.90
As	6.13	7.57	22.00	21.00	1092.50	1040.50
	SEd	CD (0.05)	SEd	CD (0.05)	SEd	CD (0.05)
Amendment	0.180	1.578	0.04	0.081	64.99	130.10
Irrigation	0.114	0.225	0.03	0.061	41.11	82.30
A × I	0.440	0.868	0.06	0.120	91.92	184.02

The number of pods and 100 kernel weights were higher in A_3 which received pressmud as amendment. The yield was lower in A_5 (TNPLS) (Table 33.6). This might be due to increased available nutrient status would have led to greater utilization of nutrients by the crops resulting in higher yield. Higher rhizophere microflora might have played a role in mobilizing the nutrients by way of solubilisation and fixation and also through growth promoting substances. The present result was in accordance with Insam *et al.* (1991), Goyal *et al.* (1992), Dutta and Biossya (1999). Treatment received TNPL sludge showed lesser yield might be due to salts added, which restricted the root growth by increasing soil osmotic pressure (Feagly *et al.*, 1994).

Effect of Effluent and Amendments Nutrient Uptake

Diluted paper mill effluent increased the chlorophyll content, plant height, shoot and root biomass, grain yield, protein, carbohydrate and lipid contents in wheat grains, while undiluted effluent caused inhibition in plant growth resulting in a sharp decline of yield. Pure soil provided better growth and yield results than that soil mixed with sand (Singh *et al.*, 2002).

Acknowledgement

The authors are extremely grateful to Tamil Nadu Newsprint and Paper Limited for granting permission to publish these data, and for providing financial assistance during the course of investigation.

References

Abdul-Baki, A.S. and Anderson, J.O., 1973. Vigour determination in soybean seed by multiple criteria. *Crop Sci.*, 3: 630–633.

Anonymous, 1965. *Standard Methods for the Examination of Water and Wastewater*. Amer. Pub. Health Assoc. and Amer. Water works Assoc. and Amer. Water Pollution Conf. Federation, Broadway, New York, p. 1193.

Benefield, C.B., 1971. A rapid method for measuring cellulase activity in soil. *Soil Biol. Biochem.*, 3: 325–329.

*Beijarano, M. and Madrid, L., 1992. Effect of Alpectin on heavy metal solubilization. Effecto del Alpechin sobrede metales pesados. *Sueloyplanta.*, 2: 541–549.

Bharathi, S.G. and Krishnamoorthy, S.R., 1988. Effect of paper mill effluent on the distribution of algal flora of river Kali around Dandeli, Karnataka. *Proc. Ind. Sci Cong.*, 3: 19–20.

Bolton, H., Jr. Elliott, L.F. and Papendick, R.I., 1985. Soil microbial biomass and selected soil enzyme activities: effect of fertilization and cropping practice. *Soil Biol. Biochem.*, 17: 297–302.

Brendecke, J.W., Axelson, R.D. and Pepper, S.L., 1994. Soil microbial activity as an indicator of soil fertility: Long-term effects of municipal sewage sludge on an arid soil. *Soil Biol. Biochem.*, 25: 751–758.

Chauhan, U.K. and Kaur, P., 1991. Impact of pulp and paper mill effluent on soil microflora. *Proc. Ind. Sci. Congr. Ass.*, Indore, 3(8): 24–25.

Chopra, S.L. and Kanwar, J.S., 1982. *Analytical Agricultural Chemistry*. Kalyani Publishers, New Delhi, p. 162.

Datta Roy, S. and Malty, B.R., 1984. Pollution load of paper mill effluents discharged into the Hoogly river at Kalyani, West Bengal. *Env. Eco.*, 1: 29–34.

Dhevagi, P., Rajannan, G. and Oblisami, G., 2000. Effect of paper mill effluent on spermosphere microflora. *J. Ecobiol.*, 12: 149–152.

Dhevagi, P. and Oblisami, G., 2000. Effect of paper mill effluent on germination of agricultural crops. *J. Ecobiol.,* 12: 243–249.

Dhevagi, P., Maheswari, M., Ilamurugu, K. and Oblisami, G., 2003. Influence of paper mill effluent on soil microflora. *J. Ecotoxicol. Environ. Monit.,* 13: 103–109.

Dhevagi, P., Udayasoorian, C. and Oblisami, G., 2004. Fly ash: Can it be used as nursery mix? *J. Ecobiol.,* 16: 33–36.

Dhevagi, P., Maheswari, M. and Oblisami, G., 2006. Effect of paper mill effluent on soil microflora of Maize. *Indian Journal of Environmental Science* (Accepted for Publication).

Dhevagi, P. Maheswari, M. and Oblisami, G., 2007. Effect of paper mill effluent on soil microflora of Maize. *Indian Journal of Environmental Science* (Accepted for Publication).

Dobereiner, J., 1980. Forage grasses and grain crops. In: *Methods for Evaluating Biological Nitrogen Fixation,* (Ed.) F.J. Bergerson. John Wiley and Sons, New York, pp. 535–555.

Dutta, S.K. and Biossya, C.L., 1999. Effect of paper mill effluent on chlorophylls, leaf area and grain number in transplanted rice (*Oryza sativa* L. Var Masuri). *Eco. Env. Conserv.,* 5: 369–372.

Epstein, E., Taylor, J.M. and Chaney, R.L., 1976. Effects of sewage sludge compost applied to soil on some soil physical properties. *J. Environ. Qual.,* 5: 423–426.

Feagly, S.E.,Valdez, M.S. and Hudnall, W.H., 1994. Paper mill sludge, phosphorus, potassium and lime effect on clover grown in mine soil. *J. Environ. Qual.,* 23: 759–765.

Foley, B.J. and Cooperband, L.R., 2002. Paper mill residuals and compost effects on soil carbon and physical properties. *J. Environ. Qual.,* 31: 2086–2095.

Galstyan, A.Sh., 1965. A method of determining the activity of hydrolytic enzymes in soil. *Soviet Soil Sci.,* 2: 170–175.

Gomathi, V. and Oblisami, G., 1992. Effect of pulp and paper mill effluent on germination of tree crops. *Indian J. Environ. Hlth.,* 34: 326–328.

Goyal, S., Mishra, M.M., Hooda, I.S. and Singh, R., 1992. Organic matter microbial biomass relationships in field experiments under tropical conditions: Effects of inorganic fertilization and organic amendments. *Soil Biol. Biochem.,* 24: 1081–1084.

Hasbe, A., Kanazava, S. and Takai, V., 1985. Microbial biomass in paddy soil II. Microbial biomass C measured by Jenkinson's fumigation method. *Soil Sci. Plant Nutr.,* 31: 349–359.

Indira Raja, M. and Raj, D., 1981. Influence of pressmud application on yield and uptake of nutrients by finger millet in some soils in Tamil Nadu. *Madras Agri. J.,* 58: 314–322.

Ilyaltdinov, A.N., Kurmambaev, A.A., Sandanov, A.K. and Klysheva, A.L., 1990. Biological activity of secondary saline soils in the kzyl-ordinsky irrigation areas as a result of the introduction of pulp and paper mill wastes. *Seriya Biologicheskaya,* 3: 50–54.

Insam, H., Mitchell, C.C. and Dormaar, J.F., 1991. Relationship of soil microbial biomass and activity with fertilization practice and crop yield of three ultisols. *Soil Biol. Biochem.,* 23: 459–464.

Jackson, M.L., 1973. *Soil Chemical Analysis.* Prentice Hall of India (Pvt.) Ltd., New Delhi, p. 103.

Juwarkar, A., Dutta, S.A. and Pandey, R.A., 1988. Impact of domestic waste water on soil chemical properties and microbial populations. *J. Micro. Biotechnology,* 14: 169–177.

Kallas, J. and Munter, R., 1994. Post-treatment of pulp and paper industry waste waters using oxidation and adsorption processes. *Wat. Sci. Tech.*, 29: 259–272.

Kannan, K. and Oblisami, G., 1990a. Influence of paper mill effluent irrigation on soil enzyme activities. *Soil Bioi. Biochem.*, 22: 923–926.

Kannan, K. and Oblisami, G., 1990b. Influence of irrigation with pulp and paper mill effluent on soil chemical and microbiological properties. *Biol. Fertility Soils*, 10: 197–201.

Kannan, K. and Oblisami, G., 1992. Effect of raw and treated paper mill effluent irrigation on vigour indices of certain crop plants. *Madras Agric. J.*, 79: 18–21.

Kannapiran, S., 1995. Studies on the effect of solid and liquid wastes from paper and pulp industry on forest species. *M.Sc. (Env. Sciences)*, Tamil Nadu Agricultural University, Coimbatore.

Kannapiran, S., Dhevagi, P., Ponniah, C. and Oblisami, G., 1997a. Influence of treated paper mill effluent on germination of tree crops. *Indian J. Environ. Health*, 39: 330–332

Kannapiran, S., Dhevagi, P., Ponniah, C., Oblisami, G. and Udayasoorian, C., 1997b. Response of *Pongamia pinnata* to solid waste. In: *Proceeding of the 6th National Symposium on Environment*, January 7–9, 270–275.

Killham, K., 1994. *Soil Ecology*. Cambridge University Press, New York, p. 242.

Krueger, C.L. and Sheikh, W., 1987. A new selective medium for isolating *Pesudomonas* spp. from water. *Appl. Environ. Microbiol.*, 53: 895–897.

Lahdesmaki, P. and Piispanen, R., 1988. Degradation products and the hydrolytic enzyme activities in the soil humification processes. *Soil Biol. Biochem.*, 20: 289–292.

Mahmood, T., Azam, F. and Malik, K.A., 1985. Decomposition and humification of plant residues by some soil fungi. *Biotechnol. Lett.*, 7: 207–212.

Manikya Reddy, P. and Venkateswarlu, B., 1985. Ecological studies in the paper mill effluents and their impact on the river Tungabhadra: Heavy metals and algae. *Proc. Indian Acad. Sci.*, 95: 139–146.

Mariappan, N., Nagappan, M. and Sadanand, A.K., 1983. Pressmud in the reclamation of alkaline soils in Kothari sugar factory area. In: *Proc. National Seminar on Utilization of Organic Wastes*, TNAU, Coimbatore.

Moura, J.B., 1987. Treatment of liquid effluents from a cardboard factory by a spray-irrigation system. *Wat. Sci. Tech.*, 19: 147–153.

Narashimha Rao, P. and Narashimha Rao, Y., 1992. Quality of effluent water discharged from paper board industry and its effect on alluvial soil and crops. *Indian J. Agric. Sci.*, 62: 9–12.

Niemelae, S.I. and Vaeaetaenen, P., 1982. Survival in lake water of *Klebsiella penumoniae* discharged by a paper mill. *Appl. Environ. Microbial.*, 44: 264–269.

Oblisami, G. and Palanisami, A., 1991. Studies on the effect of paper and sugar factory on soil microflora and agricultural cropping system–Report. Tamil Nadu Agricultural University, Coimbatore, p. 297–305.

Olsen, S.R., Cole, L.L., Watanable, F.S. and Dean, D.A., 1954. Estimation of available phosphorus in soils by extraction with sodium bicarbonate. *USDA* Cric. 939.

Palaniswami, C. and Sree Ramulu, U.S., 1994. Effects of continuous irrigation with paper factory effluent on soil properties. *J. Indian Soc. of Soil Sci.*, 42: 139–140.

Panse, V.G. and Sukhatme, P.V., 1985. *Statistical Methods for Agricultural Workers*. ICAR Publ., New Delhi, p. 359.

Piper, C.S., 1966. *Soil and Plant Analysis*. Inter Science Publications, New York, p. 368.

Pramer, D. and Schmidt, E.L., 1965. *Experimental soil Microbiology*. Burges Publishing Co., Minneapolis, Mimnesota, p. 107.

Reddy, M.R., Jivendra, S. and Jain, S.C., 1991. Paper mills effluent for sugarcane irrigation. *IAWPC Technol. Annu.*, 8: 129–146.

Sandhya, S., Joshi, S.R. and Parhad, N.M., 1990. Simultaneous disposal and resource recovery from lignocellulosic waste of paper mills. *J. Microb. Biotechnology*, 5: 5–9.

Sandana, K.M.C. and Oblisami, G., 1996. Effect of paper factory effluent on soil characteristics and growth of tomato. *Abst. National Conference on Environmental Biotechnology Utilization of Industrial Wastes Potentials and Limitations* held at Bharadhidasan University, Trichy, p. 65.

Singh, A., Agrawal, S.B., Rai, J.P. and Singh, P., 2002. Assessment of the pulp and paper mill effluent on growth, yield and nutrient quality of wheat (*Triticum aestivum* L.). *J. Environ. Biol.*, 23: 283–28.

Stanford, S. and English, L., 1948. Use of flame photometer in rapid soil tests of K and Ca. *Agron. J.*, 41: 446–447.

Subba Rao, N.S., 1988. *Biofertilizers in Agriculture*. Oxford and IBH Publishing Company Private Limited, New Delhi, p. 65–190.

Subbiah, B.V. and Asija, G.L., 1956. A rapid procedure for estimation of available nitrogen in soils. *Curr. Sci.*, 25: 259–260.

Subrahmanyam, P.V.R., Juwarkar, A.S. and Sundaresan, B.B., 1984. Utilization of pulp and paper mill waste water for crop irrigation. In: *Proc. Asian Chemical Conference on Priorities in Chemistry in Development of Asia*, Priochem Asia, Kulalumpur, Malaysia, pp. 26–31.

Sun, H.G. and Lie, W.Q., 1992. The effects of straw paper pulping wastes using ammonium sulfate on the microstructural characteristics of calcareous drab soil and paddy soil. *Acta Agriculturae*, 18: 181–188.

Udayasoorian, C., Dhevagi, P. and Ramasami, P.P., 1990a. Case study on the utilization of paper and pulp mill effluent irrigation for field crops. In: *Proceedings of the Workshop on Bioremediation of Polluted Habitats*, p. 71 –73.

Udayasoorian, Sagaya, Alfred, C., and Ramasami, P.P., 1990b. Long-term effect of bagasse based paper mill effluents irrigation on soil and groundwater. In: 2nd *International Conference on Contaminants in Soil Environment in the Australia–Pacific Region*, p. 80–81.

Verma, A.M., Dudani, V.K., Kumari, B. and Kargupta, A.N., 1988. Algal population in paper mill waste water. *Indian J. Environ. Hlth.*, 30: 388–390.

Waksman, S.A. and Fred, E.B., 1922. A tentative outline of the plate method for determining the number of micro organisms in soil. *Soil Sci.*, 14: 27–28.

Yingming, L. and Corey, R.B., 1993. Redistribution of sludge borne cadmium, copper and zinc in a cultivated plot. *J. Environ. Qual.*, 22: 1–8.

Chapter 34

Cyanobacteria can Efficiently Detoxicate Herbicides

Gyanendra Kumar Dwivedi[1], *A.K. Pandey*[1]* *and K.N. Mishra*[2]

[1]*Biological Research Laboratory, Department of Botany, Kutir P.G. College, Chakkey, Jaunpur (U.P.)*
[2]*Department of Botany, T.D. College, Jaunpur (U.P.)*

ABSTRACT

Cyanobacterium *Nostoc linckia* was tested for detoxication of paddy field herbicide alachlor and propanil. A gradual loss in the toxicity of alachlor and propanil was noticed by repeated incubation and removal of the cyanobacterium from the herbicide medium as algicidal doses started to support the cyanobacterial growth, pH of the medium was maintained 7.5 in each inoculation. This indicated that herbicides are either accumulated or detoxified in the cyanobacterial cells.

Keywords: *Herbicides, Detoxication, Cyanobacteria, Nostoc linckia.*

Introduction

The growth and activity of cyanobacteria in submerged soil, some of which possessing efficient system for fixing atmospheric nitrogen, are widely recognized in tropical paddy field soil (Singh, 1961, Stewart *et al.*, 1978). The photoautotrophic habit of these organisms and their luxurious growth in aquatic environments explain their agricultural and ecological importance. The extensive use of various types of herbicides at any level in ecosystem may lead to much greater emphasis on the possibility of serious environmental contamination through food chain from primary producers to

* Corresponding Author.

high trophic level. (Woowell *et al.*, 1967) and may affect the community composition of the ecosystem by disturbing the diversity and number of microbes. The accumulation of toxicants and their residues in an organism by adsorption and absorption resulted increased concentration in cell of microscopic plants and higher aquatic flora (Audus, 1964). There are only few reports are available on biodegradation/detoxification and accumulation of pesticides including herbicides by cyanobacteria and green algae (Kruglov, 1970; Valentine, 1973; Fitzerald, 1975; Das and Singh, 1977; Kar and Singh, 1979; Tiwari and Pandey, 1981), although their biological effect on cyanobacteria have been reported (Dasilva *et al.*, 1974; Tiwari and Pandey, 1981; Tiwari *et al.*, 1984 Singh and Tiwari, 1988; Pandey, *et al.*, 1984; Pandey, 1985; Vaishampayan *et al.*, 2000; Pandey and Singh, 2001; Pandey *et al.*, 2006). Since herbicides are often used in rice cultivation, an extensive study on their detoxification by cyanobacteria may be significant for biomonitoring studies and could be useful for phytoremediation technology to restore the paddy soil quality by harvesting such cyanobacteria. Therefore, the present communication deals with the ability of paddy soil nitrogen fixing cyanobacterium *Nostoc linckia* in defoliating rice field herbicide alachlor and propanil.

Materials and Methods

A paddy soil nitrogen fixing cyanobacterium *Nostoc linckia* cultured axenic condition from clonal culture, in modified Chu–10 medium (Safferman and Morris, 1964) free of combined nitrogen sources (C–N) at 24±2°C in a culture room under day light fluorescent illuminatiom (light intensity approximately 2200 lux) for 14 hrs daily.

Alachlor (50 per cent EC) obtained from Mansanto Chemicals Chennai and propanil (35 per cent EC) from BPM Pvt. Ltd. Mumbai in a liquid containing 0.5 g and 0.35g chemicals per ml respectively. Freshly prepared aquous solution of the herbicide was filter sterilized separately before use and its stock solution was prepared by approximately diluting it with sterilized growth medium.

Exponantially grown cyanobacterial cultures were centrifuged, washed thrice and homogenized with neutral glass beads and resuspended into the basal medium. Aliquots of 0.2 ml were inoculated into 50 ml culture medium containing various concentrations of herbicide.

Growth of the cyanobacterium was measured torbidimetrically at 650 nm with spectrophotometer (Systronics 166). The cyanobacterium was removed by centrifugation on every third days and a fresh culture of the cyanobacterium with 0.01 OD was inoculated into the herbicide containing supernatent medium. The above process was repeated for every three days till 18 days. In due course of study the pH of the medium was maintained 7.5.

Results and Discussion

Cyanobacterial growth was measured on the third day of first inoculation and it was observed that 10 μg ml^{-1} of alachlor and 20 μg ml^{-1} of propanil were algistatic whereas 30 and 40 μg ml^{-1} of alachlor and propanil were lethal respectively (Table 34.1) to cyanobacterium *Nostoc linckia* as evidenced by loss of turbidity and absence of filaments under microscopic observation. After removal of cyanobacterial population by centrifugation and reinoculation of cyanobacterium into the resulting supernatants and incubation for next three days, no growth was observed at lethal concentration till second inoculation where as in third incubation (after 9 days), it was observed that medium containing, algicidal/lethal dose of 30 μg ml^{-1} (alachlor) and 40 μg ml^{-1} (propanil) started to support cyanobacterial growth as evidenced by turbidity and appearance of cyanobacterial cells (Table 34.1). After fourth and fifth subsequent removal and reinoculation of the fresh cyanobacterium into the herbicide containing supernatants, it was observed that algicidal/lethal concentration of alachlor and propanil supported growth of the cyanobacterium. The cyanobacterium exhibit reduced growth in this medium which may be on account of exhaustion of some essential nutrients. The present observations suggest that the repeated cyanobacterial inoculation and growth in the lethal dose must have caused the gradual removal and detoxification of the herbicide from the medium. Although there is no direct evidence to demonstrate their metabolic degradation by cyanobacteria and exact site of accumulation in cells is not known. The accumulation of 2–8 per cent of herbicide C^{14}–Simazine and binding to protein fraction by green algal *Chlorosarcina* sp. and *Ankistrodesmus braunii* within 20 days from nutrient solution was reported by Kruglov (1970). Herbicides mediated cyanobacterial growth is possible either biodegradation during metabolism or simply taken up and accumulated into the cells as reported earlier in green algae (Valentine, 1975). There is no direct evidence to demonstrate that metabolic degradation by the cyanobacteria/green algae but based on the present observation it may be concluded that photosynthetic algae may accumulate these herbicides in their cell and subsequently metabolise them either as energy source (as carbon or nitrogen) or as waste product in order to create a detoxified environment for their better survival.

Table 34.1: Effect of Repeated Cyanobacterial Growth in Detoxification of Rice Field Herbicides Alachlor and Propanil by *Nostoc linckia* (Mean value of three independent experiments)

Incubation Period (3 days interval)	*Control*	*O.D. 650 hm*			
		Alachlor (μg ml)		*Propanil (μg ml⁻¹)*	
		*10 **	*20*	*20*	*40*
Zero	0.010	0.010	0.010	0.010	0.010
First	0.025	0.020	*	0.050	*
Second	0.045	0.030	*	0.035	*
Third	0.040	0.050	0.015	0.025	0.020
Fourth	0.040	0.065	0.030	0.015	0.035
Fifth	0.035	0.075	0.045	0.055	0.050

*: No growth.

Acknowledgement

The authors are grateful to Professor D.N. Tiwari, ex. Professor and Head, Department Botany, Banaras Hindu University, Varanasi for helpful discussion. We express our gratitude to Shri S.N. Mishra, Principal, Kutir P.G. College Chakkey, Jaunpur and Dr. U.P. Singh, Principal, T.D. College, Jaunpur for providing facilities.

References

Audus, L.J., 1964. Herbicide behaviour in soil. In: *The Physiology and Biochemistry of Herbicides*. Academic Press, London.

Das, B. and Singh, P.K., 1977. Detoxification of the pesticides bengene-hexachloride by blue-green algae. *Microbios Letters*, 1: 99–102.

Dasilva, E.J., Henriksson, L.E. and Henrikson, E., 1974. Effect of pesticides on blue-green algae and nitrogen fixation. *Arch. Envi. Contaminant. Toxicol.*, 1: 197–24.

Fitzgerald, G.P., 1975. *Are Chemicals Used in Algae, Water and Sewerage Works*, pp. 82–85.

Kar, S. and Singh, P.K., 1979. Detoxification of pesticides carbafuran and hexachlorocyetohexane by pesticide carbafuran and hexachlorocyetohexane by blue-green alage *Nostoc muscorum* and *Wollea bhardwajae. Microbios. Letters,* 10: 111–114

Kruglov, Yu.V., 1970. Detoxification of simazine by microscopic algae. *Microbiology*, 39: 139–142.

Pandey, A.K., Srivastava, V. and Tiwari, D.N., 1984. Toxicity of the herbicides stam f–34 (Propanil) on *N. calcicola.* Zeit. Allg. *Microbiol.*, 26(6): 369–376.

Pandey, A.K., 1985. Effect of propanil on growth and cell constituents of *N. calcicola* Pest. *Biochem Physiol.,* 23: 157–162.

Pandey, A.K., Dongare, P.N., Singh, Y.K. and Tiwari, S.N., 2006. Regulation of growth and Biomolecule of *Alphanothece sp.* by phenoxy herbicide 2,4-D. *Biosci. Biotech. Res. Asia,* 2(a) (In press).

Pandey, A.K. and Singh, H.N., 2001. Factors regulating, 2,4-D toxicity in the cyanobacterium *N. calcicola. Ind. Environ. Ecoplan.,* 5(3): 539–543.

Saffermann, R.S. and Morris, M.E., 1964. Growth characteristics of blue-green algae virus LPP–1. *J. Bact.,* 88: 771–775

Singh, R.N., 1961. *Role of Blue Green Algae in the Nitrogen Economy of Indian Agriculture*. ICAR Publ., New Delhi.

Singh, L.J. and Tiwari, D.N., 1981. Effect of selected rice field herbicides on photosynthesis, respiration and nitrogen assimilatory enzyme systems of paddy soil diazotrophic cyanobacteria. *Pesticides Biochem. Physiol.,* 31: 120–128.

Stewart, W.D.P., Sampio, M.J., Isicheii, A.O. and Sylvester Brodley, 1978. In: *Limitation and Potentials of Biological Nitrogen Fixation in the Tropics,* (Eds.) Dobereiner, J. Burris, R.S. and Hollander, A. Plenum Press, New York.

Tiwari, D.N. and Pandey, A.K., 1981. Biodetoxification of different groups of herbicide by blue-green algae *Nostoc calcicola. Ind. J. Microbiol*., 21(2): 163–164.

Tiwari, D.N. and Pandey, A.K., 1981. 2,4-D resistant mutant strain of *Anacystis nidulans* and filament formation. *Ind. J. Exp. Bio.,* 19: 988–990.

Tiwari, D.N., Pandey, A.K. and Mishra, A.K., 1984. Toxicity of 2,4-Dichloro phenoxyacetic acid on growth and nitrogen fixation of blue-green algae *Anabaena cylindrica. Pesticides*, 11: 16–18.

Vaishampayan, A., Sinha, R.P., Gupta, A.K. and Hader, D.P., 2000. A cyanobacterial mutant resistant against a bleaching herbicide. *J. Basic Microbiol.*, 40(4): 279–288.

Valentine, J.P., 1973. The influence of four algae on herbicide residue in water. *Diss Abstra Inten. B.,* 38: 1251–1252.

Wood Well, G.M., Inluister Jr. C.F. and Isaaeson, P.A., 1967. DDT residues in an east coast estuary: A case of biological concentration of a persistant insecticides. *Science*, 156: 821–823.

Chapter 35

Immunomodulation of Humoral Immune Response by Carbosulfan in Swiss Albino Mice

P. Dhasarathan, Lighty George, A.J.A. Ranjit Singh, C. Padmalatha and M. Madhumitha

Department of Biotechnology, Sri Kaliswari College, Sivakasi – 627 412, Tamil Nadu, India
Department of Biology, Sri Paramakalyani College, Alwarlrurichi, Tamil Nadu, India
Department of Zoology, Rani Anna Govt. College, Tirunelveli

ABSTRACT

Carbosulfan, which is a broad-spectrum organochlorine, insecticide against insects. It is an commercial products and so affects man through food-chain. As immunotoxicology is less explored, it's a serious concern now to explore the adverse effect of insecticide on animals and human beings. In the present study, carbosulfan is given with diet to Swiss albino mice so as to examine the effect of chemical changes on immune system. In present study, show clear exploration on immunological changes.

Keywords: *Modulation, Immune response, Carbosulfan, Toxicology and Swiss Albino mice.*

Introduction

The immune system is a versatile defense system to protect the animals and human beings. It will apparently recognize the foreign invaders and kill them (phagocytosis). The impact of xenobiotics on mammalian immune system is less studied. So, in this article humoral immune response of chemical carbosulfan in action of Swiss albino mice has been revealed. The immunotoxicology can be defined as the toxicological assessment of drugs, chemica1s and biological/xenobiotics for human risk (Bick, 1985 and Akthar *et al.*, 1996). Long term exposure to modem insecticides results in adverse effects like cancer, birth defects etc. (Lochner *et al., 2002*). That's why the development of biologically natural methods of insects control is preferential today.

Materials and Methods

Animal and Treatment

For this experiment, Swiss albino mice were selected as candidate species. It's reared in standard lab condition of light and darkness (12/12 hrs) and temperature (22°C±2°C). It fed with standard mice pellet feed and water *ad libitum* to the animals. Using standard toxicological test (Spraque, 1965) LD50 doses for the carbosulfan was found out. The sublethal doses of pesticide dissolved in water and given to the animals. Food consumption, general condition and other symptoms were observed up to 3 weeks and body weights were recorded.

Immunological Assays

The following immunological parameters were analysed in both control and carbosulfan treated mice. They are B-cell EAC rosette assay, antibody titration and plaque forming cell assay.

B-cell EAC Rosette Assay

Blood is collected nom pesticide treated and control mice using a heparin pretreated vials. B and T cell counts in the blood are carried out by the following method.

5–10 ml of blood was collected and it was introduced into sterile conical flask/beaker containing (4–5) sterile glass beads. It was then continuously swirled until no sounds were heard from the beads. This indicates that all the fibrins have adhered to the beads. This blood was considered as defibrinated blood. This defibrinated wood was taken and diluted with equal volume of physiological saline. 3 ml of the lymphoprep solution was taken in a centrifuge tube. The tube was kept in slanting position and 9 ml of diluted blood was slowly added along the sides of the centrifuge tube using Pasteur pipette. Care was taken so that the FICON layer of the lymphoprep solution present in the centrifuge tube was not disturbed. The content of the centrifuge tube was then centrifuged at 1600 rpm for 20 min. The interphase (containing lymphocytes) was removed using pipette. The cells were washed with 1 ml saline and excess FICON was removed. The sample was again washed with 1ml of saline after centrifugation the supernatant was decanted by inverting the tube over a filter paper after all saline was drained; the pellet was then resuspended in 300 µl of RPMI 1640 medium.

12–14 cm of drinking straw was cut. One end of the straw was slantly cut and sealed by slightly heating the tip in a flame. Nylon wool fibres were finely teased using a pair of forceps and the teased fibres were packed (loosely) into the straw. Adding 5 ml of physiological saline washed the packed nylon wool column. A small opening was made at the sealed end of the straw to drain the physiological saline. After washing with physiological saline, the nylon wool was then filled with 3 ml of RPMI 1640 medium in a horizontal position. The nylon wool column was kept in the incubator (at 370 C for 30 minutes) in horizontal position. This process activates the nylon wool column.

Resuspended lymphocytes were loaded into the activated nylon wool column. Then the column was held vertically above an eppendorf tube, now hot saline (about 600°C) was slowly dripped into the column. The hot saline passing out of the column was collected in the eppendorf tube, which contain T lymphocytes. After hot saline elution, cold saline in then dripped was through the column.

The column in gently squeezed to release the adhered B-cell (repeat twice). The cold saline dripping out of the column was collected in another eppendorf tube. 0.2 ml of the saline containing B lymphocyte (from the eppendorf tube containing B cells) was taken in a separate eppendorf tube. To this 0.2 ml of 1 per cent SRBC was added and then the mixture was centrifuged for 12 minutes at 1600 rpm. After centrifugation the sample were incubated in an icebox or refrigerator (at 40°C) for 5 minutes. After cold incubation, the pellet in the eppendorf tube was re-suspended by gentle flushing with a Pasteur pipette. Then a drop of it was taken in a clean dry slide, observed and enumerated B-cell under the microscope (20x/40x) for rosettes. Number of B-cell rosettes formed were observed among hundred lymphocytes observed was tabulated.

Antibody Titration

The total amount of antibody production was carried out by Log_2 titre plate method. In this method, 50 µl of physiological saline is added using a 50 µl dropper into all the wells of a clean microtitre plate. Then take 50 µl of the antiserum in a pipette man the antiserum is serially diluted in the wells of the first row till the 11th well of the microtitre plate leaving the 12th well as a positive control. 25 µl per cent SRBC in saline is added to all the wells of the microtitre plate. The microtitre plates hand shaken for effective mixing of reagents. The reagents are incubated for an hour at 37°C and for one more hour at 10°C. The highest dilution of the serum samples which shows detectable agglutination is recorded and expressed a Log_2 antibody titre of the serum.

Plaque Forming Cell Assay

The direct Plaque Forming Cell Assay (PFCA) was carried out in the pesticide treated and control mice, after immunization as mentioned in Section 3.4 by the method of Jeme and Nordin (1963). From the control and pesticide treated mice (stimulated and non-stimulated), spleen was aseptically and carefully removed. The spleen was separately teased with a needle to prepare the cell suspension in RPMI–1640 medium. The splenic cell suspension was prepared from different groups and carefully marked for further analysis. Cells were washed twice with the medium and suspended to a density of 1×10^6 cells/ml. Petri dishes (2.5 cm diameter) were layered with 12 per cent agarose in 0.15 M NaCl. A mixture of 2 ml of 0.6 per cent SRBC (v/v packed cell volume) and 1×10^6 spleen cells in 0.1 ml was poured over the base layer. The Petri dishes were incubated at 37°C for 90 minutes. Two ml of 1 : 10 diluted fresh guinea pig serum were added to each Petri-dishes as a complement. Incubation was further continued for 45 minutes. Plaques formed on the agarose surface were then counted and the values were expressed as counts per 10^6 spleen cells.

Results and Discussion

B Lymphocyte Counts

B-lymphocytes counts using rosette techniques revealed significant changes in pesticide treated mice. When compared to control (Table 35.1). Carbosulfan treated mice show the decrement in B-cell number in the first week. The decrease in B-cell number in pesticide exposed mice due to impact on pesticide molecules. Synthesis proliferation and activation of B-cell. B-cell have population potentially auto reactive clones at most of B-cell maturation (Vos, 1977 and Sindemman, 1983) and transforming growth factors. Pesticides Used in the present study were found to be genotoxic and damages DNA. Thus, Ig gene rearrangement in B-cell and TFG-β might be affected by the pesticides leading to the reduction in B-cell count. Rapid GSH depletion excess Oxygen Free Radical (OFR) production, increased rapid peroxidations. progressive uncoupling of oxidative phosphorylation DNA fragmentation and apoptosis due to pesticide suppress the functioning of the immune system.

Table 35.1: Enumeration of B-cells Using Rosette-forming Assay in Mice Exposed to Different Sublethal Concentrations of Carbosulfan for Three Weeks

Sl.No.	Test Chemicals	Number of B Cells Rosette Formed in 100 Lymphocytes Observed		
		I Week	II Week	III week
1.	Control	24±]	24±0.6	24±1.0
2.	1/10th LD_{50} Conc. Carbosulfan	8±1.5	6±0.4	9±05
3.	1/20th LD_{50} Conc. Carbosulfan	10±0.3	8±0.2	11±05

Plaque Forming Cell Assay

Effect of pesticide carbosulfan on direct spleenic plaque forming cells shows a reduction in secondary plaque forming cells in the first 3 weeks (Table 35.2). The peak antibody forming occurred in the 4th week of primary response and in 3rd week in the secondary response in control as well as in treated animals indicated that there was be delay in antibody formation. The cytotoxicity of pesticides reported in various workers (Cassale *et al.*, 1984; Dean *et al.*, 1994 and Dhasarathan *et al.*, 2005), it may be mechanism responsible for the reduced PFC response observed in the present study.

Table 35.2: Carbosulfun Exposed Animals PFC Assay at Different Time Intervals

Sl.No.	Test Chemicals	Distribution of PFC/10^6 Spleen Cells		
		I Week	II Week	III week
1.	Control	36.4±0.2	36.5±0.4	37.1±0.1
2.	1/10th LD_{50} Conc. Carbosulfan	10.3±0.5	11.5±0.3	12.8±0.4
3.	1/20th LD_{50} Conc. Carbosulfan	11.3±0.6	12.1±0.3	16.8±05

Conclusion

In the present investigation, carbosulfan was tested for immune toxicity. Toxicity to immune system was assessed directly by quantifying immunological factors that governs cells and humoral immune response and indirectly by assessing changes in blood cells and in DNA of lymphocytes. This study strongly recommends the necessity for monitoring the concentration of farm chemicals in the run off water from the crop fields and applies appropriate control measures to ensure the safety of terrestrial animals and the ecosystem at large scale.

Acknowledgement

The authors are grateful to the management Sri Kaliswari College, Sivakasi for providing facility to complete this work.

References

Akthar, N., Kayam, S.A., Ahmad, M.M. and Shahab, M., 1996. Insecticide induced changes in secretary activity of the thyroid glands in rats. *Journal of Applied Toxicology*, 16(5): 397–400.

Bick, P.R., 1985. The immune system: Organization and function. In: *Immunotoncology and Immunopharmacology*, (Eds.) Dean, M.I., Luster, Munson, A.E. and Amos, R. Raven Press, New York, pp. 1–10.

Cassale, G.P., Steven, D.C. and Richand, A.D., 1984. Parathion induced suppression of humoral immunity in inbred mice. *Toxico. Lett.*, 23: 239–248.

Dean, K.H., Comacoff, J.B., Rosenthal, G.J. and Luster, M.J., 1994. *Principles and Method of Toxicology*, (Ed.) Hayes, A.W. Karen Press Ltd., New York, p. 1065–1085.

Dhasarathan, P., Ranjitsingh, A.J.A. and Sukumaran, N., 2005. Humoral and cellular immunomodulation induced by endosulfan in Swiss albino mice. *J. Curr. Sci.*, 7(1): 111–116.

Jerne, N.K. and Nordin, A.A., 1963. Plaque formation in agar by single antibody producing cells. *Science*, 140: 405.

Lochner, M., Wagner, H., Classen, M. and Forster, I., 2002. Generation of neutralizing mouse anti mouse. IL–18 antibodies for inhibition at inflammatory responses *in vivo*. *Journal of Immunological Methods*, 259: 149–157.

Sindemman, C.J., 1983. An examination of relationships between pollution and diseases. *Const. Int. Explor. Mor.*, 183: 37–43.

Sprague, J.B., 1973. The ABC's of pollutant bioassay using fish biological methods for the assessment of water quality. *ASTM STP* 528. American Society for Testing and Materials, p. 6–30.

Vos, J.C., 1977. Immune suppression as related to toxicology, *CRC critical. Rev. Toxicol*, 5: 67–101.

Chapter 36

Antioxidant Potential of a Brassinosteroid in Alloxan Induced Diabetic Male Rat

P. Muthuraman and K. Srikumar*

Department of Biochemistry and Molecular Biology, School of Life Sciences, Pondicherry University, Kalapet, Pondicherry – 605 014, India

ABSTRACT

Diabetes mellitus is a clinical disorder that manifests in a variety of complications. The etiology of the disease is multi-factorial. Lipid peroxidation being considered a hallmark of cellular degeneracy, the degree of lipid peroxidation and antioxidant status in liver, kidneys, heart and testis of alloxan induced diabetic rats used as controls were initially investigated. Further, the effect of a brassinosteroid phytohormone isoform homobrassinolide administered intra peritoneal to a group of diabetic male rats was also probed tissuewise. The diabetic rats exhibited an increase in lipid peroxidation and consequently a decrease in their antioxidant status when compared to control rats. However administration of homobrassinolide to diabetic rats significantly reduced the level of lipid peroxidation in 2hrs in all the tissues studied but not in rat testis. The phyto-oxysterol homobrassinolide clearly prevented the oxidation of lipids and thereby protected the antioxidant status of the experimental diabetic rats. Homobrassinolide therefore remains a critical natural dietary constituent that prevents lipid peroxidation in diabetic animals.

Keywords: *Diabetes mellitus, Homobrassinolide, Antioxidant, Lipid peroxidation, MDA, 4-HNE.*

* E-mail: kotteazeth_srikumar@excite.com; Tel: +91-413-265-4422.

Introduction

Diabetes mellitus is a clinical disorder that exhibits multiple symptoms and manifests in a variety of complications. Cellular lipids and proteins undergo a wide variety of structural modifications under a variety of disease conditions. Peroxidative change of lipids is the result of one such structural modification brought about by oxygen free radicals generated through the course of a disease progression. The Lipid Peroxidation (LPO)/antioxidant status of normal and animal tissues under duress was generally estimated by monitoring the endogenous content of malondialdehyde (MDA) and 4-hydroxy nonenol(4-HNE) generated in a tissue. Oxidative stress induced by hyperglycemia, hyperlipidemia or by the generation of advanced glycated end products is regarded as an important indicator of diabetes related complications. The generation of oxygen free radicals and the reduction in the anti oxidative defense mechanisms observed in diabetic patients is thought to cause the late diabetic complications (Godlin *et al.,* 1998). Alloxan acts as a diabetogenic agent due to its ability to destroy pancreatic cells via free radical mechanism. A state of increased oxidative stress is thus indicated in diabetes and reflects through the observed increase in Lipid Peroxidation (LPO) and decrease in antioxidant reserves found in animal and human models (Palanivel *et al.,* 1998). Diabetic levels of glucose content in the plasma and tissues of a subject was generally found to correlate with an increased production of reactive oxygen metabolites in a subject (Giugliano *et al.,* 1996). Consequently, the reactive oxygen metabolites caused damage to cellular proteins, lipids, and the DNA. Furthermore, activation of the stress-sensitive signaling pathways that regulated specific gene expression also caused cellular damage (Droge, 2001). Since increased levels of lipid peroxidation combined with vascular complications, had been reported in type 1 and type 2 diabetes (Griesmacher *et al.,* 1995; Jennings *et al.,* 1991), and since our earlier work had reported an antidiabetic effect for the phytooxysterol brassinosteroid isoform homobrassinolide (Muthuraman and Srikumar, 2007), studies were therefore initiated to investigate the effect of homobrassinolide on lipid peroxidation in male rats.

Brassinosteroids in general influenced growth, seed germination, rhizogenesis, flowering and senescence in plants. The hormone is reportedly found at very low concentration in pollen, seeds and young vegetative tissues of plants (quantity per gram tissue weight unknown). Isomeric forms of this compound are known as epi and homo brassinolides. They exhibited a polyoxygenated steroid structure. They confered resistance to plants against various abiotic stresses (Zollo, 2003). A leucine-rich protein (BRL1), was identified as binding the polyoxygenated steroid in Arabidopsis thaliana and was considered a brassinolide receptor (Wang and Chory, 2006). Receptor binding by brassinolides caused activation of a receptor kinase domain and led to the phosphorylation of additional kinases within plant cells, similar to the cell signaling process found in animal cells. It was hypothesized by us therefore that either the epi or the homobrassinolide isoforms may potentially have receptors within animal cells, possibly mediating its biological potency in animal cells while yielding recognizable circulatory responses. The study therefore investigated the effect of homobrassinolide in the reduction of lipid peroxidation in alloxan induced diabetic rats. Of the 44 and odd bras sino steroid varieties available, homobrassinolide [(6-oxa-7-oxo-28 homobrassinolide) $C_{28}H_{48}O_6$] with molecular weight 480.69 was used for the study. We evaluated the level of malondialdehyde (MDA) measured as Thiobarbituric Acid Reactive Substances (TBARS) and 4-hydroxy 2-nonenal (4-HNE), both compounds having been considered as indices of lipid peroxidation in alloxan-induced experimental diabetic control rats and in diabetic rats supplemented with homobrassinolide (50 µg), 2hrs post intraperitonial administration of the compound. Results of our study is reported here.

Materials and Methods

Male Wistar strain-albino rats weighing 150–200 g, were used for the investigation. The animals were housed under controlled temperature and hygiene conditions in propylene cages. Commercial rat chow and with free access to drinking water ad libitum were provided for the animals. Experiments were carried out in accordance with internationally accepted ethical guidelines for the care of laboratory animals. Eighteen rats, were divided into three groups, each group consisting of six animals. Group 1 was designated normal control, group 2 diabetic control and group 3 comprised of diabetic rats treated using homobrassinolide. Rats that were fasted for 16 hrs (Mustafa *et al.*, 2007). Diabetes was induced by the intraperitoneal (i.p.) injection of alloxan in physiological saline at a dose of 200 mg/kg body weight (Vasily *et al.*, 1998). Fifteenth day following the injection, blood glucose level in control and diabetic rats was measured. Animals with blood glucose concentration > 250 mg/dl was considered diabetic, and were used for the experiment. Following induction of diabetes, homobrassinolide (50 μg) was administered in 95 per cent ethanol, blood samples and tissues were collected from rats after 2hr of treatment. Hypoglycemia occurring within the 24 hrs following alloxan administration was prevented by feeding a 5 per cent glucose solution orally to the diabetic rat.

Preparation of Tissue Homogenate

Rats were anesthetised by exposure to a minimum dose of diethylether. The liver, kidney, heart and testis of each rat was immediately excised and chilled in ice-cold 0.9 per cent NaCl and 1.0 g of each wet tissue was homogenized in 9 ml of 0.25 M sucrose using a potter-elvehjam glass-teflon homogenizer to obtain a 10 per cent suspension. The cytosolic fraction was obtained by centrifugation of the suspension initially at 1500 × g for 10 min and later at 20000 × g for 30 min at 4°C and used for the following assays.

Determination of Tissue Malondialdehyde

Lipid peroxidation in tissue samples was measured by the method of Ohkawa *et al.* (1997) as modified by Jamall and Smith (1985). An aliquot of the homogenate (0.20 ml) was transferred to a vial and was mixed with 0.2 ml, 8.1 per cent (w/v) sodium dodecyl sulphate solution, 1.50 ml of 0.8 per cent (w/v) TBA solution and the final reaction volume was adjusted to 4.0 ml with distilled water. Each vial was tightly capped and heated in a boiling water bath for 60 min. The vials were then cooled under running water. An equal volume of tissue blank or test sample was mixed with 10 per cent TCA and was transferred into a centrifuge tube. The mixed sample was centrifuged at 1000 × g for 10 min. The supernatant was collected and its absorbance was measured at 532 nm. Control tubes were processed similarly but without use of the TBA solution.

Determination of Tissue 4-hydroxy-2-nonenal

2 ml of homogenized solution was treated with 1.5 ml of 10 per cent TCA, and centrifuged at 3000 rpm for 30 min and filtered. The filtrate (2 ml) was treated with 1 ml of 2,4 dinitrophenyl hydrazine (1 mg/ml of 0.5 M HCl), and allowed to stand for 1 hr at room temperature. The samples were then extracted with hexane, and the extract was evaporated at 40°C. After cooling to room temperature, 2 ml of methanol was added to each sample mixed thoroughly and the absorbance of the sample was measured at 350 nm against methanol as blank (Kinter, 1996). A series (300, 400, 500 and 600 ml) of primary standard of 4-HNE were diluted with 4.70, 4.60, 4.50 and 4.40 ml of phosphate buffer. From each diluted sample 2 ml was pipetted out, and transferred into a stoppered glass tube. 1 ml of DNPH solution (2,4 dinitro phenylhydrazine) was added to each sample, and the samples were kept at room temperature for 1 h. Each sample was then extracted with 2 ml of hexane three times. All extracts were

collected in stoppered test tubes. The extract was evaporated to dryness under argon at 40°C and the residue was reconstituted in 1 ml methanol. The absorbance of the reconstituted residue was at 350 nm against a blank. The nonenol content of the test sample was estimated employing the relationship 4-1-INE= (A_{350}–0.005603185)/0.003262215. A standard curve was prepared.

Statistical Analysis

Statistical differences between the treated and control samples were evaluated by one-way analysis of variance (ANOVA). A difference between control and treated was found to be statistically significant. Values are presented as mean±SEM.

Results and Discussion

Liver, kidney and heart tissue MDA and 4-HNE content of the diabetic controls were dramatically greater than the normal control but not that in diabetic testis control. Administration of homobrassinolide in ethanol led to a reduction in the content of MDA and 4-HNE in the liver, kidney and heart tissues of these animals, but to lesser degree in the testis. The restorative effect of HB on the liver, kidney and heart content of MDA and 4-HNE was found statistically significant (Figure 36.1).

There has been considerable debate over the extent to which increased oxidative stress contributed to the development of diabetic complications. Increased membrane rigidity, decreased cellular

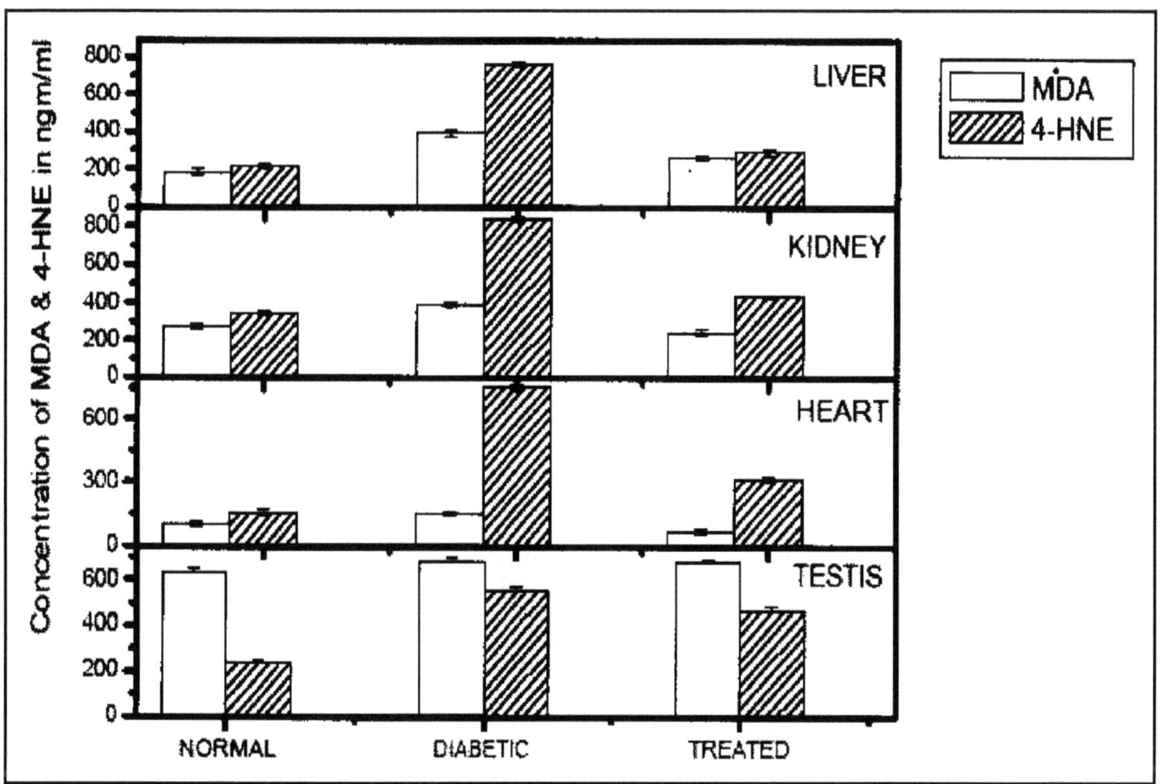

Figure 36.1: Malondialdehyde (MDA) and 4-hydroxy-2-nonenal (4-HNE) Levels in Normal, Diabetic and Treated Male Albino Wistar-Strain Rats

deformability, reduced erythrocyte survival and lipid fluidity due to peroxidation of membrane lipids has been implicated in diabetes mellitus (Selvam and Anuradha, 1988). Hyperglycaemia resulted in the generation of free radicals that exhausted antioxidant defenses within a cell, and contributed to the disruption of cellular function through oxidative damage to cell membranes and through enhanced susceptibility to lipid peroxidation (Wall *et al.*, 1987; Giugliano *et al.*, 1986; Van Dam *et al.*, 1995). So far, no study examining the effect of homobrassinolide *in vivo* has been made on normal and diabetic rat tissues MDA, 4-HNE content. An increase or decrease in the antioxidant defense may result in an increase of superoxide radicals. The increased content of MDA and 4-HNE may result from an increase of hydroxyl radicals (˙OH) (19) while MDA and 4-HNE are major oxidation product of peroxidized polyunsaturated fatty acids. Increased MDA and 4-HNE contents are an important indices of lipid peroxidation (20). Doyotte *et al.* (1997) pointed out that a decreased response may accompany a first exposure to pollutants, which can be followed by an induction of antioxidant systems. We demonstrate that homobrassinolide is significantly inhibitory to the generation of MDA and 4-HNE in the liver, kidney and heart of diabetic rats. Our experimental observations suggested that dietary levels of the brassinosteroid isoform homobrassinolide, potentially offered a natural beneficial protection against oxidative damage to animal tissues and consequently for the alleviation of certain related diabetic complications.

Acknowledgment

The authors gratefully acknowledge the financial support received as grants from the UGC [F.3-31/2004 (SR-1)] and DST (SR/SO/AS-16/2004).Technical help from Mr. S. Ravikumar is thankfully appreciated.

References

Doyotte, A., Cossu, C., Jacquin, M.C., Babut, M. and Vasseur, P., 1997. Antioxidant enzymes, glutathione and lipid peroxidation as relevant biomarkers of experimental or field exposure in the gills and the digestive gland of the freshwater bivalve *Unio tumidus*. *Aquat. Toxicol.*, 39: 93–110.

Droge, W., 2001. Free radicals in the physiological control of cell functions. *Physio. Rev.*, 82: 47–95.

Freeman, B.A. and Crapo, J.D., 1981. Hyperoxia increases oxygen radical production in rat lung and lung mitochondria. *J. Biol. Chem.*, 256: 10986–10992.

Giugliano, D., Ceriello, A. and Paolisso, G., 1986. Oxidative stress and diabetic vascular complications. *Diabetes Care*, 19: 257–267.

Giugliano, D., Ceriello, A. and Paolisso, G., 1996. Oxidative stress and diabetic vascular complications. *Diabetes Care*, 19: 257–267

Godin, D.V., Wohaieb, S.A. and Garnett, M.E., 1998. Antioxidant enzyme alterations in experimental and clinical diabetes. *Mol. Cell. Biochem.*, 84: 223–231.

Griesmacher, A., Kindhauser, M., Andert, S.E., Schreiner, W., Toma, C. and Knoebl, P., 1995. Enhanced serum levels of thiobarbituric-acid-reactive substances in diabetes mellitus. *Am. J. Med.*, 98: 469–475.

Halliwell, B. and Gutteridge, J.M.C., 1984. Oxygen toxicity, oxygen radicals, transition metals and disease. *Biochem. J.*, 219: 1–14.

Jamall, I.S., and Smith, J.C., 1985. Effects of cadmium on glutathione peroxidase, superoxide. dismutase, and lipid peroxidation in the rat heart: A possible mechanism of cadmium cardiotoxicity. *Toxicology and Applied Pharmacology*, 80: 33–42.

Jennings, P.E., McLaren, M., Scott, N.A., Saniabadi, A.R. and Belch, J.J., 1991. The relationship of oxidative stress to thrombotic tendency in type 1 diabetic patients with retinopathy. *Diabet. Med.*, 8: 860–865.

Kinter, M., 1996. In: *Free Radicals: A Practical Approach,* (Eds.) N.A. Punchard and F.J. Kelly. Oxford University Press, Oxford.

Mustafa, Aslan, Orhan, Didem Deliorman, Orhan, Nillifer, Sezik, Ekrem and Yesilada, Erdem, 2007. *In vivo* antidiabetic and antioxidant potential of *Helichrysum plicatum* ssp. *Plicatum capitulums* in streptozotocin-induced-diabetic rats. *Journal of Ethnopharmacology*, 109: 54–59.

Muthuraman, P. and Srikumar, K., 2007. A brassinosteroid as an antihyperglycaemic in alloxan induced diabetic male rats. *Journal of Current Science*, 10 (Article in press).

Ohkawa, H., Ohkawa Ohishi, N. and Vagi, K., 1997. Assay for lipid peroxides in animal tissues by thiobarbituric acid reaction. *Analytical Biochemistry,* 95: 351–358.

Palanivel, R., Ravichandran, P. and Govindasamy, S., 1998. Biochemical studies on peroxidation in ammonium para-tungstate-treated experimental diabetic rats. *Med. Sci. Res.*, 26: 759–762

Selvam, R. and Anuradha, C.V., 1988. Lipid: Peroxidation and antiperoxidative enzyme changes in erythrocyte in diabetes mellitus. *Indian J. Biochem. Biophys.*, 25: 268–272.

Van Dam, P.S., Van Asbeck, B.S., Erkelens, D.W., Marx, J.J.M., Gispen, W.H. and Bravenboer, B., 1995. The role of oxidative stress in neuropathy and other diabetic complications. *Diabetes Metabolism Reviews,* 11: 181–192.

Vasily Popov, N., Volvenkin, V. Sergei, Eprintsev, T. Alexander and Igamberdiev, U. Abir, 1998. Glyoxylate cycle enzymes are present in liver peroxisomes of alloxan-treated rats. *FEBS Letters*, 440: 55–58.

Wall, R.K., Jaffe, S., Kumar, D., Songenete, N. and Kalra, V.K., 1987. Increased adherence of oxidant treated human and bovine erythrocyte to cultured endothelial cells. *J. Cell. Physiol.*, 113: 25–36.

Wang, Xuelu and Chory, Joanne, 2006. Brassinsteroids regulate dissociation of BK11, a negative regulator of BRI 1 signaling, from plasma membrane. *Science*, 313: 1118–1121.

Zollo, Marco Antonio Texeira and Burgos, Mariangela de, 2003. Some notes on the terminology of Brassinosteroids. *Plant Growth Regulation*, 39: 1–11.

Chapter 37

External Morphology of the *Pars Stridens*, the Strigil and the Pala of *Agraptocorixa hyalinipennis* Fabricius (Heteoptera : Corixidae) as Revealed by Scanning Electron Microscopy

Susmita Gupta

Department of Ecology and Environmental Science,
Assam University, Silchar – 788 001, Assam, India
E-mail: susmita_au@rediffmail.com

ABSTRACT

A scanning electron microscopic study of the *pars stridens*, the pala, and the strigil of *Agraptocorixa hyalinipennis* (Heteroptera : Corixidae) collected from Ward Lake, Shillong, Meghalaya, North East India has revealed that the *pars slridens*, consisting of rows of pegs are better developed in the males than in the females though palae are same in size both in males and females, distinctive palar pegs are present only in the males. Strigil is also present in the males only. The role of the different cuticular sensilla in the functioning of these structures is discussed.

Keywords: *Pars stridens, Pala, Strigil, Corixidae.*

Introduction

The family Corixidae comprises the largest family of aquatic bugs. The essence of their success lies in their prolonged copulation for sure insemination and species specific acoustic communication systems. Underwater stridulation has been demonstrated to occur only in the sub families of Corixinae and Micronectinae (Jansson, 1989). It has been recorded that stridulating males of *Sigara slriata* (Corixidae, Heteroptera) are able to synchronize the time structure of their calls (pulse train synchronous stridulation) (Finke and Prager, 1980) which help them in different behavioural activities like mate selection, territory marking, agonistic signaling etc. Through the survival of individuals carrying successful genes natural selection has moulded all aspects of their biology (Mc Gavin, 1993). It has been observed that in Corixidae, besides reproductive organs, the process of reproduction is aided by a few specialized morphological structures either in the male or in both the sexes. The *pars slridens* in the fore femur, the pala in the fore tarsi, and the strigil in the last abdominal tergum of the male are such structures. Early studies on corixids concentrated mainly on the method of sound production (Ball, I 846; Kirkaldy, 1901; Von Mitis, 1936; Schaller, 1951; Finke, 1968) until Jansson (1972, 1978) described the morphology of the *pars stridens* and palar pegs of several species of corixids. *Pars stridens* is the stridulatory apparatus of most of the members of the subfamily Corixinae (Heteroptera : Corixidae) consisting of rows of pegs on the inner side of the fore femora which is rubbed against a plectrum formed by the edge of the head or that of the maxillary plate to produce sound (Yon Mittis, 1936; Hungerford, 1948; Jansson, 1972, 1973a,b). The pala or the fore tarsi, both in the males and females serve as a food scoop and in the males of some species, pegs on the anterior side are species specific and serve to attach the two sexes firmly together during copulation (Popham, 1961; Jansson, 1978). The strigil is located on the dorsum of abdominal tergum VII in the subfamily Micronectinae and are said to be another sound producing organ (Jansson, 1972). While males of most species in the sub-family Corixinae also have the strigil, Von Mitis (1936) observed that this was not associated with sound production. Although several species of Corixinae (Corixidae : Heteroptera) have been described from India (Tonapi, 1959; Julka, 1977) no detailed description of their morphology exists till date. In this paper an attempt has been made to study the morphological structures of *pars stridens*, pala, and strigil of male and female *Agraptocorixa hyalinipennis* Fabricius (sub-family Corixinae) in relation to their function and ecology by using scanning electron microscopy. *A. hyalinipennis* is a common corixid species of lentic systems of Meghalaya, Northeast India (Ahmed, 1984).

Specimens of *A. hyalinipennis* were collected from Ward Lake Shillong (lat. 25°34′N; 90°52′E), Meghalaya State, India. The fore femora, the fore tarsi and the whole body of males and females were fixed in 2.5 per cent Glutaraldehyde in 0.1 M Na-Cacodylate buffer for 2–4 hours, followed by washing in buffer, dehydration in graded concentration of acetone and drying (Wolley and Vossbrinck, 1977). They were examined in a JSM-35 scanning electron microscope operated at 15 KV.

The *pars stridens* area on the inner femora of male *A. hyalinipennis* is comprised of two types of sensory structures (Figure 37.1). The first type, being sensilla trichoidea (St. 1) comprises short pegs located in a round cup shaped depression having a distinctively swollen base. They are collared, and more or less abruptly tapered to a slightly curved flexible hair like structure. Their number ranges from 70–75. They are 28–36 μm long with a basal thickness of 4–5 μm (Figure 37.2). Ten to twelve rows of such pegs are present on each fore femur. The second type–also sensilla trichoidea (St. 2)–are long (34–40 μm) and socketed with a basal thickness of 1.3–1.5 μm (Figure 37.3). In the *pars stridens* area of female, which has the same size as that of the male, peg like *Sensilla trichoidea* (St. 1) are totally absent.

Both in the male and female, the fore tarsi or the pal a is large and the antero-Iateral sides are densely covered with a third type of *Sensilla trichoidea* (St. 3) that are 7.7–42.3 μm long, socketed and collared. The male is distinctive with short (1.5–3.8 μm) palar pegs, 36–38 in number, and arranged in a single row on the ventral surface (Figure 33.4), which arc absent in the female.

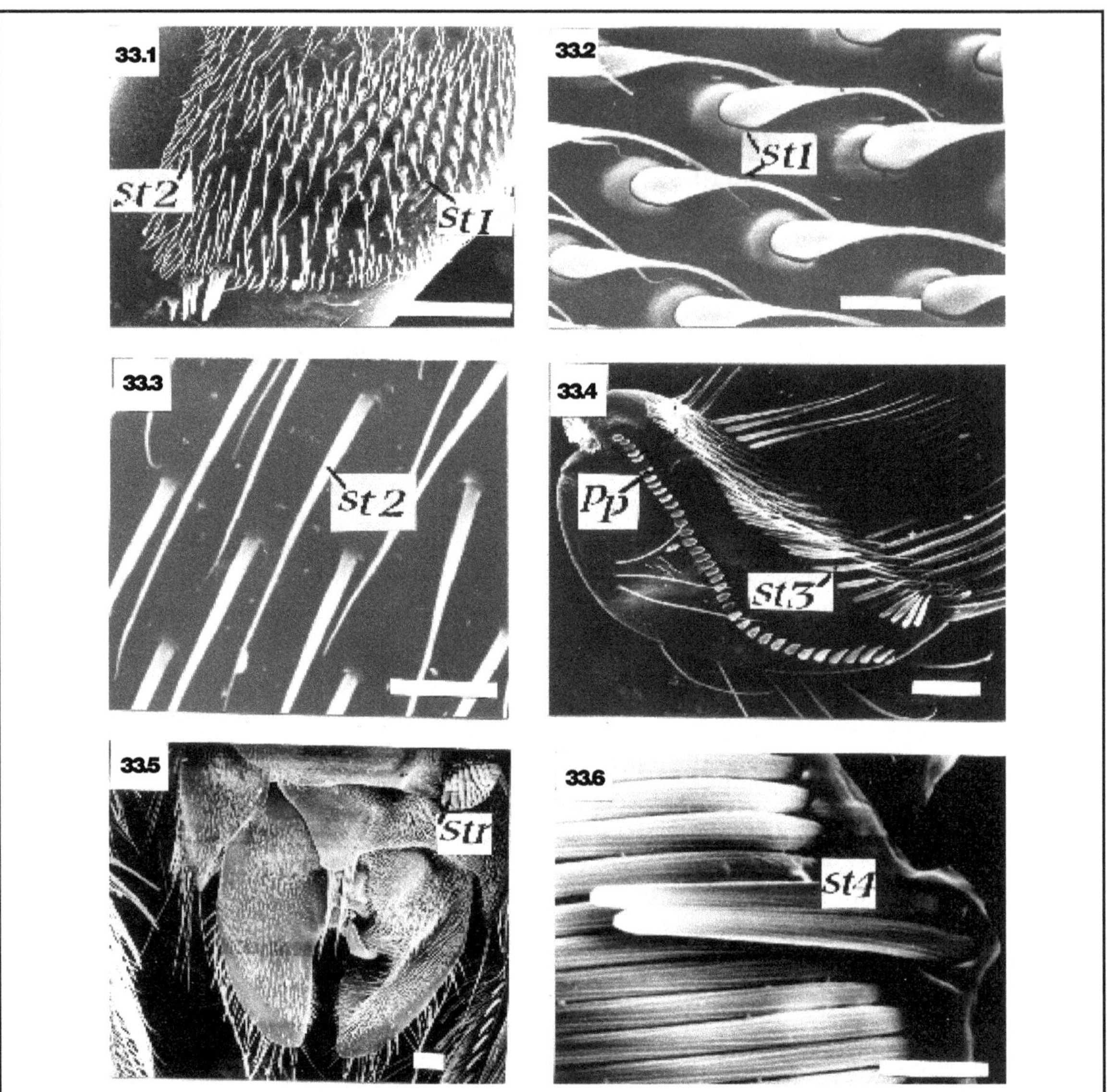

Plate 37.1: External Morphology of the *Pars stridens*, the Strigil and the Pala of *Agraptocorixa hyalinipennis* Fabricus (Heteroptera : Coreixidae) are Revealed by Scanning Electron Microscopy by Susmita Gupta

Figure 33.1: Pars Stridens on the Femora of Male *A. hyalinipennis*. Scale bar–100 □ m St. 1: *Sensilla trichoidea* Type 1, St. 2: *Sensilla trichoidea* Type 2.

Figure 33.2: St. 1 Magnified. Scale bar–10 □ m

Figure 33.3: St. 2 on Femora of Female *A. hyalinipennis*. Scale bar–10 □ m

Figure 33.4: The Pala of Male. Pp: Palar pegs; St. 3: *Sensilla trichoidea* Type 3. Scale bar–100 □ m

Figure 33.5: The Strigil on the Abdomen of Male *A. hylinipennis*. Strigil: Str. Scale bar–100 □ m

Figure 33.6: Strigil Magnified. St. 4: *Sensilla trichoidea* Type 4. Scale bar–10 □ m

The strigil is present only in the last abdominal tergum of male *A. hyalinipennis*. It is an oval structure comprising ten comb like rows of striated densely packed *Sensilla trichoidea* (St. 4). They are socketed but not collared, uniformly thick, and strong with slightly rounded tips. Their length ranges from 31–35 µm (Figure 33.5).

The *pars stridens*, palar pegs and strigil are found to be specialized structures evolved in the body of corixids for aiding their process of reproduction either by stridulating or by attaching and holding the female during copulation. From the morphological structures of *pars stridens*, amplitude of stridulation can be predicted and in most species, the thickness of the peg was found to be directly proportional to the amplitude of the signal produced (Jansson, 1972). In the dung beetles of the genus *Trypocopris* (Coleoptera, Geotrupidae), sub-species and populations within each species were differentiated with regard to body size and stridulatory organ, and the length of the *pars stridens* was found to be positively correlated with the width of the coxa. The duration of sound emission was also positively correlated with the length of the apparatus and the sub-pulse rate was negatively related to the distance between two consecutive crests (Carisio *et al.*, 2004). In corixids also, the other factors which work in symphony are the size of the plectrum and the speed of the movement of the *pars stridens* (Yon Mitis, 1936; Hungerford, 1948; Jansson, 1972). The same signal served in two or even three behaviourally different situations like spontaneous calling, agonistic signaling between the males or premating signaling to attract females (Jansson, 1973b; 1976). In the present study stridulatory signals of *A. hyalinipennis* could not be recorded. Comparison of the morphological structure and thickness of the first type of peg (St. 1) of male *A. hyalinipennis* with that of the males of *Cenocorixa* spp. belonging to the same sub-family Corixinae revealed that males with distinctive sensory peg is expected to stridulate louder than the females, which do not posses this peg (St. 1) at all. The presence of only second type of *Sensilla trichoidea* (St. 2) in the pars stridens area of females reveals that stridulatory apparatus is poorly developed in females as *Sensilla trichoidea* (St. 2) can make only faint sound. Standard recording also showed that the female signals are always much fainter than the male signals, although lack of well developed stridulatory pegs does not necessarily mean that a species or sex is unable to stridulate, as it is not known whether the sound itself is formed by the vibration of the pegs or the maxillary plate (Jansson, 1972).

The number and arrangement of palar pegs is the most important identifying character which is species specific and always included in the species description. While in most of the species the palar pegs are arranged in two rows apical and basal, with aberrantly arranged palar pegs (Jansson, 1978), in *A. hyalinipennis* the palar pegs are arranged in a single row in a definite pattern. This arrangement is significant and might be species specific as when mounting a female the male stretches the front legs sideways and grabs the female on the basal part of the forewings so that the palar pegs catch on the ridges of the embolium *i.e.*, the lateral flange of hemielytron (Jansson, 1978). Although Dumortier (1963) explained that some *Sigara* and *Corixa* species (Sub-family: Corixinae) stridulate by scratching the transverse ridges of the clypeus with the point of the fore tarsus (Bruyant, 1894) or with the palar pegs (Weber, 1933; Leston and Pringle, 1963), Jansson (1972) is of the view that the palar pegs have nothing to do with stridulation. In *A hya/inipennes* also, the palar pegs are not likely to be suitable for sound production as continuous scratching of the palar pegs would probably damage the long row of socketed, collared, and pliable *Sensilla trichoidea* (St. 3) present in both the edges of male and female pala which otherwise have a great role in cleaning their mouthparts and other areas as observed in several aquatic insects (Gupta *et al.*, 1999). This is in agreement with the study of Hsu (1937) who said that densely covered areas with sensory setae will not be suitable for sound production. In the sub-family Micronectinae and also in some species of subfamily Corixinae, strigil is located on the dorsum of abdominal tergum VI, and for the mechanism of mounting signals two possible structures exist, *viz.*, the strigil and the pegs on the abdominal tergum VI. In *A. hyalinipennis* strigil is located in the last abdominal segment. Moore (1961) and Finke (1968) have reported that certain Corixinae produce sounds by rubbing their hind legs against the forewings or the abdomen. Jansson (1989) argued against the role of srigil as part of the stridulatory mechanism in the Finnish species of the genus *Micronecta* because it is located underneath a plate, and is too long, soft and with less number of pegs. Contrary to the Finnish species, the strigil of *A. hyalinipennis* is prominently placed in the last abdominal segment without any protection or plate, and with 10 comb-like rows of thick, striated and strong pegs (Figures 37.5 and 37.6). Hence, in this species the role of strigil as sound emitter cannot be ruled out. This structure also can be associated with the attachment of male to female during copulation and also for the mechanism of producing mounting signal which is common in the subfamily Corixinae, as exchange of signals between males and females has an important function in enabling the two sexes to meet (Larsen, 1938; Jansson, 1972; 1973b).

Scanning electron micrographs of *pars stridens,* strigil, and pal a in *A. hyalinipennis* (Corixidae: Heteroptera) reveal that these structures might play an important decisive role in mate selection and copulation. However a detailed investigation with sound recording, audiospectrography and behavioural studies are required for explaining the morphological structures, the mechanism as well as the functioning of the signals.

Acknowledgement

The author is thankful to Dr. D.T.K. Khathing, Head, R.S.I.C., NEHU, Shillong (India) for the encouragement and also for providing SEM facilities during the course of investigation.

References

Ahmed, A.K.Z., 1984. Studies on the ecology of aquatic insects with special reference to fish pond. *Ph.D. Thesis,* North-Eastern Hill University, Shillong, India.

Ball, R., 1846. On noise produced by one of the notonectidae. *Report Brit. Adv. Sci.,* p. 64–65

Bruyant, C., 1894. Sur un Hemiptere aquatique stridulant *Sigara minutissima* Lin. *C.R Acad. Sci.,* 118: 299–301.

Carisio, L., Palestrini, C. and Rolando, A., 2004. Stridulation variability and morphology: An examination in dung beetles of the genus *Trypocopris* (Coleoptera, Geotrupidae). *Population Ecology,* 46(1): 27–37.

Dumortier, B., 1963. Morphology of sound emission apparatus in Arthropoda. In: *Acoustic Behaviour in Animals,* (Ed.) R.G. Busnel. Elsevier Co., Amsterdam, p. 277–345.

Finke, C., 1968. Lautausserung und verhalten von Sigara striata und Callicorixa praeusta (Corixidae Leach., Hydrocorisae Latr.). *Zschr, vergl. Physiol.,* 58: 398–422.

Finke, C. and Prager, J., 1980. Pulse-train Synchronous pair stridulation by male sigara striata (Heteroptera, Corixidae). *Cellular and Molecular Life Sciences (CML.), Birkhauser Basel,* 36(10): 1172–1173.

Gupta, S., Gupta, A. and Meyer-Rochow, V.B., 1999. Cuticular microstructures of abdominal tergites and sternites of *Cloeon* sp. (Ephemeroptera : Baetidae) during post embryonic development. *Entomologica Fennica,* 10: 51–59.

Hsu, F., 1937. Structure de lappareil soit-distant phonateu de Corixa (c.falleni Fieb.). *Ann. Soc. Sci. Bruxelles,* 57(2): 128–136.

Hungerford, H.B., 1948. The corixidae of the western hemisphere (Hemiptera). *Univ. Kansas Sci. Bull.,* 32: 1–87.

Jansson, A., 1972. Mechanism of sound production and morphology of the stridulatory apparatus in the genus *Cenocorixa* (Hemiptera : Corixidae). *Ann.Zool. Fennici,* 9: 120–129.

Jansson, A., 1973a. Diel periodicity of stridulating activity in the genus *Cenocorixa* (Hemiptera : Corixidae). *Ann. Zool. Finnici,* 13: 48–62.

Jansson, A., 1973b. Stridulation and its significance in the genus *Cenocorixa* (Hemiptera : Corixidae). *Behaviour,* 46: 1–36.

Jansson, A., 1976. Audiospectrographic analysis of stridulaory signals of some North American Corixidae (Hemiptera). *Ann. Zool. Finnici,* 13: 48–62.

Jansson, A., 1978. Aberrant arrangement of male palar pegs in some *Callicorixa* species (Heteroptera : Corixidae). *Notulae Entomologica,* 58: 15–17.

Jansson, A., 1989. Stridulation of Micronectinae (Heteroptera : Corixidae). *Annales Enlomologici Fennici,* 55: 161–175.

Julka, J.M., 1977. On possible seasonal fluctuations in the population of aquatic bugs in a fish pond. *Oriental Insects,* 11: 139–149.

Kirkaldy, G.W., 1901. The stridulation of *Corixa* (Rhynchota). *Entomologist,* 34: 9.

Larsen, O., 1938. Untersuchungen uber den Geschlechtsapparat der aquatilen Wanzen. *Opuscula Entomol., Suppll.,* p. 1–388.

Leston, D. and Pringle, J.W.S., 1963. Acoustic behaviour of Hemiptera. In: *Acoustic Behaviour of Animals,* (Ed.) R.G. Busnel. Elsevier Co., Amsterdam, p. 391–411.

McGavin, G.C., 1993. *Bugs of the World.* Blandford, London, 193 p.

Moore, T.E., 1961. Audiospectrographic analysis of sounds of Hemiptera and Homoptera. *Ann. Entomol. Soc. Amer.,* 54: 273–291.

Popham, E.J., 1961. The function of paleal pegs of corixidae (Hemiptera : Heteroptera). *Nature,* London, 190: 742–743.

Schaller, F., 1951. Lauterzeugung und Horvermogen von corixa (Callicorixa) striata (L.). *Zeitschr. Vergl. Physiol.,* 33: 476–486.

Tonapi, G.T., 1959. Studies on the aquatic insect fauna of Poona (aquatic Heteroptera). *Proc. Natn. Inst. Sci.,* India, 25: 321–332.

Von Mitis, H., 1936. Zur Biologie der Corixiden Stridulation. *Zeitschr. Morphol., Okol. Tiere,* 30: 479–495.

Weber, H., 1933. Lehrbuch der Entomologie. *Fisher Verlag,* Jena, 726 p.

Woolley, T.A. and Vossbrinck, C.R., 1977. Scanning electron microscopy as a tool in arthropod taxonomy. In: *SEMI* 1977 / IIT Research Institute, p. 645–652.

Chapter 38

Effect of Endosulfan on Blood Glucose of Freshwater Fish in *Channa gachua*

***D.R. Deshmukh*[1] *and S.R. Sonawane*[2]**

[1]*Department of Zoology, Pratishthan Mahavidyalaya Paithan*
[2]*Department of Zoology, Dr.Babasaheb Ambedkar Marathwada University, Aurangabad (MS)*

ABSTRACT

In the present investigation the freshwater fish *Channa gachua* were exposed to LC_{50} concentration (0.0035 ppm) of endosulfan pesticide for short term (96 hrs.). The blood glucose levels were increased after exposure to endosulfan toxicity and compare to control group.

Keywords: *Endosulfan, Blood glucose, Channa gachua.*

Introduction

Pesticides used in excessive amount have caused potential health hazards not only to human but also to all forms of aquatic life. An exposure to pesticides for a considerable length of time is known to adversely affect a number of vital functions. The study of haematology of fish has contributed significantly to an understanding of comparative physiology, phylogenetic relationship, mode of animal life, food selection and other significant ecological parameters.

Several attempts have been made to study the effect of a number of pesticides on different parameters of the blood of fishes. Mukhopadhaya and Oehadri (1981) observed that *Clarias batrachus* exposed to malathion showed that blood glucose level increased significantly. It has been reported that there is increase in glucose level in carps exposed to various pollutants by Hanke *et al.* (1982). Ceron *et al.* (1997) reported that there is increase in glucose in *Anguilla anguilla* exposed to sublethal concentration of diazinon for 96 hrs. Srinivas *et al.* (200 I) also reported that there was significant increase in blood glucose level in *Catla catla* exposure to malathion and dichlororovos.

The present investigation has been designed to study the effect of endosulfan pesticide on blood glucose level of *Channa gachua* after 96 hrs exposure.

Materials and Methods

In the present investigation, live specimens of *Channa gachua* were collected from Kham river near Aurangabad and were brought to the laboratory without any mechanical injury. The fishes were maintained in glass aquaria and were allowed to acclimatize for nearly about four weeks before being used for the test.

To determine the effect of acute treatment of endosulfan pesticide on haematological parameters a separate set of experiment was specially run for the short term of 96 hours. Ten fishes were exposed for a period of 96 hours to the LC_{50} values, the 96 hours LC_{50} values of endosulfan level is 0.0035 ppm. After completion of short-term exposure the blood from the caudal peduncle of fish was taken with the help of sterile disposable syringe. The blood was taken in bulb and heparin was used as an anticoagulant. After taking the blood, the blood glucose parameters were calculated. Simultaneously, a control tank was also maintained. The blood glucose was estimated by using phenol sulphuric acid method.

Results and Discussion

Values of blood glucose of experimental and control fishes are presented in Table 38.1. Fishes under osmatic stress show increase in blood glucose level after exposure to 96 hrs.

Table 38.1: Change in Blood Glucose Level in Freshwater Fish *Channa gachua* Exposure for Short Term (96 hrs)

Parameters	*Short Term Exposure*	
	Control	*Endosulfan Pesticide (0.0035 ppm)*
Blood glucose mg/100 ml.	63.0±4.404	88.0±4.123

Each value is a mean of ten observation±S.D.

Dalela *et al.* (1981) showed that the blood glucose was increased in *Mystus villatus* was exposed to three pesticides thiotox, dichlorovos and carbofuran exposed for 30 days. Bhattacharya *et al.* (1987) observed that the blood glucose increased in *Channa Punctatus* exposed to industrial pollutant. Bhoopathy and Gunasegar (1999) showed that the blood glucose in exotic fish *Oreochromis mossambicus* (Trewaves) exposed to sublethal concentration of potassium dichromate for long term toxicity, blood glucose level was increased gradually.

Koundinya and Ramamurthy (1978) reported that toxicity stress induced increase hyperglycemic condition to energy demand. The observed hyperglycemic condition suggests that during anoxia or hypoxia the utilisation of carbohydrate may be increased was reported by Dezwaun and Zandee (1972). Larson (1973) observed that the increase in glucose level may be due to increase in the catacholamines which deplete glycogen reserves to form glucose in the fish. There is another chance for increase in glucose in blood that is due to the inability of the tissues to utilize the glucose in blood for energy demands, resulting in the accumulation of glucose in the blood.

Srinivas *et al.* (2001) observed that the–fish *Catla catla* exposed to malathion and dichlorovos pesticide showed that the blood glucose level was increased. Rangaswamy (1984) observed that there is a continuous breakdown of glycogen reserve to meet the energy demand of the fish as a result of pesticide stress, thus increasing the blood glucose level. Another reason for the hyperglycemic condition might be the hypoxic condition to which the animal has been exposed where oxygen consumption of the fish has been reduced after exposure to endosulfan. Thus endosulfan stress also might be induced hyperglycemia through glycogenolysis

Acknowledgements

The author is thankful to the Professor and Head Department of Zoology, Dr. Babasaheb Ambedkar Marathwada University, Aurangabad for providing the necessary laboratory facilities required for this work.

References

Bhattacharya, Tapati, Ray, Arun Kumar and Bhattacharya, Shelley, 1987. Blood glucose and hepatic glycogen interrelationship in *Channa Punctatus* (Bloch): A parameters of nonlethal toxicity bioassay with industrial pollutants. *Ind. J. of Exp. Biol.,* 25: 539–541.

Bhoopathy, S. and Gunasegar, N., 1999. Potassium dichromate induced physiological changes at sublethal concentration in the exotic fish *Oreochromis mossambicus* (Trewavas). In: *Proc. Nat. Acad. Sci., India,* 69(B) III and IV.

Ceron, U., Sancho, E., Ferrando, M.D., Gutierrez, C. and Andreu, E., 1997. Changes in carbohydrate metabolism in Eel *Anguilla anguilla,* during short-term exposure to diazinon. *Toxicol. Environ. Chem.,* 60: 201–210.

Dalela, R.C., Rani, S., Kumar, V. and Verma, S.R., 1981. *In vivo* haematological alterations in a freshwater teleost *Mystus vittatus* following subacute exposure to pesticide and their combinations. *J. Environ Biol.,* 2(2): 79–86.

Dezwaun, A. and Zandee, D.I., 1972. The utilization of the glycogen and accumulation of some intermediates during an aerobiosis in *Mytilus edulis. Comp. Biochem. Physiol. A&B.*

Hanke, W., Bittner, A., Horn, G., Muller, R. and Keppler, R., 1982. *Jul. Spez.,* 163: 42.

Koundinya and Ramamurthy, 1978. Haematological abnormalities of the fish, *S. mossambicus* exposed to sumithon and sevin. *Curr. Sci.,* 48: 877.

Larson, A., 1973. Clinical methods applied to fish blood with reference to effect of chlorinated hydrocarbon. In: *Comparative Physiology,* (Eds.) J. Botis K. Schmidt-Nelson and S.H.P. Maddrel. North Holland Publishing Co., Amesterdam.

Mukhopadhya, P.K. and Dehadri, P.V., 1981. Biochemical changes in the air breathing catfish *Clarias batrachus* (Linn) exposed to malathion. *Environ. Pollut.,* (Sera, a) 2(2): 149–158.

Rangaswamy, C.D., 1984. Impact of endosulfan toxicity on some physiological properties of the blood and aspects of energy metabolism of afresh water fish, *Tilapia massambica* (Peters). *Ph.D. Thesis,* Sri Venkateswara University, Tirupati.

Srinivas, A., Venugopal, G., Pisca, R.S. and Waghray, S., 2001. Some aspects of haemato biochemistry of Indian major carp, *Catla catla* influenced by malathion and dichlorovos (DDVP). *J. Aqua. Biol.,* 16(2): 53–56.

Chapter 39

Host Based Intrusion Detection for Enforcing Kernel Level Security in Desktop Grids

S. Mary Saira Bhanu N.P. Gopalan***

National Institute of Technology Tiruchirappalli – 620 015, India

ABSTRACT

The swelling number of applications and consequent increase in the amount of critical data over the Desktop grids has considerably raised stakes for efficient security architecture. The Desktop Grid Security System must adequately protect the desktop machine from malfunction or malicious applications. Intrusion detection mechanisms play frontline role in detecting the system vulnerabilities and policy violations. Intrusion detection mechanisms rely on wide variety of observable data to distinguish between legitimate and illegitimate activities. An approach for detecting insider misbehavior is to monitor the system call activity and watch for unusual behavior. In this chapter, we present an extended security layer based on host based intrusion detection that identifies intrusions by analyzing system calls at the kernel level using kernel loadable modules in Desktop grids. This proposed security layer does not impose any hermetic environment restrictions (in policy selection and enforcement) like in the case of sandboxes.

***Keywords:** Desktop grid, Intrusion detection, Security layer.*

Introduction

Today the largest computing system in the world is based on distributed computing through the assembly of a large number of PC's over the Internet. There are millions of desktop systems around the

* E-mail: *msb@nitt.edu; **gopalan@nitt.edu.

world. Even though they are resource rich they are under utilized. Projects like SETI @ home (Anderson *et al.*, 2002) harvests the computing power of over a million CPU's using cycle stealing technology. Grid computing is a new way of distributed computing which tights up all physically distributed resources such as computational power, memory, useful data or hard disk space (Foster and Kesselman, 2004). In grid similar to SETI @ home, PC's are connected to some servers to form a large computing infrastructure called a Desktop Grid (DG) (Kacsuk and Podhorszki, 2005). DG utilizes the unused processing power of desktop resources under a secure environment. It enables the availability of the processing power to an end user in a transparent way by hiding the complexity of managing jobs. In DG, the locus of security concern moves from network perimeter to PC's. The idle PC's are configured to execute programs rather than to transfer data. This makes DG potentially a fertile ground for malicious programs like Viruses, Trojan horses etc. The DG security system must adequately protect the desktop machine from malfunction or malicious applications. The integrity of distribution computation should also be protected.

The security is a vital part in DG and it raises additional security challenges including the establishment of a trustworthy system and automatic handling of changes in configuration. Preventing the attacks using intrusion detection can reduce the holes in the DG security. In this paper, we present an extended security layer based on host based intrusion detection that identifies intrusions by analyzing system calls at the kernel level using kernel loadable modules in DG. The proposed work is designed and developed in Windows platform, which is used by almost all desktops. Windows prop up for kernel loadable modules and once loaded; the modules run within the kernel space and have full access to kernel data structure.

Related Work

In host based intrusion detection systems, an agent located on a single host identifies intrusions by analyzing system calls, application logs and file system modifications and auditing data related to the dynamic behavior or state of the host computers. The key design issue in IDS is to identify a set of dynamic behavioral characteristics of the system. Forrest *et al.* (1994, 1996) analyzed the sequence of system class to discriminate between processes. The total number of mismatches over an arbitrary length of the trace was used to quantify the number of system calls in anomalous position. This work was extended by Hofmeyer (1998) to achieve intrusion detection by counting the mismatches within small fixed length section of the trace and comparing the recorded sequences against a database of normal sequences. Applications can be protected using kernel hypervisors (Mitchen *et al.*, 1997). The hypervisors provide a set of virtual system calls for selected system components. They provide application specific policy decision-making and enforcement. Harmony (Naik *et al.*, 2003)–a desktop grid for delivering enterprise computations uses Virtual machines based on hypervisor to limit the intrusions. The proliferation of mobile code via the internet can also be prevented by using sandboxing. In Entropia (Caler *et al.*, 2005) a sandbox called Windows device driver provides security against malicious applications. This device driver runs as part of the as and every single call that grant access to windows resources by all processes/threads is examined by the driver code. Other systems, which uses sandbox, are BlueBox (Chari and Chen, 2002), MAP box (Acharya and Raje, 2000), Windowbox (Kacsuk and Podhorszki, 2005), etc. These systems have no provisions for recording audit trails. Furthermore, specifying security policies is labor intensive, as the sandbox needs to be programmed into applications.

Add-on Security Layer Operation View

In order to provide security services with fewer overheads, the new security architecture is proposed. The main objectives of the proposed architecture are:

System Security

The system must be protected from possible hostile acts performed by applications.

Transparency

The application does not have to know about the underlying system. For the application each system access must occur as usual, and the security sub-system should take care of each possible violation.

The proposed security layer shown in Figure 39.1 includes multiple interception, monitoring and analysis stages. At initialization a system wide monitor is first loaded into the kernel as a kernel extension module, which monitors the new process creation in the operating space of the application. It also takes into account of number of child processes created by a parent application to detect a form of Denial of Service (DOS) attack. If the number of child processes created by an application exceeds the prescribed threshold value, that process is stopped from further execution and the application properties are added to the in-memory database. The normal execution of a system call instruction is as follows

1. The control switches to kernel mode.
2. Using the system call number, the address of the requested service from a predefined pointer table is retrieved.
3. The call is executed.

In our work, when a system call instruction is executed, the control is transferred to the Call Analysis Module (CAM, which is a kernel module). The CAM compares the incoming parameters against the database of trusted signatures. If the call is found to be malicious, the monitoring fails its execution. Alternatively, when the request is legitimate, execution proceeds normally. The layer is transparent and effectively prevents an attack before any damage occurs. The detection module controls any interaction with the OS. It is activated at the right time just before the system service is executed. When it gains control, it has the appropriate context information as the system call parameters provide the relevant information regarding the service request. In most cases, it is possible to accurately determine the user identity as well as the module that initiated the call. As a result, the detection of malicious requests is accurate and false positives are rare. Most of the system faults are detectable precisely because of the pattern of failures that they produce, but human attackers are motivated to cover their tracks and keep their intrusions inconspicuous. To achieve fine-grained security an autonomic system element called Trace Distribution Broker (TDB) is used. This has the capability of decision-making and works in a non-hermetic test environment. TDB has the modules (1) Call Trace Analyzer (CTA); (2) Temporal Signature Analyzer (TSA); (3) Intense Call Argument Analyzer (ICAA).

TDB passes control to Call Trace Analyzer (CTA) to analyze the system calls to the finer level. CT A module does the raw system call trace analysis by taking into account the number of occurrences of system calls in the trusted application and the application under test. This produces finer level of control in the application execution and detects any system vulnerability at the initial stage itself. Any imbalance in this stage will make the TDB to pass control to the Temporal Signature Analyzer (TSA) of this security architecture. This stage concentrates on the timing values between the system calls as the principle tool in detecting vulnerabilities and the policy violations. It consists of two sub-modules: signature generator and run time monitor. The signature generator generates the rich signature for future use and the run time monitor observes the execution of the application to find an anomalistic

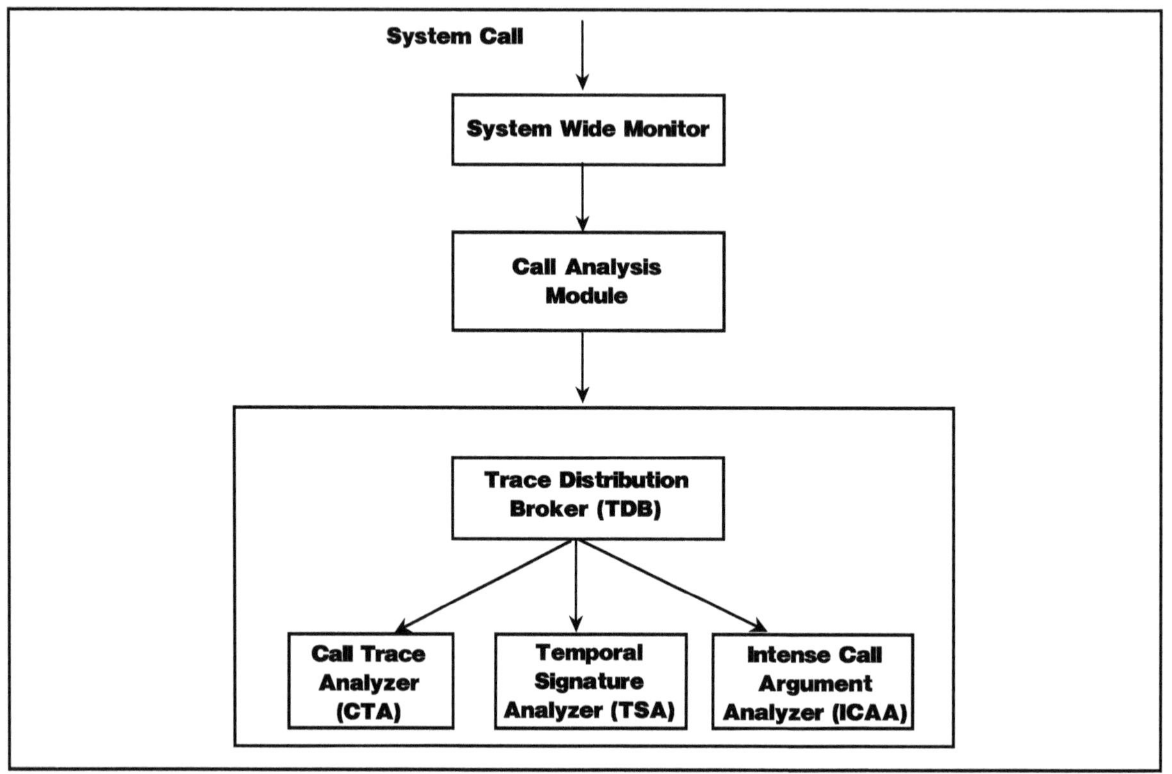

Figure 39.1: Add on Security Layer

behavior associated with inter-call timing measurements. The ICAA stage of intrusion detection concentrates on the in-call analysis rather than the inter-call analysis as in the case of CTA and TSA. Policy violations through specific system calls are monitored through the ICAA module. It provides minimal overhead compared to other security layers. It also incorporates the work ground monitor module, which monitors the entire file system and reports about malicious actions back to the user. It works in an active mode unlike passive mode, by recording the actions into the security log file. This analyzer module detects most of the file system violations.

Experimental Analysis

The on-kernel (loadable kernel) module is used to intercept the system calls and the intercepted calls are then processed. As a real time analysis, the number of kernel calls logged in a Desktop Grid client (in 60 sec) is considered. From the experiment, it is clear that the proposed CT A distinguishes the applications based on the number of occurrences of the system calls. The observed result is shown in Table 39.1. The TSA component takes into account of system calls generated by the application along with the time elapsed between subsequent system calls in the execution of the application. Time measurements are on a per process basis and only the time duration (between calls) for which the process is actually executing is measured. Context swap and process sleep are elided. Since the elapsed time between two sequence calls can vary, the signature of an application is generated by one or more executions of an application. We monitor the application execution multiple times both in a real time environment and also synthetically simulating the application, with the variety of inputs. We have taken the sequence length of 6 for our analysis.

Table 39.1: Comparison of Calls Generated by the Trusted Applications and the Application Under Test

System Calls	*Original Application (Number of Occurrences)*	*Masqueraded Application (Number of Occurrences)*
Open	18	22
Close	31	38
Commit	0	0
Read	11	25
Write	1	4
Unlink	0	0

It is necessary to remove the high variance data from the observed data to obtain a rich signature for the application based on timing measures. To determine criteria for excluding such data, we use a notation for the variance of such interval, called the *z-score* (Table 39.2).The unique signature for the normal application is identified by considering various sequence lengths and different normal databases and if any discrepancies occur during comparison with the application under test will be termed as malicious. This is a one-time check for an application and hence imposes a less overhead. The core module of ICAA is concerned with the monitoring of Low-level I/O calls like the open system call. It monitors the file system modification through I/O calls. The events and the change notification will be made by monitoring the work ground and the associated work area (unlike the closed environment restrictions in the case of sandboxes) and the file system change notification will be made by FILE_NOTIFY events. When the event occurs, the application is notified. To be useful, this stage is lightweight with little overhead and flexible.

Table 39.2: Z-score Calculation from the Calculated Mean and Standard Deviation

Cases	*Open*	*Close*	*Read*	*Create*	*Write*	*Z*
Case 1	67	27	31	33	25	1.813
Case 2	65	28	31	33	22	0.734
Case 3	71	27	32	37	21	1.088
Case 4	81	30	33	34	22	1.981
Case 6	65	27	31	33	22	0.734

Conclusion

The presence of autonomic element in security layer provides flexibility and reduces the overhead in testing the trustworthiness of an application. We believe our work can be implemented on any operating system and in any system architecture with ease. In our work the policies are hard-coded in the architectural components, which complicate their dynamic modification. It would be much easier to add and modify policies by using a constraint-based language. However integrating a constraint based language interpreter in an operating system is not feasible due to the performance cost it would induce. It would therefore be interesting to specify a constraint-based language, which would ease the specification of protection policies without hindering the execution of the system.

References

Acharya, Anurag and Raje, Mandar, 2000. MAPbox: Using parameterized behavior classes to confine applications. In: *The Proceedings of the 9th USENIX Security Symposium*.

Anderson, D.P., Cobb, J., Korpela, E.M. Lebofsky and Werthimer, D., 2002. SETI@home: An experiment in public resource computing. *Communications of the ACM*, 45(11): 56–61.

Balfanz, Dirk and Simon, Daniel R., 2000. Window box: A simple security model for the connected desktop. In: *The Fourth Usenix Windows Systems Symposium*, pp. 37–48.

Calder, Brad, Andrew, A. Chien, Ju Wang and Yang, Don, 2005. The entropia virtual machine: Machine for desktop grids. *VEE'05, ACM.*

Chari, Suresh N. and Cheng, Pau-Chen, 2002. Blue box: A policy-driven, host-based intrusion detection system. In: *The Proceedings of the ISOC Symposium on Network and Distributed System Security*.

Dasgupta, D. and Forrest, S., 1994. Novelty detection in time series data using ideas from immunology. In: *The Fifth International Conference on Intelligent Systems*, Reno, Nevada.

Forrest, Stephanie, Hofmeyr, Steven A., Somayaji, Anil and Longstaff, Thomas, A., 1996. A sense of self for unix processes. In: *The Proceedings of the 1996 IEEE Symposium on Research in Security and Privacy*, pp. 120–128.

Foster, I. and Kesselman, C., 2004. *The Grid: Blueprint for a New Computing Infrastructure*. Morgan-Kaufman Publishing, San Francisco.

Hofmeyr, Steven, A., Forrest, Stephanie and Somayaji, Anil, 1998. Intrusion detection using sequences of system calls. *Journal of Computer Security*, 6(3): 151–180.

Kacsuk, P. and Podhorszki, N., 2005. *Scalable Desktop Grid System*. Core Grid Technical Report Number TR-0006, http://www.coregrid.net.

Mitchen, Terrence, Lu, Raymond and O'Brien, Richard, 1997. Using kernel hypervisors to secure applications. In: *The Proceedings of the 13th Annual Computer Security Applications Conference*, p. 175.

Naik, Vijay K., Subramanian, Swaminathan, Bantz, David and Krishnan, Sriram, 2003. Harmony: A desktop grid (or delivering enterprise computations). In: The *Proceedings of the Fourth International Workshop on Grid Computing (GRID'03).*

Chapter 40

On Fuzzy Variable Liner Programming in Duality

D. Stephen Dinagar[1], *A. Nagoor Gani*[2], *P. Martin Deva Prasath*[3]* *and R. Sangeetha*[1]

[1]*P.G. and Research Department of Mathematics, T.B.M.L. College, Porayar*
[2]*P.G. and Research Department of Mathematics, Jamal Mohamed College, Trichy, India*
[3]*P.G. Department of Chemistry, T.B.M L. College, Porayar, Tamil Nadu, India*

ABSTRACT

In this paper the important notion fuzzy variable linear programming is discussed. The notion called the dual of fuzzy variable linear programming problem is introduced. Some important duality results with the aid of ranking function are deduced. A numerical example is also included.

Keywords: *Duality, Fuzzy number, Fuzzy variable linear programming, Ranking functions.*

Introduction

Bellman and Zadeh (1979) have precisely defined the notion decision-making in a fuzzy environment. Linear programming problems with interval coefficients have been studied by several authors *viz.*, Sengupta *et al.*, (2001), Bitran (1980) and Ishibuchi and Tanaka (1980). Different methods of comparison of fuzzy numbers can be found as in Sengupta and Kumar Pal (2000). Mahdavi-Amiri and Nasseri (2005) have described the duality in fuzzy variable linear programming. This study focuses on Duality in Fuzzy Variable Linear Programming (DFVLP) problem. Some important results

* E-mail: martinprasath@rediffmail.com

and theorems in DFVLP are proved. Finally we solve a DFVLP problem by use of linear ranking functions.

Basic Definitions

Definition

A convex fuzzy set A on R is a fuzzy number if the following conditions hold

(*a*) The membership function is piecewise continuous.

(*b*) There exist three intervals [a, b] and [c, d] such that the membership function is increasing on [a, b], equal to 1 on [b, c], decreasing on [c, d] and equal to 0 elsewhere.

Definition

Let $\mathbf{a} = (a_1, a_2, a_3, a_4)$ denotes the trapezoidal fuzzy number, such that $a_1 \le a_2 \le a_3 \le a_4$ where $[a_1, a_4]$ is the support of **a** and $[a_2, a_3]$ is its modal set.

Ranking Functions

A convenient method for solving fuzzy linear programming problem is based on the concept of comparison of the fuzzy numbers, by use of ranking functions (Maleki, 2002). We define a ranking functions $\Re$: F(R) $\to$ R, which maps each fuzzy number in to the real line, where the real order exists and F(R) denotes the set of all trapezoidal fuzzy numbers.

Remark

Suppose that **a** and **b** be two trapezoidal fuzzy numbers We define orders on F(R) as

$\mathbf{a} \underset{\Re}{\geq} \mathbf{b}$ if and only if $\Re(\mathbf{a}) \ge \Re(\mathbf{b})$

$\mathbf{a} \underset{\Re}{\gneqq} \mathbf{b}$ if and only if $\Re(\mathbf{a}) > \Re(\mathbf{b})$

$\mathbf{a} \underset{\Re}{=} \mathbf{b}$ if and only if $\Re(\mathbf{a}) = \Re(\mathbf{b})$

Also, we write $\mathbf{a} \underset{\Re}{\leq} \mathbf{b}$ if and only if $\mathbf{b} \underset{\Re}{\geq} \mathbf{a}$.

Definition

A linear ranking function of **a** E F(R) is defined as follows:

$\Re(\mathbf{a}) = a_2 + a_3 + ½ [(a_4+a_1) - (a_3+a_2)]$

Arithmetic Operations

Let $\mathbf{a} = (a_1, a_2, a_3, a_4)$, $\mathbf{b} = (b_1, b_2, b_3, b_4) \in$ F(R).

We now define the arithmetic operations as:

Addition

$\mathbf{a} + \mathbf{b} = (a_1 + b_1, a_2 + b_2, a_3 + b_3, a_4 + b_4)$

Subtraction

$\mathbf{a} - \mathbf{b} = (a_1 - b_4, a_2 - b_3, a_3 - b_2, a_4 - b_1)$

Scalar Multiplication

$x > 0, x \in R : x\,\mathbf{a} = (xa_1, xa_2, xa_3, xa_4)$

$x < 0, x \in R : x\,\mathbf{a} = (xa_4, xa_3, xa_2, xa_1)$

Multiplication

$a \bullet b =$

$$\left(\frac{a_1}{2}(b_1+b_2+b_3+b_4), \frac{a_2}{2}(b_1+b_2+b_3+b_4), \frac{a_3}{2}(b_1+b_2+b_3+b_4), \frac{a_4}{2}(b_1+b_2+b_3+b_4)\right)$$

if $\Re(\mathbf{b}) > 0$

$$= \left(\frac{a_4}{2}(b_1+b_2+b_3+b_4), \frac{a_3}{2}(b_1+b_2+b_3+b_4), \frac{a_2}{2}(b_1+b_2+b_3+b_4), \frac{a_1}{2}(b_1+b_2+b_3+b_4)\right)$$

if $\Re(\mathbf{b}) < 0$

Division

$$\mathbf{a/b} = \left(\frac{2a_1}{b_1+b_2+b_3+b_4}, \frac{2a_2}{b_1+b_2+b_3+b_4}, \frac{2a_3}{b_1+b_2+b_3+b_4}, \frac{2a_4}{b_1+b_2+b_3+b_4}\right)$$

if $\Re(\mathbf{b}) > 0, \Re(\mathbf{b}) \neq 0$

$$= \left(\frac{2a_4}{b_1+b_2+b_3+b_4}, \frac{2a_3}{b_1+b_2+b_3+b_4}, \frac{2a_2}{b_1+b_2+b_3+b_4}, \frac{2a_1}{b_1+b_2+b_3+b_4}\right)$$

if $\Re(\mathbf{b}) < 0, \Re(\mathbf{b}) \neq 0$

Definition

A *Fuzzy Variable Linear Programming* (FVLP) problem is defined as

$$\text{Min } \mathbf{z} \underset{\Re}{=} \mathbf{c}\,\mathbf{x}$$

$$\text{s.t } \mathbf{A}\,\mathbf{x} \underset{\Re}{\geq} \mathbf{b}$$

$$x \underset{\Re}{\geq} 0$$

where, $\mathbf{b} \in [F(R)]^m$, $\mathbf{x} \in [F(R)]^n$, $\mathbf{A} \in R^{m \times n}$, $c^T \in R^n$.

Duality in Fuzzy Variable Linear Programming Problem

Formulation of the Dual Problems

For the FVLP problem

$$\text{Min } \mathbf{z} \underset{\Re}{=} \mathbf{C}\,\mathbf{x}$$

$$\text{s.t} \qquad \mathbf{A\,x} \underset{\Re}{\geq} \mathbf{b} \tag{3.1.1}$$

$$\mathbf{x} \underset{\Re}{\geq} \mathbf{0}$$

define its dual, the DFVLP problem as:

$$\text{Max} \qquad \mathbf{w} \underset{\Re}{=} \mathbf{b\,y}$$

$$\text{s.t} \qquad \mathrm{Ay} \underset{\Re}{\leq} \mathrm{c} \tag{3.1.2}$$

$$\mathrm{x} \underset{\Re}{\geq} 0$$

For the FVLP problem

$$\text{Max } \mathbf{z} \underset{\Re}{=} \mathbf{c\,x}$$

$$\text{s.t} \qquad \mathbf{A\,x} \underset{\Re}{\leq} \mathbf{b} \tag{3.1.3}$$

$$\mathrm{x} \underset{\Re}{\geq} 0$$

define its dual, the DFVLP problem as

$$\text{Min w} \underset{\Re}{=} \mathbf{by}$$

$$\text{s.t} \qquad \mathbf{Ay} \underset{\Re}{\geq} \mathrm{c} \tag{3.1.4}$$

$$\mathbf{y} \underset{\Re}{\geq} \mathbf{0}$$

Theorem (Weak Duality Theorem)

If **x** be any feasible solution to the primal (3.1.3) and **y** be any feasible solution to the dual then $\mathbf{cx} \underset{\Re}{\leq} \mathrm{by}$

Proof

Let the primal problem (3.1.3) be

$$\text{Max } \mathbf{z} \underset{\Re}{=} \mathbf{cx}$$

$$\text{s.t} \qquad \mathbf{A\,x} \underset{\Re}{\leq} \mathbf{b} \tag{3.2.1}$$

$$\mathbf{x} \underset{\Re}{\geq} \mathbf{0}$$

Multiplying both sides of the constraints (3.2.1) by **y**, we have

$\mathbf{Axy} \underset{\mathfrak{R}}{\leq} \mathbf{b\,y}$ (3.2.2)

Now the dual problem (3.1.4) be

Min $\mathbf{w} \underset{\mathfrak{R}}{=} \mathbf{b\,y}$

s.t $\mathbf{Ay} \underset{\mathfrak{R}}{\geq} \mathbf{e}$ (3.2.3)

$\mathbf{y} \underset{\mathfrak{R}}{\geq} \mathbf{0}$

Multiplying both sides of the constraints (3.2.3) by **x**, we have

$\mathbf{Ayx} \underset{\mathfrak{R}}{\geq} \mathbf{cx}$ (3.2.4)

From (3.2.2) and (3.2.4) we have

$\mathbf{cx} \underset{\mathfrak{R}}{\leq} \mathbf{Axy} \underset{\mathfrak{R}}{\leq} \mathbf{by}$

Hence $\mathbf{cx} \underset{\mathfrak{R}}{\leq} \mathbf{by}$.

Proposition

Suppose that **x*** and **y*** are feasible solutions to the primal (3.1.3) and the dual (3.1.4) respectively, such that **cx*** = **by***, then **x*** and **y*** are optimal solutions to the primal and dual problems respectively.

Proof

Let $\mathbf{cx^*} \underset{\mathfrak{R}}{=} \mathbf{b\,y^*}$.

From the weak duality theorem, we have

$\mathbf{c\,x^*} \underset{\mathfrak{R}}{=} \mathbf{b\,y^*} \underset{\mathfrak{R}}{\geq} \mathbf{c\,x}$ for all $\mathbf{x} \in X = \{\mathbf{x} \underset{\mathfrak{R}}{=} (x_1, x_2, \ldots\ldots x_n): \mathbf{Ax} \underset{\mathfrak{R}}{\leq} \mathbf{b}, \mathbf{x} \underset{\mathfrak{R}}{\geq} \mathbf{0}\}$

$\mathbf{c\,x^*} \underset{\mathfrak{R}}{\geq} \mathbf{c\,x}$ for all $\mathbf{x} \in X$.

Similarly $\mathbf{b\,y^*} \underset{\mathfrak{R}}{=} \mathbf{c\,x^*} \underset{\mathfrak{R}}{\geq} \mathbf{b\,y}$ for all $\mathbf{y} \in Y = \{\mathbf{y} \underset{\mathfrak{R}}{=} (y_1, y_2, \ldots\ldots y_n\}: \mathbf{Ay} \underset{\mathfrak{R}}{\geq} \mathbf{c}, \mathbf{y} \underset{\mathfrak{R}}{\geq} \mathbf{0}\}$

$\mathbf{by^*} \underset{\mathfrak{R}}{\leq} \mathbf{b\,y}$ for all $\mathbf{y} \in Y$.

Hence the proof.

Theorem (Strong Duality Theorem)

If **x** is an optimal solution to the primal (3.1.3), then there exist a feasible solution **y** to the dual (3.1.4) such that $\mathbf{cx} \underset{\mathfrak{R}}{=} \mathbf{b\,y}$.

Proof

We convert the primal problem (3.1.3) to its standard form by adding slack variables as follows:

$$\text{Max } \mathbf{z} \underset{\Re}{=} \mathbf{cx}$$

$$\text{s.t} \qquad \mathbf{Ax} \underset{\Re}{=} b \qquad (3.4.1)$$

$$\mathbf{x} \underset{\Re}{\geq} \mathbf{0}.$$

Suppose that $\mathbf{x}B \underset{\Re}{=} \mathbf{B}^{-1}\mathbf{b}$ is an optimal basic feasible solution to (3.4.1), where $\mathbf{B}$ is the corresponding basis matrix, $\mathbf{c}_B$ is the cost vector corresponding to $\mathbf{x}_B$ and $\mathbf{z}_j \underset{\Re}{=} \mathbf{c}_B\,\mathbf{y}_j$

$$\mathbf{a}_j \underset{\Re}{=} \sum_{i=1}^{m} y_{ij} b_i \underset{\Re}{=} y_n B$$

$$\mathbf{y}_j \underset{\Re}{=} \mathbf{B}^{-1}\mathbf{a}_j.$$

Also

$$(\mathbf{z}_j - \mathbf{c}_j) \underset{\Re}{=} (\mathbf{c}_B\mathbf{y}_j - \mathbf{c}_j) \underset{\Re}{=} \begin{cases} c_B B^{-1} a_j - c_j, \text{ for } j = 1,2,\ldots.n. \\ c_B B^{-1} e_j - 0, \text{ for } j = n+1,\ldots.n+m. \end{cases}$$

Since $\mathbf{x}_B \underset{\Re}{=} \mathbf{B}^{-1}\mathbf{b}$ is an optimal basic feasible solution to (3.4.1), we have

$$(\mathbf{z}_j - \mathbf{c}_j) \underset{\Re}{\geq} \mathbf{0} \text{ for all } j$$

$$\text{so, } (\mathbf{z}_j - \mathbf{c}_j) \underset{\Re}{\geq} \mathbf{0} \Rightarrow (\mathbf{c}_B\, \mathbf{B}^{-1}\mathbf{a}_j - \mathbf{c}_j) \underset{\Re}{\geq} \mathbf{0} \text{ and } (\mathbf{c}_B\, \mathbf{B}^{-1}\mathbf{e}_j - 0) \underset{\Re}{\geq} \mathbf{0}.$$

$$(\mathbf{c}_B\, \mathbf{B}^{-1}\mathbf{a}_j - \mathbf{c}_j) \underset{\Re}{\geq} \mathbf{0} \text{ and } (\mathbf{c}_B\, \mathbf{B}^{-1}\mathbf{e}_j - 0) \underset{\Re}{\geq} \mathbf{0}$$

$$(\mathbf{c}_B\, \mathbf{B}^{-1}\mathbf{A} - \mathbf{c}) \underset{\Re}{\geq} \mathbf{0} \text{ and } \mathbf{c}_B\, \mathbf{B}^{-1} \underset{\Re}{\geq} \mathbf{0}$$

$$\mathbf{c}_B\, \mathbf{B}^{-1}\mathbf{A} \underset{\Re}{\geq} \mathbf{c} \text{ and } \mathbf{c}_B\, \mathbf{B}^{-1} \underset{\Re}{\geq} \mathbf{0}.$$

Suppose that $\mathbf{y} \underset{\Re}{=} \mathbf{c}_B\mathbf{B}^{-1}$.

Then $\mathbf{y} \underset{\Re}{\geq} \mathbf{0}$ and $\mathbf{c}_B\, \mathbf{B}^{-1}\mathbf{A} \underset{\Re}{\geq} \mathbf{c}$

$$\mathbf{yA} \underset{\Re}{\geq} \mathbf{c}$$

$\mathbf{Ay} \underset{\Re}{\geq} \mathbf{c}$

y is feasible solution to the dual (3.1.4)

Also, $\mathbf{by} \underset{\Re}{=} \mathbf{yb} \underset{\Re}{=} \mathbf{c_B B^{-1} b} \underset{\Re}{=} \mathbf{c_B x_B} \underset{\Re}{=} \mathbf{cx}$, whenever **x** is an optimal solution to the primal.

Hence $\mathbf{cx} \underset{\Re}{=} \mathbf{by}$.

Numerical Example

FVLP

$$\text{Min. } \mathbf{z} \underset{\Re}{=} (3, 5, 7, 9)\, \mathbf{x_1} + (9, 10, 12, 17)\, \mathbf{x_2}$$

$$\text{s.t } (-2, 1, 2, 3)\, \mathbf{x_1} + (1, 3, 5, 7)\, \mathbf{x_2} \underset{\Re}{\geq} (2, 6, 8, 12)$$

$$(-2, 2, 3, 5)\, \mathbf{x_1} + (1, 2, 4, 5)\, \mathbf{x_2} \underset{\Re}{\geq} (2, 4, 6, 8)$$

DFVLP

$$\text{Max } \mathbf{w} \underset{\Re}{=} (2, 6, 8, 12)\, \mathbf{y_1} + (2, 4, 6, 8)\, \mathbf{y_2}$$

$$\text{s.t } (-2, 1, 2, 3)\, \mathbf{y_1} - (-2, 2, 3, 5)\, \mathbf{y_2} \underset{\Re}{\leq} (3, 5, 7, 9)$$

$$(1, 3, 5, 7)\, \mathbf{y_1} + (1, 2, 4, 5)\, \mathbf{y_2} \underset{\Re}{\leq} (9, 10, 12, 17)$$

$$\mathbf{y_1}, \mathbf{y_2} \underset{\Re}{\geq} \mathbf{0}$$

we may rewrite

$$(-2, 1, 2, 3)\, \mathbf{y_1} + (-2, 2, 3, 5)\, \mathbf{y_2} + (-2, 0, 1, 3)\, \mathbf{y_3} + (-2, -1, 1, 2)\, \mathbf{y_4} \underset{\Re}{=} (3, 5, 7, 9)$$

$$(1, 3, 5, 7)\, \mathbf{y_1} + (1, 2, 4, 5)\, \mathbf{y_2} + (-2, -1, 1, 2)\, \mathbf{y_3} + (-2, 0, 1, 3)\, \mathbf{y_4} \underset{\Re}{=} (9, 10, 12, 17)$$

$$\mathbf{y_1}, \mathbf{y_2}, \mathbf{y_3}, \mathbf{y_4} \underset{\Re}{\geq} \mathbf{0}.$$

}_ we obtain the first tableau as follows:

Basis	y_1	y_2	y_3	y_4	*R.H.S*	*ℜ (R.H.S)*
$\mathbf{y_3}$	(–2, 1, 2, 3)	(–2, 0, 1, 3)	(–2, –1, 1, 2)	(–72, –24, 36, 84)	12	
$\mathbf{y_4}$	(1, 3, 5, 7)	(1, 2, 4, 5)	(–2, –1, 1, 2)	(–2, 0, 1, 3)	(–72, –12, 36, 96)	24
w	(–32, –18, 4, 18)	(–28, –16, 6, 18)	(–4, –2, 2, 4)	(–4, –2, 2, 4)	(–72, –36, 36, 72)	0

From the above tableau, we obtain

$(\mathbf{z}_1 - \mathbf{c}_1, \mathbf{z}_2 - \mathbf{c}_2) \underset{\Re}{=} (-14, -10) \Rightarrow (\mathbf{z}_1 - \mathbf{c}_1) \underset{\Re}{\leq} (\mathbf{z}_2 - \mathbf{c}_2)$

Hence related fuzzy non basic variable $\mathbf{y}_1$ is an entering variable

$$\text{Now min}\left(\frac{(-72, -24, 36, 84)}{(-2, 1, 2, 3)}, \frac{(-72, -12, 36, 96)}{(1, 3, 5, 7)}\right) \underset{\Re}{=} \min(6, 3) = 3.$$

Hence $\mathbf{y}_4$ is a leaving variable. After pivoting the new tableau becomes

Basis	y_1	y_2	y_3	y_4	*R.H.S*	$\Re$ *(R.H.S)*
$\mathbf{y}_3$	1/4 (–15, –1, 5, 11)	1/4 (–13, 4, 1 0, 19)	(–5/2, –1/4, 5/4, 7/2)	(–11/4, –5/4, 1, 10/4)	(–96, –33, 39, 102)	6
y_1	1/8 (1, 3, 5, 7)	1/8 (1, 2, 4, 5)	1/8 (–2, –1, 1, 2)	1/8 (–2, 0, 1, 3)	1/8 (–72, –12, 36, 96)	24
w	(–10, –2, 2, 10)	Y_4 (–46, –16, 18, 48)	(–4, –2, 2, 4)	1/4 (–15, –5, 12, 22)	(–6, 12, 30, 48)	42

According to the above tableau, for fuzzy non basic variables $\mathbf{y}_2, \mathbf{y}_4$, we have, all $\mathbf{z}_j - \mathbf{c}_j \underset{\Re}{\geq} 0$

}_The optimal fuzzy solution is obtained, $\mathbf{y}_1{}^* \underset{\Re}{=} 1/8\,(-72, -12, 36, 96)$

$\mathbf{y}_2{}^* \underset{\Re}{=} (0, 0, 0, 0)$

$\mathbf{y}_3{}^* \underset{\Re}{=} (-96, -3339, 102)$

$\mathbf{y}_4{}^* \underset{\Re}{=} (0, 0, 0, 0)$

and

$\mathrm{w} \underset{\Re}{=} (-6, 12, 30, 48)$ with $\Re\,(\mathbf{w}) = 42$

Conclusion

The fuzzy basic feasible solution for the notion fuzzy variable linear programming problem is defined. The duality concept in fuzzy variable linear programming problem is introduced. Relevant results are discussed. An illustration of solving these problems by the help of linear ranking function is also included.

References

Baldwin, F.F. and Guild, N.C.F., 1979. Comparison of fuzzy sets on the same decision space. *Fuzzy Sets and Systems*, 2: 213–231.

Bellman, R.E. and Zadeh, L.A., 1970. Decision making in a fuzzy environment. *Management Science*, 17(4): B141– B164.

Bitran, G.R., 1980. Linear multiple objective problems with interval coefficients. *Management Science*, 26: 694–706.

Ganesan, K., 2004. Duality in interval number linear programming. In: *Proceeding of the Annual Conference of KMA and National Seminar on Fuzzy Mathematics and Applications*, p. 271–283.

Ishibuchi, H. and Tanaka, H., 1980. Formulation and analysis of linear programming problem with interval coefficients. *Journal of Japan Industrial Management Association*, 40: 320–329.

Mahdavi-Amiri, N. and Nasseri, S.H., 2005. Duality in fuzzy number linear programming. In: *35th Annual Iranian Mathemetics Conference*, January 26–29. Ahvaz, Iran.

Maleki, H.R., 2002. Ranking functions and their applications to fuzzy linear programming. *Far East J. Math. Sci.*, 4: 283–301.

Maleki, H.R., Tata, M. and Mashinchi, 2000. Linear programming with fuzzy variables. *Fuzzy Sets and Systems*, 109: 21–33.

Nagoorgani, A. and Stephen Dinagar, D., 2005. Duality in close interval approximation of fuzzy number linear programming. In: *National Symposium on Mathematical Methods and applications*, IIT Chennai on December.

Sengupta, Atanu and Pal, Tapan Kumar, 2000. Theory and methodology: On comparing interval numbers. *European Journal of Operational Research*, 127: 28–43.

Sengupta, Atanu, Pal, Tapan Kumar and Chakraborty, Debjani, 2001. Interpretation of inequlity constraints involving interval coefficients and a solution to interval linear programming. *Fuzzy Sets and Systems*, 119: 129–138.

Zadeh, L.A., 1965. *Fuzzy Sets: Information and Control*, 8: 338–353.

Chapter 41

Observations on Venation Pattern in Pteridaceae of Thoubal District, Manipur: A Phylogenetic Study

Y. Sanatombi Devi and P.K. Singh

Enthnobotany and Plant Physiology Laboratory, Department of Life Sciences, Manipur University, Canchipur, Imphal – 795 003 Manipur, India

ABSTRACT

Ten species from the family Pteridaceae found in Thoubal District, Manipur have been investigated for morphological features of their fronds with specific reference to placement and pattern of venation. Two types of venation patterns are seen namely, free and anastomosis type. *Ceratopteris thalictroides* (L.) Brongn. shows anastomosing veins and is regarded as the most advanced group among the ten species and *Pityrogramma calomelanos* (L.) Link as most primitive one.

Keywords: *Pteridaceae, Venation, Phylogenetic.*

Introduction

The Pteridaceae is a large and diverse family with 35 genera. It is represented by six tribes *viz., Taenitideae, Platyzomateae, Cheilantheae, Ceratopterideae, Adiantae* and *Pterideae* (Tryon and Tryon, 1982). These six tribes of Pteridaceae are mainly based on the venation pattern and soral morphology. The venation pattern of fronds has long been considered as one of the most important feature in the classification of higher leptosporangiate ferns, and often serve as the sole taxonomic parameter. Studies of leaf venation in the genera *Pteris* and *Adiantum* have been reported by Nair and Das (1974a, b) and Tryon and Tryon (1982). Hara (1964) investigated venation pattern in *Onoclea sensibilis*. Britto *et al.*

(1994) studied the venation pattern in Thelypteroid ferns of the Western Ghats of South India. Bambie and Madan (1982) studied the venation pattern of Ophioglossaceous ferns. However, our knowledge on the venation pattern in the taxa Pteridaceae is meager, and so far, no complete report is available on the details of these structures. The present investigation is a part of a comparative morphological study with a view to seek evidence for understanding the taxonomic and phylogenetic significance of venation pattern in Pteridaceae.

Materials and Methods

Materials for the present investigation are collected from Thoubal District, Manipur. The taxa Pteridaceae often have a heteroblastic leaf development, and for this reason only the character of large and matured leaves were employed. To study the venation pattern, the laminal segments were cleared in 5 per cent sodium hydroxide followed by chloral hydrate (Arnott, 1959). The cleared material was washed thoroughly in water and stained with safranin. Identification of species were based on various literatures (Bedddome, 1883; Baishya and Rao, 1982; Vasudeva and Chhibber, 1989 and Borthakur *et al.*, 2001). Diagrammatic representation on the venation type of each species is also incorporated (Figure 41.1).

Observations

Adiantum capillus-veneris L.

Fronds bipinnate, 6–15 × 5–8 cm, spreading, firm, deltoid-ovate; pinnae stalked, alternate, 3–5 on each side, basal the largest, flabellate; pinnules shortly stalked, obovate-cuneate, lower margin straight or concave and entire, upper margin 4–5 lobed, finely dentate, texture pellucid, herbaceaous, both surfaces glabrous, glacous green; veins free, dichotomously branched, terminating up to the tips (Figures 40.1a, b).

Adiantum lunulatum Burm.

Fronds caepitose, rooting at apex, unipinnate, 10–25 × 3–8 cm; pinnae 10–20 on each side, alternate, long stalked, basal ones larger, diminished towards the apex, pinnae sub-orbicular to obliquely oblong-ovate, lower margin entire, upper margin inciso-lobate, texture herbaceous, both surfaces glabrous, pale-green; veins distinct on both surfaces, dichotomously branched, free, reaching the margin (Figures 41.1e, f).

Adiantum caudatum L.

Fronds unipinnate, rachis prolonged bearing apical vegetative buds, 10–30 × 2–3 cm, pinnae 10–30 on either side, alternate, sessile to subsessile, pinnae largest in the middle, gradually diminishing towards the apex, oblong-lanceolate, truncated at base, upper margin deeply incised into 3–6 lobes, lower margin slightly concave, texure herbaceous, both surfaces hairy, rachis hirsute, pale green; venation free, dichotomously branched, reaching the margin (Figures 41.1c, d).

Pteris quadriaurita Retz.

Fronds upto 70 × 30 cm, bipinnatifid, deltate, 3–7 pairs of lateral pinnae, basiscopic pinnae forked once at base; pinnae 20–30 × 5–9 cm, sessile, opposite to sub-opposite; margin entire, lanceolate, congested, coriaceous to sub-coriaceous, dark green, both surfaces glabrous; veins prominent, spinules present on upper surface of the segments along mid-veins, vein forked, free, raised on the ventral surface (Figures 41.1m, n).

Figure 41.1: *Adiantum capillus-veneris*

(*a*) Habit, (*b*) Pinna Showing Venation; (*c–d*) *A. caudatum*, (*c*) Habit, (*d*) Pinna Showing Venation, (*e–f*) *A. lunulatum*, (*e*) Habit, (*f*) Venation of Pinna, (*g–i*). *Pityrogramma calomelanos*, (*g*) Habit, (*h*) Venation of pinna, (*i*) Fertile pinna, (*j–l*) *Ceratopteris thalictroides*, (*j*) Habit, (*k*) A portion of pinna with adventitious bud, (*l*) Anastomosing veins, (*m–n*) *Pteris quadriaurita*, (*m*) Habit, (*n*) Venation, (*o–p*) *P. biaurita*, (*o*) Habit, (*p*) Arc formation, (*q–s*) *P. vittata*, (*q*) Habit, (*r*) Basal auricle pinna, (*s*) Venation of lateral pinna, (*t–u*) *P. cretica*, (*t*) Habit, (*u*) Venation in fertile frond, (*v–x*) *P. ensiformis*, (*v*) Habit, (*w*) Venation in laterial pinna, (*x*) Fusion offertile and sterile fronds.

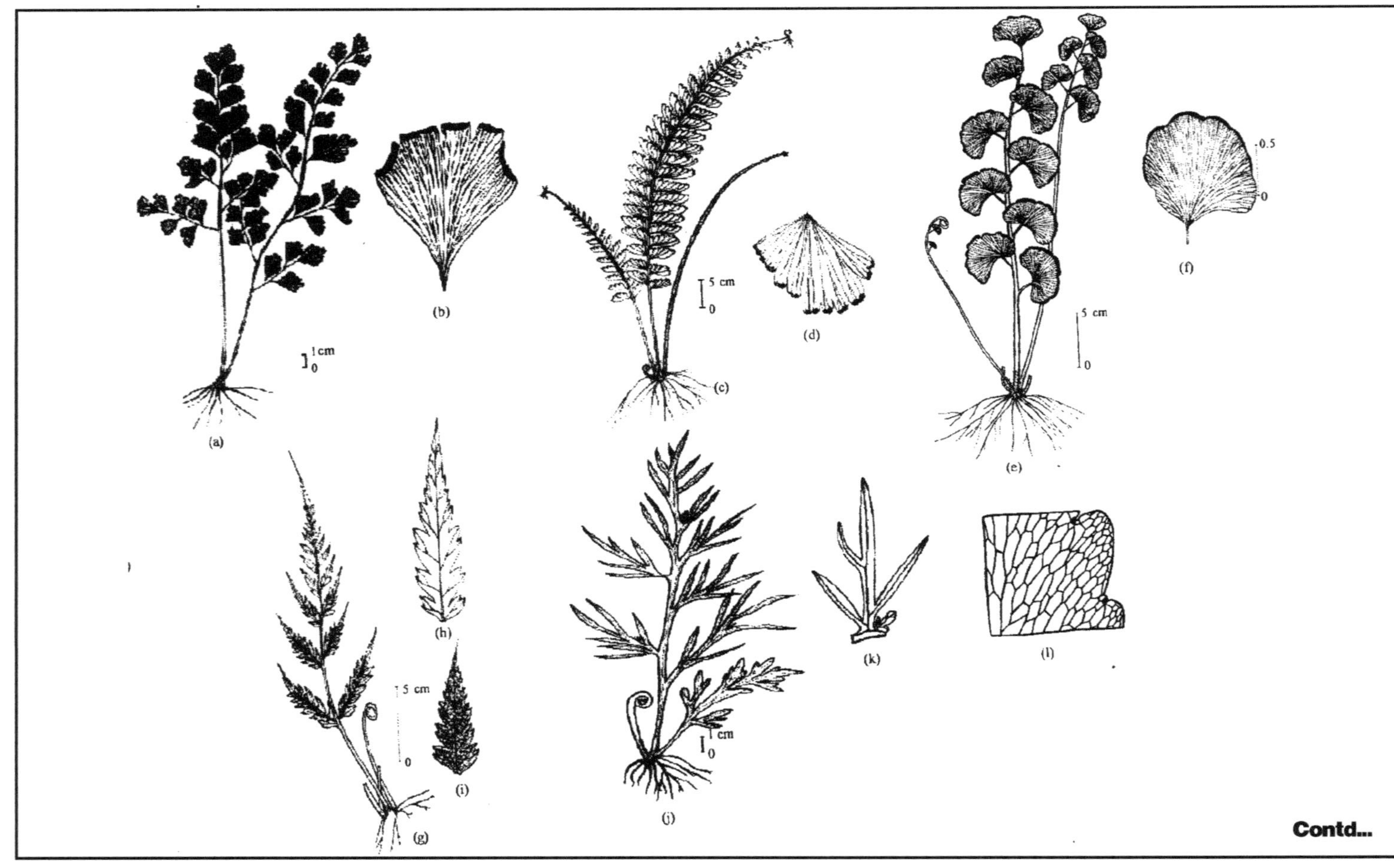

Contd...

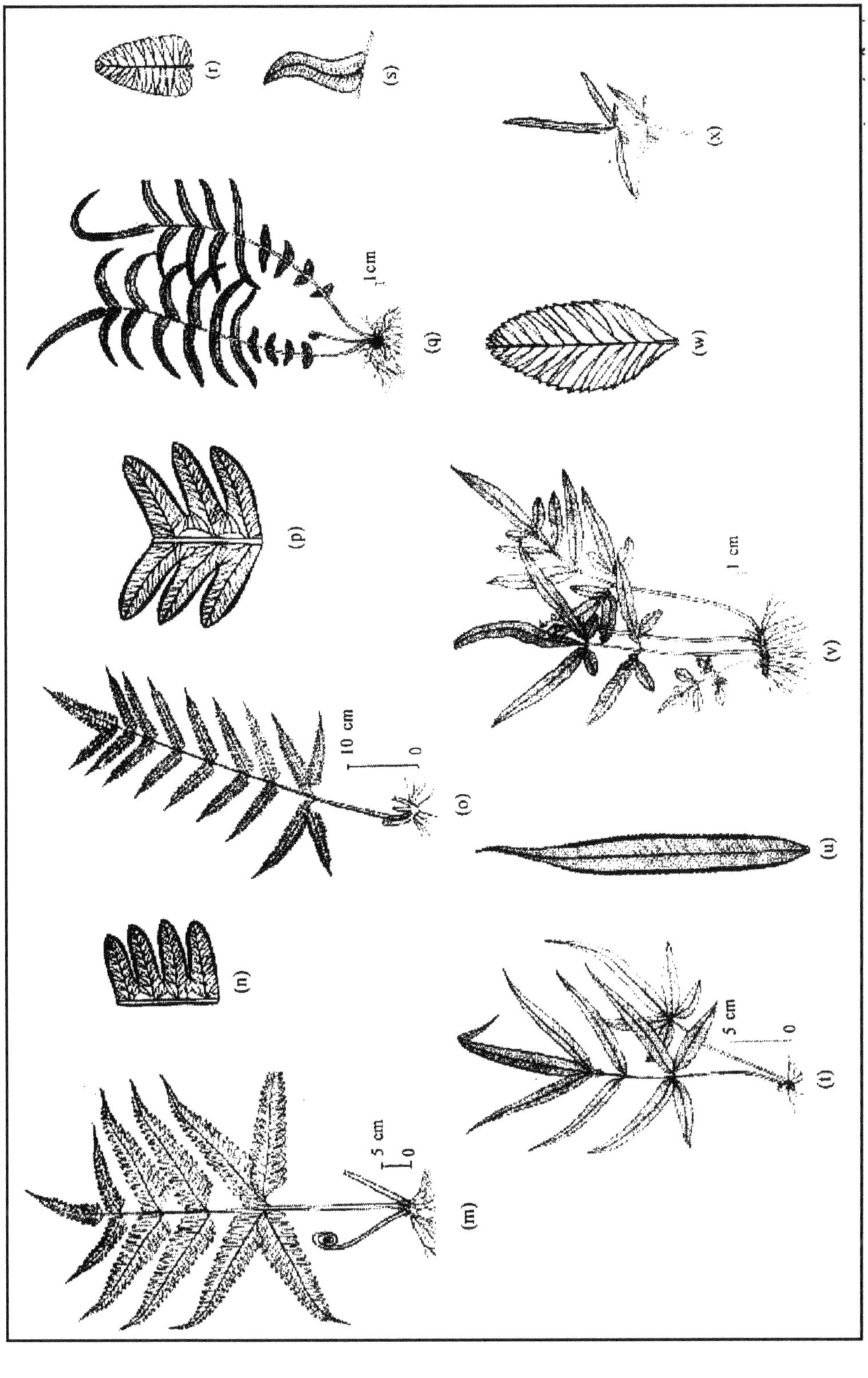

Figure 41.1–Contd...

Pteris biaurita L.

Fronds large, 40–70 × 40–45 cm, bipinnatifid, lanceolate, 7–10 pairs of lateral pinnae, basiscopic pinnae fork once downwards at base, sessile, segments opposite to sub-opposite; pinnae cut down nearly two-third distance to costae, margins entire, pale-green, both surfaces glabrous, thin texture; veins prominent, dichotomously branched except the basiscopic basal vein from each leaflet joins with the aeroseopic basal vein of the leaflet next below forming an arc along costa with 4–6 veinlets passing to the base of sinus (Figures 41.1o, p).

Pteris vittata L.

Fronds large, 20–90 × 10–15 cm, simple pinnate with an elongated large terminal pinna, linear, deltoid-cordate at base, opposite to sub-opposite, middle pinnae largest, upper and basal ones gradually reduced, 5–10 pairs of basal pinnae reduced to auricle-like appendages, nearly all pinnae fertile except reduced basal ones, pinnae sessile, herbaceous, densely covered with whitish hair, thin texture, dark-green; veins prominent, simple or forked once, free (Figures 40.1q, r, s).

Pteris cretica L.

Fronds dimorphic, 25–40 cm long, sterile frond smaller than fertile ones, simple pinnate, 2–5 pairs of lateral pinnae with a terminal one, terminal pinnae 20 × 3 cm; barren pinnae broader, shortly stalked or sessile, opposite to subopposite, margins sharply serrated; fertile pinnae linear, entire; pinnae dark green, glabrous, coriaceous texture, one or two pairs of basal pinnae forked at base, deflexed, unequal; veins prominent, forked once, free (Figures 41.1t, u).

Pteris ensiformis Burm.

Fronds dimorphic; fertile fronds 40 × 20 cm, lateral pinnae about 2–5 pairs, terminal pinnae longest, lateral pinnae forked or trifid, narrower than sterile pinnae, entire, shortly stalked; sterile fronds shorter, 20 × 12 cm, lateral pinnae 2–5 pairs, oblique, oblong-lanceolate, margins sharply dentate, basiscopic one pair of pinnae lobed, lobes oval to elongate-obovate, serrated, sometime both the sterile and fertile fronds fused together; lamina herbaceous, green, glabrous; veins prominent, forked, oblique, free (Figures 41.1v, w, x).

Pityrogramma calomelanos (L.) Link

Fronds tufted on rhizome, 30–50 × 5–20 cm, bipinnate, oblong triangular, gradually shortened towards apex, glabrous above, lower surface covered with white waxy powder, pinnae linear-subulate, deeply lobed, margin crenato-lobate, shortly petiolate; veins dichotomously radiated, free rarely reaching the margin (Figures 41.1g, h, i).

Ceratopteris thalictroides (L.) Ad. Brongn.

Fronds dimorphic; sterile lamina 20–25 × 10–15 cm, bipinnate, pinnae variously lobed, glabrous, membranous, apex acute, base cuneate, petiolate, margin entire, pale-green, texture soft, veins distinct above and below, copiously anastomosing without included veinlets, adventitious buds ore often present in marginal sinuses. Fertile lamina with much branched pinnae, 15–30 × 10–15 cm, bipinnatifid-tripinnatifid, pinnae alternate, apex acute and acicular, margin reflexed and covered sporangia on the lower surface, dichotomously branched (Figure 41.1j, k, l).

Results and Discussions

Two main types of venation patterns are seen in the taxa Pteridaceae. These are free and anastomosis types. Free and forked veins are seen in the genus *Pityrogramma, Adiantum* and *Pteris,* while *Ceratopteris* shows anastomosing venation. Various authors like Tryon and Tryon (1982), Holtum (1973), Vashista (1995) and others pointed out a number of morphological characters in sporophytes that can decide the phylogenetic position of a taxon. Specis with dimorphic fronds having indeterminate simple lamina with entire margins and anastomosis type of venation are considered more specialized. The presence of decompound laminae with many flabellate to cuneate segments and dichotomously forked veins are considered least derived or primitive.

In the present context, *Pityrogramma calomelanos* is regarded as the most primitive species due to the presence of monomorphic and highly bipinnatifid fronds. The margins are deeply crenato-Iobated and veins are dichotomously branched and also slightly radiated from one another. Fronds are monomorphic with indeterminate growth in the genus *Adiantum.* The veins are free, dichotomous and with finely dentate leaf margins. The lamina is bipinnatifid in *A. capillus-veneris* and margins 4–5 lobed. It is placed as least advanced amongst the three species of *Adiantum.* Both the lamina in *A. lunulatum* and *A. caudatum* are unipinnate and rachis often prolonged producing apical vegetative buds and rootings. But the margin of *A. caudatum* are deeply incised with 3-6 lobes and therefore can be reassessed as less specialized than *A. lunulatum.*

The genus *Pteris* exhibits both frond monomorphism (*P. quadriaurita, P. biaurita, P. vittata*) and dimorphism (*P. cretica, P. ensiformis*). The leaf architecture in *P. quadriaurita* and *P biaurita* almost resemble to one another, except in having more pairs of lateral pinnae in *P. biaurita* (7–10 pairs) and presence of an arc along the costa (Figures 41.1o, p). This very manifestation shows the level of specialization in the orientation of leaf venation. In *P. qudriaurita,* the number of lateral pinnae ranges between 3–7 pairs and no arc formation takes place (Figures 41.1m, n). So, it is least specialized amongst the five species of *Pteris. P. vittata* also lacks frond morphism, but there is functional differentiation among the pinnae which is an advanced character. The basal pinnae are reduced to auricle like appendages and usually incapable to bear spores (Figures 41.1r). Except for the basiscopic pinnae, all the segments are potentially sporophyll. Hence *P. vittata* can be regarded as slightly advanced than *P. quadriaurita* and *P. biaurita.* Fronds are dimorphic in both *P. cretica* and *P. ensiformis.* However, the degree of specialization is more reflected towards *P. ensiformis* than in *P. cretica.* In *P. cretica,* one or two (rare) pairs of basal pinnae are forked while in *P. ensiformis,* the lateral pinnae are usually trifid or at times even more than trifid condition. Sometimes both the sterile and fertile fronds may fused together in *P. ensiformis* (Figure 41.1x). Tryon and Tryon (1982) treated *P. ensiformis* under the *P. cretica* group. But after critical analysis, *P. ensiformis* is treated here as more specialized group than *P. cretica,* and hence most advanced group amongst the genus *Pteris.*

In *Ceratopteris thalictroides,* fronds are dimorphic as well as hydrophytic in nature which is an advanced character than the terrestrials or lithophytes (Saxena and Saxena, 2001). The veins are distinct and anastomosing without any included veinlets. This manifestation is considered as most advanced character compared to other types of venation (Holttum, 1973).

It is evident from the present investigation that venation pattern in leptosporangiate ferns is of great taxonomic importance. Amongst the ten species of the taxa Pteridaceae, *Ceratopteris thalictroides* and *Pityrogramma calomelanos* can be regarded as most advanced and primitive respectively. The trend of specialization can be formulated as follows:

Pityrogramma calomelanos < *Adiantum capillus-veneris* < *A. caudatum* < *A. lunulatum* < *P. quadriaurita* < *P. biaurita* < *P. vittata* < *P. cretica* < *P. ensiformis* < *C. thalictroides.*

It may be thus, concluded that the degree of architectural complexities in the venation pattern and other morphological characters of a sporophyll may help to correlate the advance nature of various ferns. Therefore, further critical studies in this field is much awaited and needed.

References

Arnott, H.J., 1959. Leaf clearings. *Turlox News*, 37: 192–194.

Bashya, A.K. and Rao, R.R., 1982. *Ferns and Fern-allies of Meghalaya State, India*. Scientific Publishers, Jodhpur.

Beddome, R.H., 1883. *Handbook to the Ferns of British India, Ceylon and the Malay Peninsula with Supplement*. Today and Tomorrow's Printers and Publishers, New Delhi.

Borthakur, S.K., Deka, P. and Nath, K.K., 2001. *Illustrated Manual of Ferns of Assam*. Bishen Singh Mahendra Pal Singh, Dehra Dun, India.

Bhambie, S. and Madan, P., 1982. Observation on the venation pattern in *Ophioglossum, Botrychium and Helminthostachys. Fern Gaz.*, 12: 215–223.

Britto, A.J., Manichkam, V.S. and Gopalakrishnan, S., 1994. Venation patterns in the *Thelypteroid* ferns of the Western Ghats of South India. *Indian Fern J.*, 11: 124–129.

Hara, N., 1964. Ontogeny of the reticulate venation in the pinna of *Onoclea sensibils. Bot. Mag., Tokyo*, 77: 381–387.

Holttum, R.E., 1973. The family *Thelypteridaceae* in the Old World. *Bot J. Linn. Soc.*, 67: 173–189.

Nair, N.C. and Das, A., 1974a. Studies on the venation pattern in Ferns I. Anastomoses in *Adiantum incisum* Forsk (*A. caudatum* sensu Bedd. Non Lin. Proparte). *Bull. Bot. Surv., India,* 15: 108–117.

Nair, N.C. and Das, A., 1974b. Studies on the venation pattern in Ferns II. Further observations on *Adiantum lncisum* Forsk. *Acta. Bot., India,* 6: 148–153.

Saxena, N.B. and Saxena, S., 2001. *Plant Taxonomy*. Pragati Prakashan Publications, Meerut, India.

Tryon, R.M. and Tryon, A.F., 1982. *Ferns and Allied Plants with Special Reference to Tropical America.* Springer-Verlag, New York Inc., USA.

Vashishta, P.C., 1995. *Pteridophyta*. S. Chand and Company Ltd., Ram Nagar, New Delhi.

Vasudeva, S.M. and Chhibber, S., 1989. Taxonomic studies on some species of *Pteris* Linn. in India. *Indian Fern J.*, 6: 204–216.

Chapter 42

Effect of Nomadism and Grazing on the Phytodiversity of Kalakote Range, Rajouri (J&K)

Jagbir Singh and Shashi Kant

Department of Botany, University of Jammu, Jammu – 180 006

ABSTRACT

The Himalayas has been a great controlling factor for the various ecosystems throughout country. This Himalayan region is facing serious problems today due to excessive exploitation of its natural resources. Uncontrolled biotic interferences have tilted the balance of the mountain regions and have exerted an immense pressure on the biodiversity of the area. The disproportionate growth in human and livestock population has inevitably and inadvertently degraded the area.

Introduction

Kalakote Range (Rajouri) lies between 33°10′N latitude at 74°45′E longitude in district Rajouri, Jammu and Kashmir State. The area mainly constitutes inner Shivaliks. The area is undulating with moderately sloped hills. Altitude of the area ranges from 600 to 1070 m.

Biodiversity in India is under constant anthropogenic stress and threat of extinction due to a hoard of reasons. The protection and conservation of the country's biodiversity is a Herculean task. Considering the livestock population alone, which is around 450 million, while the carrying capacity of the grazing land and forest is not more than 50 million, the Himalayan ecosystems are degrading faster and a great biodiversity loss is being witnessed. Over grazing and browsing by the livestock of nomads and local inhabitants and over exploitation of natural resources are causing great damage to the biodiversity.

Observations

It is believed that the primitive man started his life as a food gatherer, but when he switched over to food producers, he started leading as a pastoral community, which with course of time changed to transhumance and then to sedentrization. Nomadism as a system of grassland management for maintaining socioeconomy of pastoral and transhumance communities in different parts of his country, whose sustenance and life style is directly related to the existence of pasture and the livestock.

Every year during winter season nomadic graziers with their livestock migrate to Kalakote Range from high mountains and stay there for about six months. During rainy season Gujjars and Bakerwals migrate along with their Buffaloes and cattle to this area °and leave their animal for open grazing in the forest area. Besides open grazing, graziers buy leaf fodder and grasses from private land as well.

There are three major passes in the area through which the migration of these nomads takes place. Seir-Kalakote-Sarhanoo; Mahogala-Khawas-Bhadal; and Lamberi-sail sui-Dhangri and many small routes which nomads with their livestock pass and thus, study area becomes the prey of nomadic activities like grazing, browsing, and lopping during rainy season, winter and autumn.

The effects of the activities of these different nomadic communities on the biodiversity of the area is as under:

Gujjars

They rear predominantly buffaloes, cattle and poultry birds. They come through all the three passes from Thanamandi, Darhal and Rajouri area. They reach the study areas in June-July and stay during rainy season in forest area and return back by the end of September every year. The economy of these Gujjars primarily depends upon the sale of milk products and meat.

Bakerwals

They rear predominantly goats, sheep, and horses, and cattles. They reach in the study area during August–September, stay during winter season in the forest area and return back by March–April every year.

Local Inhabitants

Local inhabitants are permanent settlers or those who have encroached forest area, rear cattles, buffaloes, sheep and goats.

In the present study it has been recorded that the Bakerwals and Gujjars are nomadic migratory grazing communities in the area. It has also been recorded that three main grazing systems exist at present *i.e.*, nomadic, pastoral and transhumance.

In the area analysed, about 1,71,437 heads of the animals remain presents for grazing round the year (Tables 42.1 and 42.2). Animals have been observed to feed on the grasses, herbs and shrubs, in the forest area. Unrestricted and uncontrolled grazing by the cattle and buffaloes of nomads and local inhabitants in the forest area during rainy season has been observed. The pressure of local livestock grazing remains throughout the year.

Continuous over-grazing by a large herd of animals destroy the seedlings and recruits due to grazing and browsing of young succulent plant parts and trampling. This adversely affects the whole regeneration process. Grazing is destructive to the better strains of grasses which are palatable because they are being overeaten. It leads to deterioration of grass ecosystem. Unpalatable bushes such as

Lantana, Parthenium, Euphorbia sps. Dodonaea and *Polygonum* have invaded such areas in large numbers, thereby changing the whole ecosystem.

Table 42.1: Livestock Population of Rajouri District

	1972	*1977*	*1982*	*1988*	*1992*
Cattle	118652	114457	126204	123512	133210
Buffaloes	78411	78545	89811	88739	88959
Sheep	24444	80003	78212	146726	202103
Goats	53120	71194	86395	207456	245850
Others	6862	6447	8270	8790	11108
Total	281480	32046	388892	572223	681230

Sources: (i) Chief Animal Husbandry Office, Distt. Rajouri

(ii) District Sheep Husbandry Office, Rajouri.

Table 42.2: Livestock Population of Kalakote Block

	1998	*1999*	*2000*
Cattle	28000	30100	29200
Buffaloes	35000	37200	36800
Sheep	43738	49613	43917
Goats	70870	68005	57444
Horse	–	–	2908
Others	–	–	1168
Total	182608	184918	171437

Sources: (i) Chief Animal Husbandry Office, Distt. Rajouri

(ii) District Sheep Husbandry Office, Rajouri.

Soil compaction is mainly caused by unrestricted, unregulated grazing and trampling. Camping of nomads in forest area also contribute, to compaction of soil. Continuous trampling reduces the ability of soil to recover, due to the decrease in abundance of active roots. Consequences of compaction include impeded drainage, which leads to increased run-off and erosion, decreased water and air availability to plant roots and soil organisms causing alteration in soil organism populations and plant death, and decreased abundance of larger pore spaces. There is decrease in water absorbing capacity, soil becomes susceptible to erosion. There is increase in surface run-off and dislodging of soil particles also takes place.

The congregations of nomadic and local animals inevitably degrade and finally desertify the area around the settlements and then extend to adjoining areas. Confronted with pastoral shortage, the nomads leave the settlement after every six months and so and move on to the greener pastures elsewhere with the result that they contribute to the desertification process to the adjoining or other places.

The problem of grazing takes a serious turn in this area because of so many factors, like very high unregulated grazing pressure during rainy and spring season, when rain and cattle grazing are synchronous and soil gets exposed by heavy hooves, puddling of ground surface, moist soil and erosion of the top soil by rain water. The problems get aggravated because of the hilly terrain of the area, which accelerates the process of soil erosion.

The livestock of nomads, goats, sheep, horses and mules exert an immense pressure on herb and shrub flora of the region by grazing and browsing. Horses and mules cause a lot of damage to topsoil with their strong and sharp hoofs. Sheep grazing is more harmful as it effects soil properties more adversely than cattle grazing. Browsing by goats is even more serious. Besides local goat that remain in the forests year round, migratory goat population swells the total number many folds. They have been found eating up the twigs and vegetative buds voraciously. Bakerwals axe is far more destructive than his goat.

When there is less or nothing on the ground, the sheep and goats take to the tree growth and graziers resort to the lopping of the trees and branches for feeding their animals. Goats, sheep, oxen and cattle are harmful in that order.

Lopping affect has also been observed in the study area. Nomads as well as local inhabitants have been found to lop the trees. *Acacia catechu, Ziziphus mauritiana, Albizia, lebbeck, Bauhinia variegata, Celtis australis, Grewia optiva* and *Morus alba* have been found to be exploited for stall feeding of cattle, goats and their young ones which are generally kept in pens. Repeated lopping, year after year, adversely affects the growth and form of these trees. Insect-pests and diseases easily attack such trees.

Over grazing and open grazing are the major causes of poor regeneration/degradation of forests areas. This view is substantiated by the fact that livestock density per unit of land in Himalaya is much higher than in low lands and lack exclusive fodder crops farming in mountains.

Biological spectrum of area has come to a Thero-chamaephytic phytoclimate and normal expected biological spectrum (Raunkiaer's, 1934) of the area is phanerophytic. The dominance of therophytes is due to biotic interferences coupled with heavy grazing pressure.

The nomadic as well as local inhabitants totally depend upon the forests for meeting their requirements of timber, agricultural implements, house building and repairs, firewood and fodder. The people are also totally dependent on forests for grazing requirements of their cattle, sheep goats etc which leads to deterioration and degradation of ecosystem.

Discussion

Biodiversity is under constant stress due to disproportionate growth in human and livestock population coupled with deforestation and widening gap between demand and supply of biotic resources.

Continuous unregulated and overgrazing by a large herd of animals destroys the seedlings and recruits due to grazing and browsing of young succulent plant parts and trampling. This adversely effects the whole regeneration process. These observations find supports from the works of Gupta *et al.* (1994) and Babu *et al.* (1984) who also observed similar results.

Milchunas (1992) pointed out that indirect effect of herbivorous mediated through the effects of grazing on microenvironment and plant competition were more important than direct effects of consumption and trampling on the invasion capacity of the grassland by opportunistic species. In the present study area, similar effects of grazing have been observed.

Unrestricted grazing and browsing by the livestock of local inhabitants and nomads, exploitation of natural resources, encroachment and forest fires are causing great damage to biodiversity of the area. Similar observations were observed by Kant and Sharma (1999). Dhaliwal *et al.* (2000) attributed the reasons of degradation of forests to indiscriminate deforestation by the local population to meet their fuel wood and timber requirement. It is true for the present study area also.

Singh (1981) reported that loss is not entirely due to illicit/over felling but is the total loss in productivity due to indiscriminate grazing, lopping, hackings, over felling, fires etc. Present studies are in conformity with above studies since the pressure on the forests are from local inhabitants, Government, nomadic population and illicit smugglers etc.

Myers (1982) reported that the depletion of tropical moist forest are due to timber exploitation, high population pressure and cattle raising in the area. This is true for our study area also. Similar observations have also been reported by Srivastava and Kumar (1995).

Patnaik and Kumar (1996) observed that productivity of grazing lands of Jammu region has decreased considerably because the livestock far exceed land's carrying capacity and the number is ever increasing. Present finding are in consonance with above findings.

In the present study area a thero-chameophytic phytoclimate has been worked out. The normal expected biological spectrum of area is phanerophytic. The dominance of therophytes is due to biotic interferences coupled with heavy grazing pressure. Similar observations were observed by Gupta and Singh (1990), Sharma and Dhakre (1993) and Cain (1950)

India has large livestock population. It accounts for about 15 per cent of the livestock population of the world and for only about 2 per cent of the total worlds geographical area. It possesses a livestock of about 450 millions. India supports more than half of world's buffaloes, 15 per cent goats, 4 per cent sheep and a large percentage of other livestock. The reason for such a huge livestock is rural nature of settlements. About 80 per cent population of India live in villages, which directly or indirectly, is linked with agriculture. These people graze their cattle in grazing lands/pastures and in the forests. Though grazing as a necessary evil has to be tolerated yet over grazing and unrestricted grazing has to be curbed immediately.

References

Babu, C.R., Gaston, A.J., Chauduri, A. and Khandwa, R., 1984. Effects of human disturbance in three areas of west Himalaya moist deciduous forest. *Environment Conservation*, 11(1): 55–60.

Cain, S.A., 1950. Life form and Phytoclimate. *Bot. Rev.*, 16: 1–32.

Dhaliwal, S.S., Litoria, P.K., Singh, C. and Sharma, P.K., 2000. Degradation of reserved protected forests in district Patiala (Punjab): A vegetation change detection study using remote sensing technology. *Ecol. Env. and Cons.*, 6 (2): 153–157.

Gupta, B. and Singh, R., 1990. Phenology and biological spectrum of grazed and ungrazed grassland vegetation in Gambhar Catchment, Himachal Pradesh. *Range Mnag. Arof.*, 11(2): 123–134.

Gupta, B., Singh, R. and Verma, R.K., 1994. Biomass fluctuations in grazing lands around Shimla, Himachal Pardesh. *Indian Forester*, 120(6): 497–499.

Kant, S. and Sharma, K.K., 1999. Effect of nomadism and grazing on the phytodiversity of Patnitop and Sanasar Hills J&K (India). In: *Proc. Acad. Environ. Biol.*, 8(2): 165–175.

Milchunas, D.G., 1992. Effect of grazing on grassland community structure. *Ph.D. Thesis,* Colorada State University, USA, pp. 177.

Myers, N., 1982. Depletion of tropical moist forests: A comparative review of rates and causes in the three main regions. *Acta Amazonica*, 12(4): 745–757.

Patnaik, P. and Kumar, A., 1996. Graziers problems: A solution (In Jammu region). *J&K Forest News Letter*, p. 39–43.

Raunkiaer, C., 1934. *The Life Forms of Plants and Statistical Plant Geography: Collected Papers of C. Raunkiaer.* Clarendon press, Oxford, 639 pp.

Sharma, M. and Dhakre, J.S., 1993. Life form classification and biological spectrum of the flora of Shahjahanpur District, Uttar Pradesh (India). *Indian J. Forestry*, 16(4): 366–371.

Singh, V.P., 1981. Effect of biotic interference on forests: A case study in Udaipur forest division of Rajasthan. *Indian Forester,* 107(11): 693–697.

Srivastava, A.K. and Kumar, S., 1995. The disturbed mountain ecosystem: A case study of Salari village in Kumaon Himalaya. *Indian Forester,* 121(2): 103–109.

Chapter 43

Water Quality Study of Selected Dug Wells of Chirayinkeezhu Grama Panchayat in Kerala

K.S. Binu and V.R. Prakasam

Department of Environmental Sciences, University of Kerala, Thiruvananthapuram – 695 581

ABSTRACT

A study was conducted to evaluate the water quality of selected dug wells of Chirayinkeezhu Grama Panchayat, Kerala, during November 2003 to October 2004. The physico-chemical parameters using standard methods showed that all the parameters except turbidity were within the limits for drinking water specified by BIS. Season-wise differences were also negligible except in the case of total dissolved solids, nitrate and potassium. The amounts of TDS and potassium were greater during rainy season whereas nitrate was greater during non-rainy season, but within limits of BIS. So it was concluded that the dug well water was safe for drinking and domestic needs

Keywords: *Dug well, Water quality, Kerala panchayat.*

Introduction

A perusal of literature showed that studies on water quality of dug wells at panchayat level are rare in Kerala. Chirayinkeezhu is one of the panchayats located in Thiruvananthapuram district of the state. People of the panchayat depend on dug well and pipelines for their domestic water needs. In a recent report from the panchayat (Progress Report, 2002–2007) doubts have been cast on the quality of dug well water for drinking purpose. Therefore, the study was proposed with the objective of monitoring the quality of groundwater available in selected dug wells of Chirayinkeezhu grama panchayat.

Materials and Methods

Chirayinkeezhu grama panchayat is located in the coastal region of Thiruvananthapuram district in Kerala. In the panchayat lies the brackish Anjengo lake, the Vamanapuram river and several ponds. Several wells have also been dug for meeting domestic requirements.

Seven dug wells were randomly selected for water quality analysis. These were Kottappuram well (KKW), Pazhanchira well (PW), Pulumthuruthi welll (PTW 1) Pulumthuruthi well II (PTW II) Inthikadavu well (IKW), Anchakadavu well (AKW), and Karunthakadavu well (KKW). Monthly collection of water samples was done from November 2003 to October 2004. The various parameters as indicated in Table 43.1, were analysed by following standard methods given in APHA (1985) and Trivedi and Goel (1986). The results are presented season wise (rainy season and non-rainy season). Considering the data of all the wells together annual mean values were calculated and presented as mean±SD. The values obtained were then compared with the specification of drinking water of BIS (1993).

Table 43.1: Physico-chemical Parameters of Dug Wells of Chirayinkeezhu Grama Panchayat during 2003–2004

Sl.No.	*Parameter (Units)*	*Rainy Season (Mean±SD)*	*Non-Rainy Season (Mean±SD)*	*Annual Mean (Mean±SD)*
1.	Temperature (°)	28.33±0.31	29.33±0.27	28.83±05
2.	Turbidity (NTU)	8.09±3.02	11.74±12.33	9.92±1.83
3.	pH	7.16±0.67	6.64±0.64	6.9±0.26
4.	Conductivity (mS)	0.06±0.08	0.06±0.07	0.06±0.0
5.	Total Solids (mg/l)	599.29±551.14	394.29±165.05	496.79±102.5
6.	Total Dissolved Solids (mg/l)	372.67±460.30	163.83±25.85	268.25±104.42
7.	Total Suspended Solids (mg/l)	70.95±24.81	68.67±22.28	69.81±1.14
8.	Total Hardness (mg/l)	73.86±129.44	72.17±124.52	73.02±0.85
9.	Calcium Hardness (mg/l)	26.11±16.03	26.34±15.67	26.23±0.12
10.	Magnesium Hardness (mg/l)	72.57±114.56	71.96±112.33	72.27±0.37
11.	Dissolved Oxygen (mg/l)	4.65±1.09	3.56±0.36	4.29±0.51
12.	Biological Oxygen Demand (mg/l)	3.72±0.64	2.54±0.48	3.13±0.59
13.	Chloride (mg/l)	56.36±8.73	56.80±9.48	56.58±0.22
14.	Nitrate (mg/l)	0.12±0.08	0.40±0.18	0.26±0.14
15.	Nitrite (mg/l)	0.02±0.01	0.08±0.06	0.05±0.03
16.	Phosphate (mg/l)	0.13±0.67	0.11±0.62	0.12:t0.01
17.	Silicate (mg/l)	0.20±0.15	0.22±0.14	0.21±0.01
18.	Fe (mg/l)	0.06±0.15	0.05±0.14	0.06±0.01
19.	Sodium (mg/l)	32.04±21.65	31.63±20.77	31.84±0.021
20.	Potassium (mg/l)	16.18±9.46	10.89±12.22	13.54±2.65

Results and Discussion

The results of the analysis of the physico-chemical properties of water are given in Table 43.1. It could be noted that the mean difference in temperature between the rainy season and non-rainy season was 1.0°C and the annual mean temperature was 28.83°C. Turbidity value of rainy season was less by a slight margin of 2.65 NTU than non-rainy season. These values were above the desirable or permissible limits set by BIS.

The pH of rainy season was 7.16 and that of non-rainy season was 6.64. Both values were within the limits of BIS. Though the pH has no direct effect on human health, all biochemical reactions are reported to be sensitive to the variation of pH (Jayakumar *et al.*, 2003). However, a pH value below 4 produced sour taste and high value above 8.5 produced bitter taste to water (Naik and Rajendra Prasad, 2004).

Water conductivity of both the seasons was same at 0.06mS. According to Davis and De Wiest (1996), normal groundwater showed the conductance from 30 to 2000 micromhos and the present value is within this range.

The quantity of total solids of rainy season was 599.29 mg/l, and that of non-rainy season was 394.29 mg/l, which indicated that the value was greater during rainy season. A limit of 1000 mg/l, of total solids has been imposed by WHO (1984) for drinking water and above this level it is known to cause gastro-intestinal irritations (ICMR, 1975).

According to BIS, desirable value TDS is 500 mg/l, and the permissible value is 2000 mg/l. The dug well water during both the seasons showed TDS values below the desirable and permissible limits of BIS as given in Table 43.1.

Total hardness of rainy season was 73.86 mg/l, and that of non- rainy season was 72.17 mg/l, the difference being negligible. As per BIS desirable value is 300 mg/l and the permissible value is 600 mg/l, and accordingly during both the seasons the mean value was below the desirable level. WHO (1984) stated that water hardness was not a health hazard but its value should remain within permissible limit to restore the taste to water. Besides, hardness makes water unsuitable for several domestic operations such as washing, cooking etc.

Dissolved oxygen of rainy season was 4.65 mg/l, and that of the non-rainy season was 3.56 mg/l, with an annual mean value of 4.29 mg/l. The low content of DO might be due to the underground source of the water and its low photosynthetic activity. The recommended DO limit for all domestic purposes is 4–5 mg/l (Naik and Rajendra Prasad, 2004).

Biological Oxygen Demand values of rainy and non-rainy season were 3.72 mg/l, and 2.54 mg/l, respectively. It indicated that the groundwater was not organically polluted. Chloride values during rainy season (56.36 mg/l) and non-rainy season (56.80 mg/l) remained almost same. As per specification of BIS the desirable value of chloride is 250 mg/l and the permissible value is 1000 mg/l, and accordingly all the values were below the desirable limit.

Nitrate values of rainy season (0.12 mg/l) and non-rainy season (0.40 mg/l) indicated that these were lower than the limits recorded by BIS (desirable value of nitrate is 45 mg/l, and the permissible value is 100 mg/l). It could be further noted that all the other nutrients (nitrite, phosphate, silicate) were present in low amounts in the dug well water.

The mean value of iron in dug well during the rainy season was 0.06 mg/l and in the non-rainy season was 0.05 mg/l. Desirable limit of Fe is 0.3 mg/l and permissible limit is 1.0 mg/l (WHO, 1984). The present values were below the limit.

With reference to Na, the seasonal difference was negligible, but K showed an increase during rainy season.

It may be concluded from the data that the dug well water of Chirayinkeezhu grama panchayat is suitable for drinking and other domestic purposes.

References

APHA, 1985. *Standard Methods for the Examination of Water and Wastewater,* 15th Edition, American Public Health Association, Inc., New York.

BIS, 1993. *Indian Standard: Specification for Drinking Water*. Bureau of Indian Standards, New Delhi.

Progress Report, 2001. *Chirayinkeezhu Grama Panchayat*, Chirayinkeezhu, Kerala.

Trivedi, R.K. and Goel, P.K., 1986. *Chemical and Biological Methods for Water Pollution Studies*. Environmental Publications, Karad.

Mahesha, N.K. and Prasad, N.R. Rajendra, 2004. Physio-chemical characteristics of bore well water. *Arikere Taluk, Harran, I.J.E.P.*, 24(12): 897–904.

Davis, S.N. and Wiest, R.J.M. De, 1996. *Hydrogeology*. John Wiley and Sons, Inc., New York.

Indian Council for Medical Research, 1975. *Manual of Standards of Quality for Drinking Water Supplies*, 2nd edn. Special Report, Series, No. 44.

Jayakumar, S., Indira, P. and Arasu, Thillai, 2003. Status of groundwater quality and public health around Thiruchendur. *IJEP*, 23(3): 256–260.

WHO, 1984. *Guidelines for Drinking Water Quality,* Vol. 122: Recommendations and World Health Organization, Geneva.

Chapter 44

Influence of Zinc Toxicity on Some Selected Biomarkers in a Freshwater Teleost Fish, *Cyprinus carpio* var. *communis*

J. Vinoliya[1]*, M. Manavalaramanujam*[1] *and M. Ramesh*[2]

[1]*Department of Zoology, Holy Cross College, Nagercoil, Kanyakumari*
[2]*Department of Zoology, Bharathiar University, Coimbatore – 46, India*

ABSTRACT

The present investigation deals with the changes in plasma cortisol, prolactin, Glucose, Na^+K^+ATPase levels of the fish *Cvprinus carpio* when exposed to sublethal concentrations of zinc sulphate for 28 days. Plasma cortisol level was seen increased throughout the experimental period, whereas plasma prolactin level declined upto seventh day and significantly increased thereafter and remained so during the rest of the study period. A significant hyperglycemic condition and inhibition of Na^+K^+ATPase was observed. The reasons for the above parameters are discussed and can be used as logical candidates in pollution monitoring programmes.

***Keywords**: Cyprinus carpio, Zinc sulphate, Cortisol, Prolactin, Glucose,* Na^+K^+*ATPase.*

Introduction

Polluted habitat poses a severe challenge to the organisms physiological integrity and the homeostasis of the animal are mediated to a large extent by the endocrine system (Brouwer *et al.*, 1990). Osmoregulation, energy metabolism, growth or reproduction are hormonally regulated physiological functions that may be adversely affected by pollutants (Hontela *et al.*, 1993).

Mayer *et al.* (1992) reported that during acute stress, the endocrine system (*i.e.*, catecholamines and glucocorticoids) generally controls energy mobilization and during chronic stress both endocrine and tissue level system interact to affect energy mobilization which is required to maintain homeostasis during chemical challenge.

Hormones have been included as they are measurable in blood and circulating levels of these hormones can be altered by exposure to xenobiotic chemicals as suggested by Folmar (1993). He further reported that for aquatic toxicologists, measurement of circulating levels of hormones can provide additional information of the sublethal effects of many chemicals.

Cortisol, one of the glucocorticosteriod hormone is considered an indicator of primary stress in many fishes (Donaldson, 1981; Passino, 1984; Matty, 1985). The importance of the pituitary hormone prolactin for regulating

the water and ion balance in fish has been well established (Yada *et al.*, 1991; Flik *et al.*, 1989; Sivakumari, 1997 and Waring, 1996). Blood glucose level can be utilized as a parameter of stress response as it is rapid, practicable and quantitative (Chavin *et al.*, 1973).

The rationale for studying an enzyme was to see if they could be used as biochemical parameter in diagnosing sub-lethal metal toxicity or stress (Eugene, 1974). Na^+K^+ATPase is an important component of active transport in teleost gills and is involved in osmoregulation and intercellular functions, *e.g.*, the "sodium pump" are sensitive indicators of trace metal toxicity (Haya and Waiwood, 1983).

Zinc when in excess in the environment may enter the fish body through nutrients and general surface of body and gills. It may get accumulated in tissues and result in general enfeeblement, retardation in growth and cause metabolic and pathological changes in various organ tissues (Spehar *et al.*, 1978). The present investigation was carried out to study the alterations in the levels of the hormones cortisol and prolactin, Glucose and Na^+K^+ATPase in *Cyprinus carpio* var. *communis* when exposed to sublethal concentrations of zinc.

Materials and Methods

Specimens of *Cyprinus carpio* var. *communis*, a fresh water teleost fish were collected from Tamil Nadu Fisheries Development Corporation, Aliyar, Tamil Nadu, without any physical or mechanical injury and acclimatized to laboratory conditions for about 20 days. During this period fish were fed with rice bran and ground nut oil cake in the ratio of 3 : 1. The tap water was analyzed for physico-chemical features as per APHA (1971) and was as follows: Temperature 26.0±2°C; pH 7.2±0.1; Salinity 0.6±0.01 ppt; Dissolved oxygen 6.2±0.02 mg/l; Total hardness 18.0±0.5 mg/l and Total alkalinity 21.00±10.00 mg/l. Water in the tank was changed daily and aerated to ensure sufficient oxygen supply. For experimental purpose healthy fish with an average weight of 5 to 6 gm and length of 7 to 8 cm were selected. The median lethal concentration (LC_{50}) of zinc sulphate was determined for 24 hrs and 96 hrs by probit analysis method of Finney (1971) which was 81.75 and 69.82 ppm, respectively. 1/10th of the 24 hr LC_{50} value (8.1 ppm) was taken for sub-lethal studies according to Sprague (1971). In the present study fishes were exposed to sub-lethal concentration of zinc sulphate for a period of 28 days. A common control was maintained. At the end of every seven days fish from control and sub-lethal group were sacrificed and the plasma cortisol, prolactin, glucose and Na^+K^+ATPase levels were measured. Cortisol and prolactin was estimated by solid phase competitive Enzyme Linked Immunosorbent Assay (ELISA) following the method of Skelly *et al.* (1973). Plasma glucose was estimated by O-Toludine method (Cooper and McDaniel, 1979). The specific activities of Na^+K^+ATPase were estimated following the method of Shiosaka *et al.* (1971). The data was subjected to statistical analysis.

Results

Tables 44.1 and 44.2 shows the plasma cortisol and prolactin level of *Cvprinus carpio* exposed to zinc sulphate for 28 days. During the above treatment period plasma cortisol level increased throughout the study period and it was directly proportional to the exposure period showing a per cent increase of 3.80, 9.45,15.71 and 14.13 at the end of 7, 14, 21and 28 days, respectively. The plasma prolactin level decreased showing a percent decrease of 5.94 at the end of the 7th day. After the 7th day plasma prolactin level recovered showing a per cent increase of 3.40, 3.50 and 7.41 at the end of 14, 21 and 28 th day respectively.

Table 44.1: Changes in the Plasma Cortisol Level of *Cyprinus carpio var. communis* Exposed to Sub-lethal Concentration of Zinc Sulphate for 28 Days

Exposure Period (in days)	*Cortisol ng/ml*		*Per cent Change*	*Calculated 't' Value*
	Control	*Experimental*		
7	4.84±0.128	5.024±0.040	+3.80	1.373
14	5.50±0.200	6.021±0.037	+9.45	2.561*
21	6.441±0.186	7.642±0.093	+15.71	5.764*
28	6.321±0.188	7.360±0.093	+14.13	4.976*

Values are mean±S.E. of five individual observations.

+: Denotes percent increase over control.

*: Values are significant at 5 per cent level.

Degrees of freedom at $8_{t0.05}$ = 2.306.

Table 44.2: Changes in the Plasma Prolactin Level of *Cyprinus carpio var. communis* Exposed to Sub-lethal Concentration of Zinc Sulphate for 28 Days

Exposure Period (in days)	*Prolactin ng/ml*		*Per cent Change*	*Calculated 't' Value*
	Control	*Experimental*		
7	3.70±0.230	3.48±0.086	–5.945	0.898
14	6.46±0.263	6.68±0.086	+3.405	0.797
21	7.42±0.265	7.68±0.066	+3.500	0.962
28	7.82±0.128	8.40±0.070	+7.410	3.920*

Values are mean±S.E. of five individual observations

–: Denotes percent decrease over control.

+: Denotes percent increase over control.

*: Values are significant at 5 per cent level.

Degrees of freedom at $8_{t0.05}$ = 2.306

Changes in plasma glucose level of *Cyprinus carpio* treated with sub-lethal concentration of zinc sulphate showed a gradual increase as the exposure period was extended showing Ii minimum percent increase of 23.59 at the end of the 7th day and a maximum percent increase of 96.30 at the end of the 28th day (Table 44.3).

Table 44.3: Changes in the Plasma Glucose Level of *Cyprinus carpio var. communis* Exposed to Sub-lethal Concentration of Zinc Sulphate for 28 Days

Exposure Period (in days)	*Plasma Glucose (mg/100 ml)*		*Per cent Change*	*Calculated 't' Value*
	Control	*Experimental*		
7	38.23±0.164	47.25±0.213	+23.59	33.65*
14	30.55±1.795	51.88±1.154	+69.81	9.99*
21	45.71±0.871	80.94±0.233	+77.07	39.23*
28	53.13±1.11 0	104.29±0.145	+96.30	45.72*

–: Denotes percent decrease over control.

+: Denotes percent increase over control.

*: Values are significant at 5 per cent level.

Degrees of freedom at $8_{t0.05}$ = 2.306

Table 44.4 gives the data on Na^+K^+ATPase activity in the gills of *Cyprinus carpio.* During the treatment period, the enzyme activity decreased in the treated fish. The significant decrease in enzyme activity gradually increased upto the 14th day showing a percent decrease of 61.93 at the end of 28th day.

Table 44.4: Changes in the Na^+K^+ATPase Activity in Gills of *Cyprinus carpio var. communis* Exposed to Sub-lethal Concentration of Zinc Sulphate for 28 Days

Exposure Period (in days)	*Na^+K^+ATPase Activity (g/h/g)*		*Per cent Change*	*Calculated 't' Value*
	Control	*Experimental*		
7	128.96±5.48	61.0 1±3.31	–50.37	10.15*
14	13 0.08±2.47	29.76±3.18	–77.12	24.89*
21	134.2±0.99	42.20±1.50	–68.55	51.39*
28	120.98±0.33	46.16±1.89	–61.93	20.95*

Values are mean±S.E. of five individual observations

–: Denotes percent decrease over control.

*: Values are significant at 5 per cent level.

Degrees of freedom at $8_{t0.05}$ = 2.306

Discussion

Changes in the concentration of hormones particularly those regulating vital such as osmoregulation, energy metabolism, functions reproduction or growth may have potential as early warning indicators of toxic stress in fish.

The capacity of an animal to interact with its environment is reflected by changes in corticosteroid levels in response to a specific stress (Weiss *et al.*, 1979). Plasma cortisol has been monitored as a general index of stress. Secretion of the steroid hormone cortisol by inter renal tissue is a characteristic reaction of teleost fish to almost all forms of environmental stress (Donaldson, 1981).

According to Wendelaar Bonga (1997) an elevation of plasma cortisol is the most widely used indicator of stress in fish. During chronic stress cortisol levels may remain elevated, although well below peak levels and such cortisol responses have been reported for many fish species after many different treatments ranging from handling and disturbances, heavy metals, organic pollutants, rapid temperature changes and acid waters, to confrontations with predators (Barton and Iwama, 1991; Brown, 1993).

Lorz *et al.* (1978) and Schreck and Lorz (1978) reported a significant increase in plasma cortisol level in coho salmon *Oncorhynchus kisutch* exposed to cadmium, copper and several metals which may be due to the presence of metals as potential harmful substances in the ambient water. Lidman *et al.* (1979) in European eel *Anguilla* (*L.*) and Sheridan (1986) in coho salmon *Oncorhynchus kisutch* observed stimulation of proteolysis and lipolysis for gluconeogenesis leading to elevation of plasma glucose and also appears to be related to increased cortisol levels under stressful condition.

A possible benefit of cortisol mobilization in response to stressors may be its documented effect in stimulating chloride cell proliferation on the secondary lamellae as reported by Doyle and Epstein (1972) and Perry and Wood (1985). Elevation in plasma cortisol levels also triggers osmotic and ionic regulation along with prolactin by increasing branchial $Na^{+}K^{+}$ATPase and Ca^{2+}ATPase activities (Flik and Perry, 1989; Laurent and Perry, 1990).

In the present study, significant increase in plasma cortisol levels in *Cyprinus carpio* exposed to sublethal zinc treatment may be due to the recognition of the presence of a noxious or potential harmful substance or abnormal plasma chloride level or the process of trying to restore the values to normal or stimulation of proteolysis and lipolysis for gluconeogenesis, leading to the elevation of plasma glucose level.

Prolactin with cortisol is one of the main osmoregulatory hormones in fish maintaining the plasma electrolyte levels mainly by controlling permeability of the gill epithelium (Clark and Bern, 1930). Wendelaar Bonga and Pang (1991) reported that the main function of prolactin in fish is the inhibitory control of the permeability of the integument to water and ions in fresh water environment.

The key osmoregulatory hormone prolactin of fresh water teleost maintains plasma Na^{+} and Cl^{-} concentrations by its action on osmoregulatory tissues (Bern and Madsen, 1999). According to Rosseland and Staurnes (1994) prolactin reduces the permeability of gill epithelium which is time dependant and it is a mechanism of resistance.

Wendelaar Bonga and Van der meig, (1981) and Flik *et al.* (1989) observed that reduction in gill permeability by an increased response of prolactin in cadmium treated fish clearly indicate the maintenance of a stable Na^{+} and water content since prolactin has been shown to decrease branchial efflux and osmotic water influx.

Such compensatory response by prolactin could account for the apparent undisrupted Na^{+} and water balance in metal exposed fish and such a response apparently enables the fish to maintain homeostasis (Fu *et al.*, 1989; Fu and Lock, 1990). In the present study the increased prolactin response under sub-lethal concentration exposure may be a step to re-establish ionic disturbances supporting the views of the above authors.

Decline in plasma prolactin levels in brown trout *Salmo trutta* may be due to a more stressful environment, where mortalities are common (Waring *et al.*, 1996). Notter *et al.* (1976) suggested that severe stress may also cause atrophy of pituitary prolactin cells thereby inhibiting their biosynthetic activity as observed in brook trout *Salvelinus fontnalis* under acute acid and aluminum exposure. In the present study also decline in prolactin level during zinc treatment at the end of 7th day may be due to the destruction of prolactin cells and/or high acute stress due to zinc toxicity.

Glucose is released by corticosteroids whose elevation has been described as a primary response to most stressors including heavy metals (Donaldson and Dye, 1975). Chan and Woo (1978) reported that cortisol has been shown to promote the catabolism of peripheral tissues which through increased gluconeogensis, leads to hyperglycemia. The hyperglycemic condition of plasma in the present study may be attributed to enhanced liver glycogenolysis or induced activation of adrenal pituitary glucocorticoid hormone which stimulates the hepatic glucose production thereby elevating glucose level or it may be a physiological response to meet the critical energy under toxic stress as suggested by Varley (1975), Sastry and Gupta (1978) and Shaffi (1978).

Towle (1981) reported that Na^+K^+ATPase is the prime indicator of ion transport across cellular membranes and play an important role in whole body ion regulation. According to Banks (1965) Na^+K^+ATPase is responsible for the regulation of membrane polarization in addition to playing a vital role in the release and uptake of biogenic amines in the central nervous system. In the present study the inhibition of Na^+K^+ATPase activity of gills of *Cyprinus carpio* during sub-lethal treatment with zinc sulphate may be due to change in physical properties of the cell membrane or alterations in the lipid content or disruption of oxidative phosphorylation within the cells.

References

APHA, 1971. *Standard Methods for the Examination of Water and Wastewater*, 13th edn. American Public Health Association, New York, USA, pp. 874.

Banks, P., 1965. *Biochem. J.*, 95: 490.

Barton, B.A. and Iwama, G.K., 1991. Physiological changes in fish from stress in aquaculture with emphasis on the responses and effects of corticosteroids. *Ann. Rev. Fish. Dis.*, 1: 3–26.

Bern, H.A. and Madsen, S.S., 1992. A selective survey of the endocrine system of the rainbow trout, *Oncorhynchus mykiss* with emphasis on the hormonal regulation of ion balance. *Aquaculture*, 100: 237–262.

Brouwer, A., Murk, A.J. and Koeman, I.H., 1990. Biochemical and physiological approaches on ecotoxicology. *Funct. Ecol.*, 4: 75–281.

Brown, J.A., 1993. Endocrine responses to environmental pollutants. In: *Fish Ecophysiology*, (Eds.), I.C. Rankin and F.B. Jensen. Champman and Hall, London, pp. 276–296.

Chan, D.K.O. and Woo, N.Y.S., 1978. Effect of cortisol on the metabolism of the eel, *Anguilla japonica. Gen. Compo Endocrinol.*, 35: 205–215.

Chavin, W., 1973. In: *Response of Fish to Environmental Changes,* (Ed.) W. Chavin. Springfield, Illinois, pp. 199.

Clark, W.C. and Bern, H.A., 1980. Comparative endocrinology of prolactin. In: *Hormonal Proteins and Peptides,* (Ed.) C.H. Li. Academic Press, New York, pp. 105–297.

Cooper, G.R. and McDaniel, V., 1970. *Standard Methods of Clinical Chemistry*, pp. 159.

Donaldson, E.M. and Dye, H.M., 1975. *J. Fish. Res. Bd. Can.*, 32: 533.

Donaldson, E.M., 1981. The pituitary-internal axis as an indicator of stress in fish. In: *Stress in Fish,* (Ed.) A.D. Pickering. Academic press, London, pp. 11–47.

Doyle, W.L. and Epstein, F.H., 1972. Effects of cortisol treatment and osmotic adaptation on the chloride cells in the eel, *Anguilla rostrata. Cytobiologie*, 6: 58–73.

Eugene, Jackim, 1974. Enzyme responses to metals in fish. In: *Pollution and Physiology of Marine Organisms*, (Eds.) F.J. Vernberg and W.B.Vernberg. Academic Press, New York, pp. 59–65.

Finney, D.J., 1978. In: *Statistical Methods in Biological Assay*, 3rd edn. Griffin Press, London, pp. 508.

Flik, G. Van den Velden, J.A., Seegers, H.M., Kolar, Z. and Wendelaar Bonga, S.E., 1989. Prolactin cell activity and sodium fluxes in tilapia *Oreochromis mossambicus* after long-term acclimation to acid water. *Gen. Comp. Endocrinol.*, 75: 39–45.

Flik, G. and Perry, S.F., 1989.Cortisol stimulates whole body calcium uptake and the branchial calcium pump in fresh water rainbow trout. *J. Endocrinol.*, 120: 75–82.

Folmar, L.C., 1993. Effects of chemical contaminants and blood chemistry of teleost fish: A bibliography and synopsis of selected effects. *Environ. Toxicol. Chem.*, 12: 337–375.

Fu, H. and Lock, R.A.C., 1990. Pituitary response to cadmium during the early development oftilapia, *Oreochromis mossambicus. Aquat. Toxicol.*, 16(1): 9–18.

Fu, H., Lock, R.A.C. and Wendelaar Bonga, S.E., 1989. Effect of cadmium on prolactin cell activity and plasma electrolytes in the fresh water teleost, *Oreochromis mossambicus. Aquatic. Toxicol.*, 1/14: 295–306.

Haya, K. and Waiwood, B.A., 1983. Adenylate energy charge and ATPase activity: Potential biochemical indicators of sub-lethal effects caused by pollutants in aquatic animals. In: *Aquatic toxicology*, (Ed.) J. Nrigau. John Wiley and Sons, New York, 13: 525.

Hontela, A., Rasmussen, J.B. and Chevalier, G., 1993. Endocrine responses as indicators of sub-lethal toxic stress in fish from polluted environments. *Water Poll. Res. J. Can.*, 28(4): 767–780.

Laurent, J. and Perry, S.F., 1990. Effects of cortisol on gill chloride cell morphology and ionic uptake in the fresh water trout, *Salmo gairdneri. Cell Tissue Res.*, 259: 429– 442.

Lidman, V., Dare. G., Johansson-sjbeck, M.L., Larsson, A. and Lewander, 1979. Metabolic effects of cortisol in the European eel, *Anguilla anguilla* (L). *Comp. Biochem. Physiol.*, 63A: 339–344.

Lorz, H.W., Williams, R.H. and Fustish, A.C., 1978. Effect of several metals on smooting of coho salmon *Oncorhynkus kisutusch. U.S. Environ. Prot. Agency, Environ. Res. Lab.*, Corvalli.

Matty, A.J., 1985. In: *Fish Endocrinology*. Croon Helm, London, pp. 267.

Mayer, F.L., Versteeg, D.J., Mckee, M.J., Folmar, L.C., Graney, R.L., McCume, D.C and Rattner, B.A., 1992. Physiological and non-specific biomarkers. In: *Biomarkers, Biochemical, Physiological and Histological Markers of Anthropogenic Stress*, (Eds.) Huggett, R.J., R.A. Kimerle., Jr. P.M. Mehrle and H.L. Bergman. In: *Proceedings of the 8th Pellston Workshop*, Keystone, Colorado, July 23–28, 1989. Lewis Publishers, Boca Raton, USA, p. 5–85.

Notter, M.F.D., Mudge, J.E., Neff, W.H. and Anthony, A., 1976. Cytophotometric analysis of RNA changes in prolactin and stannous corpuscle cells of acid stressed brook trout. *Gen. Comp. Endocrinol.*, 30: 273–284.

Passino, D.R.M., 1984. Biochemical indicators of stress in fishes: An overview. In: *Contaminant Effects on Fishes*, (Eds.), V.M. Cairns, P.V. Hodson and J.O. Nriagu. John Wiley and Sons Ltd., New York, pp. 37–48.

Perry, S.F. and Wood, C.M., 1985. Kinetics of branchial calcium uptake in the rainbow trout: Effects of acclimation to various external calcium levels. *J. Exp. Biol.*, 116: 411–433.

Rosseland, B.O. and Staurnes, M., 1994. Physiological mechanism for toxic effects and resistance to acidic water: An ecophysiology andecotoxicological approach. In: *Acidification of freshwater Ecosystem: Implication for the Future*, (Eds.) Steinberg, C.E.W. and R.F. Wright. John Wiley and Sons Ltd., pp. 227–246.

Sastry, K.V. and Gupta, P.K., 1978. Chronic mercuric chloride intoxication in the digestive system of *Channa punctatus*. *J. Environ. Pathol. Toxicol.*, 2: 443–446.

Shaffi, S.A., 1978.Variation in tissue glycogen content, serum lactate, glucose level due to copper intoxication in three fresh water teleost. *Curr. Sci.*, 47: 955–966.

Shiosaka, T., Okuda, H. and Fuji, S., 1971. Mechanism of phosphorylation of thymidine by the culture filtrate of *Clostridium perfringens* and rat liver extract. *Biochem. Biophys. Acta*, 246: 171–183.

Sivakumari, K., 1977. Modifying effects of acidified waters on cadmium toxicity with reference to blood chemistry of a freshwater fish *Cyprinus Carpio* var. *communis*. *Ph.D. Thesis*, Bharathiar University, Coimbatore, India.

Schreck, C.B. and Lorz, H.W., 1978. Stress response of coho salmon, *Onchorhyncus kisutusch* elicited by cadmium and copper and potential use of cortisol as an indicator of stress. *J. Fish. Res. Bd. Can.*, 35: 1124–1129.

Schreck, C.B., 1981. Stress and compensation in teleostean fishes: Response to social and physiological factors. In: *Stress in Fish*, (Ed.) A.D. Pickering. Academic Press, London, pp. 77–82.

Sheridan, M.A., 1986. Effect of thyroxine, cortisol, growth hormone and prolactin on lipid metabolism of coho salmon, *Onchorhyncus kisutusch*, during smoltification. *Gen. Comp. Endocrinol.*, 64: 220–238.

Skelly, D., Brown, L. and Besch, P. 1973. Radioimmunoassay. In: *Clin. Chem.*, 19(2): 146.

Spehar, R.L., Anderson and Fiandt, T.J., 1978. Toxicity and bioaccumulation of cadmium and lead in aquatic invertebrates. *Environ. Pollut.*, 15: 195–208.

Sprague, J.B., 1971. Measurement of pollution toxicity to fish–III. Sub-lethal effects and safe concentration. *Water Res.*, 5: 245–266.

Towle, D.W., 1981. Role of $Na^{+}K^{+}$ATPase in ionic regulation by marine and estuarine animals. *Mar. Biol. Lett.*, 2: 107–122.

Varley, H., 1975. *Practical Clinical Biochemistry*, 4th edn. Arnold Heinemann Publication (India) Pvt. Ltd., New Delhi, pp. 88–89, 309–311.

Waring, C.P., Brown, J.A., Collins, J.E. and Prunet, P., 1996. Endocrine response of brown trout, *Salmo trutta* L., exposed to lethal and sub-lethal levels of aluminium in acidic soft water. *Gen. Comp. Endocrinol.*

Weiss, M., Oddie, C.J. and McCance, T., 1979. The effect of ACTH on adrenal steroidogenesis and blood corticosteroid levels in the echidna, *Tachyglosus acuteatus*. *Comp. Biochem. Physiol.*, 64B: 65–70.

Wendelaar Bonga, S.E. and Vander Meij, J.C.A., 1981. Effect of ambient osmolarity and calcium on prolactin cell activity and osmotic water permeability of the gills in the teleost, *Sarotherodon mossambicus. Gen. Comp. Endocrinol.*, 43: 432–442.

Wendelaar Bonga, S.E. and Pang, P.K.T., 1991. Control of calcium regulating hormones in the vertebrates: Parathyroid hormone, calcitonin, prolactin and stanniocalcium. *Int. Rev. Cytol.*, 128: 139–213.

Wendelaar Bonga, S.E., 1997. The stress response in fish. *Physiol. Rev.*, 773: 592–625.

Yada, T., Takahashi, K. and Hirano, T., 1991. Seasonal changes in sea water adaptability and plasma levels of prolactin and growth hormone in land locked sockeye salmon, *Onchorhynchus nerka* and amago salmon *Oncorhynchus rhodurus. Gen. Comp. Endocrinol.*, 82: 33–44.

Chapter 45

Protoplast Fusion Between *Bacillus thuringiensis kurstuki 3a3b* and *Bacillus thuringiensis thompsoni H12* for Improved Biocontrol Activity

U.S. Bagde[1], B.V. Bilolikar[1] and R.S. Pandit[2]

[1]*Applied Microbiology Laboratory, Department of Life Sciences, University of Mumbai, Vidyanagari, Mumbai – 400 098, Maharashtra*
[2]*Department of Zoology, Sathaye College, Vile Parle (E), Mumbai – 400 057, Maharashtra., India*

ABSTRACT

Protoplasts of *Bacillus thuringiensis kurstuki* (*3a3b*) the toxic endospore producing organism were obtained by lysozyme treatment (7 mg/ml) and *Bacillus thuringiensis thompsoni* (*H12*) were obtained by lysozyme treatment (15 mg/ml). Almost equal number of protoplasts of both the organisms were mixed and allowed to fuse in the fusion fluid containing 40 per cent PEG. The fusants were allowed to regenerate their cell walls in regeneration medium. Regenerated fusants were selected on PAA (Penicillin Assay Agar) with both the markers *i.e.*, chloramphenicol and kanamycin. Among the many fusants with these desired characteristics few showed inproved biocontrol activity against fish pathogens. These hybrids were different in colony and cell morphology from their parents.

***Keywords**: Protoplast fusion, Bacillus thuringiensis, Biocontrol activity.*

Introduction

In the present study the two strains of *Bacillus thuringiensis 3a3b* and *Bacillus thuringiensis H12* were used for the study of protoplast fusion. Protoplast fusion is a versatile technique for inducing genetic recombination in a variety of prokaryotic and eukaryotic organisms, to produce novel strains of organisms with various capabilities. Protoplast fusion, is useful in generation of novel combination of genes even from organisms of different kingdoms. Unlike other methods of recombination among microorganisms, it gives the unique opportunity of bringing together two or more complete genomes instead of fractions of cell DNA. Considering positive aspects of protoplast fusion technique, it is being increasingly used for improving the abilities of microorganisms by carrying interstrain (Rosenberg and Breiter, 1989), interspecific (Schaeffer *et al.*, 1976) and intergeneric crosses (Chen *et al.*, 1987; Gokhale and Deobagkar, 1989; Bagde and Paranjpe, 2000) that resulted into creation of many novel hybrids.

The term protoplast is used to denote an osmotically sensitive cell only when it has been established that the entire cell wall has been removed (Bagde and Singh, 2000). The protoplast fusion is unique as a mode of genetic exchange in prokaryotes, because the transfer is bidirectional and the entire genomes are combined in the same cytoplasm at high frequency. Often within even the same species there are barriers to the successful establishment of heterokaryones, leave apart genetic recombination. Protoplast fusion remains to be a powerful tool in this regard.

The aim of the present investigation was to obtain genetic recombination between two bacterial strains *Bacillus thuringiensis kurstuki* (*3a3b*) and *Bacillus thuringiensis* (*H12*), using protoplast fusion technique to develop novel strains with improved antagonistic/Biocontrol activity against fish pathogens.

Materials and Methods

Bacillus thuringiensis sub. sp. *Kurstuki* (*3a3b*) and *Bacillus thuringiensis thompsoni* (*H12*) were obtained from NCIM Pune, India and maintained on Nutrient agar medium and Cellulose agar medium (Rosenberg and Breiter, 1989) and were used for protoplast fusion study. For the selection of markers, these strains were screened for antibiotic sensitivity by disk diffusion method. They were screened for sensitivity to kanamycin, chloramphenicol, etc. They were grown in Luria broth at 30°C and used to prepare protoplasts. Approximately Iml of *B. thuringiensis k* (*3a3b*) protoplast and 1 ml of *B. thuringiensis thompsoni* (*H12*) protoplast suspensions were mixed. The mixture was then centrifuged at 2000 ×g for 15 minutes at 10°C. Supematent was discarded and pellet was suspended in 0.2 ml Sodium Malate Medium (SMM). It was then diluted ten times with 40 per cent Polyethylene Glycol (PEG) 6000 (Himedia, India) and allowed to react at 30°C for 2 minutes. After 2 minutes PEG was diluted to 10 ml by adding smm. Then it was centrifuged at 3000 × g for 15 minutes. The pellets were resuspended in 1 ml of protoplast buffer and serially diluted 0.1 ml of the diluted fusion mixture was spread on Luria agar plates, incubated at 30°C for three to four days with daily counting of colonies.

The colonies on the Luria agar were replicated first on plane L-agar, using replica plate technique. After eight to ten generation replication on plain L-agar, the colonies were replicated on Penicillin Assay Agar (PAA) plate with chloramphenicol and were selected as novel fusants. This novel strain was further studied for morphological and cultured characteristics as well as for endospore formation and a Novel fusant was analyzed for its antagonistic activity against isolated fish pathogens.

Results and Discussion

Bacillus thuringiensis was rod shaped bacteria. Addition of lysozyme to the actively growing culture of *Bacillus thuringiensis* (*3a3b*) and *B. thuringiensis H12* in diluted Luria broth resulted in protoplast formation.

When equal amount of protoplasts of *Bacillus thuringiensis* (*3a3b*) and protoplasts of *Bacillus thuringiensis* (*H12*) were mixed in presence of 40 per cent PEG, protoplasts have undergone fusion (Figures 45.1–45.3). After PEG treatment either two or more protoplasts undergo fusion. When more than two protoplasts have undergone fusion, it resulted in the formation of multiple fusions (Figure 45.4).

When added in semisolid medium the fusants got regenerated (Figures 45.5 and 45.6). Many regenerated fusants were selected on PAA (Penicillin assay agar) with both the markers *i.e.*, chloramphenicol and kanamycin. Fusants having stability were selected for evaluation of probiotic activity.

Evaluation of Antagonistic Activity of Fusant Cultures

When checked for their efficiency to control isolated fish pathogens *viz.*, *Aeromonas salmonicida, Vibrio anguillarum biotype II, Vibrio anguillarum biotype I, Aeromonas hydrophila* sudspecies *hydrophila, Aerognmas hydrvhila, anaerogenes, Pseudomonas putida,* the fusants showed effective zone of inhibition on agar plates, by well diffusion method.

For *Aeromonas salmonicida* diameter of zone of inhibition was 39 mm. Against *Vibrio anguillarum biotype II* and *Vibrio anguillarum biotype I*, the diameters of zones of inhibitions were 37 mm and 39 mm respectively. For both *Aeromonas hydrophila* strains the zone of inhibition was 48 mm in diameter and for *Pseudomonas putida* the zone was 38 mm (Figure 45.7).

The novel strain showed increase in diameter of zone of inhibition by 69 per cent, 20 per cent, 27.58 per cent, 77.27 per cent, 37.14 per cent and 31.03 per cent towards *Aeromonas salmonicida, Vibrio anguillarum biotype II, V. anguillarum biotype I, Aeromonas hydrophila, subspecies hydrophila, Aeromonas hydrophila anaerogenes and Pseudomonas putida,* respectively when compared with *Bacillus thuringiensis* (*3a3b*) parent strain (Figure 45.7).

When compared with parent strain *Bacillus thuringiensis* (*H12*), the increase in diameter of zone of inhibition was 25.8 per cent, 100 per cent, 44.44 per cent, 23.07 per cent, 23.07 per cent and 80.95 per cent for *Aeromonas salmonicida, Vibrio anguillarum biotype II, Vibrio anguillarum biotype I, Aeromonas hydrophila, hrdrophila, Aeromonas hydrophila anaerogenes* and *Pseudomonas putida* respectively (Figure 45.7).

The fusant cultures were found to be highly antagonistic against fish pathogens when tested on plates. Thus, through protoplast fusion the antagonistic activity of *Bacillus thuringiensis* culture was improved.

Evaluation of Probiotic Activity of Novel Strain of *Bacillus thuringiensis* in Fish Aquarium Experiment

After testing for antagonistic effect on pathogens *viz.*, *Aeromonas salmonicida, Vibrio anguillarum biotype II, Vibrio anguillarum biotype I, Aeromonas hydrophila* subspecies *hydrophila, A.hydrophila* sub. sps. *anaerogenes* and *Pseudomonas putido* the novel strain was tested in vivo for its disease control activity by testing it in fish aquarium experiment. In fish aquarium the experiment was set up to evaluate the probiotic activity.

Figure 45.1: Fusion of Protoplasts after PEG Treatment

Figure 45.2: Fusion of Protoplasts of *B. thuringiensis* (3a3b) and *B. thuringiensis* (H12)

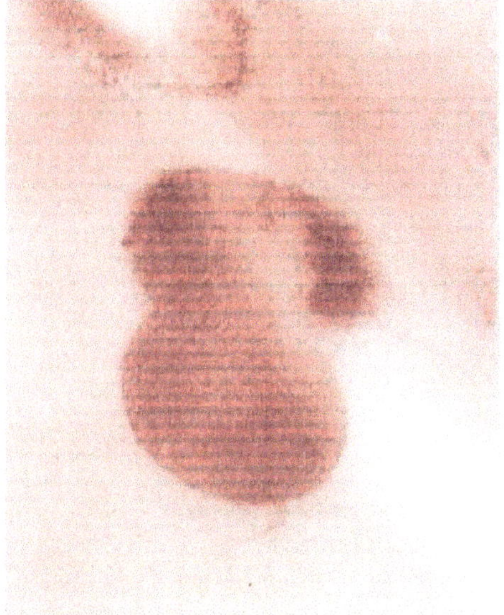

Figure 45.3: Protoplasts Undergoing Fusion

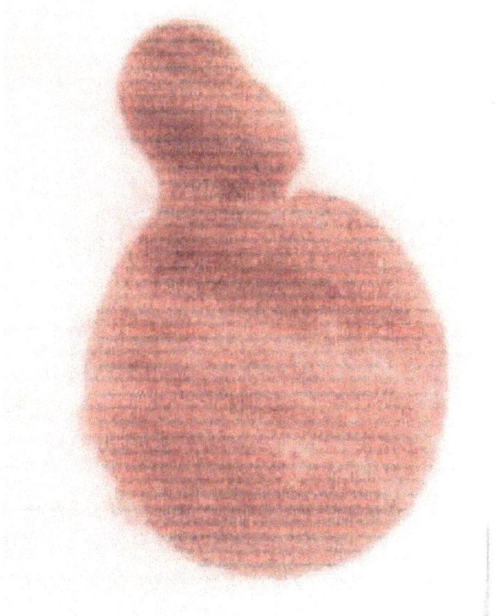

Figure 45.4: Multiple Fusion of Protoplasts

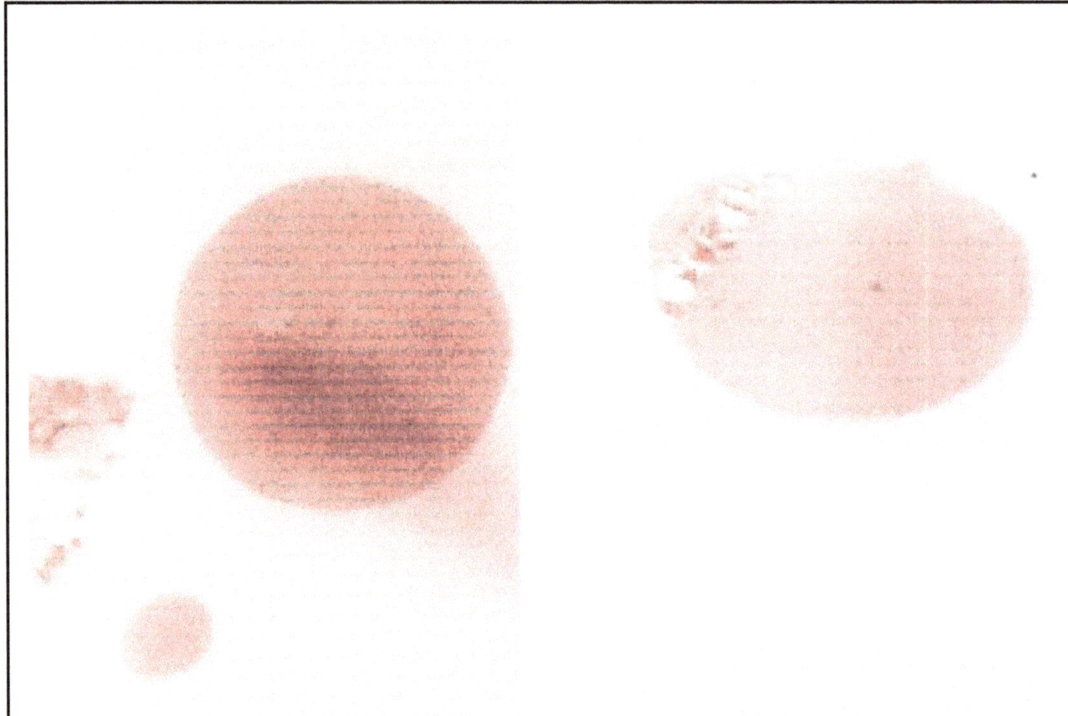

Figure 45.5: Regenerating Cell after Fusion

Figure 45.6: Fusant Cell Undergoing Regeneration

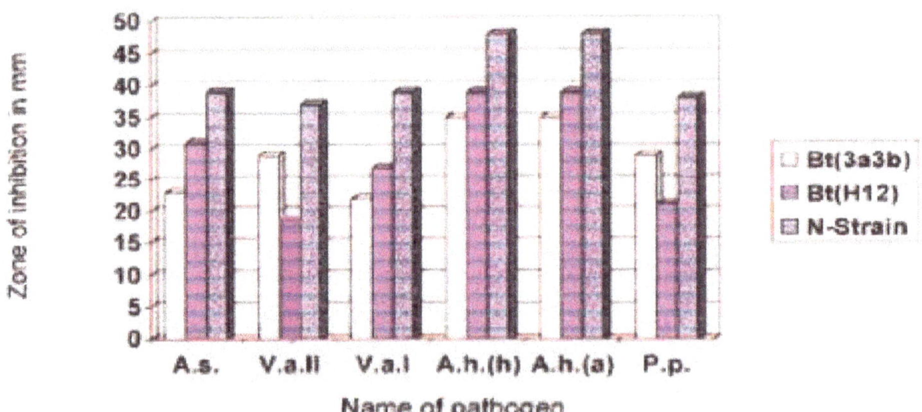

Figure 45.7: Comparative Study of Biocidal Affect of Regenerated Fusants and Parent Culture ***B. thuringiensis*** **(3a3b)**

Keywords: Bt(3a3b): *Bacillus lhuringiensis* **(3a3b); Bt(H 12):** *Bacillus thuringiensis* **(H12); N-Strain: Novel strain developed by protoplast fusion; A.s.:** *Aeromonas salmonicida;* **V.a.II:** *Vibrio anguillarum biotype II;* **V.a.I:** *Vibrio anguillarum biotype I;* **A.h.(h):** *Aeromonas hydrophila hydrophila;* **A.h(a):** *Aeromonas hydrophila anerogenes;* **P.p:** *Pseudomonas pulida.*

Fishes when challenged with *Aeromonas salmonicida,* in control tank: (without treatment with probiotic strain) developed disease in a week with symptoms like tail fin rot, obscaling patches, and sluggishness in movements. The first fish died in eight days showing hemorrhagic patches on liver in control tank (Table 45.1). The pathogenic bacterial count did not show sharp decrease (Count decreased from 24×10^6 to 9×10^4) and 100 per cent morality was observed in 18 days in control tank while it was nil (0 per cent) in probiotic treated tank. There was a sharp decrease from 47×10^6 to 880 in pathogenic bacterial count in treatment tank while probiotic count was observed to remain constant from 88×10^4 to 24×10^3 in 45 days (Table 45.2). The fishes were normal with average SGR (specific growth rate) of 30 per cent per day.

Table 45.1: *In vivo* Experiment to Check the Biocontrol Activity of Novel Strain of *Bacilus thuringiensis* Developed by Protoplast Fusion in Fish Aquarium using *Tilapia* Species Fishes

Organism Name	*Tank Name*	*Weight of Fish in gm.*	*Time*	*Death Rate*	*Symptoms Developed*
Aeromonas sa/monicida	Control	6 to 11	18 days	100 per cent	Tail fin was deterorated ulcerative patches
	Treatment	8 to 10	45 *days	NIL	NIL
Vibrio anguillarum biotype II	Control	8 to 10	25 days	100 per cent	Ulcerative patches and body fins were degraded
	Treatment	8 to 10	45 *days	100 per cent	NIL
Vibrio anguillarum biotype I	Control	5 to 8	30 days	100 per cent	Red patches on trunk mucus from mouth and gills
	Treatment	6 to 10	45 *days	NIL	NIL
Aeromonas hydrophila hydrophila	Control	8 to 10	7 days	100 per cent	Fins are deteriorated, red patches on trunk
	Treatment	9 to 12	45 *days	20 per cent	No diseased symptoms were observed
Aeromonas hydrophila anaerogenes	Control	10 to 12	10 days	100 per cent	Tail fins were weathered away
	Treatment	9 to 12	45 *days	NIL	NIL
Aeudomonas putida	Control	10 to 15	9 days	100 per cent	Body and tail fins were deteriorated
	Treatment	8 to 10	45 *days	NIL	NIL

*: Denotes last day of the experiment not the death of the fish.

In presence of novel probiotic strain when fishes were challenged with *Vibrio anguillarum biotype II* in treatment tank, the mortality rate was found to be 20 per cent after 35 days of experiment with SGR of 0.025 per cent per day. In control tank the fishes developed ulceration patches in 18 days of incubation with pathogen (Table 45.1). The mortality started after 20 days and reached 100 per cent mortality in 25 days, in case of control tank and there was no sharp decrease in pathogenic bacterial count (from 29×10^6 to 29×10^4 in 18 days). There was a sharp decrease in pathogenic count from 52×10^7 cfu/ml to 300 cfu/ml in treatment tank where as the probiotic count was maintained from 36×10^4 to 69×10^2 in 45 days (Table 45.2).

In the challenged trial, using *Vibrio anguillarum biotype I* in control tank, the fishes developed de scaling patches on the trunk, redness and ulcerative patches at the base of body fins and the first fish died on 15th day of inoculation and 100 per cent mortality was achieved in 28 days (Table 45.1). There was no sharp decrease (from 68×10^6 to 56×10^4 cfu/ml) in pathogenic count in control tank, but in case of treatment tank, there was a sharp decrease from 39×10^5 to 20 cfu/ml, while probiotic count was observed to remain more or less constant from 90×10^3 cfu/ml to 8×10^2 cfu/ml (Table 45.2). In probiotic treatment, the mortality was nil and fishes were healthy with average SGR of 4.0 per cent per day after 45 days.

Table 45.2: Biochemical Analysis of Fish Aquarium Used for Evaluation of Novel Strain of *Bacillus thuringiensis* as Probiotic Strain

A. salmonicida	*Control Tank*		*No. of Days*		*Treatment Tank*		*No. of Days*	
	0	*7*	*14*	*18*	*0*	*15*	*30'*	*45*
pH	8.3	8.3	8.6	8.1	8.4	8	8.1	7.8
Temp. °C	22	20	21	23	23	21	22	20
D.O.	8.1	6.2	5.3	4.9	8.3	6.4	5.8	4.9
Pathogen	24×10^6	18×10^4	12×10^4	9×10^4	47×10^6	22×10^3	11×10^2	880
Bt. Count cfu/ml.	–	–	–	–	88×10^4	34×10^4	2×10^3	24×10^3
V. anguillarum Biotype II	*Control Tank*		*No. of Days*		*Treatment Tank*		*No. of Days*	
	0	*15*	*25*	–	*0*	*15*	*30*	*45*
pH	8.2	8.1	7.9	–	8.2	7.9	8.1	8
Temp. °C	22	21	22	–	20	23	24	21
D.O.	8.1	6.5	5.5	–	7.8	7.6	5.2	4.4
Pathogen	29×10^6	28×10^5	29×10^4	–	52×10^7	41×10^4	58×10^2	300
Bt. Count cfu/ml.	–	–	–	–	36×10^4	31×10^4	5×10^4	69×10^2
V. anguillarum Biotype I	*Control Tank*		*No. of Days*		*Treatment Tank*		*No. of Days*	
	0	*15*	*30*	–	*0*	*15*	*30*	*45*
biotype I								
pH	8.2	8.1	7.9	–	8.2	8.2	8	8
Temp. °C	22	23	22	–	20	21	22	24
D.O	7.9	6.4	5.1	–	8.2	6.4	5.9	5.2
Pathogen	68×10^6	36×10^5	56×10^4	–	39×10^5	85×10^3	27×10^2	20
Bt. count cfu/ml.	–	–	–	–	90×10^3	9×10^2	34×10^2	8×10^2
A. hydrophila hydrophila	*Control Tank*		*No. of Days*		*Treatment Tank*		*No. of Days*	
	0	*7*	–	–	*0*	*15*	*30*	*45*
pH	8.1	7.9	–	–	8.1	7.8	7.9.	7.8
Temp. °C	24	22	–	–	23	22	22	24
D.O.	8.3	5.3	–	–	7.8	6.2	5.5	4.8
Pathogen	28×10^5	16×10^5	–	–	43×10^4	51×10^4	32×10^2	500
Bt. count cfu/ml.	–	–	–	–	84×10^4	55×10^3	57×10^2	22×10^2

Contd...

Table 45.2–Contd...

A. hydrophila anaerogenes	*Control Tank*		*No. of Days*		*Treatment Tank*		*No. of Days*	
	0	*15*	*10*	*45*	*0*	*15*	*30*	*45*
pH	8.2	8.0	8.1	–	8.1	8.3	8.2	8.1
Temp. °C	21	23	22	–	20	22	23	23
D.O.	8.1	6.6	5.7	–	8.2	6.5	5.3	4.6
Pathogen	57×10^7	39×10^6	11×10^5	–	56×10^6	56×10^4	31×10^2	105
Bt. count cfu/ml.	–	–	–	–	59×10^3	25×10^3	79×10^2	52×10^2
P. putida	*Control Tank*		*No. of Days*		*Treatment Tank*		*No. of Days*	
	0	*9*	*–*	*–*	*0*	*15*	*30*	*45*
pH	8.2	8	–	–	8.2	8.1	8.1	7.8
Temp. °C	21	20	–	–	21	22	24	22
D.O.	8.1	6.7	–	–	8.1	6.3	5.6	4.3
Pathogen	82×10^6	44×10^5	–	–	57×10^7	24×10^4	10×10^2	560
Bt. count cfu/ml.	–	–	–	–	71×10^3	28×10^3	58×10^2	15×10^2

D.O.: Dissolved oxygen; Bt. Count: *Bacillus thuringiensis* (*3a3b*) count; cfu/ml: Colony forming units per ml, Temp.: Temperature.

Aeromonas hydrophila sub. sp. *hydrophila* inoculated fishes showed fm determination, and red patches on trunk after three days of inoculation with pathogen in control tank (Table 45.1) on third day, the mortality was 20 per cent and in seven days it was 100 per cent in control tank with no sharp decrease in bacterial count *i.e.*, 28×10^5 cfu/ml to 16×10^5 cfu/ml. but in treatment tank a sharp decrease from 43×10^6 cfu/ml to 500 cfu/ml was observed in pathogenic bacterial count and the probiotic bacterial count was found to be stable from 84×10^4 to 22×10^2 (Table 45.2). In treatment tank the mortality was 20 per cent after 35 days of inoculation with pathogen in the presence of probiotic.

In the control tank, inoculated with *Aeromonas hydrophila* sub. sp. *anaerogenes*, 100 per cent mortality was observed. The fishes showed weathering of tail fins, rejection of food, sluggish movements in control tank. In treatment (Probiotic treated) tank the mortality was nil and specific growth rate was 3.6 per cent per day in 45 days (Table 45.1). Also the pathogenic bacterial count was found to remain constant from 57×10^7 to 1×10^5 in 10 days. In treatment tank there was a sharp decrease in pathogen count from 56×10^6 cfu/ml to 10^5 cfu/ml while probiotic cell count was more or less constant from 59×10^3 cfu/ml to 52×10^2 cfu/ml (Table 45.2).

Pseudomonas putida inoculated fishes showed body and tail fins determination and rejection of feed with 100 per cent mortality in 9 days. The symptoms and mortality was nil in treatment tank even after 45 days (Table 45.1). The average SGR value was found to 2.8 per cent per day in 45 days. In control tank the pathogenic bacterial count was found to be 2.8 per cent per day in 45 days. In control tank the pathogenic bacterial count was found to remain constant from 82×10^6 cfu/ml to 44×10^5 cfu/ml in nine days. While it was found to decrease sharply from 57×10^7 cfu/ml to 560 cfu/ml in 45 days, but probiotic cell count was found to remain constant from 71×10^3 cfu/ml to 15×10^2 cfu/ml (Table 45.2).

For all the control and treatment tanks the temperature variation was from 20 to 25°C depending on the room temperature and tank conditions. The pH showed variation in the range of 7.8 × 8.4, dissolved oxygen decreased from the range of 7.9 to 8.4 to the range of 4.3 to 5.1 depending on the growth of bacterial mass in the tank (Table 45.2).

Today India is one of the foremost countries in inland fish production, probably next to China. However, diseases caused by bacteria generally result in large scale mortalities of fishes and consequently reduce the economic benefits of the fish farmers (Lakshman *et al.*, 1986). The use of antibiotics in aquaculture may introduce potential hazards to public health and to the environment by the emergence of drug resistant microorganisms and antibiotic residues. Disease control of fishes needs a new approach, which is both cost effective and environmentally safe.

Tilapia spp. Fishes were collected from Lakes of Thane district Maharashtra. These water bodies were polluted with high bacterial load and had very less dissolved oxygen thereby the, fish like Tilapia spp. Suffered from diseases. The diseased fish sample of Tilapia was collected, dissected and from the diseased parts five bacterial were pathogens. The details of these isolates were studied and were identified as *Aeromonas salmonicida, Vibrio anguillarum* biotype II, *Vibrio anguillarum* biotype I, *Aeromonas hydrophila* sub.sps. *hydrophila* and *Aeromonas hydrophila* sub.sp. *anaerogenes* using Bergey's manual. These pathogens proved, postulates of Koch using Tilapia fishes.

References

Alikhanian, S.I., Ryabchenko, N.O., Bukanov N.O. and Sakahyan, V.A., 1981. Transformation of *Bacillus thuringiensis* subsp. *Galleria* protoplasts by plasmid pBC16. *J. Bacteriol.*, 146(1): 7–9.

Bagde, U.S. and Paranjape, 2000. Protoplast fusion between *Cellulomonas fimi* and *Brevibacterium divaricatum. Indian J. Environ and Ecoplan.*, 3(3): 573–578.

Chen, W., Nagashima, K. and Kajino, T., 1988. Intergeneric protoplast fusion between *Ruminococcus albus* and an anaerobic recombinant FE7. *Appl. Environ. Microbiol.*, 54: 124–125.

Chen, W., Ohmiya, K. and Shimizu, S., 1987. Inter generic protoplast fusion between *Fusobacterium varium* and *Enterococcus faecium* for enhancing dehydrodevanillin degradation. *Appl. Environ. Microbial.*, 53: 542–548.

Ferenczy, L., 1981. Microbial protoplast fusion. In: *Genetics as a Tool in Microbiology*, (Eds.) Glovan, S.W. and K. Hopwood. Cambridge University Press, Cambridge, p. 1–67.

Gokhale, D.V. and Deobagkar, D.N., 1989. Essential expression of Zylonaseb and endoglucanases in the hybrid derived from intergeneric protoplast fusion between a *Cellulomonas* sp. and *Bacillus subtilis. Appl. Environ. Microbiol.*, 55: 2675–2680.

Lakshmanan, M., Sundar, K. and Lipton, A.P., 1986. Isolation and characterization of *Aeromonas hydrophila* subsp. *hydrophila* causing haemolytic disease in Indian major carp, *Labeo rohita* (Ham.). *Curr. Sci.*, 55(21): 1080–1081.

Martin, P.A.W., Lohr, J.R. and Dean, D.H., 1981. Transformation of *Bacillus thuringiensis* protoplast by plasmid deoxyribonucleic acid. *J. of Bacteriol.*, 195(32): 980–983.

Putambekar, U.S. and Ranjekar, P.K., 1989. Intergeneric protoplast fusion between *Agrobacterium tumerfaciens* and *Bacillus thuringiensis* subsp. *kurstuki. Biotechnology Letter*, 11: 717–722.

Rodriques, H., Garcia, B., Ancheta, O. and Sipiizki, M., 1991. Formation, regeneration and fusion of protoplasts in *Cellulomonas* strain. *Curr. Microbiol.*, 23: 265–270.

Rosenberg, F.A. and Breiter, H., 1989. Role of cellulolytic bacteria in digestive processes of shipworm. *Material and Organism*, 4: 147–149.

Scaeffer, P., Cami, B. and Hotchkiss, R.D., 1976. Fusion of bacterial protoplasts. In: *Proc. Nat. Acad. Sci., U.S.A.*, 73: 2151–2155.

Sung, Nackie, Duch Hua Chung and Young, Lee Mu, 1988. Development of L-lysine producing strains from cellulosic substrates by intergeneric protoplast fusion: Conditions for formation and regeneration of protoplasts. *Koren J. Appl. Microbiol. Bioeng.*, 18: 150–155.

Chapter 46

Hydrography of a Visapur Dam, Near Ahmednagar, Maharashtra

***A.K. Pandarkar*[1] *and U.H. Mane*[2]**

[1]*Department of Zoology, New Arts, Commerce and Science College, Ahmednagar – 414 001, Maharashtra*
[2]*Department of Zoology, Dr. Babasaheb Ambedkar Marathwada University, Aurangabad – 431 004, Maharashtra*

ABSTRACT

Visapur dam is situated on Hanga River. This river originates from Pamer taluka (about 30 km) away from north west of the dam. The river receives water from many other natural sources before reaching Visapur village. This dam was completed in June 1927, mainly to fulfill drinking water supply to British colony and later for irrigation and fishing purpose. From December 1986 the dam receives water from Kukadi canal arriving from Pune district. This dam is completely filled throughout the year due to inflow from Kukadi canal. In this dam fishes are abundant. This study forms a part to approach principal aspects of fish culture. Several parameters such as water temperature, dissolved oxygen, alkalinity, hardness, pH, chlorides, sulphates, phosphates and nitrates have been studied. The seasonal variations in the above parameters were studied over a period of one year beginning from October 1995. The present study was initiated so as to monitor the base line data on the suitability of water for the fish and fisheries practices, so that changes if any in the future could be evaluated.

Keywords: *Visapur dam, Physico-chemical conditions, Fish culture.*

Introduction

Various abiotic-biotic factors in ponds and lakes play an important role for augmenting the yield capacity significantly and therefore, it is necessary to determine the dynamic effects of environmental factors on fish growth (Sarkar, 1991). The environmental variability arso strongry influences the fish population. (Rekhow *et al.*, 1987 and Freeman *et al.*, 1988). In seasonally-breeding fish many environmental factors and physico-chemical characteristics of water have been implicated in the initiation of maturational and reproductive events and of these, photoperiod, temperature and nutritional status appear to be the most important with subsidiary influences imparted by salinity, rainfall and pheromones in some species (whitehead *et al.*, 1978).

Visapur dam was chosen for the study where several edible fish species occur and fishing is done regularly. There is tremendous scope for enhancing inland fish production in this water body through scientific management. In the view to popularize the research work on edible fish species and to bring to the notice to those concerned in the cultural aspects of the fishes in Maharashtra state, the present study was undertaken.

Materials and Methods

The dam is located at village Visapur, Tal Shrigonda of Ahmednagar district Maharashtra state. The dam is 38 km from south of Ahmednagar city. It is located between 18° 501 to 18°551 North latitude and 74°351 to 74°401 East longitude, Southern part of Ahmednagar district. The dam is 841 deep (max) and the length is 7940 with a capacity 1136 million cubic feet. This dam was constructed on Hanga river in June 1927, mainly to fulfill drinking water supply to British colony and later for irrigation and fishing purpose;. From December 1986 the dam receives water from Kukadi canal arriving from Pune district. This dam is completely filled not only in monsoon season but also mostly in rest of the seasons due to inflow from Kukadi canal. To get a base line data on the dam hydrography, water samples were collected every fortnight in the morning between 9 10 am and analyzed for temperature, dissolved oxygen, alkalinity, hardness, pH, chlorides, sulphates, phosphates and nitrites, for a period of one year from October 1995 to September 1996 using standard methods described by Golterman *et al.* (1977) and APHA (1985).

Results and Discussion

Visapur dam is fed by water coming from Kukadi irrigation canal at every month and Hanga river during rainy season. The dam remains full of water not only in monsoon season but also mostly in rest of the seasons, but in summer water level decreases due to less amount of water coming from Kukadi canal.

The yearly data obtained on water parameters was divided into four seasons, representing post-monsoon (October–November), winter (December–February), summer (March–May) and monsoon (June–September) (Table 46.1 and Figure 46.1). The water analysis revealed that the water temperature throughout year varied from 24 to 34°C, being minimum in winter and maximum in summer, indicating that the water and air temperature were found to go more or less hand in hand. Dissolved oxygen varied from 1.25 to 9.88 ml/l, being minimum in summer and maximum in monsoon. Low DO during summer was probably due to two reasons, in summer at high temperature rate of oxidation of organic matter in water increased and oxygen was consumed in the *process*, secondly at higher temperature the water had a lesser oxygen holding capacity and surplus oxygen was lost to the atmosphere (Munawar, 1970; Singh and Sahai, 1979; Saha and Pandit, 1986; Yeole and Patil, 2005; Pawar and Mane, 2006). High DO observed in monsoon was probably due to circulation and mixing of atmospheric oxygen during down pour. Gupta *et al.* (1992), Subbamma and Ramasarma (1992) and Kulkarni *et al.* (1995) reported such high level of dissolved oxygen in fresh water. The annual range of alkalinity was 50–250 mg/l, being low in monsoon and high in summer season. Decrease in alkalinity during monsoon was obviously due to dilution (Mishra and Yadav, 1978), High total alkalinity values indicated the higher trophic status (Sarwar and Wazir, 1991). The annual total hardness was ranged between 24–129.3 mg/l, being low in monsoon season and high in winter. These values are comparatively low than those reported by many other workers from India. Low hardness values were also noticed by Forsyth and McColl (1975) and Jayaramaraju and Kalavati (1986) attributed high

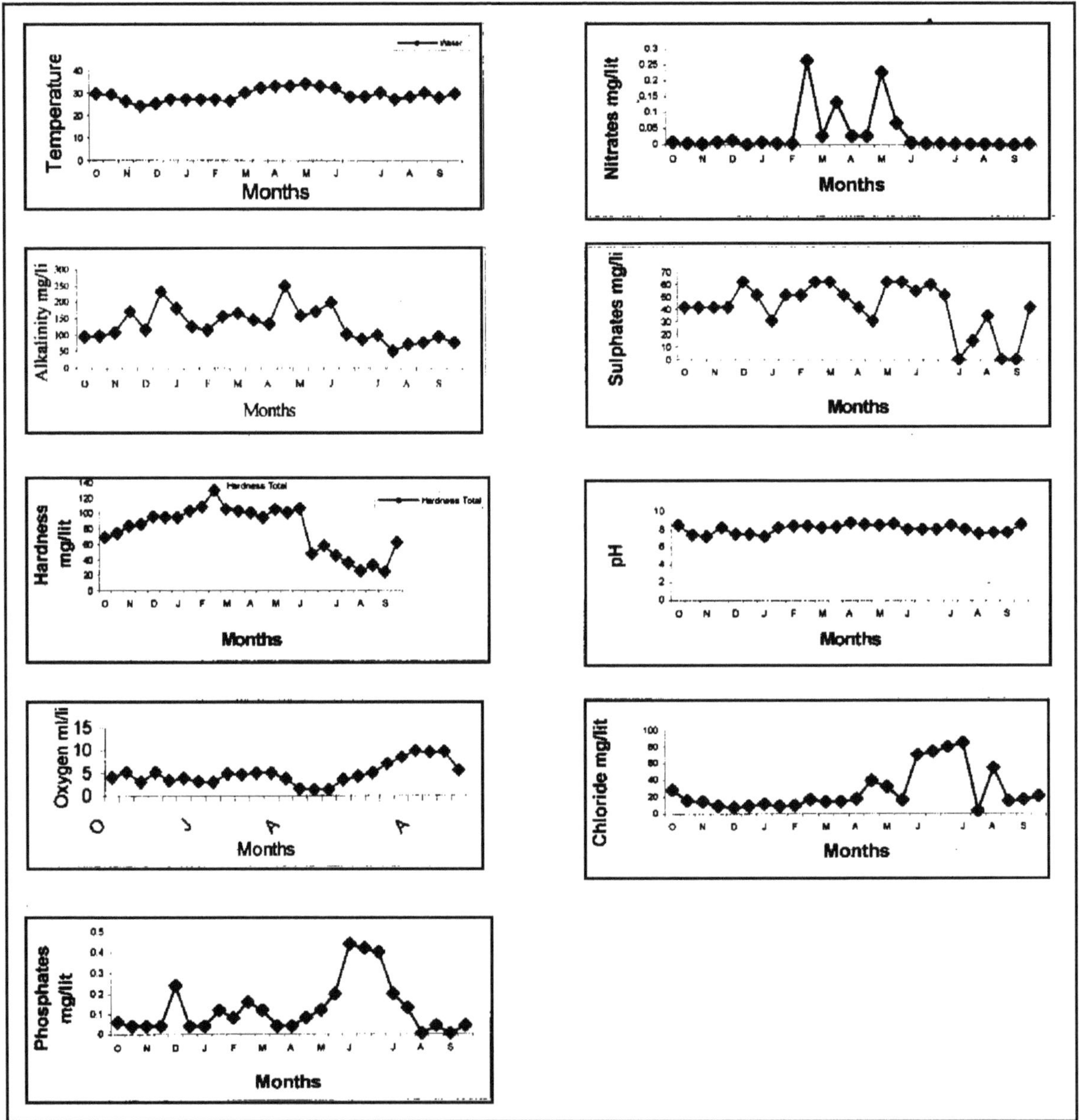

Figure 46.1: Monthly Variations in the Physico-chemical Characteristics of Water of Visapur Dam

Calcium values to the use of agricultural fertilizers like lime and super phosphate which might also be contributing towards total hardness in the water body as fertilizers are used in the catchments to enrich the vegetable fields. Higher chloride content of water was cited as an index of pollution of animal origin (Munawar, 1970), in summer due to the high temperature and high rate of evaporation. Saha and Pandit (1986), Pawar and Mane (2006) have reported lower chloride values in unpolluted ponds. In the present study throughout study period chloride values were comparatively low ranged between 7.09 to 84.97 mg/l being minimum in winter and maximum in monsoon season. Low chloride values might be due to the absence of pollution from animal or human origin, and also likely to be due to continuous availability of water in Visapur dam by feeding Kukadi canal. The annual sulphates range of Visapur dam was found to be 00 to 61.72 mg/lit. Its average minimum value was observed in monsoon, while maximum value recorded in winter and summer seasons. High values of sulphates during summer might be due to the high temperature and rate of evaporation Kulkarni *et al.* (1995). Phosphate ranged between 00 to 0.44 mg/lit, being average minimum 0.045 mg/l in post-monsoon and average maximum 0.157 mg/lit in monsoon, which probably due to influx through rain water (Michael, 1969; Munawar, 1970). The annual range of nitrites was 0.00110–0.2632 mg/lit. The average minimum value was 0.00248 mg/lit in monsoon and a maximum of 0.1105 mg/l. in winter. This is in agreement with findings of Sarwar and Wazir (1991), Gupta *et al.* (1992) and Subbamma and Ramasarma (1992).

Table 46.1: Seasonal Variations in the Physico-chemical Conditions of Visapur Dam

Sl.No.	*Parameters*	*Post–monsoon (Oct–Nov)*	*Winter (Dec–Feb)*	*Summer (Mar–May)*	*Monsoon (June–Sept)*
1.	Temperature °C (Water)	27.13±1.97 (24.0–29.5)	26.92±1.46 (25.0–28.0)	32.5±1.68 (30.0–34.0)	28.89±1.88 (27.0–32.0)
	Temperature °C (Air)	29.88±2.79 (24.0–35.0)	27.33±2.3 (25.0–32.5)	27.17±2.38 (22.0–30.0)	30.94±2.3 (26.5–34.0)
2.	Dissolved Oxygen (ml/lit)	4.33±1.28 (2.87–5.18)	3.76±1.14 (2.94–4.78)	2.93±1.62 (1.25–4.97)	6.95±2 (4.2–9.88)
3.	Alkalinity (mg/lit)	117.33±7.33 (94–170)	153.4±9.13 (114–232)	170.3±9.13 (132–250)	94.93±10.2 (50–197.2)
4.	Hardness mg/lit (Total)	77.4±3.5 (68–85.3)	103.67±5 (93.6–129.3)	101.05±2.8 (94–105)	48.17±7.6 (24–105.6)
5.	Hardness mg/lit (Carbonate)	6±2.8 (3–13.7)	9.13±3.14 (3.6–17.6)	8.67±1.9 (5–10)	5.27±2.7 (0.3–10.7)
6.	pH	7.83±0.96 (7.2–8.5)	7.88±o.9 (7.25–8.4)	8.52±0.7 (8.2–8.8)	8.02±1.2 (7.57 –8.6)
7.	Chloride (mg/lit)	16.84±3.7 (9.22–28.36)	10.63±2.7 (7.09–17.02)	22.22±4.2 (14.18–38.99)	49.64±7 (14.99–84.97)
8.	Sulphates (mg/lit)	41.15±0 (41.15–41.15)	51.43±4.7 (30.86–61.72)	51.43±4.7 (30.86–61.72)	32.06±5.7 (14.40–60.27)
9.	Phosphates (mg/lit)	0.045±0.12 (0.04–0.06)	0.113±0.3 (0.04–0.16)	0.10±0.34 (0.04–0.20)	0.157±0.53 (0.04–0.44)
10.	Nitrites (mg/lit)	0.005±0.06 (0.001–0.007)	0.111±0.43 (0.003–0.263)	0.095±0.38 (0.026–0.227)	0.003±0.06 (0.001–0.006)

Seasonal means±S.D. are given. (Bracket values represent range of variations).

In the present study the physico-chemical condition of the Visapur dam described above, exhibits some basic characteristics, which suggest suitability of water for pisciculture, High values of total alkalinity (above 50 mg/lit) alkaline pH, high values of dissolved oxygen (6.95 mg/lit) in monsoon and the total hardness (24–129.3 mg/lit), indicate that the water is suitable for fish Culture. The values of chloride (7.09–84.97 mg/lit), Sulphates (00–67.72 mg/lit) and phosphates (00–0.44 mg/lit) showed that the water is quite suitable for fish survival and growth.

The comparison of water quality of Visapur dam with limits laid down by fresh water quality criteria for fish and fisheries practiced by Subbamma and Ramasarma (1992) and Chandra Prakash (2001) suggest that the water parameters of Visapur dam are with in the permissible limits for fish and fisheries practices. Since, the dam at Visapur is almost free of any visible pollution and also devoid of any industry in the vicinity it is possible to intensify the cultural activities of the edible fish species.

Acknowledgement

The author Pandarkar, A.K. is thankful to the Management, A.J.M.V.P.S, Ahmednagar, Principal, New, Arts, Commerce and Science College, Ahmednagar, Dr. Aher S.K., Head of the Zoology Department and Professor Khose R.G. for providing the laboratory facilities and encouragement for research

References

APHA, 1985. *Standard Methods for the Examination of Water and Wastewater*, 15th edn. American Public Health Association, Washington.

Chandra Prakash, 2001. Status of soil and water quality parameters in brood stalk management. In: *Course Manual CAS Training Programme on Brood Stock Management and Genetic Selection in Fish Seed Production* (February–March), CIFE, Mumbai.

Forsyth, D.H. and McColl, R.H.S., 1975. Limnology of lake Nagahewa, North Island, New Zealand. *NZ. J. Freshwat. Res.*, 9 (3): 311–322.

Freeman, M.C., Crawford, M.K., Barett, J.C., Facey, D.E., Flood, M.G., Hill, J., Stducer, D.J. and Grossman, G.D., 1988. Fish assemblage stability in a southern application stream (USA). *Can. J. Fish Aqua. Sci.*, 45(1): 1949–1958.

Golterman, H.L., Chlymo, R.S. and Ohanstand, M.A.M., 1978. *Methods for Physical and Chemical Analysis of Freshwater*, 2nd edn. IBP Handbook No. 8, Blackwell Scientific Pub. Oxford, London, Edinburgh, Melbourne, pp. 172–178.

Gupta, S., Micheal, R.G. and Gupta, A., 1992. A preliminary investigation on physico-chemical factors, periphyton and invertebrate communities in a protected water works of Shillong, Meghalaya. In: *Proc. Nat. Acad. Sci., India*, 62(B): 3.

Jayaramaraju, P. and Kalavati, C., 1986. Ecological studies on the ciliates associated with hydrophytes in a fresh water lake at Kondakarla (Visakhapatnam). *Uttar Pradesh J. Zool.*, 6(2): 201–206.

Kulkarni, S.D., Mokashe, S.S. and Patil, R.P., 1995. Diurnal changes in physico-chemical characteristics of Sadatpur reservoir. *J. Aqua. Biol.*, 10(1): 21–23.

Mishra, G.P. and Yadav, A.K., 1978. A comparative study of physico-chemical characteristics of river and lake water in central India. *Hydrobiol.*, 59(3): 275–278.

Munawar, M., 1970. Limnological studies on freshwater ponds of Hyderabad, India. *Hydrobiologia*, 31: 101–128.

Pawar, B.A. and Mane, U.H., 2006. Hydrography of a Sadatpur lake, near Pravaranagar, Ahmednagar District, Maharashtra. *J. Aqua. Biol.*, 21(1): 101–104.

Rekhow, K.H., Robert, W.B., Thomas, B.S., and Udithwood, J., 1987. Empherical models of fish response to lake identification. *Can J. Fish. Aqual. Sci.*, 44(8): 1432–1442.

Saha, L.C. and Pandit, B., 1986. Comparative limnology of Bhagalpur ponds. *Comp. Physiol. Ecol.*, 11(4): 213–216.

Sarkar, S.K., 1991. Dynamics of aquatic ecosystem in relation to fish growth exposed to ammonium sulphates. *J. Environ. Biol.*, 12(1): 37–43.

Sarwar, S.G. and Manzoor, A. Wazir, 1991. Physico-chemical characteristics of a freshwater pond of Srinagar (Kashmir). *Poll. Res.*, 10(4): 223–227.

Subbamma, D.V. and Ramasarma, D.V., 1992. Studies on the water quality characteristics of a temple pond near Machilipatnam, Andhra Pradesh. *J. Aqua. Biol.*, 7: 22–27.

Whitehead, C., Bromage, N.R. and Forster, J.R.M., 1978. Seasonal changes in reproductive function of rainbow trout (Salmo-gairdneri). *J. Fish Biol.*, 12: 601–608.

Yeole, S.M. and Patil, G.P., 2005. Physico-chemical status of Yedshi lake in relation to water pollution. *J. Aqua. Biol.*, 20(1): 41–44.

Chapter 47

Polysaccharides, Proteins and Lipids from Basidiomycetes Fungi

J. Vinaya Sagar Goud, M. Jaya Prakash Goud and M.A. Singaracharya

Department of Microbiology, Kakatiya University, Warangal, A.P., India

ABSTRACT

Fruit bodies of 40 basidiomycetes species were screened for the presence of polysaccharides, proteins and lipids. Maximum amounts were observed in *Fornes fomentarious* 187.2 (m/g), *Grifola berkely* 265.4 m/g and *strobilomyces* sp. 4.1 m/g respectively. High amounts of polysaccharides and proteins are recorded in six species and lipids in nine species. The study forms the basis for selection of potent species of medicinal importance.

Keywords: *Basidiomycetes, Proteins, Polysaccharides, Human health.*

Introduction

Badidiomycetes have been considered by human kind as an edible and medicinal resource. A number of bioactive molecules have been identified in many basidiomycetes. Higher basidiomycetes represent an unlimited source to cure different human ailments with their Bio-molecules such as polysaccharides, proteins and lipids (Wasser, 2002). Potent polysaccharides and proteins are wide spread among higher basidiomycetes. Most of them have unique structure in different species. Medicinal mushrooms have been intensively investigated for medicinal effect *in vitro* and *in vivo* model systems. Many new antitumor immuno modulatory polysaccharides have been identified and put in to practiced polysaccharidied, proteins and lipid content (Wasser and Weis, 1999).

Materials and Methods

40 species of basidiomycetes were collected from forests of Andhra Pradesh. The fruit bodies were dried in shade and packed in air tight containers. They were identified based on the Friesian classification system (1874). Estimation of biomolecules were done according to standard procedures (Fokh *et al.*, 1957)

Results and Discussion

The identified basidomycetes species and their biomolecular content via, proteins, polysaccharides, lipids are depicted in Table 47.1, maximum polysachrides content was recorded in *Fomes fomentarius* 187.2 (m/g), followed by *Lenzites sepoiaria* 171.6 m/g and *Lepiota* sp. (168.4). Minimum amount (8.3) was observed in *clavatia cyathiformis.* Out of 40 species studied six species are with high polysaccharide content two with moderate amount and 32 with low polysachharide content. A vide range of polysaccharides of different chemical from basidiomycetes had been investigated (Kiho, 1992). These are the best known and most potent mushroom derived, substances with anti tumour and immunomodulating properties (Mizuno 1996). *Grifola frondosa* is one of the most popular medicinal mushrooms. Fruit body of this mushroom contain B-(1→6)-glucan, (Mizuno *et al.*, 1986). Fornes fomentarious has anti-bacterial, anti-viral properties (Brandt and Piraino, 2000). *Grifola frondosa* has Antibacterial, immunomodulatory, antitumor (Altshul, 2001).

Table 47.1: Polysaccharides, Proteins and Lipids (mg/g) in the Fruit Bodies of Higher Basidiomycetes

Macro Fungi	*Polysaccharides*	*Protein*	*Lipid*
Amuroderma rugsom	32.7	113.1	3.9
Armillaria mellea	13.5	169.2	0.9
Bolbitius vitellinus	158.0	102.5	0.4
Bovista sp.	15.6	252.4	2.4
Ceripora viridans	26.0	96.1	2.4
Clavatia cynthiformis	8.3	57.2	1.4
Coriolopsis occidentalis	21.8	37.7	0.7
Daedalea flavida	30.1	99.8	2.6
Dichomitus squalens	29.6	133.3	1.5
Fistulina sp.	26.0	169.2	2.9
Fomes fomentarius	187.2	251.6	1.4
Geaster sp.	13.5	242.0	0.6
Geastrum sp.	10.2	42.0	1.1
Grifola berkely	135.2	265.4	1.2
Innotus sp.	18.7	91.5	0.7
Lactarius volemus	26.0	96.0	1.5
Laetiporus sulphuerus	31.7	74.5	2.2
Lentinus edodes	11.4	85.5	0.6
Lenzites betulina	31.7	155.0	3.4

Contd...

Table 47.1–Contd...

Macro Fungi	*Polysaccharides*	*Protein*	*Lipid*
Lenzites sepiaria	171.6	252.0	1.8
Lepiota sp.	168.4	89.2	0.9
Lycoperdon perlatum	166.4	67.4	1.5
Merulius sp.	18.7	97.0	2.4
Mycena chlorophanus	62.4	105.9	0.3
Nidularia sp.	20.8	1.3	0.8
Oxyporus populnus	42.6	155.2	1.9
Phalous sp.	10.4	86.1	1.3
Phellinus ignaaricus	20.8	47.3	0.8
Pleurotus ostreatus	11.4	208.8	1.2
Podaxis pistillaris	39.5	158.2	1.3
Polyporus squalosas	47.8	115.9	1.6
Poria obliqua	24.9	67.4	0.9
Pycnoporus cinnabarinus	79.0	65.7	1.4
Russula sp.	10.4	46.0	1.0
Schizophyllum commune	26.0	169.2	1.4
Scleroderma	52.0	76.1	1.2
Strobilomyces	35.3	127.4	4.1
Trametes versicolar	36.4	77.2	1.5
Tulostoma sp.	82.3	36.1	1.9
Tyromyces sp.	12.4	1.4	1.2

Grifola berkely (265.4 mg/g) was recorded with high amount of protein followed by *Bovista* sp. (252.4 mg/g), *Lenzites sepiaria* 252.0 (mg/g) and least reorded in *Nidularia* sp. (1.3 mg/g). Mushrooms are an important source of edible protein for human consumption. More than 2000 species of mushroom exist in nature but only approximately 22 species are intensively cultivated fro commercial purposes, on soil or wood and utilizing particular environmental and nutritional conditions. Mushrooms are rich in protein mushrooms they can be a good supplement, mushrooms offer themselves as potential protein source the great advantage is that mushrooms has the capacity to convert nutritional value less substances in to high protein food (Yudiz *et al.,* 2005). In ancient cultures such as the Indian Greek and Roman mushrooms have been described as sophisticated delicacies associated with royal class (Manju, B 1995). Maximum amount of lipids was recorded in *strobilomyces* sp. (4.1 mg/g) *Amauroderma rugosam* (3.9 mg/g) *Lenzites betulna* (3.4 mg/g) and low amount in *Mycena Chlorophanus* (3.0 mg/g) overall, high amount of lipids are recorded in 9 moderate in 20 species and less in 11 species. In general fruiting bodies have been found to contain much higher proportions of lipids compared with Vegetative mycelia. Lipids could help basidiomycetes adapt to low growth temperature. These unsaturated fatty acids also may have a role in lignin degradation (Ana Gutierrez *et al.,* 2002).

Among the higher basidiomicetes, thoroughly studied prganisms may be around 10 per cent. The number of mushrooms with known pharmacological qualities by their biomolecules is much lower.

Recent development in biotechnology can be applied to transfer genes encoding bioactive molecules from unculturable species into culturable species. If utilized substantially the basidiomycetes containing pharmacological important molecules can be a treasure for every economy.

References

Altshul, S., 2001. Mushroom remedy for chronic yeast infections. *Prevention Magazine*, 1: 53–60.

Ana Gutierrez, Jose C., del Rio, Maria Martinez J., Maria Mardinez J. and angel Martinez, T., 2002. Production of New unsaturated lipids during wood decay by lignolytic basidiomycetes. *Applied Environmental Microbiology*, 3: 1344–1350.

Brandt, C.R. and Piraino, F., 2000. Mushroom antivirals. *Recent Research Development for Antimicrobial Agents and Chemotherapy*, 4: 11–26.

Folch, J.M., Lees and Stanely, G.H.S., 1957. A simple method for the isolation and purification of total lipids from animal tissues. *J. Biol. Chem.*, 226: 497–509.

Ikekawa, T., Ikeda, Y., Yoshikova, Y., Nakanishi, M., Yokoyama, E. and Yamazoki, E., 1982. Anti tumour polysachasides of Flammulina veludipoes 2. The structure of EA-3 and further purification of EA-5. *J. Poharmacobiol Dyn.*, 5: 576–581.

Kiho, T., Nagai, Y.S., Sakushima, M. and Ukai, S., 1992. Polysacchorides in fungi. XXIX. Structural features of two antitumour polysaccharides from the fruiting bodies of *Armillaria tabscens. Chem. Pharm. Bull.*, 40: 2212–2214.

Manju, B., Vadher, S. and Soni, G., 1995. Nutritional evaluation of *pleurotus florida. Mushroom Research*, 5: 101–104.

Mizumo, T., Ohsawa, K. Hagiwava, N. and Kuboyama, R., 1986. Fraction and characterization of antitumour polysaccharides from maitake, *Grifola frondosa. Agric. Biol. Chem.*, 50: 1679–1688.

Mizuno, M.C., 2000. Anti-tumour polysaccharides from mushrooms during storage. *Biofactors*, 12: 275–281.

Mizuno, T., 1996. Development of antitumour polysachrides from mushroom fungi. *Foods Food Ingred. J. Jpn.*, 167: 69–85.

Mizuno, T., Ashowa, K., Hagiwava, N. and Kuboyamo, R., 1986. Fractionation and characterization of antitumour polysaccharides from maitake, *Grifola frondosa. Agric. Biol. Chem.*, 50: 1679–1688.

Wasser, S.P., 2002. Medicinal mushrooms as a source of antitumour and immuno modulating polysaccharides. *Appl. Microbil. Biotechnol.*, 60: 258–274.

Wasser, S.P. and Weif, A.L., 1999. Medicinal properties of substances occurring in higher Basidiomycetes mushrooms: Current perspectives. *Int. J. Med. Mushrooms*, 1: 31–62.

Yoshida, I., Kiho, T. Usui, S. Sakushima M. and Ukai S., 1996. Polysaccharides in fungi. XXXVII. Immunomodulasing activities of carboxymethylated derivatives of linear (1→3)–alpha–D–glucans extracted from the fruiting bodies of *Agrocybe cylindracea* and *Amanita muscaria. Bio. Pharm. Bull.*, 19: 114–121.

Yudiz, A., Yesil, O.F. and Yavuz, O., 2005. Organic elements and protein in some macro fungi of south east Anatolia in Turkey. *Food Chemistry*, 89: 605–609.

Chapter 48

Studies on a Comparative Note on Physico-chemical Features of the Raw and Treated Textile Bleaching Effluent and its Heavy Metal Toxicity on *Cyprinus carpio*

M. Manimegalai, S. Swapna, S. Umavathi and T.C. Greeshma

Department of Zoology, Kongunadu Arts and Science College, Coimbatore – 641 029, Tamil Nadu, India

ABSTRACT

The effluents discharged from textile bleaching factory has shown higher values of physico-chemical parameters like Temperature, TDS, pH, Alkalinity, BOD, COD and Nutrients were found to be increased in the river Bhavani at Mettupalayam. Bio accumulation of heavy metals like copper, lead and hexavelant chromium and their LC_{50} values were detected in fish *Cyprinus carpio* maintained in sublethal concentration of textile bleaching effluent for a period of 30 days.

Keywords: *Cyprinus carpio, Textile bleaching effluent, Physico-chemical parameters, Heavy metals.*

Introduction

Clean drinking water is essential for human requisite for the sustenance of life. Clean water is also indispensable condition for the development of fishery resources. Out of total global water only three per cent in the form of freshwater, is suitable for human use. Even this limited amount has been put under tremendous pressure of pollution due to stupendous increase in population resulting in rapid urbanization, industrialization and agriculture (Chatterjee, 2000). Most of the developmental activities use natural resources as raw material and waste generated is disposed into different environmental media. The signs of stress on the natural resources are evident from the deteriorating air quality, soil degradation, polluted rivers and streams and general status of the environment in various regions. All the major rivers in India are facing severe water pollution due to rapid industrial growth, which encourages the establishment of manufacturing installations on the banks of rivers where water is readily available. The more important types of industries whose effluents contribute to the pollution of river system in India are pulp and paper, textiles, tanneries, sugar, distilleries, shellac, petrochemical and miscellaneous industries. Improper modes of effluent discharge in the water bodies are polluted and they carry deadly substances (Muthusamy and Jayabalan, 2001). In most cases huge amount of diverse nature of effluent released from varied industries is disposed in open environment causing pollution of soil and water bodies (Kumar *et al.*, 2003). Principally the human activities also add up the heavy metals to the water systems in the urban areas. The chemicals forms of these heavy metals in water are accessible to the biota through significant accumulation in the food chain that may subsequently cause harmful effects to living organisms (Godwin, 2004).

The discharge of waste which contains strong acids, alkalies render the waters corrosive to river structure such as pipelines, concrete and metal parts of structures and treatment plants. Toxic metals like chromium, copper, lead, mercury found in untreated and partially treated industrial effluents contribute massive kills of fish and other aquatic biota (Prasad, 2002). Pollution created by these industries are mainly attribute to various waste liquor coming out of operations in wet processing such as desizing, scoring, bleaching, mercerizing, dyeing, printing, and finishing. These effluents when discharged into water receiving body without adequate treatment can cause irreversible changes (Sharma and Arora, 2001).

Materials and Methods

Source of Effluent

Both treated and untreated effluents were collected from the United Bleaching Factory. This factory locally known as UBL located at Mettupalayam in Coimbatore District, Tamil Nadu. It discharges its treated effluent through an open outlet into river Bhavani, which is a major tributary of Cauvery. The rate of discharge is 8 lakhs liters per day.

Physico-chemical Features of the Effluent

Composite sample were collected in plastic bottles with optimum care from two different sites representing raw and treated effluent. The pH and temperature were recorded by using pH meter and mercury thermometer at the spot itself. Biochemical Oxygen Demand (BOD), Chemical Oxygen Demand (COD), Dissolved Oxygen (DO), Total Solids (TS), and Total Suspended Solids (TSS) were measured by the Standard methods (APHAA WW A- WPCF, 1985). Calcium and magnesium were estimated by versenate titration and sodium by flame photometer. Carbonate, Sulphate, Nitrate, Silicate, Phosphate and Chloride were determined by the methods suggested by Trivedy and Goel (1984). Total lead, $Chromium^{+6}$ and Copper were determined using atomic absorption spectrophotometer after wet digestion with HNO_3 and $HCLO_4$ (4 : 1 v/v) mixture.

Evaluation of LC_{50}

Toxicity of both treated and raw effluent to *Cyprinus carpio* was studied by employing static bioassay method as described by Trivedy *et al.* (1987). Fishes were subjected to preliminary range finding test to determine a mortality range of 0–100 per cent in various concentration of effluent. Based on the preliminary exploratory studies, concentrations of effluent ranging from 26.0 per cent to 29.0 per cent for treated and from 16.6 per cent to 19.0 per cent for raw were selected for the present toxicity studies.

Twenty liters of sample water with a particular concentration of effluent was taken in a rectangular glass jar. Ten fishes were introduced into each jar. Mortality of fish was noted at the end of 24 hours in each concentration of effluent. The effluent water in the jar was renewed for every 6 hours (with least disturbance of the fish) to maintain the dissolved oxygen content and the effluent concentration through out the experiment period. Control fishes were also maintained in effluent free unchlorinated tap water. The number of fish survived at the end of 24 hours was recorded in both control and effluent media. The percent mortality values were calculated. LC_{50} values of bleaching factory effluent to *Cyprinus carpio* for 24 hours were obtained by employing probit analysis of Finney (1964) as described in detailed by Busvine (1971).

Bioaccumulation of Heavy Metals

Heavy metals *viz.*, copper, lead, mercury and chromium^{+6} in tissues were estimated according to the methods detailed by Ayyadurai *et al.* (1994), Rao *et al.* (1999). The fish samples were dried at constant temperature of 120°C for 48 hours and made into a fine powder. Tissue digestion of the dried, powdered samples was carried out by ashing the samples in a muffle furnace at 550°C for 2 hours. The ash was further digested using concentrated nitric acid and perchloric acid. The digest was analyzed for copper, hexavalent chromium and lead by Atomic Absorption Spectrophotometer. The estimation was done at 10th, 20th and 30th day of exposure in ppm wet weight.

Results and Discussion

The textile bleaching effluent collected from united bleaching factory selected as for the present study was yellowish brown in colour and the treated effluent was pale yellow. Villegas-Navarro *et al.* (2001) absorb that wastewater from bleaching station and textile processing was beige in colour. The colour of the effluent are bacteriostatioc in nature and many ofthese are toxic to microorganism (Sharma and Arora, 2001). The raw and treated had an ammonia odour. This foul smell as a social impact upon the people near by the factory. The various physico-chemical parameter of both raw and treated effluent were measured according to APHA-AWWA-WPCF (1985) and Trivedy and Goel (1984) and are given in the Table 48.1 along with the tolerance limit (ISI, 1981).

The temperature and pH levels of raw and treated effluents were observed. The temperature of the raw effluent was 35°C while collecting and the treated effluent was 28° C. Both are below the ISI tolerance limit. The impact of effluents did not show any influence on the temperature levels of the river water. The pH of the raw effluent was 11.64 and the treated effluent it reduced to 11.31. But both the sample were highly alkaline in nature compared to ISI permit range from 5.5 to 9 pH for these effluents that could be release into natural water resources. High pH was due to the presence of chemicals like hypochlorite, chlorine, caustic soda that are used during the bleaching process (Rajagopalan, 1990). Jacob *et al.* (1999) found that pH of Noyyal river water due to the textile industries in Tirupur areas was slightly alkaline. The pH and carbonate levels are not considerably altered after the treatment. The high alkalinity was also in correlation with the amount of carbonate in raw (2560 mg/l) and treated (2490 mg/l) effluent.

Table 48.1: Physico-chemical Characters of the Raw Textile Bleaching Effluent

Parameter	*Raw Textile Bleaching Effluent*	*Treated Textile Bleaching Effluent*	*Per cent Reduction Per cent Increase* Over Raw*	*ISI Tolerance Limit*
Colour	Yellowish	Pale Yellow	–	Colourless
Odour	Ammonia	Ammonia	–	Odourless
Temperature	35	28	–	40
pH	11.64	11.31	2.84	5.5 to 9.0
Electrical conductivity (microm hos/cm)	6650	196	97.05	–
Carbonate	2560.0	2490.0	2.73	–
Chloride (Cl)	960.06	907.52	5.47	1000.0
BOD	900.0	210.0	76.67	30.0
DO	0.05	4.0	98.75*	6.0
COD	2720.0	1040.0	61.76	250.0
TS	2350.0	1678.0	28.60	500.0
TSS	265.0	184.0	30.57	100.0
Organic carbon (%)	0.03	BDL**	–	–
Ammoniacal nitrogen	7.6	5.2	31.58	50.0
Nitrate (NO_3)	5.0	2.6	48.00	–
Phosphate (PO_4)	2.96	1.98	33.11	5.0
Silcate	260.0	100.0	61.54	–
Sulphate (SO_4)	2.5	0.92	63.20	1000.0
Sodium (Na)	2806.12	437.5	84.41	6.0
Calcium	18.0	14.0	22.22	75.0
Magnesium (Mg)	7.2	2.4	66.67	50.0

All the values are expressed in mg/l, except temperature, pH, electrical conductivity and organic carbon.

**: Below Detectable Limit.

Electrical Conductivity (EC) is total parameter for dissolved and disassociated substances and indicates the concentrations of dissolved electrolytes. The permissible value of conductivity for drinking water 300 mhos/cm (USPHS). Electrical Conductivity (EC) of the raw effluent was 6650 μmhos/cm. High electrical conductivity shown more cations and anions in soluble form in the effluent. On treatment, EC reduced to 196 μmhos/cm, the percentage of reduction was 97.05. When EC was values exceed 3000 μmhos/cm, the germination of almost all the crops would be affected and it may result in much reduced yield (Lokhande *et al.*, 1996). The carbonate content of the raw effluent was 2560 mg/l. The treated effluent contained a carbonate content of 2490 mg/l. The pH and carbonate levels not altered after treatment Srivastava, *et al.* (1994) observed alkaline nature of river Tons due to discharge of carbonates. Alkalies destroy bacteria and microbes thus, inhibit the self purification and are harmful for crops as well as impair their growth (Sharma and Arora, 2001).

The raw effluent showed a chloride content of 960.06 mg/l. The amount of chloride content present in the treated effluent was 907.52. The chloride content showed a reduction of only 5.47 per cent over raw effluent. The concentration of chloride is considered to be the indicator of salinity and pollution, it will give salty taste to water. High chloride concentrations were reported in the groundwater in Tirupur area due to the input from textile industries (Jacks *et al.*, 1994).

Dissolved Oxygen (DO) are important parameter to determine water quality and also assessing the biological activities in a water system. BOD and COD are the most important parameters used to determine degree of pollution of river water. In the present study BOD and COD of the raw and treated effluent were about the tolerance limit. The BOD of the raw effluent was 900 mg/l. The treated effluent showed BOD of 210 mg/l; which was reduced by 76.67 per cent compared to that of raw effluent. The Dissolved Oxygen (DO) content of the raw effluent was 0.05 mg/l and that of treated effluent was 4 mg/l. The dissolved oxygen content was increased to 98.75 per cent on treatment. Clean surface water is normally saturated with DO (7.6 mg/l at 30°C. The COD of treated of the raw effluent was 2720 mg/l. The COD of treated effluent was 1040 mg/l; the values are higher than the standard limits of industrial as advised by ISI. The present observation is in accordance with the report of Kumar *et al.* (2001). Mohanty *et al.* (2003) evaluated the relationship between BOD and COD for river Nagavalli and Kolab and noticed their impact as the proliferation of bacteria in the river waters.

The Total Solids (TS) 2350 mg/l and suspended solids 265 mg/l was high in raw effluent and showed no remarkable decrease even after treatment (1678 mg/l and 184 mg/l respectively) and they were relatively greater than the tolerance limits (500 mg/l and 100 mg/l respectively). High suspended solids were also reported in tannery and Sago waste. High-suspended solids contribute for turbidity in the receiving water, there by affecting biological productivity by preventing light penetration and more than 350mg/l will affect the fish adversely by causing mechanical injury to the gills (Anima *et al.*, 2002; Rao and Rao *et al.*, 2002). The organic content of the raw effluent was 0.03 per cent. The organic carbon content of the treated effluent was below detectable limit. The amount of ammonical nitrogen present in the raw and treated effluent was 7.6 mg/l and 5.2 mg/l, respectively. The percent reduction was 31.58 per cent.

Heavy metals present in the raw and treated textile bleaching effluents were determined according to the methods detailed by Khan and Weis (1992), Ayyadurai *et al.* (1994). The concentration values of heavy metals enlisted in Table 48.2. The values are higher than the standard limits of industrial effluents as advised by ISI. The amount of iron present in the raw effluent was 12.95 mg/l. The amount of iron present in the treated effluent was 8.35 mg/l. The percent reduction in the amount of iron was 35.52mg/l compared to that of raw effluent. The amount of hexavalent chromium present in the raw effluent was 1.35 mg/l and in the treated effluent was 0.74 mg/l; which was lower (45.19 per cent) than that of raw effluent. The amount of copper present in the raw effluent was 2.31mg/l. The amount of copper present in the treated effluent was 0.93 mg/l. The quantity of lead content of the raw effluent and treated effluent was 4.80 mg/l and 2.90 mg/l respectively.

Toxicity of both treated and raw textile bleaching effluent to *C. carpio* was studied by employing static bioassay method as described by Trivedy *et al.* (1987). The fish were subjected to preliminary range finding tests to determine a mortality range of 0-100 per cent in various concentrations of raw and treated effluent percentage. The median lethal concentration was calculated according to Finney's method of probit analysis. The LC_{50} values of the raw and treated textile bleaching effluent at 24 hrs of exposure to the fish, *C. carpio* were 17.58 and 27.23 per cent respectively (Tables 48.3–48.4).

Table 48.2: Heavy Metal Concentration of the Textile Bleaching Effluent

Parameter	*Raw Bleaching Textile Effluent*	*Treated Bleaching Textile Effluent*	*Per cent Reduction Over that of Raw Effluent*	*ISI Tolerance Limit*
Iron (Fe)	12.95	8.35	35.52	0.3
Chromium^{+6} (Cr^{+6})	1.35	0.74	45.19	0.1
Copper (Cu)	2.31	0.93	59.74	3.0
Lead (Pb)	4.80	2.90	39.58	0.1

All values are expressed in mg/l.

Table 48.3: LC_{50}/24 Hours Value of Raw Textile Bleaching Effluent to *Cyprinus carpio*

Concentration of Effluent (Per cent)	*Correlation (Per cent) Kill*	*Log (+) Dose*	*Emprical Probit*	*Expected Probit*
19.0	90	1.28	6.28	6.25
18.6	80	1.27	5.84	5.80
18.2	70	1.26	5.52	5.40
17.8	60	1.25	5.25	4.95
17.4	40	1.24	4.74	4.50
17.0	20	1.23	4.15	4.10
16.6	10	1.22	3.71	3.70

LC_{50} = 17.58 (1.245); Fiducial limits (95 per cent confidence) UL = 17.95 (1.254)

LL = 17.23 (1.236); Values in parameters are in log dose.

Table 48.4: LC_{50}/24 Hours Value of Treated Textile Bleaching Effluent to *Cyprinus carpio*

Concentration of Effluent (Per cent)	*Correlation (Per cent) Kill*	*Log (+) Dose*	*Emprical Probit*	*Expected Probit*
29.0	95	1.46	6.64	5.9
28.5	80	1.45	5.84	5.5
28.0	70	1.45	5.52	5.5
27.5	55	1.44	5.12	5.0
27.5	35	1.43	4.61	4.55
26.5	20	1.42	4.15	4.10
26.0	10	1.41	3.71	3.65

LC_{50} = 27.23 (1.435); Fiducial limits (95 per cent confidence)

UL = 17.95 (1.443); LL = 17.23 (1.427)

Values in parameters are in log dose.

Concentration of Ca^{+2} and Mg^{+2} ions of raw and treated effluents are 18 mg/l; 14 mg/l and 7.2 mg/l, 2.4 mg/l respectively. Nitrate values found to be 5 mg/l for raw effluent and 2.6 mg/l treated effluent. The quantity of phosphate ion for raw and treated effluent has been found to be 2.96 mg/l and 1.98 respectively. The amount of silicate present in the raw effluent was seems to be 260 mg/l and for the treated effluent was 100 mg/l respectively. The high concentration of silicate in the effluent may be due to the usage of sodium silicate during the bleaching process. Thus, the above data of physico-chemical parameters and concentrations of toxic metals are higher than the acceptable limits (Borole and Patil, 2004).

In the present investigation, *Cyprimus carpio* was exposed to different sub-lethal concentrations like 1.5 per cent, 2.0 per cent and 2.5 per cent of textile bleaching effluents for a period of 30 days and in accumulation of copper, lead and hexavalent chromium was detected (Table 48.5). In the present study, copper was found to be considerably accumulated in the body of fish. A maximum accumulation of 1.6 ppm, copper 0.2 ppm, of lead and very trace amount of chromium was observed when exposed to 25 per cent effluent. Many investigators have reported the accumulation of heavy metals in fish species under polluted conditions (Rao and Patnaik, 1999; Misra *et al.*, 2002).

Table 48.5: Heavy Metal Concentrations in the Whole Body of *Cyprinus carpio* Exposed to Different Concentrations of Textile Bleaching Effluent After 30 Hrs

Concentration (%)	*Copper*	*Lead*	*Chromium*
Control	0.12	0.001	Nil
1.5	0.7	0.13	Nil
2	1.0	0.16	Nil
2.5	1.6	0.22	0.02

All values are expressed in ppm.

Singh and Rai (2003) studied the impact of the industrial effluents and domestic sewage on river Ganga at Allahabad and reported that all the pollution parameters are beyond the permissible limits and unfit for human consumption, Tiwari (2004), studied the pollution potential of river Pandu contaminated heavily by the discharges of various industrial effluents. Rajyalakshmi and Srilatha (2005) studied the physico-chemical parameters of river Goutami-Godavari flowing across Yanam reported the negative impact of agro-chemical parameters of river Ganga and Gomti with reference to human activity. Kulshresta and Sharma (2006), Chetna *et al.* (2006) observed the impact of mass bathing during festive days and analysed the bacteriological water quality in different rivers in India.

It can be suggested that the effluent discharged into the river should be properly treated using modem techniques to remove various toxic constituents present in it so as to avoid damage to fish life and prevent deterioration of down stream water supplies. If some internal measures are adopted by the factory, the pollution load can be reduced to a great extent. These measures include the use of peracetic acid instead of hypochlorite as bleaching chemical, peroxide bleaching with organic stabilizers and replacement of acetic acid by formic acid which will generate less BOD. Recently, a new German invention for one step preparatory process named Reco-yet Remisch kleinwefers is launched for textile processing. This works on the principle of water aerosol and needs less volume of water results into less pollution.

References

Anima, S. Dadhich, Avasnmaruthi, Y. and Srinivas, P., 2002. The fate of chromium in water and soil: A case study of Andhra Pradesh tanneries. *J. Ecotoxicol. Environ. Monit.*, 12: 139–145.

APHA, AWWA and WPCF, 1985. *Standards Methods for the Examination of Water and Wastewater*, 16th edn. American Public Health Association, American Water Association and Water Pollution Control Federation, Washington, D.C.

Ayyadurai, K., Swaminathan, C.S. and Krishnasamy, V., 1994. Studies on heavy metal pollution in the fin fish, *Oreochromis mossambicus* from river Cauvery. *Indian J. Environ. Hlth.*, 36: 99–103.

Borole, D.D. and Patil, P.R., 2004. Studies on Physico-chemical parameters and concentration of heavy metals in sugar industry. *Poll. Res.*, 23: 83–86.

Busvine, J.R., 1971. A critical review of the techniques for testing insecticide. *Common wealth Agricultural Bureaux,* England, pp. 267–282.

Chatterjee, C., 2000. Physico-chemical studies of water quality of the river Nunia at Asanol industrial area. *J. Environ. Poll.*, 7: 259–261.

Chetna Anand, Pratima Alolkar and Chakrabarthi, R., 2006. Bacteriological water quality status of River Yamuna in Delhi. *J. Environmental Biology*, 27: 97–101.

Finney, D.J., 1964. *Probit Analysis: A Statistical Treatment of the Sigmoid Curve*. Cambridge University Press, London.

Godwin, W.S., 2004. Bioaccumulation of heavy metals by the intertidal molluscs of Kanyakumari waters. *Poll. Res.*, 23(1): 37–40.

ISI, 1981. *Indian Standard Tolerance Limits for Industrial Effluents*. Indian Standard Institution, New Delhi, IS: 2490 (Part I), pp. 7–11.

Jacob, T.C., Azariah, J. and Viji Roy, A.G., 1999. Impact of textile industries on river Noyyal and riverine groundwater quality of Tirupur, India. *Poll. Res.*, 18: 359 – 368.

Khan, A.T. and Weis, J.S., 1992. Bioaccumulation of heavy metals in two populations of mummichog (*Fundulus heteroclitus*). *Bull. Environ. Contam. Toxicol.*, 48: 1–5.

Kulshresta, H. and Sharma, S., 2006. Impact of mass bathing during Arth Kumbh on water quality status of river Ganga. *J. Environmental Biology*, 27: 437–440.

Kumar, S.RD., Narayanaswamy, R. and Ramakrishnan, K., 2001. Pollution studies on sugar mill effluent: Physico-chemical characteristics and toxic metals. *Poll. Res.*, 20: 93–97.

Lokhande, R.S., Pokale, S.S. and Regi Thomas, 1996. Physico-chemical aspects of pollution in water in some coastal areas of Shrivarcdhan (Maharashtra) India. *Pollution Research,* p. 403–406.

Misra, S.M., Borana, K., Pani, S., Bajpai, A. and Bajpai, A.K., 2002. Assessment of heavy metal concentration in grass carp (*Ctenopharyangodon idella*). *Poll. Res.*, 21: 69–71.

Mohanty, S.K., Patnaik, D. and Rout, S.P., 2003. Evaluation of relationship between BOD and COD for river Nagavali and river Kolab in Koraput district of Orissa. *Indian J. Env. and Ecoplan.*, 7: 549–552.

Muthusamy, A. and Jayabalan, N., 2001. Effect of factory effluent on physiological and biological contents of *Gossypium hirsutum* L. *J. Environ. Biol.*, 22(4): 237–242.

Prasad, B.B., 2002. Effect of copper and zinc in the gills of *Channa marulius*. *J. Ecotoxicol. Environ. Monit.*, 12: 35–40.

Rajagopalan, S., 1990. Water pollution problem in textile industry and Control. In: *Pollution Management in Industries*, (Ed.) R.K. Trivedy. Environmental Publications, Karad, India, 21–45 pp.

Rajyalakshmi, S. and Sreelatha, K., 2005. Physico-chemical parameter of river Goutami Godavari, Yanam (Union Territory of Pondicherry). *J. Aqu. Biol.*, 20: 110–112.

Rao, L.M. and Manjula Sree Patnaik, 1999. Heavy metal accumulation in the cat fish *Mystus vittatus* (Bloch) from Mehadrigedda stream of Vishakhapatnam, India. *Poll. Res.*, 19: 325–329.

Rao, P.A. and Rao, P.P.V.V., 2002. Pollution potential of sago industry: A case study. *J. Ecotoxicol. Environ. Monit.*, 12: 53–56.

Sharma, J.K. and Arora, M.K., 2001. Environmental friendly processing for textile. *Poll. Res.*, 20: 447–451.

Singh, S.K. and Rai, J.P.N., 2003. Pollution Studies on river Ganga Allahabad district. *Poll. Res.*, 22: 469–472.

Srivastava, S.K., Kumar, R. and Srivastava, A.K., 1994. Effect of textile industry effluents on the biology of river Tons at MAU (U.P.). I. Physico-chemical characteristics. *Poll. Res.*, 13: 369–373.

Tiwari, D., 2004. Pollution Potential of the wastes polluting river Pandu. *Nature Env. Poll. Techno.*, 3: 219–221.

Trivedy, P.K. and Goel, P.K., 1984. *Chemical and Biological Methods for Water Pollution Studies*. Environmental Publication, Karad.

Trivedy, P.K., Goel, P.K. and Trisal, C.L., 1987. *Practical Methods in Ecology and Environmental Science*. Environmental Publications, Karad.

Villegas-Navarro, A., Ramirez-M.Y., Salvador-S.B., M.S. and Gallardo, J.M., 2001. Determination of waste water LC_{50} of the different process stages of the textile industry. *Ecotoxicol. Environ. Safe.*, 48: 56–61.

Chapter 49

Role of Copper and Manganese Application on Plant Height, Leaf Number and Leaf Are of Ginger (*Zingiber officinale* Rosc. L.) VAR-1

A. Ksheroda Devi and P.K. Singh

Department of Life Sciences, Manipur University, Canchipur, Imphal – 795 003, Manipur, India

ABSTRACT

Field experiments were carried out during the period of April to December 2005–06 with soil and foliar application of copper and manganese and combination of the two micronutrients in randomized block design. The response of micro nutrients was determined at 60, 90, 120, 150 and 180 Days After Plantation (DAP). The plant height and leaf number increased gradually from 60 to180 DAP. Among the treatments, maximum plant height and leaf number were found in F_2 (65.08 cm) and T_{14} (21.83) at 180 DAP. The minimum plant height and leaf number were found in T_1 and T_4 at 60 DAP. The leaf area increased gradually from 60 to 120 DAP but slightly decreased at 150 and 180 DAP. The maximum leaf area is found in T_{14} (32.38 cm^2) at 180 DAP and minimum is found in control at 60 DAP. Thus, the micronutrients play an important role in plant height, leaf number and leaf area, both in soil and foliar sprayed plant resulting an increased in yield of ginger plant.

Keywords: *Ginger, Micronutrients, Soil and foliar application.*

Introduction

Micronutrient plays an important role in maintaining the plant growth yield, and quality of crops. It also influences the physiological and biochemical parameters of plants (Wallace and Romney, 1977). In higher plants a major fraction of copper (Cu) is present in chloroplast as plastocyanin (Agarwala and Chatterjee, 1996), the mobile redox carrier in the light reaction of photosynthesis. Manganese (Mn) is absorbed by plants as manganese ion and helps in nitrate assimilation, evolution of oxygen during photosynthesis. Manganese deficiency causes inhibition of photosynthesis and nitrogen metabolism (Jones, 1980). Role of copper and manganese application on nitrate reductase and protease and pigment levels of *Zingiber officinale* was also reported by Devi and Singh (2005a; 2005b). The study aim at observing the effect of variable copper and manganese by growing ginger plants in field on plant height, leaf number and leaf area.

Materials and Methods

A field experiment was conducted in the Horticultural Experimental Field, Central Agricultural University (CAU), Iroisemba, Imphal during the period from April to December, 2006. The experiment was laid out in randomized block design with 32 treatments and replicated thrice at plot size of 1.5 sq m. Number of plots were 96 and a population of 960 plants. Micronutrients were applied both in soil and foliar at 50 and 70 DAP. Soil application was done before plantation. The treatments comprising control *i.e.*T_1, three levels of copper (0.025 g/plant, 0.075 g/plant 0.250 g/plant) *i.e.*, T_2, T_3, T_4: three levels of manganese (0.025 g/plant, 0.075 g/plant, 0.250 g/plant) *i.e.*, T_5, T_6, T_7 and combination of copper and manganese (0.025g Cu + 0.025g Mn, 0.025g Cu + 0.075g Mn, 0.025g Cu + 0.250g Mn, 0.075g Cu + 0.025g Mn, 0.075g Cu + 0.075g Mn, 0.075g Cu + 0.250g Mn, 0.250g Cu + 0.025g Mn, 0.250g Cu + 0.075g Mn, 0.250g Cu + 0.250g Mn) *i.e.*, T_8, T_9, T_{10}, T_{11}, T_{12}, T_{13}, T_{14}, T_{15} and T_{16} respectively. The same concentrations were applied in foliar and are represented as F_1 to F_{16}.

The plant height was measured from ground level upto the tip of the highest leaf and leaf number was also recorded from five sampled plants under each treatments and averaged. Leaf area was calculated by the method of Joseph and Jayachandra (1994). Representation and statistical analysis of experimental data were also worked out (Cochran and Cox, 1965; Gomez and Gomez, 1976).

Results and Discussion

The plant height in an important indirect indications of crop yield (Gill *et al.*, 1975). The application of all nutrients B, Cu, Zn, Mo and Mn together at 0.1 per cent with each at 6th leaf stages of polyhouse experiments increase plant height and leaf number per plant, number of shoot per plant of French bean cv contender (Jana and Kabir, 1987).

In the present study, the plant heights are recorded from 60 DAP to 180 DAP. The plant heights are found in an increasing tread from 60 to 180 DAP in all treatments. At 60 DAP, the maximum height is found in F_{15} (0.25g Cu + 0.075g Mn in foliar spray) with 19.40 cm. At 90 DAP, maximum heights is observed in T_9 (0.025g Cu + 0.075g Mn soil applied plant) with 34.10 cm. At 120 DAP, the highest value is recorded in T_{10} (0.025 g Cu + 0.250g Mn) with 50.60 cm. At 150 and 180 DAP, maximum heights are recorded in F_2 (0.025g Cu foliar spray) with 59.20 cm and 65.08 cm respectively. In all stages, minimum plant height is found in control except at 180 DAP. At 180 DAP, minimum height is recorded in F_4 (Table 49.1).

The present study indicates that the plant height increases from 5.30 cm to 65.08 cm as the plant gets maturity. The foliar application of Cu significantly increased the plant height as compared to control. These findings are in agreement with the finding of Singh *et al.* (2002).

Table 49.1: Effect of Copper, Manganese and Combination of Copper and Manganese on Plant Height at Different Stage of *Zingiber officinale* Rose. L. VAR-1

Treatments		Plant Height (cm)									
		60 DAP		90 DAP		120 DAP		150 DAP		180 OAP	
Soil	Foliar	Soil	Foliar	Soil	Foliar	Soil	Foliar	Soil	Foliar	Soil	Foliar
T_1	F_1	530±2.10	7.90±3.60	19.65±3.25	18.20±3.10	33.35±3.65	28.70±5.00	33.10±12.73	45.50±1.20	51.00±3.67	55.77±2.74
T_2	F_2	16.0±3.60	18.50±6.31	32.40±0.85	27.10±7.20	45.00±4.20	44.56±2.60	48.90±5.90	59.20±3.00	53.75±4.09	65.08±2.74
T_3	F_3	17.40±3.10	10.60±4.40	28.50±0.55	27.20±1.15	42.50±2.30	31.95±1.75	54.10±2.60	45.66±3.15	57.92±5.92	48.50±2.50
T_4	F	6.20±1.20	5.65±0.15	20.50±2.85	21.30±4.15	42.85±4.85	37.87±9.60	45.45±6.75	39.50±6.35	46.38±6.05	37.08±9.08
T_5	F	12.60±4.80	10.00±5.00	26.80±0.25	27.70±3.35	44.20±1.15	48.40±0.90	51.35±2.15	52.7±0.85	54.34±0.84	56.50±0.50
T_6	F	7.90±2.60	7.80±2.20	21.90±2.45	28.86±0.10	37.10±0.25	38.70±2.05	40.65±5.35	38.00±1.30	43.92±4.42	40.75±1.75
T_7	F	8.85±0.15	9.00±1.10	25.90±0.20	22.10±1.05	41.90±3.60	37.50±7.50	50.85±1.95	42.10±1.70	49.43±5.90	45.25±1.25
T_8	F_8	7.00±0.1	014.25±2.05	20.10±2.65	30.50±6.37	44.00±3.05	46.00±1.50	51.20±3.60	51.25±0.95	55.58±3.08	52.33±3.50
T_9	F_9	12.15±0.15	12.15±5.25	34.10±3.70	25.89±6.75	45.10±4.70	36.45±3.95	56.56±4.60	40.05±1.25	59.10±4.17	49.50±5.50
T_{10}	F_{10}	11.00±1.20	13.60±6.10	29.80±8.35	29.20±3.25	50.60±6.25	47.15±4.45	51.60±9.9a	52.10±3.60	63.75±2.42	60.33±1.00
T_{11}	F_{11}	9.30±2.51	6.25±0.75	23.10±0.35	20.67±3.40	37.70±1.95	36.80±2.40	50.70±0.00	43.70±0.85	50.33±4.00	46.75±2.25
T_{12}	F_{12}	17.90±0.40	9.00±5.50	22.70±2.95	26.69±3.02	39.60±4.60	37.50±5.20	47.46±8.00	44.65±5.95	50.37±7.04	44.75±1.75
T_{13}	F_{13}	7.50±0.00	5.00±0.00	19.90±0.40	23.10±2.25	39.60±4.60	41.95±3.45	45.51±2.20	46.35±0.65	52.08±5.75	50.75±1.75
T_{14}	F_{14}	6.70±0.20	12.70±3.90	19.90±3.10	30.40±5.30	44.10±2.45	39.90±0.00	48.25±4.75	50.60±10.60	57.83±1.83	53.25±8.25
T_{15}	F_{15}	11.25±2.75	19.40±8.40	24.20±0.55	27.55±1.00	42.67±0.90	41.76±3.50	40.00±1.00	37.00±1.80	46.50±4.50	37.58±1.92
T_{16}	F_{16}	7.00±0.50	7.30±2.50	26.80±0.25	21.00±4.45	39.90±1.20	36.25±1.05	45.75±3.95	35.50±0.50	51.66±1.00	37.47±1.14
CD at 5 per cent		0,481	2.216	1.206	2.165	1.873	2.131	2.922	1.693	2.506	1.844
CD at 1 per cent		1,458	3.290	1.879	3.213	2.780	3.162	4.337	2.512	3.719	2.738

Table 49.2: Effect of Copper, Manganese and Combination of Two Micronutrients Application on Leaf Number at Different Stage of *Zingiber officinale* Rosc. L. VAR-1

Treatments		Leaf Number Plant Age (Days)									
		60 DAP		90 DAP		120 DAP		150 DAP		180 OAP	
Soil	Foliar	Soil	Foliar	Soil	Foliar	Soil	Foliar	Soil	Foliar	Soil	Foliar
T_1	F_1	2.50±0.50	2.35±0.65	4.35±1.00	5.90±0.60	13.15±1.15	12.50±1.85	16.50±1.50	15.65±1.65	16.83±0.83	16.83±0.17
T_2	F_2	3.20±0.20	4.25±1.25	6.25±1.06	6.25±1.95	12.15±0.15	11.40±1.60	15.50±0.80	14.00±1.3 0	17.33±0.33	18.42±0.09
T_3	F_3	3.75±1.25	2.85±1.15	6.15±1.1 5	6.20±0.60	11.80±0.85	10.85±1.85	16.80±0.50	14.00±0.70	15.66±1.00	15.00±1.00
T_4	F_4	2.00±0.00	2.50±0.50	4.65±0.15	3.95±0.35	12.15±0.15	9.65±6.5	15.70±2.00	11.65±0.65	15.6H1.34	13.17±1.17
T_5	F_5	4.15±1.50	2.03±0.00	5.70±1.10	5.10±0.1 0	13.00±0.30	13.35±0.35	17.00±0.00	15.35±0.65	16.83±0.17	16.83±0.17
T_6	F_6	2.68±0.58	2.65±0.35	5.15±0.45	7.90±0.90	11.00±1.00	10.15±0.85	14.90±0.85	15.20±0.50	16.1H0.17	15.92±1.42
T_7	F_7	2.50±0.50	4.00±1.00	6.00±0.20	5.20±0.20	12.70±1.00	11.15±0.85	15.30±2.00	15.00±0.30	15.66±0.67	16.50±0.17
T_8	F_8	2.10±0.40	3.20±0.50	4.90±0.70	6.60±0.40	12.35±1.65	11.5±1.15	16.35±0.35	16.35±0.35	16.67±0.66	17.33±0.33
T_9	F_9	2.85±0.15	3.15±1.15	6.60±0.20	6.15±1.85	13.50±2.20	12.35±1.65	16.85±1.15	13.2 0±1.5 0	16.50±1.50	14.83±2.83
T_{10}	F_{10}	2.90±0.40	2.85±0.15	6.10±0.90	6.60±0.40	13.75±0.65	13.50±2.20	17.65±0.15	15.35±1.35	18.17±1.1 7	17.50±0.84
T_{11}	F_{11}	3.10±0.60	2.50±0.50	6.50±0.50	5.80±1.50	12.35±0.65	13.65±0.65	15.85±1.15	12.80±0.50	16.16±1.50	13.16±0.50
T_{12}	F_{121}	5.50±0.50	2.65±0.65	4.50±1.10	6.90±1.40	12.7±1.00	12.35±0.65	15.65±1.65	13.35±0.35	16.83±2.50	14.00±2.00
T_{13}	F_{13}	3.25±0.25	2.50±0.50	4.70±0.74	4.90±0.50	12.15±0.15	12.7±0.15	15.90±1.3 1	13.70±0.00	15.92±0.59	14.58±1.08
T_{14}	F_{14}	27.50±0.25	3.50±1.50	5.55±0.65	7 A0±1.40	13.30±1.00	1.3 3±1.0 0	16.15±1.15	12.33±1.48	21.83±6.17	16.00±3.00
T_{15}	F_{15}	3.00±1.00	3.50±1.50	5.1 0±0.16	6.35±0.95	11.80±0.50	11.80±0.50	16.00±1.00	11.65±1.65	14.92±1.42	10.42±.0.09
T_{16}	F_{16}	2.00±0.00	2.50±0.50	5.95±0.35	4.85±1.15	13.15H.1 5	11.25±1.70	16.85±0.15	11.50±2.00	16.00±0.34	13.42±1.09
CD at 5 per cent		0.278	0.459	0.403	0.552	0.525	0.910	0.614	0.597	0.789	0.618
C0 at 1 per cent		0.413	0.682	0.599	0.820	0.779	1.351	0.912	0.887	1.171	0.917

Table 49.3: Effect of Copper, Manganese and Combination of Two Micronutrients Application on Leaf Area at Different Stage of *Zingiber officinale* Rosc. L. VAR-1

Treatments		Leaf Area									
		60 DAP		90 DAP		120 DAP		150 DAP		180 OAP	
Soil	Foliar	Soil	Foliar	Soil	Foliar	Soil	Foliar	Soil	Foliar	Soil	Foliar
T_1	F_1	0.94±0.51	1.25±0.67	9.55±0.19	14.15±0.72	24.59±1.68	21.48±1.20	23.42±0.17	29.9±1.23	20.67±2.64	24.84±1.43
T_2	F_1	5.16±2.28	7.28±1.68	15.16±1.04	17.37±1.37	23.76±1.29	20.64±1.05	22.15±0.07	18.18±5.24	21.32±0.54	20.75±3.13
T_3	F_3	5.4±0.70	13.14±5.31	13.61±1.29	14.00±1.57	24.4H0.99	20.15±0.94	29.45±3.16	29.53±2.09	29.85±0.85	17.54±5.16
T_4	F_4	1.42±0.34	1.40±0.05	12.991±2.85	9.68±0.32	25.92±5.97	18.30±6.31	24.42±1.03	25.75±2.97	27.30±2.86	17.54±5.16
T_5	F_5	8.51±0.79	5.99±0.22	12.12±1.05	12.2±1.95	26.4±0.99	27.98±0.32	23.0±0.22	22.110±0.82	28.02±5.32	24.16±1.93
T_6	F_1	2.24±0.46	0.53±0.39	13.115±0.52	11.92±1.92	19.22±1.17	16.99±2.75	18.65±1.40	19.1±2.78	21.75±1.65	17.16±2.22
T_7	F_7	2.94±1.02	4.3±2.05	15.3±0.42	11.78±1.57	24.45±1.65	21.20±3.75	25.28±5.99	23.30±3.88	20.18±2.24	19.84±2.44
T_8	F_8	0.81±0.66	4.13±1.72	10.42±3.64	13.4±0.50	25.68±1.70	24.40±1.56	22.31±0.23	23.80±0.66	22.98±3.44	21.11±0.42
T_9	F_9	1.76±0.99	4.4±1.96	16.99±1.40	14.85±3.20	26.37±4.56	23.40±4.0	25.09±1.37	20.56±0.48	23.50±3.95	18.61±3.89
T_{10}	F_{10}	4.30±1.97	1.78±1.68	15.9±5.55	15.55±2.67	33.32±3.42	25.31±2.10	26.81±1.55	26.51±0.37	26.5±6.26	25.19±5.71
T_{11}	F_{11}	1.84±1.37	0.35±0.24	13.7±0.92	12.23±3.64	20.44±0.45	18.90±1.62	26.05±1.56	23.09±1.01	20.68±1.10	19.77±0.95
T_{12}	F_{12}	10.15±0.97	2.67±1.96	15.75±1.61	17.40±4.31	25.09±3.05	19.39±0.45	22.61±4.80	18.77±2.49	21.47±4.86	20.73±2.95
T_{13}	F_{13}	3.13±2.87	2.83±0.25	11.06±1.61	9.80±0.82	27.30±1.06	22.81±2.69	24.48±1.25	26.45±1.44	20.78±1.43	20.03±2.58
T_{14}	F_{14}	2.56±2.4	73.57±3.52	8.06±2.95	16.832±3.14	24.71±2.47	22.4±0.63	24.55±1.67	21.35±1.33	32.38±2.26	25.90±3.34
T_{15}	F_{15}	1.45±1.12	7.26±5.86	13.17±3.32	14.59±0.89	17.44±1.57	24.49±3.71	20.6±1.37	22.38±0.30	22.50±1.05	18.26±1.40
T_{16}	F_{16}	1.42±0.54	2.94±2.90	15.61±2.36	13.60±3.10	23.05±3.63	17.4±1.82	22.52±2.11	20.46±0.27	18.38±1.63	16.33±1.48
CD at 5 per cent		0.738	1.180	1.224	1.228	1.389	1.075	1.075	1.060	1.630	1.753
CD at 1 per cent		1.096	1.752	1.817	1.822	2.062	1.595	1.573	1.573	2.419	2.602

The micronutrients present in the soil are taken up by the plants at a particular growth stage. The concentration of micronutrients in plants varies widely depending upon the extent of deficiency in soils, crops and their varieties, plant age and states of the interesting elements and environment conditions (Nayyar *et al.*, 1992).

In the present study, the rate of increase in number of leaves in different stages are very rigorous. The maximum number of leave is recorded in T_{12} (0.075g Cu + 0.075g Mn foliar spray) with 5.50 at 60 DAP. At 90 DAP, the highest number is found in F_{14} (0.23g Cu + 0.025g Mn foliar spray) with 7.40. At 120 and 150 DAP, maximum numbers are found in T_{10} (0.025g Cu + 0.250g Mn soil applied plant) with 13.75 and 17.65. At 180 DAP, the highest number of leaves is found in T_{14} (0.250g Cu + 0.025g Mn soil applied plant) with 21.83 (Table 49.2).

The above finding shows that the number of leaves increases from 2.00 to 18.42 as the number of days after plantation is 60 and 180 respectively. The combination of micronutrients in foliar spray plants playa key role in increasing number of leaves. This findings was an agreement with those of Deka and Das (1975) in French bean, Amarender Reddy (1989) in Cauliflower.

In case of leaf area, maximum leaf area is found in F_3 (0.075g Cu foliar spray plant) with 13.14 cm^2 and 19.53 cm^2 at 60 and 150 DAP. At 90 OAP, highest leaf area is found in T_{12} (0.025g Cu + 0.075g Mn soil applied plant) with 17.40 cm^2. At 120 DAP, maximum leaf area is found in T_{10} (0.025g Cu + 0.250g Mn soil applied plant) with 33.32 cm^2. At 180 DAP, maximum leaf area is found in T_{14} (0.250g Cu + 0.025g Mn soil applied plant) with 32.38 cm^2 (Table 49.3).

The above finding shows that the combined application of micronutrient in soil application plays an important role in increasing plant height, leaf number and leaf area of ginger.

Acknowledgement

The authors are grateful to the Head, Department of Life Sciences, Manipur University, Canchipur, Imphal, for providing laboratory facilities.

References

Amarender Reddy, S., 1989. Effect of foliar application of urea and gibberellic acid on cauliflower (*Brassica oleraceae* var. *botrytis* L.). *J.Res. APAU*, 17(1): 79–80.

Agarwal, S.C. and Chatterjee, C., 1996. Physiology and biochemistry of micronutrient elements. In: *Advancements in Micronutrient Research*, (Ed.) A. Hemantranjan. Scientific Publishers, Jodhpur, pp. 203–365.

Cochran, W.G. and Cox, G.M., 1965. *Experimental Design*. John Wiley, New York.

Devi, A.K. and Singh, P.K., 2005a. Role of copper and manganese application on nitrate reductase and protease activities of *Zingiber officinale* Rose. L.VAR–1. *J. Curr. Sci.*, 7(1): 67–70.

Devi, A.K. and Singh, P.K., 2005b. Role of copper and manganese application on various pigments levels of *Zingiber officinale* Rosc. L. VAR–1. *J. Phytol. Res.*, 18(2): 215–218.

Deka and Das, 1975. In: *Vegetable Crops*, (Eds.) T.K. Bose, M.G. Som and J. Kabir. Naya Prakash, Calcutta.

Gill, H.S., Singh, J.P. and Swarup, V., 1975. Combining ability in cabbage (*Brassica oleraceae* var *capitata*). Veg. Sci., 2: 50–55.

Gomez, K.A. and Gomez, A.A., 1976. *Statistical Procedures for Agricultural Research with Emphasis on Rice*. IARLLOS, Banos Philippines.

Jones, B.G., 1980. *Annual Review of Plant Physiology*, pp. 257.

Joseph, A. and Jayachandra, B.K., 1994. Leaf area measurement in ginger (*Zingiber officinale* Rose). *South Indian Hort*., 42(1): 56–57.

Jana, B.K. and Kabir, J., 1987. Influence of micronutrients on growth and yield of French bean contendor under polyhouse conditions. *Veg. Sci.*, 14(2): 124–127.

Nayyar, V.K., Chhibba, L.M. and Bajwa, M.S., 1992. Know the micronutrient status of your soils and crop. *Progressive Farming*, 3: 19–20.

Singh, K.K., Singh, P.R.P. and Singh, R., 2002. Effect of copper on herbage and oil yield of *Citronell ajawa* (*Cymbopogon winterianus*). *New Agric.*, 13(1–2): 67–69.

Wallace, A. and Romney, E.M., 1 977. Synergistic trace metal effects in plants Commun. *Soil Science. Plant Anal*.

Chapter 50

Oxidation of Cobalt (III) Complexes of α-hydroxy Acids by Pyridinium Chlorochromate in the Presence of Micellar Medium

Mansur Ahmed* and K. Subramani

Department of Chemistry, Islamiah College, Vaniyambadi – 635 752

Keywords: *Pyridiniumchlorochromate, Sodiumlaurylsulphate, Cetyltrimethyl ammoniumbromide, Pentaamminecobalt (III) complex.*

Conventional spectrophotometry has been employed to study the oxidation of α-hydroxyacids such as mandelic acid, lactic acid, glycolic acid and their Cobalt (III) Complexes using pyridinium chloro chromate as oxidant in presence of surfactant. One equivalent oxidant like Ce(IV) induced electrontransfer in pentaamminecobalt (III).

Complexes of α-hydroxyacids result in nearly 100 per cent reduction at Cobalt (III) centre with synchronous Carbon-Carbon bond fission and decarboxylation. Such an electron transfer route seems to be unavailable for pyridiniumchlorochromate in its reaction with cobalt (III)

bound and unbound α-hydroxyacids in micellar medium. Pyridiniumchlorochromate oxidize cobalt (III) bound and unbound α-hydroxyacids to respective keto acid/keto acid cobalt (III) complexes in sodiumlauryl sulphate (NaLS) and cetyltrimethyl ammonium bromide

* E-mail: chem_islamiah@fastmail.us

(CTAB) possibly the transition state is more electrondeficient. Such a transition state can be envisaged only when the C-H bond fission occurs in the slow step with hydride ion transfer. The absence of formation of cobalt (II) rules out the synchronous C-C bond fission and electron transfer to cobalt (III). The rate of PCC oxidation of cobalt (III) bound and unbound α-hydroxyacid increases with increase of temperature. The thermodynamic parameters are in consistant with bimolecular reaction. The rate of pce oxidation of cobalt (III) Mandelato, Lactato and Glycolato complexes depends on the first power of pee concentration. Similarly the reaction between PCC and unbound α-hydroxyacid exhibits first order kinetics with respect to concentration of PCC. Of the three complexes lactato cobalt (III) complexes react faster than mandelato and glycolato complexes, where as in the unbound ligand similar trends follows, that is lactic acid reacts faster than mandelic and Glycolic acid.

Oxidation is an important process in organic chemistry and introduction of new economic and effective reagents for oxidation under mild and anhydrous conditions constitutes a standing challenge. PCC is oxidant which is non-hygroscopic, non-photosensity, stable yellow orange solid which is freely soluble in water, acetic acid, N,N- dimethyl formamide etc. The little work has been done on PCC as oxidant in micellar media.

Materials and Methods

The surfactants used in the present work are sodium lourylsulphate (NaLS) and cetyltrimethylammoniumbromide (CTAS). The surfactants are purified by adopting earlier procedure. The chemicals were purchased CTAB from (B.D.H, UK 99 per cent) NaLS mandelic, lactic and glycolic acids from (SD Fine chemicals. India 95 per cent) Pentaammine cobalt (III) complexes of α-hydroxyacids were prepared using Fan and Gould. Double distilled (deionised and CO_2 free) water was used as a solvent and H_2SO_4 (E. Merck India 95 per cent) was standardized using standard Sodiumcarbonate (SOH, AR) solution with methyl orange as indicator. For the PCC oxidation of Co(III) Complexes of α-hydroxyacids and unbound ligands. The rate measurement were made at 35±0.2°C in 100 per cent aqueous medium and temperature was controlled by electrically operated thermostat. The total volume of reaction mixture in the spectrophotometric cell was kept as 2.5 ml in each kinetic run. A systronic spectrophotometer fitted with recording and thermostating arrangement was used to follow the rate of the reaction. Rate of these PCC oxidant with unbound ligand and Cobalt (III) bound complexes were calculated from observed decrease in absorbance at 350 nm. The excess of the reductant was used in kinetic runs. It gives pseudo first order rate constant. It were determine from the linear plot of the InA versus time. Reproducible result obtained giving good first order plot.

The stiochiometric studies for the PCC oxidation of pentaammine cobalt (III) complexes of α-hydroxyacid and unbound ligand in the presence of micelles were carried out at 35±2°C. It was observed that the cobalt (II) formation was negligibly small.

Result and Discussion

Kinetic study of the oxidation of pentaammine cobalt (III) complexes of α-hydroxyacid by PCC in micellar medium dependence of rate on PCC concentration in micellar bound ligand. The rate of oxidation of mandelatocobalt (III) complexes depends on PCC concentration, the specific rate calculated remains constant (Table 50.1) and Graph of logarithm of PCC concentration versus time (Figure 50.1) are linear. From the slope of these graphs, the specific rate calculated agree with those obtained from integrated rate equation suggesting first order dependence on PCC concentration.

Table 50.1

$[(NH_3)_5 Co^{III}–L]^{2+} = 2.00 \times 10^{-2}$ mol dm^{-3} $[PCC] = 2.00 \times 10^{-3}$ mol dm^{-3} $[H_2SO_4] = 1.00$ mol dm^{-3} $[Na LS] = 2.00 \times 10^{-3}$ mol dm^{-3} Temperature = 35±0.2°C L = mandelato		
Time (s)	*10^3 (a-x) mol dm^{-3}*	
	NaLS	*CTAB*
300	1.62	1.64
600	1.42	1.36
900	1.21	1.12
1200	1.02	0.93
1500	0.89	0.78
1800	0.75	0.64
2100	0.64	0.51
2400	0.51	0.42
2700	0.42	0.35
3000	–	0.28

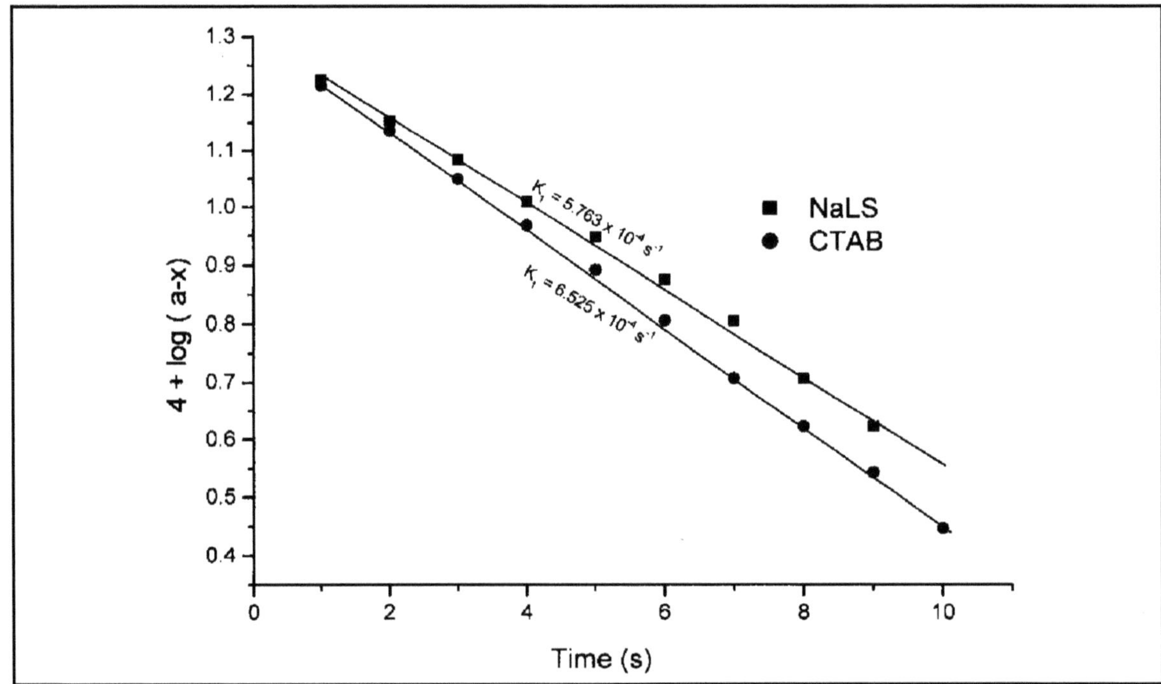

Figure 50.1: First Order Dependence Plots

When the concentration of PCC is varied from 1.00 to 3.00 × 10^{-3} mol dm^{-3} at a fixed [Cobalt (III)] and (H_2SO_4). Specific rates remains constant. Then the of rate of disappearance of Cr(VI) is given by equation–1

$$-d[Cr(VI)]/dt = k[Cr\,(VI)] \quad (1)$$

At a particular PCC concentration with increases in manlato/lactato/glycolato cobalt (III) concentration in the range 1.00 to 3.00 × 10^{-3} mol dm^{-3} there is a proportional increases in the rate of oxidation (Table 50.2).The slope of nearly unity is obtained from a linear graph of logarithm a (Figure 50.2) of specific rate (k in S^{-1}) versus logarithm of Co(III) concentration in each case suggesting first order rate dependence of rate on [Co(III)].

Table 50.2

[PCC] = 2.00 × 10^{-3} mol dm^{-3}
[H_2SO_4] = 1.00 mol dm^{-3}
[Na LS] = 2.00 × 10^{-3} mol dm^{-3}
Temperature = 35±0.2°C

10^2 [$(NH_3)_5$ Co^{III}–L] mol dm^{-3}	10^4 k_1 (s^{-1})	10^2 k_2 dm^3 mol^{-1} s^{-1}
L = Mandelacto		
1.00	2.84	2.84
1.50	4.21	2.80
2.00	5.62	2.81
2.50	7.09	2.83
3.00	8.47	2.81
L = Lactato		
1.00	3.66	3.66
1.50	5.54	3.69
2.00	7.32	3.66
2.50	9.13	3.65
3.00	11.04	3.68
L = Glycolato		
1.00	2.09	2.09
2.00	4.12	2.06
2.50	5.22	2.08
3.00	6.34	2.11

Hence the rate law for the Cr(VI) oxidation of cobalt (III) bound of α-hydroxy acids are given by equation–2

$$-d[Cr(VI)]/dt = k_2[Cr\,(VI)]\,[Co(III)] \quad (2)$$

Dependence of Rate on PCC Concentration in Micellar for α-hydroxyacid

The rate of oxidation of mandelato cobalt (III) complexes depends on pee concentration. In any specific run the change in concentration of pee, the specific rate calculated remains constant (Table

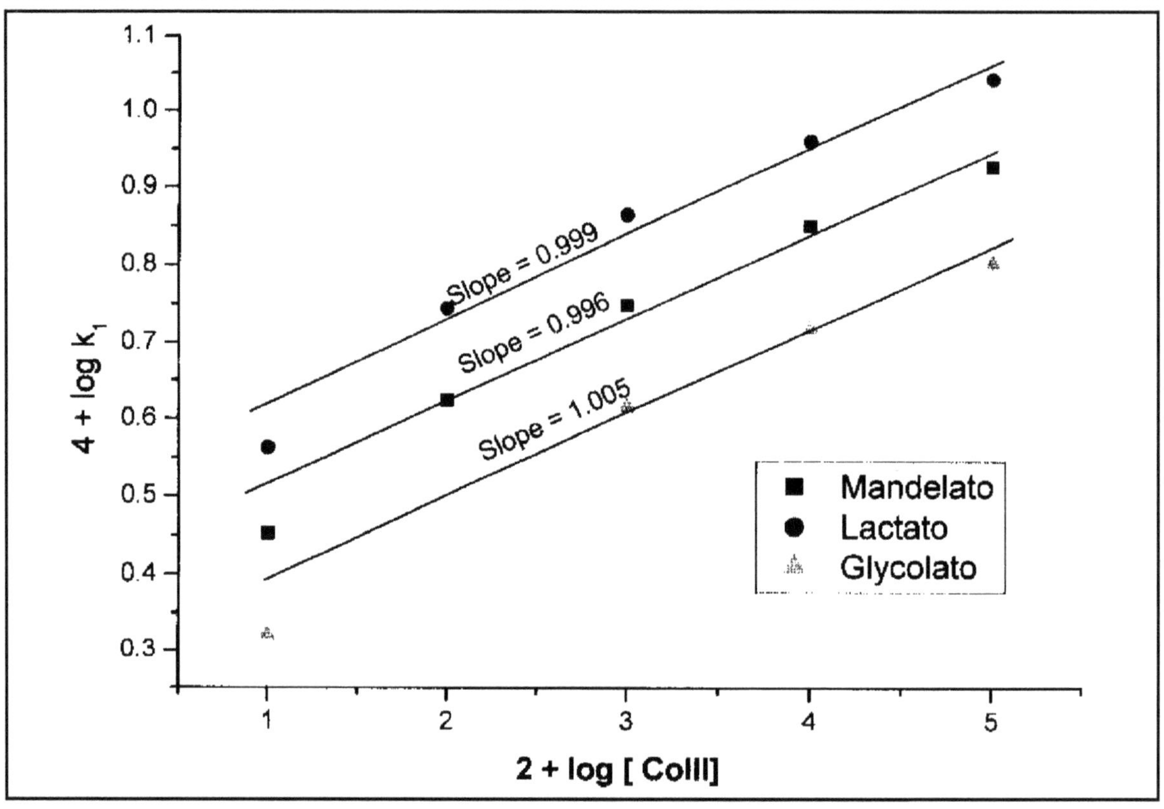

Figure 50.2: Dependence of Rate on [Cobalt (III)] in NaLS

50.3) and graphs of logarithm of pee concentration versus time are linear (Figure 50.3). From the slope of these graphs, the specific rate calculated agree with those obtained from integrated rate equation, suggesting first order dependence on pee concentration.

Table 50.3

[PCC] = 2.00×10^{-3} mol dm^{-3}

[H_2SO_4] = 1.00 mol dm^{-3}

[CTAB] = 2.00×10^{-3} mol dm^{-3}

Temperature = 35±0.2°C

10^2 [$(NH_3)_5$ Co^{III}–L] mol dm^{-3}	10^4 k_1 (s^{-1})	10^2 k_2 dm^3 mol^{-1} s^{-1}
L = Mandelacto		
1.00	3.26	3.26
2.00	6.44	3.22
2.50	8.11	3.24
3.00	9.92	3.30

Contd...

Table 50.3–Contd...

10^2 [$(NH_3)_5$ Co^{III}–L] mol dm^{-3}	10^4 k_1 (s^{-1})	10^2 k_2 dm^3 mol^{-1} s^{-1}
L = Lactato		
1.00	4.14	4.14
1.50	6.25	4.17
2.00	8.32	4.16
3.00	12.52	4.17
L = Glycolato		
1.00	2.32	2.32
1.50	3.48	2.32
2.00	4.55	2.28
2.50	5.71	2.29

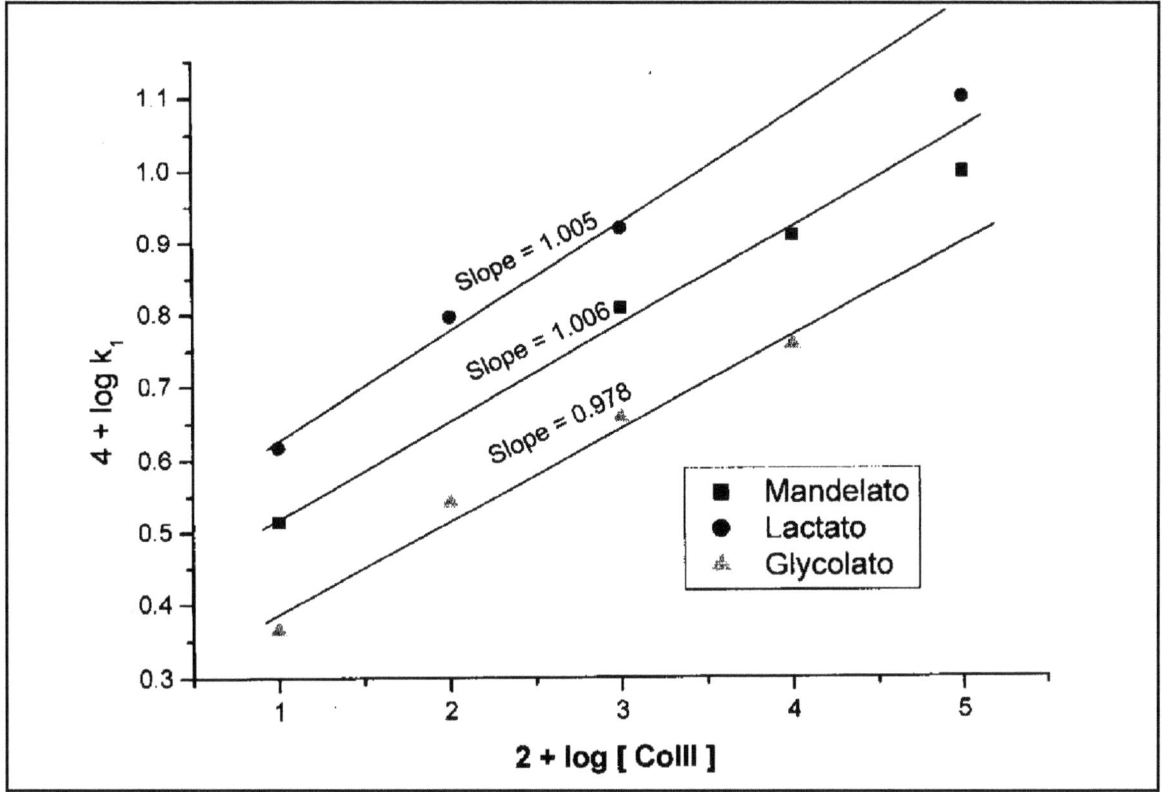

Figure 50.3: Dependence of Rate on [Cobalt (III)] in CTAB

When concentration of PCC is varied from 1.00 to 3.00 × 10^{-3} mol dm^{-3} at a fixed [Co(III)] and [H_2SO_4] specific rates remains constant. Then the rate of disappearance of Cr(VI) is given by equation–4

$$-d[Cr(VI)]/dt = k_1[Cr\,(VI)] \quad (4)$$

Dependence of Rate on the Concentration of α-hydroxyacid in NaLS and CTAB

The oxidation studies were carried out by varying initial [α-hydroxy acid] in the range 1.00 to 3.00×10^{-3} mol dm^{-3} by keeping other variable constant. The near constancy in the k_2 values (Tables 50.4 and 50.5) and the slope of nearly unity is obtained from a linear graph of logarithm of specific rate (k_1 in S^{-1}) verses logarithm of α-hydroxy acid concentration in each case suggesting first order dependence of rate on [α-hydroxy acid] (Figures 50.3 and 50.4). Hence the rate law for the Cr(VI) oxidation α-hydroxy acid of is given below equation–5

$$-d[Cr(VI)]/dt = k_2[Cr\,(VI)]\,[\alpha\text{-hydroxy acid}] \quad (5)$$

Table 50.4

[PCC] = 2.00×10^{-3} mol dm^{-3}
[H_2SO_4] = 1.00 mol dm^{-3}
[Na LS] = 2.00×10^{-3} mol dm^{-3}
Temperature = 35±0.2°C

10^2 [α-hydroxy acid] mol dm^{-3}	10^4 k_1 (S^{-1})	10^2 k_2 dm^3 mol^{-1} S^{-1}
Mandelic acid		
1.00	1.31	1.31
1.50	1.96	1.30
2.00	2.59	1.29
2.50	3.41	1.36
3.00	4.02	1.34
Lactic acid		
1.00	2.21	2.21
1.50	3.32	2.22
2.00	4.36	2.18
2.50	5.44	2.17
3.00	6.72	2.24
Glycolic acid		
1.00	1.09	1.09
2.00	2.12	1.06
2.50	2.73	1.09
3.00	3.32	1.10

Comparison of Rates on Oxidation of Pentaammine Cobalt (III) Complexes of Both Bound and Unbound α-hydroxyacid by PCC

Specific rate of the lactato complex is more compared to both the rates of oxidation of unbound ligand and mandelato complex deserves an explanation. The ligation of lactic acid to Co(III) centre has probably increased its reactivity towards PCC and this effect seems to be more specific for this ligand only. If the reaction proceeds through a preformed Chromate ester, then the rate α-C-H, fission

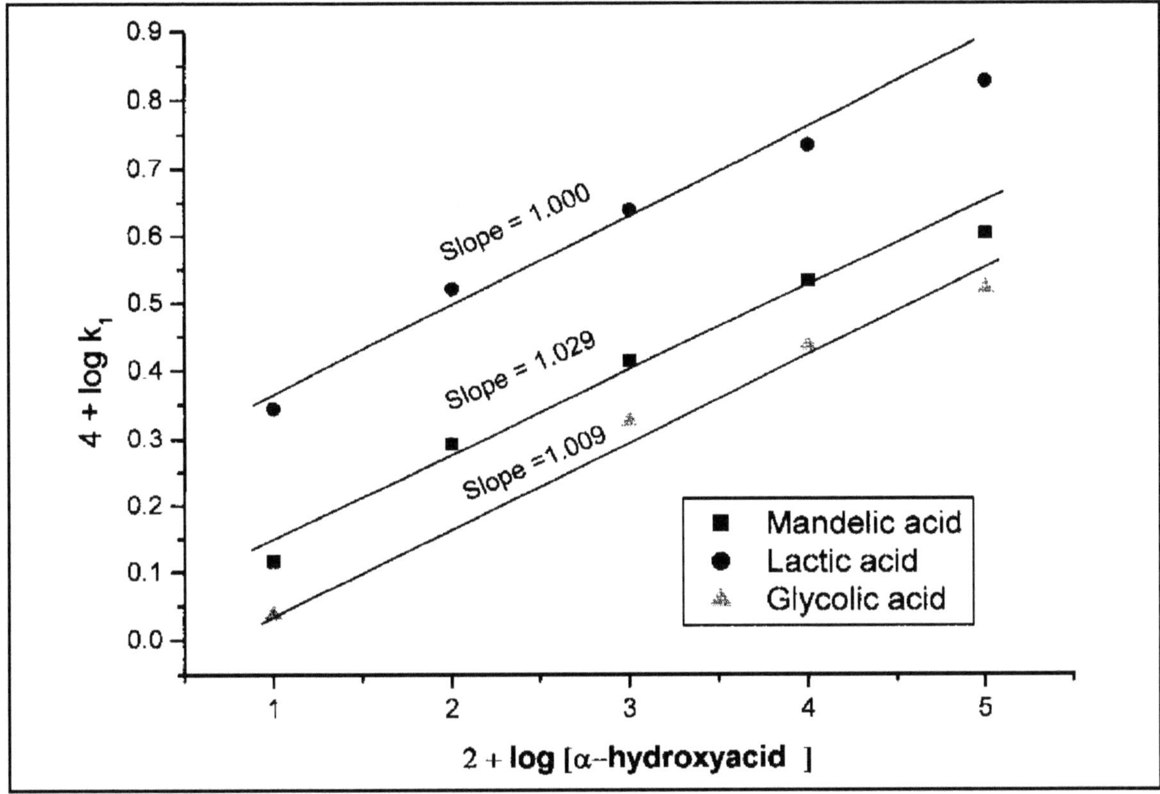

Figure 50.4: Dependence of Rate on [α-hydroxyacid] in NaLS

will been enhanced, resulting in an increased rate of oxidation of lactato complex such a precursor complex may be sterically hindered in the case of mandelato and glycolato complexes.

Table 50.5

[PCC] = 2.00×10^{-3} mol dm^{-3}

[H_2SO_4] = 1.00 mol dm^{-3}

[CTAB] = 2.00×10^{-3} mol dm^{-3}

Temperature = 35±0.2°C

10^2 [α-hydroxy acid] mol dm^{-3}	10^4 k_1 (S^{-1})	10^2 k_2 dm^3 mol^{-1} S^{-1}
Mandelic acid		
1.00	1.62	1.62
1.50	2.37	1.58
2.00	3.30	1.65
2.50	4.17	1.60
3.00	4.82	1.61

Contd...

Table 50.5–Contd...

10^2 [α-hydroxy acid] mol dm^{-3}	10^4 k_1 (S^{-1})	10^2 k_2 dm^3 mol^{-1} S^{-1}
Lactic acid		
1.00	2.39	2.39
1.50	3.56	2.37
2.00	4.75	2.38
2.50	5.97	2.39
3.00	7.18	2.39
Glycolic acid		
1.00	1.29	1.29
2.00	2.51	1.26
2.50	3.26	1.30
3.00	3.92	1.31

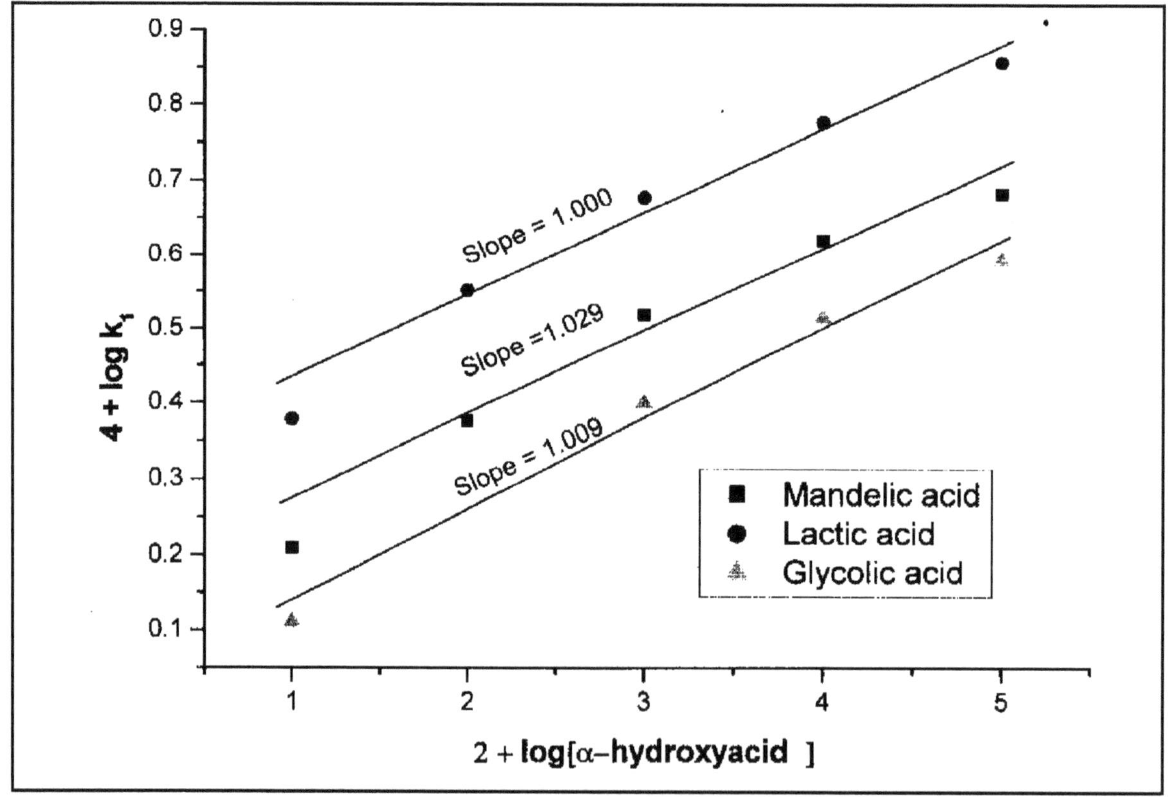

Figure 50.5: Dependence of Rate on [α-hydroxyacid] in CTAB

Mechanism

Oxidation of Pentaammine cobalt (III) complexes of both bound and unbound Ligands in micellar medium.

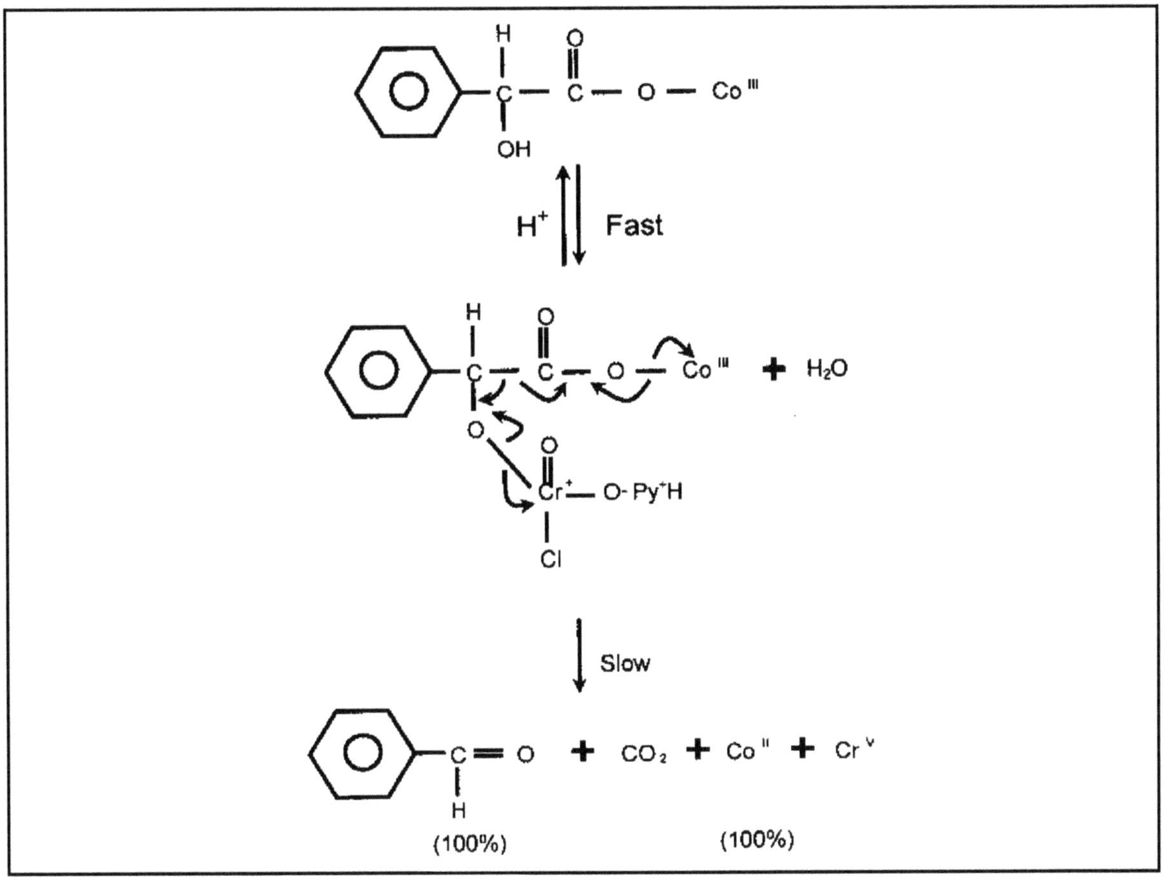

Conclusion

The kinetics of one electron transfer route seems to be unavailable for PCC with Cobalt (III) bound and unbound complexes of α-hydroxyacid in micellar medium, PCC oxidizes Cobalt (III) bound and unbound α-hydroxyacids. It rules out the synchronous C-C bond fission and electron transfer to Cobalt (III) centre. Oxidation of above complexes increases with increase of temperature. With increase in micellar concentration an increase in the rate is observed. The added CTAB enhances the rate of oxidation of a reaction much more than NaLS. Similar trends has been observed in lactato and glycolato Co(III) complexes.

Chapter 51

Extraction and Bioautography of *Aristolochia elegans* Plant for Antimicrobial Activity

M.S. Imran and U.S. Bagde

Applied Microbiology Laboratory, Department of Life Sciences, University of Mumbai, Vidyanagari, Santacruz (E), Mumbai – 400 098

ABSTRACT

A study on antimicrobial properties of leaf extract of *Aristolochia elegans* was carried out using agar cup diffusion method against clinical isolates of *Morganella morganii* and *Acinetobacter*. The ethanol leaf and methanol leaf extracts proved to be the most active against *Morganella morganii* and *Acinetobacter* and were subjected to phytochemical screening by thin layer chromatography and bioautography. The results indicated that the leaf extract contained the most efficient antimicrobial compounds.

Keywords: *Aristolochia elegans, Thin Layer Chromatography, Bioautography.*

Introduction

Medicinal plants have been used for centuries as remedies for human diseases because they contain components of therapeutic value Nostro *et al.* (2000). Since prehistoric times, people have used natural resources for medical purposes. Plant oil and extracts have been used for a wide variety of purposes for many thousands of years Jones (1996). Medicinal plants have always had an important place in the therapeutic armoury of mankind. In recent years, the use of medicinal plants and crude extracts has widely progressed. It can be assumed that this trend will not continue unless standardization methods for these plant materials become available. Numerous methods of identification for plant extracts have been proposed (Stahl, 1969) on the other hand, the quantitative determination of their biologically active constituents has been less studied, especially with regard to drugs used in phytotherapy.

Many reports have shown the effectiveness of traditional herbs against microorganisms, as a result, plants are one of the bedrocks for modem medicine to attain new principles (Evans *et al.*, 200). Moreover the increasing use of plant extracts in the food, cosmetics and pharmaceutical industries suggest that, in order to find active compounds, a systematic study of medicinal plants is very important. However, due to over exploitation, some traditionally used plants are disappearing and the sustainable usage of natural resources are currently questioned by ecologists (Nigg and Seigler, 1992).

The aim of the present study was to extract and investigate the antimicrobial activity of leaf extract of *Aristolochia elegans* plant by Bioautography.

Materials and Methods

Plant Materials

Aristolochia elegans plants were collected in February 2003, from the Mahableshwar hills in the Konkan mountain range, Maharashtra, India. The plant was taxonomically identified.

Preparation of Extracts

The leaves of the plant were separated and were shade dried for a period of 72 hours. The pulverized air dried plant material was ground and about 10 grams of the grounded leaf material was extracted with 2 × 100 ml ethanol and methanol in Soxhlet apparatus for 24 hours following the procedure described earlier (Sokmen *et al.*, 1996). The crude ethanolic and methanolic extracts were filtered through cheese cloth and cotton wool and then through a Buchner funnel with No. 4 filter paper. The filtrate was evaporated to dryness in a vacuum evaporator and reconstituted with 20 ml ethanol and methanol respectively.

Microorganisms

The test organisms *Morganella morganii* and *Acinetobacter* used in the study were obtained from Lokmanya Tilak Government Hospital, Mumbai, Maharashtra, India.

Screening for Antibacterial Activity

For bioassay, agar well diffusion method was performed (Perez and Anesini, 1993). A suspension of approximately 1.5×10^8 cells in sterile normal saline was prepared by the reported method of Forbes *et al.* (1990) and about 100 µl of each sample was uniformly seeded on Mueller-Hinton Agar prepared in glass Petri plates and left aside for 15 minutes. Two wells of 8 mm diameter and 3 cm apart were punctured in the culture plate using sterile cork borers. In one of the wells 100 µl of control (ethanol/methanol) solution while in other well test solution of leaf extract was put with the help of micro pipette and kept for diffusion in refrigerator at 4°C. Culture plates were incubated at 37°C for 24 hours. The plates were observed after 24 hours and the zone of inhibition was measured in millimeters. All tests were performed in duplicates and the antibacterial activity was expressed as the mean of inhibition produced by the plant extract.

Phytochemical Screening

The ethanol and methanol extracts obtained were selected for preliminary phytochemical screening because of their good antibacterial activity.

Identification by Thin Layer Chromatography

The ethanol and methanol extracts were subjected to Thin Layer Chromatography (TLC). Commercially available 20 cm × 20 cm TLC aluminium sheets (layer thickness 0.2 mm) precoated with silica gel 60 was obtained from Merck (Darmstadt, F.R.G.). About 5–10 µl of ethanolic and methanolic leaf extracts were applied to the TLC plates to form spots. Two different solvent systems [Upper phase of toluene: ethyl acetate : water : formic acid (20 : 10 : 1 : 1) and Benzene: Chloroform (6 : 4)] were used for the development of the plates. Separated components were visualized under visible and ultraviolet light (254 and 360 nm, Camac Universal UV lamp TL-600) or chemically visualized. The plates were made in duplicates, of which one set was used for derivitization to identify the secondary metabolite and the other set for bioautography to check the potencies of different isolated components.

Bioautography

Bioautography of extracts was performed with the cultures of *Morganella morganii* and *Acinetobacter* which showed good sensitivity to the extracts. A suspension of inoculum, at a final concentration of 106 and 0.1 per cent of 2, 3, 5-triphenyltetrazolium chloride (Merck) in Mueller-Hinton agar, maintained at 45°C, was applied to the

developed plate kept inside a petri plate. The agar plates were allowed to cool at room temperature and were kept for incubation for 24 hours. Where microbial growth got inhibited; the spots could be seen against a deep pink-red background. Duplicates of the chromatograms were used in the bioautographic method. The result of the bioautography was compared with the fIrst set to know the location of active component.

Results

The results obtained in this study indicated pronounced activity when the isolates *Morganella morganii* and *Acinetobacter* were subjected to ethanol and methanol leaf extracts. The antimicrobial susceptibility proflles of the microorganisms, obtained by the agar cup diffusion method are given in Table 51.1. The isolates of *Morganella morganii* and *Acinetobacter* showed almost similar results. The ethanol and methanol controls showed no inhibiting effect. The ethanol leaf extract showed inhibition diameters in a range of 21 mm against *Morganella morganii* and 19 mm against *Acinetobacter*, while methanol leaf extract showed inhibition diameters in a range of 22 mm against *Morganella morganii* and 21 mm against *Acinetobacter*.

Table 51.1: Antibacterial Activities of Leaf Extracts of *Aristolochia elegans*

Microorganisms	*Zone of Inhibition (mm)*			
	EL	*EC*	*ML*	*MC*
Morganella morganii	21	00	22	00
Acinetobacter	19	00	21	00

EL: Ethanol Leaf; EC: Ethanol Control; ML: Methanol Leaf; MC: Methanol Control; > 14 → highly potent, < 14 → less potent, - - → No activity.

A systematic study for identification of the potent antimicrobial from the leaf extracts was carried out using thin layer chromatography. Both ethanol and methanol leaf extracts which were found to be potent were selected on the basis of their broad spectrum of activity and largest zone of inhibition. The spot given by the ethanol and methanol leaf extracts using Upper phase of toluene: ethyl acetate:

water: formic acid (20 : 10 : 1 : 1) and *Benzene: Chloroform* (6:4) as a solvent system was circular with Rfvalue 0.83 and 0.44. The florescence color of the spot was bluish when observed under the ultraviolet light.

The component separated by TLC was checked for antibacterial activity by bioautography. The bioautography revealed clear zones of bacterial growth inhibition for *Morganella morganii* and *Acinetobacter* for the spots separated using Upper phase of toluene: ethyl acetate: water: formic acid (20 : 10 : 1 : 1) as a solvent system.

Discussion

The discovery of potent antimicrobials was one of the greatest contributions to medicine in the 20th century. Increased antimicrobial resistance presents a major threat to public health because; it reduces the effectiveness of antimicrobial treatment, leading to increased morbidity, mortality, and health care expenditure (Smith and Coasts, 2002). It is important to investigate scientifically those plants which have been used in traditional medicines as potential sources of novel antimicrobial compounds (Mitscher *et al.*, 1987). The results of the primary screening revealed that both the extracts were active against *Morganella morganii* and *Acinetobacter.* The results were in accordance with the findings of Gordana *et al.* (2005) who tested methanol extracts of aerial parts of four *Achillea* species and Kartal *et al.* (2003) who tested ethanol extracts from propolis samples for antimicrobial activity by the disk diffusion assay against five bacteria *Staphylococcus aureus, Escherichia coli, Klebsiella pneumoniae, Pseudomonas aeruginosa* and *Salmonella enteritidis* and a fungi *Candida albicans.*

Since *Aristolochia elegans* demonstrated maximum activity against the most prevalent bacteria namely *Morganella morganii and Acinetobacter,* the use of plant is validated. The results obtained indicated the existence of antimicrobial compounds in the crude ethanol and methanol leaf extracts of this plant. Very few data on chemical nature of antimicrobial principles of this plant is available.

To investigate the number of components in the extracts several TLC systems were developed. In order to obtain well separated components it was essential to select the right mobile phase and was decided to use two solvent systems. Silica gel TLC [Upper phase of toluene : ethyl acetate : water : formic acid (20 : 10 : 1 : 1) and Benzene : Chloroform (6 : 4)] of the extract showed, after visualization, differently coloured spots. The plates were made in duplicates, of which one set was used for derivitization to identify the secondary metabolite and the other set for bioautography to check the potencies of different isolated components.

The component separated by TLC was checked for its antibacterial activity by bioautography. However, a much simpler pattern was observed by bioautographic detection. Ethanol leaf extract exhibited activity against two organisms *viz., Morganella morganii* and *Acinetobacter.* Similar results were observed for methanol leaf extract against two organisms *viz., Morganella morganii* and *Acinetobacter* employed in the test. Only the spots at Rf 0.83 and 0.44 were antimicrobially active as well as the origin spot. This selectivity was a guide for the preparative TLC. Antimicrobial substances are visible on the plates as clear zones without growth of the microorganism. These observation were in accordance with the findings of Martini and Eloff (1998), who demonstrated 14 different antimicrobial components in the leaves of *C. erythrophyllum* which inhibited the growth of *Staphylococcus aureus* using bioautography technique.

References

Evans, C.E., Banso, A. and Samuel, O.A., 2002. Efficacy of some nupe medicinal plants against *Salmonella typhi:* An *in vitro* study. *J. Ethnopharmacol.*, 80: 21–24.

Forbes, B.A., Sahm, D.F., Weissfeld, A.S. and Trevino, E.A., 1990. Methods for testing antimicrobial effectiveness. In: *Bailey and Scott's Diagnostic Microbiology*, 8th edn, (Eds.) Baron, E.J., Peterson, L.R. and S.M. Finegold. Mosby Co., St Louis, Missouri, p. 171–194.

Gordana, S., Niko, R., Toshihiro, H. and Radosav, P., 2005. *In vitro* antimicrobial activity of extracts of four *Achillea* species: The composition of *Achillea clavennae* (Asteraceae) extracts. *J. Ethnopharmacol.*, 5: 1–8.

Jones, F.A., 1996. Herbs-useful plants: Their role in history and today. *European J. Gastroenterol. Hepatol.*, 8: 1227–1231.

Kartal, M., Sulhiye, Y., Serdar, K., Semra, K. and Gulacti, T., 2003. Antimicrobial activity of propolis samples from two different regions of Anatolia. *J. Ethnopharmacol.*, 86: 69–73.

Martini, N. and Eloff, J.N., 1998. The preliminary isolation of several antibacterial compounds from *Combretum erythrophyllum* (Combretaceae). *J. Ethnopharmacol.*, 62: 255–263.

Mitscher, L.A., Drake, S., Gollapudi, S.R. and Okwute, S.K., 1987. A modem look at folkloric use of anti-infective agents. *J. Nat. Prod.*, 50: 1025–1040.

Nigg, H.N. and Seigler, D., 1992. *Phytochemical Resources for Medicine and Agriculture*. Plenum Press, New York, 76 p.

Nostro, A., Germano, M.P., Angelo, V.D., Marino, A. and Cannatelli, M.A., 2000. Extraction methods and bioautography for evaluation of medicinal plant antimicrobial activity. *Lett. Appl. Microbiol.*, 30: 379–384.

Perez, C. and Anesini, C., 1993. Screening of plant used in Argentine folk medicine for antimicrobial activity. *J. Ethnopharmacol.*, 39(2): 119–128.

Smith, R.D. and Coast, J., 2002. Antimicrobial resistance: a global response. *Bull World Health Organ.*, 80(2): 126–133.

Sokmen, A., Jones, B.M. and Erturk, M., 1996. The *in vitro* antibacterial activities of Turkish medicinal plants. *J. Ethnopharmacol.*, 67: 79–86.

Stahl, E., 1969. *Thin-layer Chromatography*, 2nd edn. George Allen and Unwin, London, p. 747–750

Chapter 52

Enhanced Chlorpyrifos Pesticide Detoxification by Fusants of *E. coli* and *Achromobacter lacticum*

Chitra Verma and U.S. Ragde

Applied Microbiology Laboratory, Department of Life Sciences, University of Mumbai, Vidyanagari, Santacruz (E), Mumbai – 400 098, India

ABSTRACT

Nationwide study by Centre for Science and Environment (CSE) 2006 has reported that a Coca Cola sample manufactured in Thane, Maharashtra, India, contained the neurotoxin Chlorpyrifos, 200 times the standard. This is an alarming condition and there is a serious need to apply bioremediation technology for degradation of this pesticide. In present study we have tried to find out Chlorpyrifos degrading strains and developed them for better degrading ability with the help of protoplast fusion technique. Certain strain of *E. coli* was isolated from the specific pesticide contaminated sites of Gharda chemicals, Mumbai (India), having capabilities for degradation of Chlorpyrifos pesticide. This isolate was chosen for protoplast fusion with *Achromobacter lacticum* for enhancing its ability of degradation. The capabilities of fusants for chlorpyrofos biodegradation were assayed and compared with the parent bacteria. These fusants showed a superlative increase in Chlorpyrifos biodegradation.

Keywords: *Neurotoxin, Protoplast fusion, Biodegradation. Chlorpyrifos.*

Introduction

Chlorpyrifos is a phosphorothioate attached to a pyrimidine ring containing three chlorines, which makes this compound very hydrophobic in nature. It is used extensively throughout the world.

Chlorpyrifos is moderately toxic to humans. Poisoning from Chlorpyrifos may affect the central nervous system, the cardiovascular system, and the respiratory system. It is also a skin and eye irritant and high level of exposure results in acetylcholine accumulation, which interferes with the muscular responses, leading to the possibility of death. Repeated or prolonged exposure can cause delayed cholinergic toxicity and neurotoxicity. WHO (2001) has considered Chlorpyrifos as 'moderately hazardous'. These health and environmental concerns have led to an interest in detoxification of Chlorpyrifos in environment.

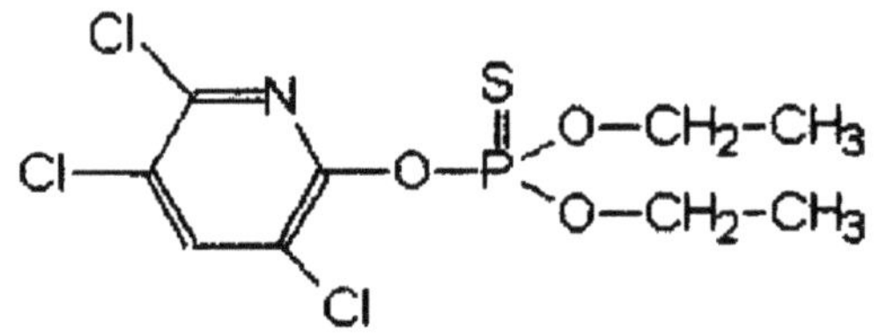

Chemical Structure of Chlorpyrifos

Biological methods or biodegradation methods are simple and could effectively reduce the risk of environmental contamination caused by commercial application of hazardous compounds. Degradation processes are constantly taking place on a large scale in the natural environment still the pesticides persists; there persistence is an example of inability of local micro flora to degrade these compounds. Considering this fact it is desirable to develop a novel organism with better degrading ability. A technology based on the use of local bacteria and developing the efficiency of isolated strains by adaptation as described in our reports could be introduced as an efficient, cheap and environmental friendly technique for pollutants decontamination (Mansee *et al.,* 2004a). In this context protoplast fusion technique has been successfully applied to carry out interspecific and intergeneric fusions by several researchers (Gokhale and Deobagkar, 1994; Bagde and Paranjpe, 2000; Rodriques *et al.,* 1991) to obtain novel strains. To enhance the capability of isolated bacteria for Chlorpyrifos biodegradation, the present study was undertaken to investigate the possibility of hybridization, through protoplast fusion, between isolated *E. coli* strains that previously indicated their capability for OP biodegradation, with *Achromobacter lacticum.* Moreover, the efficiency of resulted hybrid strains for biodegradation of Chlorpyrifos was investigated by performing CO_2 evolution test.

Materials and Methdos

Bacterial Strains

Bacterial strains *E. coli* and *Achromobacter lacticum* were isolated from the sites near Gharda chemicals, Mumbai, India. They were identified, based on cultural and physiological characteristics according to Bergey's manual of systematic bacteriology (Krieg and Holt, 1984). Chemicals used for isolation were purchased from Hi-media (Mumbai). Chlorpyrifos 20 per cent TC was provided by AIMCO pesticides Ltd., Mumbai.

Protoplast Formation, Fusion and Regeneration

Protocol followed for protoplast formation and fusion was the modification of the procedure given by Gokhale *et al.* (1984).

Protoplasts of *E. coli* were obtained in protoplast buffer containing 2M tris HCL buffer (30mM conc., pH–7.5) with 25 per cent sucrose, 0.5mM EDTA. And 0.5 M NaCl using Lysozyme enzyme, with a concentration of 2 mg/ml. It was incubated in a rotary shaker at 37°C for approx. one hour. Similarly protoplasts of *Achromobacter lacticum* were obtained by using Lysozyme (1.5 mg/ml) in protoplast buffer containing 5 ml of Tris HCL buffer (30 mM, pH–7.5) with 25 per cent sucrose and 0.5 mM EDT A and incubating it in a rotary shaker at 37°C for approx. two hours (Gokhale *et al.*, 1984)

One ml of the protoplasts of the two genera (*E. coli* and *Achromobacter lacticum*) were mixed and fused in the presence of 40 per cent polyethylene glycol 6000 (40 per cent PEG 6000). Fusants were regenerated in the regeneration medium containing Glucose 5 gm, Casamino acid 5 gm, L-Tryptophan 0.1 gm, K_2HPO_4 3.5 gm, KH_2PO_4 1.5 gm, 2 M Sodium succinate 250 ml, pH 7.2 M MgCh 10 ml, Polyvinylpyrolidone 3 gm, Agar 8 gm, D/W 1000 ml (Akamatsu and Sekiguchi, 1993).

Degradation Studies

Preparation of Bacterial Cells for Biodegradation

Pure culture of bacteria was inoculated in nutrient broth and incubated for 48 hrs at 37°C. The cells were separated after centrifugation, washed twice with saline. 6×10^8 numbers of cells (According to Mac Farland scale) were then mixed with 10 ml of mineral medium in a tube. They were further incubated for 48 hours. These pre-adapted cells in each case were used as an inoculum.

CO_2 Evolution Test

The extent to which an organic chemical is mineralized to CO_2 provides a definite measure of 'ultimate' biodegradation. For degradation studies of parent strain and fusants, CO_2 evolution test was done as per the method given by Bartha and Pramar (1965) in a biometer flask. 0.1 ml of Chlorpyrifos was mixed with 100ml of mineral meia in a biometer flask. The percentage biodegradation was calculated from the formula according to OECD guidelines, 301B(1992)

$$\text{Per cent Biodegradation} = \frac{\text{mg } CO_2 \text{ evolved} \times 100}{\text{TOC added} \times 3.67}$$

mg CO_2 produced = (CO_2 from inoculum + Test substance) – CO_2 from inoculum

3.67 is conversion factor (44/12) from carbon to carbon dioxide

TOC = Total organic carbon produced from Chlorpyrifos

Results and Discussion

High ratios of bacterial cells (about 80–90 per cent) were converted to protoplasts, under the standard conditions used. The protoplasts started coming closer after 5 minutes of 40 per cent PEG 6000 treatment. After coming in intimate contact, the cell membrane at the point of contact softened and its dissolution took place to facilitate the fusions.

Figure 52.1 shows the colonies of *E. coli*, *Achromobacter lacticurn* and their hybrid strain. Hybrids cells were obviously bigger than the parent cells. These results agreed with the results obtained by Gokhale *et al.* (1984) and Shweil *et al.* (1998). It was observed that the hybrid strain was a combination of both the parents. *For example*, like *E. coli* hybrid strain was indol positive, showed good growth on MacConkey's agar and similar to *Achromobacter lacticum* it was oxidase positive and aerobic. At the same time, the hybrid showed physiological and morphological differences too from its parent strains. It was different in shape, size, colony characteristics and growth pattern in different types of media. These differences may be due to the combination and expression of the genetic markers of the parent in the hybrids. (Gokhale and Deobagkar, 1989).

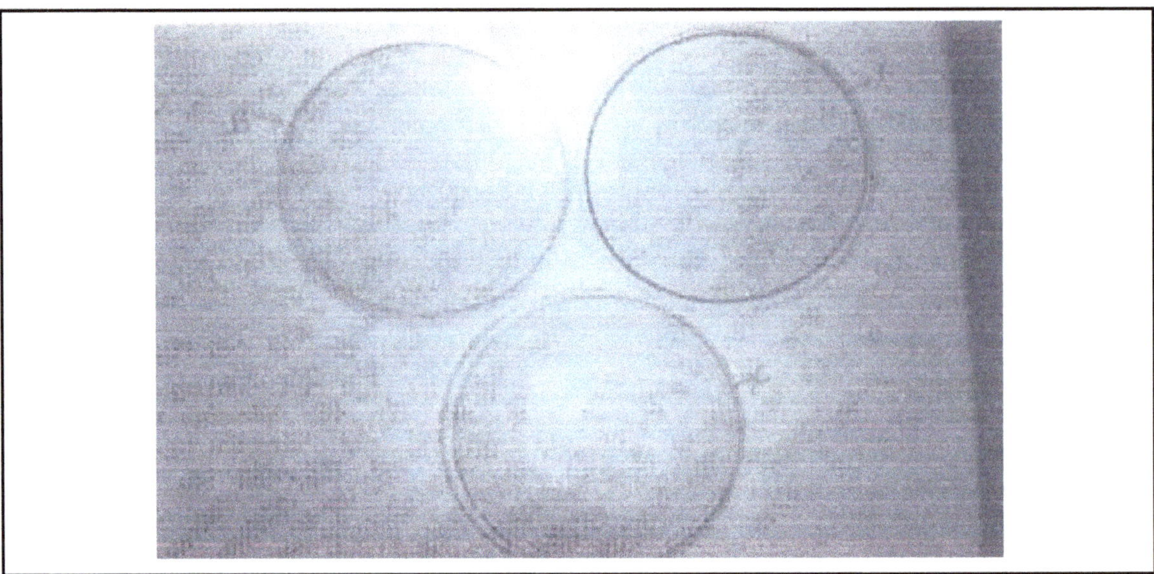

Figure 52.1: Colonies of *E. coli* (A), *Achromobacter lacticum* (B), Hybrid Strain (C)

Degradation Analysis

Degradation of Chlorpyrifos using the isolated strain *E. coli* and it's hybrid with Achromobacter lacticum was assayed and has been presented in Figure 52.2, which shows the cumulative percentage of biodegradation on 28th day by *E. coli, Achromobacter lacticum* and the fusant.

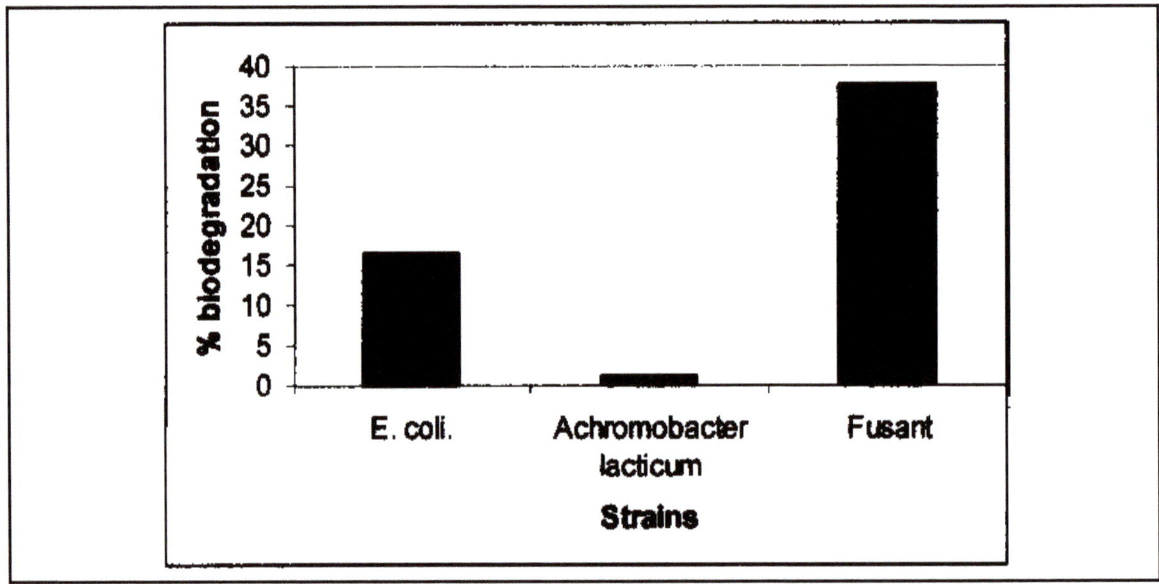

Figure 52.2: Cumulative Percentage of Biodegradation

For calculating the biodegradation percentage, first total CO_2 production in terms of mg is to be calculated. Table 52.1 shows the total amount of CO_2 production using Chlorpyrifos as a sole source of carbon, which in case of *E. coli* was 41.6 mg and the hybrid strain produced 91.9 mg CO_2. *Achromobacter lacticum* could produce only 3 mg of CO_2. Cumulative percentage of biodegradation calculated for *E. coli* was 16.6 per cent and that for hybrid strain 37.5 per cent and for *Achromobacter lacticum* only 1.1 per cent (Figure 52.2).

Table 52.1: Cumulative Value of mg CO_2 Produced after 28 Days of Incubation

Strains	*mg CO_2 Evolved*
E. coli.	41.6
Achromobacter lacticum	3
Fusant	91.9

Note: Value in each case is a mean of two replicates.

These findings imply that *Achromobacter lacticum* plays a negligible role in Chlorpyrifos degradation. However, after fusion with *E. coli* it enhances the degrading ability of hybrid strain. No reports are available till now about the degradation of Chlorpyrifos by *Achromobacter lacticum.* Although Maloney *et al.* (1998) has reported the transformation of Pyrethroid insecticides by *Achromobacter* sp. To the best of our knowledge', fusion between *Achrmobacter lacticum* and *E. coli* strains with an intention to enhance the degrading ability of Chlorpyrifos is the first of its kind.

The present results show the role *E. coli* isolates for Chlorpyrifos biodegradation either when used individually or after hybridization. Richins *et al.* (1997) and Wang *et al.* (2002) has reported the degradation of Chlorpyrifos co-metabolically in liquid media by *Escherichia coli* clone with an opd gene.

After fusion with *Achromobacter lacticum* the degrading ability increased to almost double. These findings are supported by the results of Mansee *et al.* (2004b). They enhanced the detoxification of organophophorus pesticide using intergeneric protoplast fusion between *E. coli* and two bacterial genera *Bacillus thuringiensis* D55 and *Agrobacterium tumefaciens.*

Siddavattam *et al.* (2003) cited that the efficiency of bacteria to degrade organophosphorus pesticides was generally due to their content of the enzyme organophosphorus hydrolase (OPH) that is encoded by a plasmid gene (opd). Therefore, the Chlorpyrifos (an organophosphorus) degradation enhancement by the hybrid of *E. coli* and *Achromobaceter lacticum* could be due to the DNA recombinations after protoplast fusion as well as improving the gene action of opd gene of *E. coli* in the hybrids. However, the definite changes on enzymatic level in these two strains which led to enhanced Chlorpyrifos degradation, after fusion is subject to further investigations.

Conclusion

The present work has brought about the formation of a stable recombinant fusant from two different gram negative bacteria, with the help of protoplast fusion technique. Intention of this fusion was to develop a novel strain with an ability to degrade Chlorpyrifos more efficiently. For the first time one of the genetic recombination technique *i.e.,* Protoplast fusion technique has been applied for fusion between *E. coli* and *Achromobacter lacticum* and to obtain a strain for rapid degradation of Chlorpyrifos pesticides. Results showed that after protoplast fusion pesticide degrading ability almost doubled. This method can be made commercially viable and can be adopted to develop other bacterial strains for combating pesticide's pollution.

Acknowledgements

We sincerely thank Jaslok Hospital and Research Center, Mumbai for providing technical help in electron microscopy for this work.

References

Akamatsu, T. and Sekiguchi, J., 1983. Selection methods in Bacilli for recombinants and transformants of intra and their inter specific fused protoplast. *Archive of Microbiol.*, 134: 303–308.

Bagde, U.S. and Paranjape, V.V., 2000. Protoplast Fusion between *Cellulomonas fimi* and *Brevibacterium divaricatum. Indian J. Environ. and Ecoplan.*, 3(3): 573–578.

Bartha, R. and Pramer, D., 1965. Features of a flask and method for measuring the persistence of pesticides in soil. *Soil Sci.*, 100: 68–70.

Gokhale, D.V. and Deobagkar, D.N., 1994. Isolation of intergeneric hybrids between *Bacillus subtilis* and *Zymomonas mobilis* and the production of thermostable amylase by hybrids. *Biotechnol. Appl. Biochem.*, 20: 109–116.

Gokhale, D.V., Han, E.S., Srinivasan, V.R. and Deobagkar, D.N., 1984. Transfer of DNA coding for cellulases from *Cellulomonas* sp. to *Bacillus subtilis* by protoplast fusion. *Biotechnol. Lett.*, 6: 627–632.

Gokhale, D.V. and Deobagkar, D.N., 1989. Differential expression of xylanses and endoglucanases in the hybrid derived from intergeneric protoplast fusion between a *Cellulomonas* species and *Bacillus subtilis. Appl. Environ. Microbiol.*, 55: 2675–2680.

Kreig, N.R. and Holt, J.G. (Ed.), 1984. *Bergey's Manual of Systematic Bacteriology.* The William and Wilkins Co., Baltimore, London.

Maloney, S.E., Maule, A. and Smith, A.R.W., 1998. Microbial transformation of the pyrethroid insecticide: Permethrin deltamethrin, fastac, fenvalerate and fluvalinate. *Appli. and Environ. Microbio.*, 54(2): 2874–2876.

Mansee, A.H., Montasser, M.R. and Abou Shanab, A.S., 2004a. Decontamination of pollutants in aquatic system. 1. Biodegradation efficiency of isolated bacteria strains from contaminated areas. *Pakistan. J. Biolog. Sci.*, 7 (7): 1202–1207.

Mansee, A.H., Montasser, M.R. and Abou Shanab, A.S., 2004b. Decontamination of pollutants in aquatic system. 2. Adapting certain bacteria strains for organophosphorus pesticides degradation. *The Egyptian Science Magazine*, 1(1): 23–26.

OECD Guidelines for the Testing of Chemicals 301B.(1992). Ready Biodegradability: CO_2 Evolution (Modified sturm test) Paris, Cedex.

Richnis, R., Kaneva, I., Mulchandani, A. and Chen, W., 1997. Biodegradation of organophosphorus pesticide using surface expressed organophosphorus hydrolase. *Nat. Biotechnol.*, 15: 984–987.

Rodriques, H., Gracia, B., Ancheta, O. and Sipicizki, M., 1991. Formation, regeneration and fusion of protoplast in *Cellulomonas* strain. *Curr. Microbio.*, 23: 265–270.

Shweil, S.F., Yacout, M.A. and Abou-Youssef, A.Y., 1998. Enhancement of genetic activity for toxicity against *Spodoptera littoralis* through protoplast fusion of some *Bacillus thuringiensis* strain. *Alexandria Journal of Agriculture Research*, 43: 205–211.

Siddavattam, D., Khajamohiddin, S., Manavathi, B., Pakala, S. and Merrick, M., 2003. Transposan-like organization of the plasmid-borne organophosphate degradation (opd) gene cluster found in *Flavobacterium* sp. *Appl. Environ. Microbiol.*, 69(5): 2533–2539.

Wang, A., Mulchandani, A. amd Chen, W., 2002. Specific adhesion to cellulose and hydrolysis of organophosphate nerve agents by genetically engineered *Escherichia coli* strain with surface expressed cellulose binding domain and organophophorus hydrolase. *Appl. Environ. Microbiol.*, 68: 1684–1689.

WHO, 2001. *International Programme on Chemical Safety*. Geneva.

Chapter 53

Estimation of Host Infestation and Nematode Multiplication on Two Different Crops by Root Knot Nematode in Manipur

L. Joymati Devi and W. Mema Devi

Department of Zoology, G.P. Women's College, Imphal

ABSTRACT

The host infestation rate and multliplication pattern of root knot nematode on *Allium hookeri* and *Calocasia gigantica* were studied using different inoculum level *viz.*, 0, 10, 100, 1000 and 10,000. The economic threshold level was 100 nematodes which caused 31.54 per cent reduction in the fresh root weight over control. Maximum disease incidence and retardation in plant growth was found in 1000 and 10,000 nematode.

Keywords: *Infestation, Multiplication, Meloidogyne incognita, A. hookerii, C. gigantica.*

Introduction

Root knot nematodes are important and cosmopolitan group of pest on different crops throughout the country. The first record of injury of root knot nematode to vegetables was given by Berkeley, 1855. The sites for infestation and disease incidence for this pest occur in root system and spend most of their lives in roots or soils.

In Manipur some preliminary work has been started by Joymati *et al.* (1999) and from their work these two host plant species *i.e., Allium hookerii and Calocacia gigantica* were found infested with root knot nematode and they were identified as *M. incognita* from the perennial cuticular pattern of female. *A. hookerii* is one of the important condiment plants as well as medicinal plants which grows in north eastern region of India (Sinha, 1996). This plant is commonly used for making different types of dishes. It is a herb, bulb hardly and cultivated. Local medicine men use the root of this plant is used for reducing blood pressure. Leaf juice mixed with salt is prescribed for stomach ulcers. *Calocasia gigantica* is also one of the important commonly used vegetable crops of Manipur (Sinha, 1996). This plant can be used for making different types of dishes and both corm and stem of the plant are edible. It is a herb with root stock and stolons. The juice of petioles is styptic, stimulant and rubefacient. After child birth, lactating mothers take soup of this plant for gaining extra calcium. These two plans are commonly cultivated in both plain and hilly areas for their different purposes and another important point is that the root portion of these two plants are used for their different purposes. Thus, due to their importance specially for their uses and also review to the perusal literature revealed that no detailed investigations have been done so far to determine the extent of damage caused by root know nematode, *M. incognita.* Thus, the present work has been taken up to evaluate the host infestation and nematode multiplication pattern caused by root knot nematode, *M. incognita* on *Allium hookerii* and *Calocasia gigantica.*

Materials and Methods

Experiments for assessing the effect of *Ai incognita* on *A. hookerii* and *C. gigantica* conducted in pot experiments were kept in a net house. The soil used was sand loam type and sterilized in earthen pots of 15 cm diameter by autoclaving. The test plants were propagated in nursery plots. Healthy nursery plants were collected from *M. incognita* tree healthy fields. One plant was planted in each earthen pot having 500 gm of sterilized soil. Seven days after planting of the experimental plants, the freshly hatched second stage larvae of *M. incognita* were collected from culture pots and inoculated in a logarithmic series of 10, 100, 1000 and 10.000 larvae per pot at a depth of 3 cm and covered with sterilized soil followed by light watering. After inoculation regular watering was done till the harvesting of the plants. All the treatments along with control were replicated five times.

Ninety days after inoculation of nematode, all the treated plats were carefully uprooted and observation on plant growth number of galls, larval population in both root and soil, root knot index and reproduction factor were recorded. For dry weight of shoot and root, the plant parts were cut into small pieces separately and kept in an oven at 58±2°C. The dry weights of the shoot were recorded after every 24 hours till a constant weight was obtained. Numbers of galls were recorded per plant per root system. For counting nematode population of root, the infected root was stained with phenol and acid fushcin. For soil population the entire amount of soil from each pot was processed for the extraction of nematodes by Cobb's (1918) Sieving and decanting method followed by modified Baermann's funnel technique.

Results and Discussion

The results presented in Tables 53.1(a–b), 53.2(a–b) shows influence of different inoculums levels (0, 10, 100, 1000, 10000) juveniles on *A. hookerii and C. gigantica.* The findings revealed that the increase in nematode inoculums was associated with progressive reduction in various plant growth parameters, which gave conclusive evidence that *M. incognita* is a potential pathogen for these crops. Plants treated with 10000 juveniles/500 gm of soil showed significant reduction in shoot and root lengths and shoot and root fresh weights. But, the other growth parameters like shoot and root dry weights,

Table 53.1(a): Pathogenecity of *M. incognita* on *Allium hookerii* Showing Different Plant Growth Parameters

Inoculum Level	Shoot Length (cm)	Reduction Over Control	Root Length (cm)	Per cent of Reduction	Fresh Root Wt (g)	Per cent of Reduction	Fresh Shoot Wt (g)	Per cent of Reduction	Dry Root Wt (g)	Per cent of Reduction	Dry Shoot Wt (g)	Per cent of Reduction
0	29.75	11	16.75	–	1.68	–	3.76	–	0.67	–	0.43	–
10	19.6	34.11	14.85	17.34	1.56	7.14	2.31	38.56	0.56	16.41	0.36	16.27
100	18.5	37.81	12.76	23.82	1.15	31.54	1.34	64.36	0.37	44.77	0.16	62.14
1000	15.5	47.89	9.65	42.38	0.93	44.64	1.15	49.41	0.31	53.73	0.13	69.76
10000	12	59.66	5.60	66.56	0.72	57.14	0.93	75.26	0.22	67.16	0.11	74.41
CD at 5%	1.235	–	1.175	–	0.269	–	0.069	–	0.161	–	0.144	–

Table 53.1(b): Effect of Inoculum Levels of *M. incognita* on Host Infestation and Nematode Multiplication on *Allium hookerii*

Inoculum Level	Galls/Plants	Root Knot Index	Nematode Population		Total	Reproduction Factor
			Soil	Root		
0	0	0	0	0	0	0
10	10.5	2	27	19	46	46
100	18.3	2	706	54	760	7.6
1000	25.1	3	4780	312	5092	5.09
10000	32.3	3	10400	430	10830	10.830
CD at 5 per cent	0.366		2.004	1.116	1.839	0.297
CD at 1 per cent	0.506		2.772	1.543	2.544	0.411

Table 53.2(a): Effect of Different Inoculums of *M. incognita* on Plant Growth Parameters of *C. gigantica*

Treatment	*Shoot Length (cm)*	*Per cent of Reduction Over Control*	*Root Length (cm)*	*Per cent of Reduction Over Control*	*Fresh Root Wt (g)*	*Per cent of Reduction Over Control*	*Fresh Shoot Wt (g)*	*Per cent of Reduction Over Control*	*Dry Root Wt (g)*	*Per cent of Reduction Over Control*	*Dry Shoot Wt (g)*	*Per cent of Reduction Over Control*
0 (control)	43	–	22	–	25	–	51	–	39)	–	5.42	–
10	30	3023	145	349	22	120	25	50.98	1.85	53.63	323	40.40
100	26	3953	10	5455	18	28.0	20	60.78	1.01	74.68	272	49.81
1000	22	48.8	9	59.99	10	60.0	12	76.47	0.30	92.48	1.87	65.49
10000	17	60.46	8	63.63	5	IDO	7	86.27	0.19	9523	039	92.80
CD at 5%	0.185	0.224	0.199	0.204	0.510	0.643	0.432	0.177	0.809	0.200	0.729	0.272

Table 53.2(b): Effect of Different Inoculum Levels of *M. incognita* on Host Infestation and Nematode Multiplication on *C. gigantica*

Initial Inoculum	*Galls/Plants*	*Root Knot Index*	*Nematode Population*		*Total*	*Reproduction Factor*
			Soil	*Root*		
0	0	0	0	0	0	0
10	18.3	2	131.3	324.2	455.5	45.55
100	42.1	3	376.4	613.4	989.8	9.898
1000	71.1	4	453.6	961.2	1414.8	1.4148
10000	83.1	4	898.1	1348	2246.1	0.224
CD at 5 per cent	0.516		0.601	0.466	0.509	1.748

number of leaves and number of root knot galls on the roots showed variations in different inoculums levels. Plants with inoculums levels of 1000 and 10000 juveniles showed not only reduction in plant growth but stunting, yellowing of leaves and wilting appearance.

The control plants *i.e.*, uninoculated were free from root knot galls. An increase in the level of nematode inoculums from 10 to 10000 juveniles resulted in a significant increase in host infestation as indicated by number of root knot galls. Maximum root knot galls were recorded at the inoculation level of 10000 second stage juveniles/pot which was least at the level of 10 second stage juveniles/pot. The number of root knot galls on roots, number of seconds stage juveniles in soil and number of different stages *i.e.*, eggs, J_2, J_3 and adults of *M. incognita* inside the root tissue were observed to be increasing with increase in the levels of the inoculation.

The rate of multiplication was found inversely proportional to the population density. The rate of multiplication in the population of various development stages recorded from the roots revealed that *A. hookerii* and *C. gigantica* are good host for root knot nematode *M. incognita*. The work is in conformity with Bora and Phukan (1982) in which highest nematode multiplication was recorded at the lowest inoculum levels on jute. The results can be compared with the finding of Sharma *et al.* (1999) on pathogenecity of *M. incognita* on groundnut who reported that as inoculums level increased, soil population also increased significantly. Similar results also reported by Singh and Goswami (2000) on cowpea where significant plant growth reduction over control was observed with an initial population of 1000 nematode per 500 g of soil which was established as potential pathogenic level of *M. incognita*. Haidar *et al.* (2001) also reported the effect of different inoculums levels of *M. incognita* on two species, *Carum copticum* and *Nigella sativa*. Minimum plant growth was recorded at 10000 level for both the crops. Like wise similar results were obtained during the course of study.

From the above investigation it can be concluded that these two indigenous plants of Manipur area good host for root knot nematode and there is a constant increase in the nematode population both in root and soil with increasing population that has a positive correlation with reduction in plant growth.

Acknowledgement

The author greatly acknowledge to Principal, G.P. Women's College for providing laboratory facility and D.G.C., New Delhi for giving financial assistance of Minor research project during the course of studies.

References

Bora, B.C. and Phukan, P.N., 1982. Studies on the pathogenecity of root knot nematode, *M. incognita* on jute, 1. *Res. Assam. Agric. Univ.*, 3: 176–180.

Cobb, N.A., 1918. Estimating the nema population of the soil. *Agric. Tech. Cic. Bur. Pl. Ind. Us. Dep. Agric.*, 1: 48.

Haidar, M.G., Nath, R.P. and Srivastava, S.S., 2001. Evaluation of brinjal (*Solanum melongena* L.) germplasm for resistance against *M. incognita* Race 2. *Indian J. Nematol.*, 31(1): 93–94

Joymati, L., Romabati, N. and Dhanachand, Ch., 1999. Distribution of host range studies of *M. incognita* (Kofoid and White, 1919) Chitwood, 1949 in medicinal plants of Manipur Part 1. *Indian J. Nematol.*, 19(1): 79–80.

Sitaramaiah, K., 1984. *Plant Parasitic Nematodes of India*. Today and Tomorrows Printers and Publishers, 24 B/5, Desh Bandhu Gupta Road, New Delhi, 43 pp.

Sharma, S., Siddiqui, A.U. and Parihar, A., 1999. Pathogenicity of root knot nematode *M. incognita* (Kofoid and White, 1919) Chitwood, 1949 on groundnut. *Indian J. Nematol*., 29(2): 240.

Singh, S. and Goswami, B.K., 2000. Pathogenecity of *M. incognita* on cow pea. *Indian J. Nematol*., 30(2): 249–250.

Sinha, S.C., 1996. *Medicinal plants of Manipur.* Mass and Sinha Association for Science and Society (MASS), 238 pp.

Chapter 54

Assessment of Indoor Noise Level in University of Jammu, Jammu

Raj Kumar Rampaul and Mosmi Raina

Department of Environmental Sciences, University of Jammu, Jammu – 180 006, J&K

ABSTRACT

The study of, indoor noise is very important because people spent their maximum time period indoor at home and at work place. The indoor noise level is affected by both indoor as well as outdoor sources of noise. The indoor noise level in various teaching departments, administrative block, examination block, hostels, central library, examination halls, auditorium, post office and J&K bank branch located in the Campus and Health center of the Jammu University has been estimated for the working hours except for the teaching departments where it has been recorded for working as well as non-working hours. The study revealed that indoor noise level (Leg) ranged from 61.33–68.44 dB(A) at ground floor, 64.73–77.55 dB(A) at first floor and from 63.22–70.84 dB(A) at second floor during working hours. It has been observed that the indoor noise level at all the sites far exceeds than the permissible limits. Various sources of the indoor noise has also been identified and discussed in the present paper.

Keywords: *Indoor noise, University campus,* L_{eq}.

Introduction

With rapid increase in urbanization and industrialization noise has been on the increase and it has invaded man's most private and precious possession, his mental sanctuary. The noise as a pollutant has deleterious effect on peace of mind and beauty of nature. The study of indoor noise is very important because people spent their long time period indoor at home and at work place, *i.e.*, most of the working employees in various institutes, industries, offices the entire housewives and students spent their time in indoor atmosphere. The indoor noise level is effected both by indoor as well as outdoor sources of noise.

Noise is connected with man's life from cradle to grave even before and after it. All our happy and sad moments are expressed through noise *i.e.,* man is born with noise and dies with noise. Noise affects the audiological, biological, as well as behavioral activities of man at specific noise level. Therefore it is important to evaluate the noise level or particular place to know weather the noise level is above or below permissible limits.

Several studies have been conducted at different places to evaluate the noise levels in different areas such as residential, commercial and silence zones. Persons *et al.* (1977), Sergeant *et al.* (1980), Bansal and Grewal (1990) carried out noise level studies in educational institutions.

The present study has been carried out to evaluate indoor noise level in Teaching and Non-Teaching Departments of University of Jammu, Jammu with an aim to know weather the indoor noise levels in University Campus are above or below the permissible limits as prescribed in Bureau of Indian Standards (B: 4954–1968).

Materials and Methods

To carry out the present study, the study area was divided into six sites and each site was further divided into sub-sites to cover all the Departments of the University Campus.

Site I included all the Teaching Departments located on the Ground Floor, Site II included Teaching Department located on 1st floor and Site III included Teaching Departments located on 2nd floor, Site IV, Site V and Site VI consisted of all the Non-Teaching Departments of University Campus located on Ground floor, 1st floor and 2nd floor respectively. In the Teaching Departments the noise level was recorded during both working and non-working hours whereas in the Non-Teaching Departments the noise level was recorded during working hours daily. Noise measurements were carried out in a weightage using Digital Sound Level Meter Model 8928. During each sampling of noise level, 20 readings of Sound Pressure Levels (SPL) were recorded at an interval of 30 seconds in a period of 10 minutes. From the 20 readings of SPL in 10 minutes during each time period following noise levels were calculated.

Maximum SPL dB(A), Minimum SPL dB(A).

$$L_{eq} = 10\log\left(\sum_{i=1}^{n} fi10^{Li/10}\right) dB(A)$$

where,

fi: Fraction of time for which the constant sound level persists.

I: Time interval

N: Number of observations

Li: Sound intensity at a time interval

Results and Discussion

The analysis of data revealed that the noise level in the Teaching Departments located on Ground floor ranged form 61.33 dB(A) to 68.44 dB(A) with an average L_{eq} of 65.49 ± 2.3 dB(A) during working hours and L_{eq} of 55.01 ± 2.8 dB(A) during non-working hours. The Department of Biotechnology exhibited minimum L_{eq} of 61.33 dB(A) and Department of Urdu exhibited maximum L_{eq} of 68.44 dB(A) during working hours. Whereas during non-working hours minimum Leg of 51.19 dB(A) was observed at Department of Sociology and maximum L_{eq} of 60.17 dB(A) was observed in Department of Home Sciences (Table 54.1).

Table 54.1: Range of Noise Level dB(A) in Teaching Department of University Campus

Floor	*Statistics*	*Range of Noise Level dB(A) During*	
		Working hrs.	*Non-Working hrs.*
Ground floor	Biotechnology, Botany, Geology, Home Science, MBA, Commerce, Sociology, Psychology, Urdu.	65.49±2.3 (61.33–68.44)	55.01±2.8 (51.19–60.17)
1st floor	Environmental Sciences, Zoology, Education, Mathematics, Geography, Law, Chemistry, MCA, Statistics, CEDTI, Physics, History, Dogri, Hindi	69.39±3.2 (64.73–77.55)	55.49±5.8 (48.39–73.21)
2nd floor Economics	Electronics, Political Sciences, (63.22–70.84)	67.60±3.9 (50.60–51.62)	50.96±0.5

During working hours, the average L_{eq} of 69.39 ± 3.2 dB(A) was observed in the Teaching Departments located on first floor with range of 64.73 dB(A) at Department of Mathematics and 77.55 dB(A) in Department of Law whereas during non-working hours, the average L_{eq} was observed to be 55.49±5.8 dB(A) with a range of 48.39 dB(A) at Department of Dogri to 73.21 dB(A) in Department of Law (Table 54.1).

The average L_{eq} of 67.6±3.9 dB(A) and 50.96±0.5 dB(A) was observed in the Teaching Department located on 2nd floor during working and non-working hours respectively. The minimum L_{eq} of 63.22 dB(A) and maximum L_{eq} of 70.84 dB(A) was observed in the Department of Electronics and Political Ciences during working hours and Leg of 50.60 dB(A) to 51.62 dB(A) was observed in the Department of Political Sciences and Electronics during non-working hours respectively (Table 54.1).

In the Non-Teaching Departments, the maximum average L_{eq} of 72.49 dB(A) ± 9.5 dB(A) was observed in Departments located on the ground floor. Minimum L_{eq} of 57.43 dB(A) was observed in Health Center and maximum L_{eq} of 96.72 dB(A) was observed in T.V. Hall of Girl's Hostel during working hours.

The L_{eq} ranged from 51.63 dB(A) in Vice Chancellor's Office to 78.12 dB(A) in Canteen located in Administration Block with an average L_{eq} of 64.16 ± 8.2 dB(A) in the Non-Teaching Departments located on the 1st floor. The average L_{eq} of 64.15±4.4 dB(A), was observed in the Non-Teaching Departments located on 2nd floor (Table 54.2) the L_{eq} ranged from 61.13 dB(A) in Reading Hall of Central Library to 69.27 dB(A) in Re-Evaluation section in Examination Block. .

The overall analysis of date revealed that average maximum L_{eq} of 72.49 ± 9.5 was exhibited by Non-Teaching Department located on ground floor and minimum average L_{eq} of 64.15 ± 4.4 dB(A) was exhibited by Non-Teaching Departments located on second floor during working hours. The overall value of L_{eq} (10 minutes) was observed to range from 51.63 dB(A) to 96.72 dB(A) in University Campus during working hours (Table 54.2).

Table 54.2: Range of Noise Level dB(A) in Non-Teaching Department of University during Working Hours

Floor	*Subsites*	*Range of Noise Level dB(A)*
Ground floor	Accounts section or Administration Block, Ground floor of Central Library, Ground floor of Examination Hall, Registration Section of Examination Block, P.G. Section of Examination Block, T.Y. Hall of Girl's Hostel, Mess Hall of Girl's Hostel, T.V. Hall of Boy's Hostel, Mess Hall of Boy's Hostel, Health Center, Auditorium, Post office of University Campus	72.49±9.5 (57.43–96.72)
1st Floor	Administration section of Administration Block, Vice Chancellor's Office, Canteen of Administration Block, Evaluation Section of Examination Block, Room of Girl's Hostel, Room of Boy's Hostel, 1st floor of Central Library (E_2), 1st floor of Examination Hall	64.16±8.2 (51.63–78.12)
2nd floor	Reading Hall at 2nd floor of Central Library, Development Section of Administration Block, Reevaluation Section of Examination Block.	64.15±4.4 (61.13–69.27)

The critical analysis of data revealed the indoor L_{eq} (10 minutes) value at all the sites and sub-sites exhibited values more-than 40 dB(A) which are above the prescribed limits. From the data it can also be concluded that the teaching activity adds noise to the indoor noise level which is already above 40 dB(A) during non-working hours. (Table 54.2).

Critical analysis of noise level Non-Teaching Departments of University further revealed that indoor noise level (L_{eq}) exhibited decreasing values with increase in status of floor *i.e.,* maximum values was at ground floor followed by 1st floor and minimum value was observed in 2nd floor. But Teaching Departments located on first floor exhibited maximum value followed by Departments at 2nd floor and Departments at ground floor exhibited minimum value during working hours. The noise was observed to be more during working hours. From the data it can also be concluded that the teaching activity adds noise to the indoor noise level which is already above 40 dB(A) during non-teaching hours. Sargent *et al.* (1980) estimated a mean noise level of 61 dB(A) in class rooms of a school which are having a clear view of the road. They concluded that high level of traffic is responsible for higher value of indoor noise in the classrooms.

This showed that indoor noise level is affected by outdoor sources like vehicles, public noise (*i.e.,* students, visitors etc.) which were responsible for higher L_{eq} at ground floor as compared with 1st floor. The critical analysis of the data revealed the indoor L_{eq} (10 minutes) value at all the sites and sub sites exhibited values more than 40 dB(A) which are above the prescribed limits.

From this analysis it can be concluded that the sources of indoor noise in University Campus are outdoor sources as the rooms which are away from main as well as inner roads of the University exhibited low value of indoor L_{eq} as compared to those which are nearer to the roads network. Beside this, student noise in the corridors, noise from ill maintained fans and exhaust fans are also responsible for indoor noise in the University Campus. To minimize indoor noise levels in the classrooms the plying of vehicles within Campus during working hours should be restricted as far as possible.

References

Bansal, A.S. and Grewal, P.S., 1990. Noise pollution: Its awareness and control. In: *Environmental Management,* (Eds.) G.S. Daliwal and V.K. Dilawari. Punjab Agricultural University, Ludhiana, pp. 5–11.

Sergeant, J.W., Gidman, M.I., Humphreys, M.A. and Utley, W.A., 1980. The disturbance caused to school teachers by noise. *J. Sound. Vibr.,* 70(1): 557–572.

Persons, K.S., Bennet, R. and Fidells, 1977. Speech level in various noise environments. *U.S. Environmental Protection Agency, Report* EPA-600/1-77-25, Washington, D.C. 20460.

Chapter 55

Effect of Copper Sulphate on Lipid Peroxidation: An *in vitro* Study on Fish and Chick Liver

M. Sahara, P. Mahato and J. Dandapat*

Department of Zoology, North Orissa University,
Takatpur, Baripada – 757 003, Orissa, India

ABSTRACT

Reactive Oxygen Species (ROS) are continuously generated during normal cellular functions in aerobic organisms and are neutralized by antioxidant defence components of the cell. If ROS production exceeds than the capacity of ROS scavenging it leads to oxidative stress. Lipid peroxidation (LPX) is the oxidative deterioration of membrane lipids and considered as an index of oxidative stress. Though metal ions are essential components of many cellular functions, their over exposure to organisms may induce LPX level. In the present study *in vitro* effect of different concentration of $CuSO_4$ on LPX of liver of fish (*L. rohita*) and chick (*Gallus gallus domesticus*) was compared. The result of the present investigation clearly indicate that in case of fish liver $CuSO_4$ at 100 µM concentration increases the endogenous LPX value *i.e.*, acts as prooxidant, however at higher concentration the formation of MDA is inhibited resulting in lowering of LPX value *i.e.*, act as an antioxidant. But in case of chick model $CuSO_4$ inhibited the process of LPX in all the concentrations studied here.

Keywords: *Lipid peroxidation, Copper sulphate ($CuSO_4$), Labeo rohita, Broiler chicken.*

* Corresponding Author: E-mail: jagneshwar2002@yahoo.com.

Introduction

In aerobic organisms reactive oxygen species (ROS), like Superoxide radicals (O_2^-), hydroxyl radicals ($^{\cdot}OH$) and hydrogen peroxide (H_2O_2) etc., are constantly generated during normal metabolism (Halliwell and Gutteridge, 1999). Under physiological state ROS are effectively disactivated by the antioxidant enzymes and small antioxidant molecules present in the cells (Winston and Di Giulio, 1991; Gille and Sigler, 1995). ROS are potent oxidants and excessive generation or ineffective neutralization of ROS leads to "Oxidative stress" in an individual. Short or long term oxidative stress results in protein degradation, enzymatic inactivation, lipid peroxidation (LPX) and other degenerative events, which ultimately leads to cell death (Katoch *et al.*, 2002). Membrane lipids, especially the Poly-unsaturated Fatty Acids (PUFA) are more prone to ROS attack, which is popularly known as lipid peroxidation (Elstner, 1991) and generally referred to as an index of oxidative stress (Kappus and Sies, 1981).

Earlier studies on different animal model suggest that xenobiotic compounds induce over production of ROS (Winston and Di Giulio, 1991). It has been presumed that toxicity caused to aquatic organisms by heavy metals, particularly the transitional group is partly due to excessive generation of ROS (Thomas and Wofford, 1993; Doyotte *et al.*, 1997). Heavy metals are stable, non-biodegradable, and can accumulate in amplified manner in the tissues of aquatic and terrestrial animals through food chain and cause physiological disorders.

The third most abundant trace mineral in the body, copper, is emerging as one of the most important mineral in the animal diet (Hogstrand *et al.*, 2004). Adequate amount of copper is essential for normal functioning of the immune system in laboratory and domestic animals. However, number of studies on a wide range of animal model support the fact that though copper is an essential metal it can induce the level of LPX (Radi and Markovics, 1988; Viarengo *et al.*, 1998; Doyotte *et al.*, 1997; Romeo *et al.*, 2000, Manzi *et al.*, 2004). In contrast, Dandapat *et al.*, 1999, studied the *in vitro* effect of $CuSO_4$ on LPX in the hepatopancreas of a fresh water prawn *Macrobrachium rosenbergii* and opined that $CuSO_4$ exhibits dual effects on LPX. At low concentration it acts as a prooxidant and at high concentration it behaves as an antioxidants. Thus the role of copper in the induction of LPX is quite interesting and variable in respect to dose, animal model and experimental protocol (*in vivo* or *in vitro*). Therefore the present investigation is designed to compare the kinetics of LPX in two different animal models *i.e.*, broiler chicken (*Gallus gallus domesticus*) and the major carp (*Labeo rohita*). In response to different concentration of $CuSO_4$, under *in vitro* condition. The tissue was chosen liver, because it is physiologically an active organ and an important site of oxyradical formation (Malik *et al.*, 1987) and main homeostatic organ in vertebrate copper metabolism (Blanchard *et al.*, 2004).

Materials and Methods

Chemicals

Thiobarbituric Acid (TBA), was procured form sigma, Bovine Serum Albumin (BSA), Folin clocalteu phenol reagent and all other chemicals were procured from SRL, Mumbai and were of analytical grade.

Collection of Animals and Tissue Preparation

Healthy individuals of Rohu (*L. rohita*) and adult broiler chicken were collected from the local market. The animals were sacrificed and the liver was rapidly removed with proper care. After removing the peritoriial membrane, a piece of liver was washed thoroughly in ice-cold physiological saline, wiped dry using blotting paper and weighed with the help of an electronic monopan balance and processed immediately for biochemical analysis.

Experimental Protocol

Tissues were minced and 10 per cent homogenates of the tissues were prepared in 1.15 per cent ice-cold KCl solution with the help of a motor driven glass Teflon homogeniser. Homogenates were centrifuged at 1000 x g for 10 minutes at 4°C to get Post Nuclear Supernatant (PNS). 150 µl of the PNS containing approximately 1 mg of protein was taken for the estimation of endogenous as well as $FeSO_4$-stimulated LPX. Identical tissue samples were incubated with various concentrations of freshly prepared $CuSO_4$ solutions (50 µl) for 30 min. at 37°C to study the *in vitro* effect of $CuSO_4$ on LPX. After the end of incubation time, LPX level of the samples were estimated as described below.

Lipid Peroxidation Assay and Estimation of Protein Content

Lipid peroxidation of the tissue samples was estimated according to the method of Ohkawa *et al.* (1979) by monitoring the formation of malondialdehyde (MDA), a product of LPX. In brief, the reaction mixture containing approximately 1 mg of protein, 0.3 ml of 8.1 per cent Sodium Dodecyl Sulphate (SDS), 1.5 ml of acetic acid (pH 3.5), 1.5 ml of 0.8 per cent aqueous solution of TBA and 0.6 ml of double distilled water was heated at 95°C for 60 min, then cooled at room temperature and centrifuged at 4000 rpm for 10 min. The absorbance of the supernatant was read at 532 nm in a UV-VIS Spectrophotometer. The amount of MDA formed was calculated by using an extinction coefficient of $1.56 \times 10^5 M^{-1} cm^{-1}$ (Wills, 1969), and expressed as nmol malondialdehyde (MDA) formed per mg protein.

Protein content of the samples was measured according to the method of Lowry *et al.* (1951) using BSA as standard.

Statistical Analysis

Results are presented as means±Standard Error of Mean (SEM). Significant differences between the control and individual treated means were analyzed by student t-test and were considered statistically significant when at least $P < 0.05$.

Results

Endogenous and $FeSO_4$-induced lipid peroxidation in the Post Nuclear Supernatant (PNS) of the chick liver was depicted in Figure 55.1. It is observed that when the PNS of the tissue homogenate was incubated for 30 minutes at 37°C the LPX value remain unchanged in comparison to the endogenous LPX (0 hr). However, when the identical tissue sample was incubated for 30 minutes at 37°C in the presence of 1000 µM of $FeSO_4$ a 20 per cent augmentation in LPX level was observed. Similar result was observed in case offish liver after 30 minutes of incubation at 37°C, which was presented in Figure 55.2. A 7 per cent increment in LPX level was observed when the sample was incubated for 30 minutes at 37°C in the presence of 1000 µM of $FeSO_4$. Effect of $CuSO_4$ on LPX in the PNS of the chick liver was presented in Figure 55.3. Incubation of the tissue sample with 1000 µM of $CuSO_4$ did not alter the LPX level in comparison to control. However, significant decrease (!) in LPX value in the PNS of chick liver was observed in response to 500 (20 per cent!), 1000 (33 per cent!), 2000(35 per cent!) and 4000 (48 per cent!) µM of $CuSO_4$. But in case of fish liver incubation of the tissue sample with 1000 µM of $CuSO_4$ results in 17 per cent increase in LPX level in comparison to control, however the decrease in LPX value in the PNS offish liver is observed in response to 1000 (29 per cent!), 2000 (45 per cent!) and 4000 µM (69 per cent!) of $CuSO_4$. It is interesting to note that incubation of tissue homogenate with 500 µM of $CuSO_4$ has no effect on LPX in the fish tissue, which was shown in Figure 55.4.

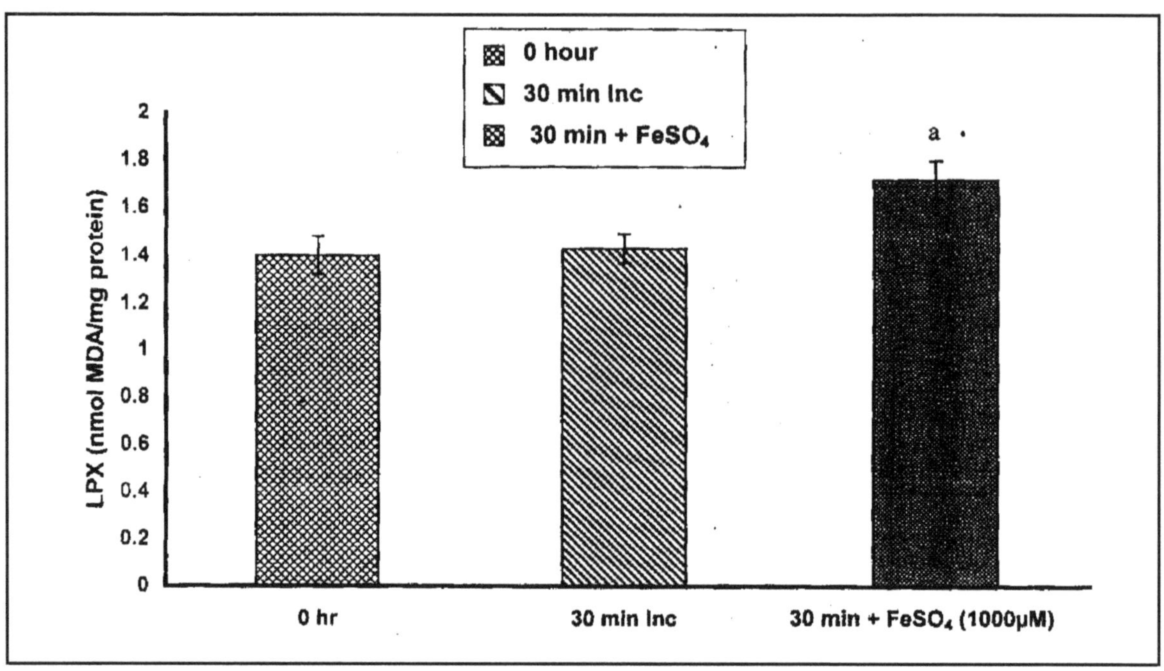

Figure 55.1: Endogenous and $FeSO_4$-induced LPX in PNS of Chick (Broiler) Liver. Data are mean±S.E.M. (n = 4), superscript indicate significant difference ([a]p < 0.05) from control.

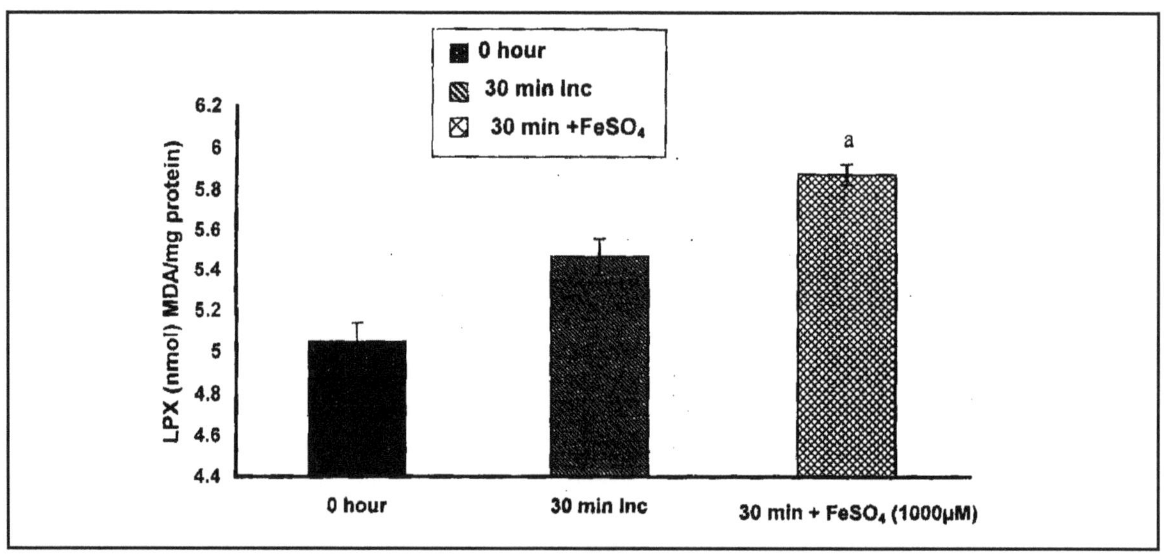

Figure 55.2: Endogenous and $FeSO_4$-induced LPX in PNS of Fish (*Labeo rohita*) Liver. Data are mean±S.E.M. (n = 4), superscript indicate significant difference ([a]p < 0.05) from control.

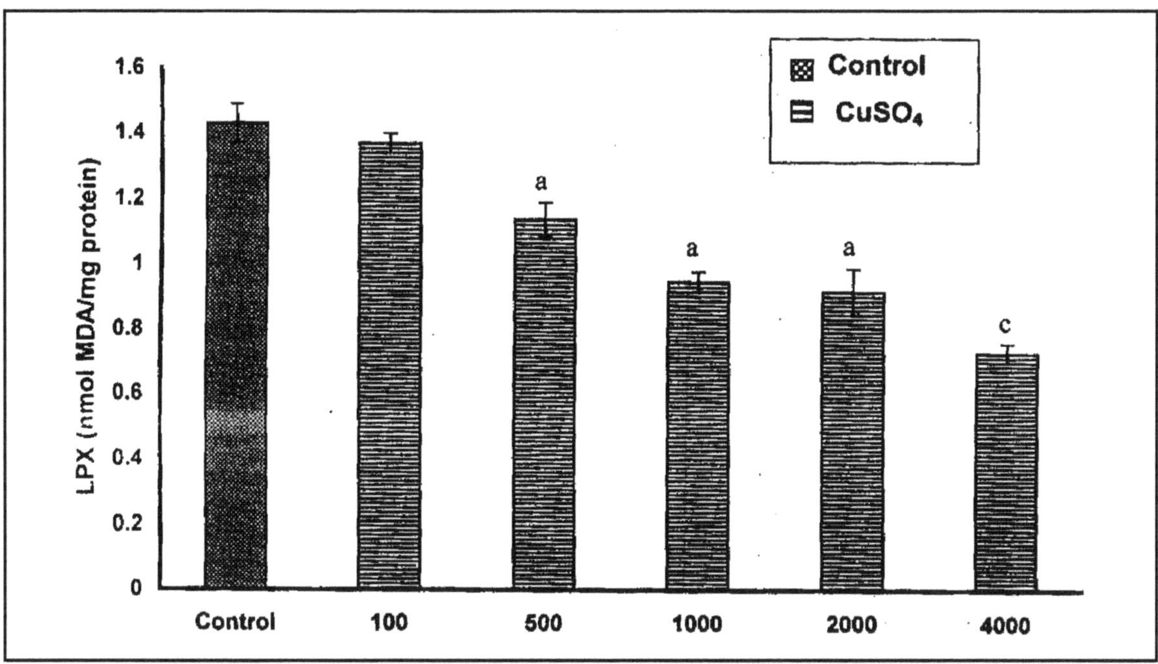

Figure 55.3: ***In vitro*** **Effect of $CuSO_4$ on LPX in PNS of Chick (Broiler) Liver. Data are means±S.E.M. (n = 4), superscripts a and c indicate significant differences from control ($^ap < 0.05$, $^cp < 0.001$).**

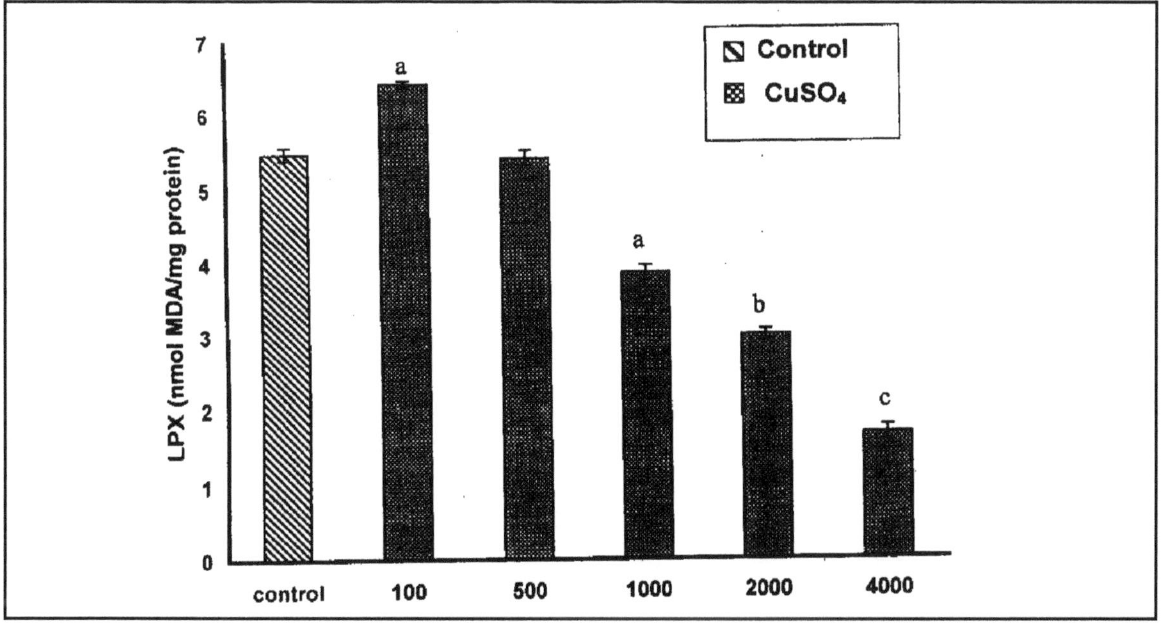

Figure 55.4: ***In vitro*** **Effect of $CuSO_4$ on LPX in PNS of Fish (*Labeo rohita*) Liver. Data are mean±S.E.M. (n = 4), superscripts a, band c indicate significant differences from control ($^ap < 0.05$, $^bp < 0.01$, $^cp < 0.001$).**

Discussion

Lipid peroxidation is induced by iron in both the animal models. Minotti and Aust (1992) reported that iron induces LPX in biological system. It is also an established fact that both Fe^{2+} and Fe^{3+} iron induce oxidative stress by facilitating the decomposition of lipid proxides and formation of hydroxyl radicals from hydrogen peroxide (Halliwell and Gutteridge, 1985).

The results of the present investigation also reveals that in case of fish liver $CuSO_4$ at 100 µM concentration (except 500 µM) increases the endogenous LPX value, however, at higher concentration the formation of MDA is inhibited resulting in lowering of LPX value. Effect of $CuSO_4$ on LPX in various animal models has been reported earlier by several authors (Radi and Markovics, 1988; Dyotte *et al.*, 1997; Varengo *et al.*, 1998). Dandapat *et al.*, 1999, opined that $CuSO_4$ at low concentration (100 µM and 500 µM) exhibit a prooxidant effect on LPX in the crude homogenate of the hepatopancreas of freshwater prawn, *Macrobranchium rosenbergii*. However, at higher concentration (1000 and 2000 µM) it probably inhibits the formation of MDA resulting the decrease in LPX value. Findings of the present investigation are in good agreement with the earlier findings of Dandapat *et al.*, 1999, in prawn model and also further strengthen the fact that $CuSO_4$ at comparatively lower concentration acts as prooxidant and at higher concentration it inhibit LPX. But in case of chick model $CuSO_4$ inhibited the process of LPX in all the concentrations studied here, however the inhibition of LPX by 100 µM of $CuSO_4$ is not significant.

Copper is a transitional metal and has been reported to induce lipid peroxidation by virtue of its redox cycling behaviour between Cu^{+2} and Cu^{+3} oxidation state. Manzi *et al.*, 2004, opined that *in vitro* toxicity of Copper in the isolated hepatopancreatic cells of snail is mediated through the formation of ROS near the site of cell membrane. However, it is also an essential trace element and a integral part of number of copper containing enzymes including Cu–Zn Superoxide dismutase, one of the important antioxidant enzyme that limit the oxidative stress through the elimination of superoxide radical. Besides, Copper has been reported to induce metallothionein synthesis in animal model, which can probably chelate copper (Bouskill *et al.*, 2004). To explain the exact mechanism of antioxidant role of copper in animal model, it needs further *in vivo* study. This will certainly throw some light to understand the metal physiology and biochemistry in relation to copper *i.e.*, from toxicity concept to protection by copper.

Acknowledgements

The authors wish to thank Dr. S.K. Dutta, Professor and Head of the Department of Zoology, North Orissa University, Baripada for providing necessary laboratory facilities.

References

Blanchard, J., Whitehead, A., Oleksiak, M., Crawford, D. and Grosell, M., 2004. (A2.3). Effects of copper on hepatic gene expression in *Funqulus heteroclitus. Abstracts/Comparative Biochemistry and Physiology,* Part A137: S7–S15.

Bouskill, N., Handy, R., Ford, T. and Galloway, T., 2004. (A2.17). Using biochemical biomarkers to differentiate models of arsenic and copper toxicity: Na^+K^+ ATPase, TBA–RS and Metallothionein Induction in the *Zebra mussel, Oreissena polymorpha* and freshwater isopod, *Asellus aquaticus. Abstracts/Comparative Biochemistry and Physiology,* Part A137: S7–S15.

Dandapat, J., Roo, K.J. and Chainy, G.B.N., 1999. An *in vitro* study of metal ion induced lipid peroxidation in giant freshwater prawn *Macrobrachium rosenbergii* (de MAN). *Bio Metal,* 12: 89–97.

Doyotte, A., Cossue, C., Jacquin, M.C., Babut, M. and Vasseur, P., 1997. Antioxidant enzymes, glutathione and lipid peroxidation as relevant biomarkers of experirnental and field exposure in the gills and the digestive gland of the freshwater bivalve unio tumidus. *Aqua Toxicol.*, 39: 93–110.

Elstner, E.F., 1991. Oxygen radicals: Biochemical basis for their efficacy. *Klin Wochenschr*, 69: 949–956.

Gille, G. and Sigler, K., 1995. Oxidative stress and living cells. *Folia Microbial*, 40(2): 131–152.

Halliwell, B. and Gutteridge, J.M.C., 1985. *Free Redicals in Biology and Medicine*. Clarendon Press, Oxford.

Halliwell, B. and Gutteridge, J.M.C., 1999. In: *Free radicals in Biology and Medicine*, 3rd Edn. Oxford University Press, NY.

Hogstrand, C., Feeney, G., Hill, N. and Kille, P., 2004. (A2.2). Essential metal ion pathways and regulatory networks. *Abstracts/Comparative Biochemistry and Physiology*, Part A137: S7–S15.

Katoch, B., Sebastian, S., Sahdev, S., Padh, H., Hasnain, S.E. and Begum, R., 2002. Programmed cell death and its clinical implication. *Indian Journal of Experimental Biology*, 406: 513–524.

Kappus, H. and Sies, H., 1981. Toxic drug effects associated with oxygen metabolism, redox cycling and lipid peroxidation. *Experimentia*, 37: 1223 –1241.

Lowry, O.H., Resebrough, N.J., Farr, AL. and Randall, R.J., 1951. Protein measurement with the Folin Phenol reagent. *J. Biol. Chem.* 193: 165–275.

Malik, Z., Jones, C.J.P. and Connock, M.J., 1987. Assay and sub-cellular localization of H_2O_2 generating mannitol oxidase in the terrestrial slug, Arion ater. *J. Exp. Zool.*, 242: 9–15.

Manzi, C., Krumschnabel, G., Schwarzbaum, P.J. and Dallinger, R., 2004. (A2.15). *In vitro* toxicity of cadmium and copper to snail hepatopancreas cells. *Abstracts/Comparative Biochemistry and Physiology*, Part A137: S7–S 15.

Minatti, G. and Aust, Steven O., 1992. Redox cycling of iron and lipid peroxidation. *Lipids*, 27: 219–225.

Ohkawa, H., Ohisi, N. and Vagi, K.,1979. Assay of lipid peroxides in animal tissues by thiobarbituric acid reaction. *Anal Biochem.*, 95: 351–358.

Radi, A.A.R. and Markovics, B., 1988. Effects of metal ions on the antioxidant enzymes activities, protein contents and lipid peroxidation of carp tissue. *Comp. Biochem. Physiol.*, 90c: 69–72.

Romeo, M., Bennani, N., Barelli-Gnassia, M., Lafaurie, M. and Girard, J.P., 2000. Cadmium and copper display different responses towards oxidative stress in the kidney of the sea bass, *Dicentrarchus labrax. Aquat. Toxicol.*, 48: 185–194.

Thomas, P. and Wofford, H.W., 1993. Effects of cadmium and aroclor: On lipid peroxidation, glutathione peroxidase activity, and selected antioxidants in Atlantic croaker tissues. *Aquat Toxicol.*, 27: 159–178.

Viarengo, A., Pertica, M., Canesi, L., Biasi, F., Cecchini, G. and Orenesu, M., 1988. Effects of heavy metals on lipid peroxidation in mussel tissues. *Mar. Environ. Res.*, 24: 354 (Abst.).

Wills, E.O., 1967. Lipid peroxide formation in microsomes: General considerations. *Biochem. J.*, 113: 315–324.

Winston, G.W. and Di Giulio, R.T., 1991. Prooxidant and antioxidant mechanisms in aquatic organisms. *Aquat. Toxicol.*, 19: 137–161.

Chapter 56

Floristic Study on Epiphytic Pteriodophytes of Thoubal District, Manipur

Y. Sanatombi Devi, Kh. Sandhyarani Devi and P.K. Singh

Ethnobotany and Plant Physiology Laboratory, Department of Life Sciences, Manipur University, Canchipur, Imphal – 795 003, Manipur, India

ABSTRACT

A survey programme of epiphytic pteridophytes distributed naturally in Thoubal District, Manipur was conducted during the period 2005–2007. A total of 20 species under 10 families and 16 genera were collected. Out ofthese plants, 10 species were found growing in extremely rare condition. Disturbance to the natural habitats may caue total elimination of these taxa and needs conservation steps to preserve them.

Keywords: *Epiphytic, Pteridophytes, Thoubal, Distribution.*

Introduction

Epiphytic pteridophytes constitute an important supplement to the terrestrial ferns. It lends a distinct charm and attracts ones attention because of their beautiful and graceful foliage. These species usually grow on the trunks, stems and branches of trees laden with leafy liverworts, mosses with humus, which often form the requisite factors for luxuriant growth of the epiphytic pteridophytes. Most of the epiphytes flourish during rainy season and usually dries up with the onset of winter.

Significant literatures pertaining to pteridophytes of Manipur are still inadequate and missing as this group of plants has been ignored compare to other aspects. Bir *et al.* (1989, 1990, 1991) and

Vasudeva *et al.* (1990) studied the Pteridophytic flora of North-Eastern India, but main emphasis has been laid on Assam plains, Khasi and Jaintia Hills, Lakhimpur District, Aka Hills and Abor Hills. Sinha (1987) studied the ethnobotanical uses of29 pteridophytes from Manipur. Singh (1990) listed 35 species of pteridophytes from Tamenglong District, Manipur. A survey on hydrophilous pteridophytes of Thoubal District, Manipur was conducted by Devi and Singh (2007).

As epiphytic pteridophytes constitute an important supplement to the terrestrial ferns, a comprehensive floristic study is needed. Many regions are yet to be explored practically and extensively. The approach of this paper is to highlight the diversity of epiphytic pteridophytes distributed naturally in Thoubal District, Manipur. The present study also encompasses distribution pattern, categorization on the position of attachment on repository trees, host epiphytic association and other important notes.

Materials and Methods

The study site Thoubal District occupies the eastern half of Manipur valley. It has an area of 514 sq.km and assumes the shape of an irregular and elongated triangle with its base facing north. It lies between 23°45′N and 24°45′N latitude and 93°45′E and 94°15′E longitude. Geologically, the whole district except some small portion around Sikhong–Sekmai area has alluvium having high base status soil. The district is dotted by few hillocks and lakes. The annual rainfall is 109.60 cm with a relative humidity of 56–95 per cent. The rainy season starts in June and continues up to mid October.

A survey on epiphytic pteridophytes distributed naturally in Thoubal District, Manipur was conducted extensively during the period from 2005–2007. Morphological characters (fronds, scales, rhizomes and spores), ecological adaptations and host-epiphytic elements were studied comparatively from live specimens as well as herbarium. The epiphytes are also categorized into 3 types based on the position of attachment on their respective repository plant (host), *viz.*, 1. Base epiphyte 2. Trunk epiphyte and 3. Branch epiphyte. Classification and identification were made by referring to various literatures (Beddome, 1883; Pichi Sermolli, 1977; Baishya and Rao, 1982; Deb, 1983; Borthakur *et al.*, 2001). The specimens are also lodged in the Herbarium, Department of Life Sciences, Manipur University.

Results and Discussion

A total of 20 species representing 16 genera and 10 families were collected and studied. Both the genera and species under each family are arranged alphabetically in Table 56.1 along with their respective repository host trees and area/localities of distribution.

Out of the 20 species of epiphytic pteridophytes that abode in Thoubal District (Manipur), 2 species *viz.*, *Plegmariurus plegmaria* and *Huperzia squarrosa* are fern allies. Many pteridophytes are found growing in diverse habitat. *Nephrolepis biserrata, Drynaria quercifolia, Pyrrosia adnacens* and *P lanceolata* can grow as lithophytes or in terrestrial or epiphytic condition. Four families *viz.*, Huperziaceae, Aspleniaceae, Platyceriaceae and Vittariaceae are exclusively epiphytic in this region. However, it is interesting to note that 3 families (Adiantaceae, Pteridaceae and Lindsaceae) are found growing in epiphytic condition, which otherwise are usually terrestrial in habitat. Polypodiaceae is the largest epiphytic fern family with a maximum of 7 species from the area. Members of Polypodiaceae are found growing together as a dense mat or as a single species on the repository tree. *Pyrrosia lanceolata, P. adnacens* and *Microsorium punctatum* can grow on any repository trees ranging from gymnosperms to angiosperms.

Table 56.1: List of Epiphytic Pteridophytes Found in Thoubal District, Manipur

Sl.No.	Epiphytic Species	Host(s) and Type	Family	Areal Locality
1.	*Adiantum caudatum* L.	*Toona ciliata* Roem Base epiphyte	Adiantaceae (Presl.) Ching	Chandrakhong. Kaina; very rare.
2.	*Asplenium nidus* L.	*Mangiftra indica* L.. *Gmelina arborea L.* Branch epiphyte	Aspleniaceae Mett. ex Frank	Waithou hill; very rare and planted as ornamentals.
3.	*Drynaria propinqua* (Wall. ex Mett.) J. Sm	*Phoenix sylvestris* Roxb., *Mangifera indica* L., *Delonix regia* (Hook.) Raf. Branch epiphyte	Drynariaceae Ching	Thoubal, Lilong, Wabagai; Common.
4.	*D. querc ifoUa* (L.) J.Sm.	*Phoenix sylvestris* Roxb., *Mimusops elengi* L. Branch/Trunk epiphyte	Drynariaceae Ching	Pallel, Thoubal, Khongjom; Common.
5.	*Huperzia squarrosa* (Frost). Trev.	*Ficus religiosa* L., *Mimusops elengi* L. Branch/Trunk epiphyte	Huperziaceae Rothm.	Kakching; very rare, also planted on pots.
6.	*Plegmariurus plegmaria* (L.) Holub	*Stereospermum personatum* (Hassk.) DC. Branch epiphyte	Huperziaceae Rothm.	Wangjing; very rare, planted as ornamentals.
7.	*Sphenomeris chinensis* (L.) Maxon	*Tectona grandis* L., *Phoenix sylvestris* Roxb. Base epiphyte	Lindsaceae Pic. Ser. very common.	Throughout the district;
8.	*Nephrolepis biserrata* (Sw.) Scott	*Phoenix sylvestris* Roxb., *Phoenix humilis* Royle Base epiphyte	Nephrolepidaceae (Ching) Pic. Ser.	Thoubal; not common.
9.	*N. cordifolia* (L.) Presl. *Trunk/*Branch/Base	*Phoenix sylvestris* Roxb. (Ching) Pic. Ser. epiphyte	Nephrolepidaceae	Throughout the district; very common, cultivated for decoration.
10.	*Platycerium alcicorne* Desv.	*Delonix regia* (Hook.) Raf., *Phoebe lanceolata* Nees., *Dipterocarpus tuberculatus* Roxb. Branch/trunk epiphyte	Platyceriaceae (Nayar) Ching	Wabagai; very rare and grown as ornamental.
11.	*Lepisorus subconfluens* Ching	*Citrus maxima* Merr., *Ficus benghalensis* L., *Cinnamomum camphora* (L.) Nees and Eberm. Branch/Trunk epiphyte	Polypodiaceae Bercht. and Presl.	Throughout the district; common.
12.	*Microsorium punctatum* (L.) Copel.	*Phoenix sylvestris* Roxb., *F. glomerata* Roxb., *F. religiosa* L., *Tamarindus indica* L. Branch/Base/Trunk epiphyte	Polypodiaceae Bercht. and Presl.	Throughout the district; common.
13.	*P araleptochilus decurrens* (Bl.) Copel.	*Mimusops elengi* L., *Stereospermum personatum* (Hassk.) DC. Base epiphyte	Polypodiaceae Bercht. and Presl.	Waithou Hill; very rare.

Contd...

Table 56.1–Contd...

Sl.No.	Epiphytic Species	Host(s) and Type	Family	Areal Locality
14.	*Pseudodrynaria coronans* (Wall ex Mett.) Ching	*Phoenix sylvestris* Roxb., *Pinus insularis* Endl.; Trunk epiphyte	Polypodiaceae Bercht. and Presl.	Thoubal, Wangjing; very rare.
15.	*Pyrrosia adnacens* (Sw.) Ching	*Phoenix sylvestris* Roxb., *Phoenix humilis* Royle, *Tamarindus indica* L., *Mangifera indica* L., *Kigelia pinnata* DC., *Ziziphus jujube* Mill. etc. Base/Trunk epiphyte	Polypodiaceae Bercht. and Presl.	Throughout the district; very common.
16.	*P. heteractis* (Mett. ex Khun) Ching	*Araucaria excelsa* R. Br. Base/Trunk epiphyte	Polypodiaceae Bercht. and Presl.	Throughout the district; very common.
17.	*P. lanceolata* (L.) Farewell	*Dalbergia sisoo* Roxb., *Plumeria acuminata* Ait., *Phoenix sylvestris* Roxb., *Ziziphus jujube* Mill. etc. Base/*Trunk/*Branch epiphyte	Polypodiaceae Bercht. and Presl.	Throughout the district; common.
18.	*Onychium siliculosum* (Desv.) C.	*Toona ciliata* Roem., *Phoenix sylvestris* Roxb. Base epiphyte	Pteridaceae Ching	Lilong; very rare.
19.	*Pteris quadriaurita* Retz.	*Phoenix sylvestris* Roxb. Base/Trunk epiphyte	Pteridaceae Ching	Kakching, Lilong, Sikhong- Sekmai; very common.
20.	*Vittaria elongata* Swartz.	*Phoenix sylvestris* Roxb. Trunk epiphyte	Vittariaceae(Presl.) Ching	Lilong; very rare.

Most of the epiphytic species seem to have locally disappeared. There has been regular destruction of natural habitats. Ten species *viz., Adiantum caudatum, Asplenium nidus, Plegmariurus plegmaria, Huperzia squarrosa, Platycerium alcicorne, Pseudodrynaria coronans, Paraleptochilus decurrens, Pyrrosia heteractis, Onychium siliculosum* and *Vittaria elongata* are found growing in extremely rare condition localized ill a particular area. There has been continuous clearing of forest and ruthless destruction of mountains and hills. This disturbance to the natural habitats may even cause total elimination of the taxa from this very region. It is therefore, high time to locate and evaluate the status of these epiphytic elements within the area so as to chalk out various measures for conservation.

Acknowledgements

The authors are thankful to the Head, Department of Life Sciences, Manipur University for laboratory facilities and to Rajeev Gogoi, Botanical Survey of India, Eastern Circle, Shillong for the encouragement and suggestions.

References

Baishya, A.K. and Rao, R.R., 1982. *Fern and Fern-allies of Meghalaya State, India*. Scientific Publisher, Jodhpur.

Beddome, R.H., 1883. *Handbook to the Fern of British India, Ceylon and Malay Peninsula with Supplement*. Today and Tomorrow Printers and Publishers, New Delhi.

Bir, S.S., Vasudeva, S.M., Kachroo, P. and Singh, Y., 1991. Pteridophytic flora of North-eastern India. V. Ecological, distributional and phytogeographical account. *Indian Fern J.*, 8: 98–139.

Bir, S.S., Vasudeva, S.M. and Kachroo, P., 1990. Pteridophytic flora of N.E. India. IV. (Families: Davalliaceae–Salviniaceae). *Indian Fern J.*, 7: 219–226.

Bir, S.S., Vasudeva, S.M. and Kachroo, P., 1989. Pteridophytic flora of North–Eastern India. I. (Families: Huperziaceae–Sinopteridaceae). *Indian Fern J.*, 6: 30–55.

Borthakur, S.K., Deka, P. and Nath, K.K., 2001. *Illustrated Manual of Ferns of Assam*. Bishen Singh and Mahendra Pal Singh, Dehra Dun, India.

Deb, D.B., 1983. *Flora of Tripura State*. Today and Tomorrow Printers and Publishers, New Delhi, India.

Devi, Y.S. and Singh, P.K., 2007. Hydrophilous pteridophytes of Thoubal District (Manipur). *The Indian Forester*, 133: 561–566.

Pichi Sermolli, R.E.G., 1977. Tentamen Pteridophytorumgenera in taxonomicum Ordinem redigende. *Webbia*, 31(2): 313–512.

Singh, O.K., 1990. Floristic study of Tamenglong district, Manipur with ethnobiological notes. *Ph.D. Thesis*, Manipur University, Canchipur, India.

Sinha, S.C., 1987. Ethnobotany of Manipur-medicinal plants. *Frontr. Bot.*, 1: 123–152.

Vashudeva, S.M., Bir, S.S. and Kachroo, P., 1990. Pteridophytic flora of North–Eastern India. III. (Families: Aspleniaceae–Oleandtaceae). *Indian Fern J.*, 7: 66–85.

Chapter 57

Studies on the Primary Productivity in Two Perennial Tanks from Kolhapur District (Maharashtra) India

Milind S. Hujare[1] *and M.B. Mule*[2]

[1]*Zoology Department, S.M. Dr. Bapuji Salunkhe College, Miraj – 41 641, Maharashtra, India*
[2]*Environmental Science Department, Dr. B.A.M. University, Aurangabad – 431 004, Maharashtra, India*

ABSTRACT

The primary productivity of two perennial tanks (Talsande and Vadgaon) Kolhapur District (Maharashtra) India has been studied for two years (May 1999 to April 2001) by following light and dark bottle method. The gross primary productivity in Talsande tank ranged from 0.329 g.c/m^3/h to 0.935 g.c/m^3/h during first year and 0.32 g.c/m^3/h to 1.132 g.c/m^3/h during second year. The gross primary productivity in vadgaon tank during first year ranged from 0.133 g.c/m^3/h to 0.629 g.c/m^3/h and during second year it was 0.187 g.c/m^3/h to 0.601 g.c/m^3/h. The seasonal variations in primary productivity along with some physico-chemical parameters and phytoplankton were studied. In both the tanks productivity showed bimodal peak. The primary peak of productivity was observed in summer and secondary in winter. The productivity was lowest in monsoon. Significant relationships were observed in case of water temperature, phytoplankton, transparency, EC, pH, dissolved oxygen and alkalinity. However relationship of GPP with dissoived oxygen was significant only in Talsande tank.

Keywords: *Primary productivity, Perennial tank, Kolhapur District.*

Introduction

The primary production of the plankton is one of the most important sources of energy input in the aquatic ecosystem. The primary productivity studies are essential for estimating the fish production potential of water body. It is well know that the primary productivity is influenced by the interaction of various factors.

Several attempts have been made to assess the primary productivity of ponds, tanks, lakes, and reservoirs of India, notable among those are of Sreenivasan (1965), Ganapati and Phatak (1969), Rao *et al.* (1981), Paulsamy *et al.(1993)* Vijay Kumar (1994), Birasal (1996), Gurusamy and Ramadoss (2000).

The Talsande and Vadgaon tanks are being used by local fishermen for pisciculture activities; however there is lack of baseline date pertaining to the primary productivity of these tanks. There fore the present investigation has been carried out, which is the part of hydrobiological studies conducted during May 1999 to April 2001.

Materials and Methods

Primary productivity studies were conducted for a period of two years (May 1999 to April 2001) by employing the light and dark bottle method (Gaarder and Gran, 1927.) The operation was carried out *in situ* at two different sites of each tank and results were expressed as average of two sites. Two sets of light and dark bottles were filled with tank water. For estimation of initial dissolved oxygen the water sample was fixed. The set of light and dark bottles were suspended in sub-surface water for the duration of one hour. Meanwhile the initial dissolved oxygen was measured. After completion of incubation period of one hour the dissolved oxygen from light and dark bottles was measured.

The physico-chemical parameters were determined forth nightly from same sites. The water temperature was measured by using mercury thermometer. The pH was measured by using pH meter (Hanna model champ). The transparency of water to light was measured by using secchi disc. The electrical conductivity was measured by using control dynamics digital conductivity meter, model APX 185-E The chemical parameters were determined by following standard methods described by APHA (1985), Trivedy *et al.* (1998). The nitrates (NO_3-N) and phosphate (PO_4-P) were determined by using systronic UV. visible spectrophotometer 118.

The plankton samples were collected fortnightly by filtering hundred liters water through a plankton net made up of bolting silk No. 125. The concentrated samples were preserved by adding 4 per cent formalin and 1 ml of lugols iodine. The quantitative analysis of phytoplankton was done by Lacky's drop count method (Lacky, 1938). The results were expressed as No. 1. Phytoplankton were identified following Fritch (1944) Adoni *et al.* (1985) and Cox (1996).

Study Area

Kolhapur is agriculturally well developed district (17°17′ to 15°43′N and 70°40 to 70°42′E) of south western Maharashtra. The Sahyadri ranges on west and river Warana in north forms natural boundaries of district. The climate is tropical monsoon. The Talsande and Vadgaon tanks are minor irrigation tanks of Hatkanangle tahsil of Kolhapur District. The area receives average annual rainfall about 900 to 1100 mm. The Talsande tank is situated in densely populated area. The maximum depth is 5.0 m. The catchment area of tank is about 8.24 km^2. The basin slope is gentle. The water depths around edges are less. The tank shows vast littoral zone which sustains profuse growth of aquatic macrophytes. The tank water is greenish due to permanent bloom of *Microcystics.* The anthropogenic activities such as washing, bathing, cattle wading are seen in and around the tank. The Vadgaon tank is located away from human locality. It has maximum length 1362.5 m, maximum width 750.0 m. The maximum depth is about 12.0 m. The water depth around the edges is greater. The tank sustains poor growth of aquatic macrophytes. No human activities except fishing are seen in Vadgaon Tank.

Results and Discussion

The monthly values of primary production (GPP and NPP) and community respiration of Talsande and Vadgaon tank are depicted in Table 57.1.

Table 57.1: Primary Productivity (Gross and Net) and Respiration of Talsande and Vadgaon Tank

Month	*Talsande Tank*			*Vadgaon Tank*		
	G.P.P g.c./m³/h	*N.P.P. g.c./m³/h*	*C.R. g.c./m³/h*	*G.P.P. g.c./m³/h*	*N.P.P. g.c./m³/h*	*C.R. g.c./m³/h*
May 1999	0.9.3	0.63	0.28	0.401	0.273	0.123
June	0.795	0.58	0.215	0.259	0.183	0.075
July	0.496	0.31	0.185	0.133	0.145	0.03
Aug	0.597	0.41	0.18	0.168	0.127	0.041
Sept	0.329	0.23	0.0795	0.139	0.08	56
Oct	0.596	0.347	0.248	0.298	0.202	0.0955
Nov	0.583	0.342	0.240	0.386	0.246	0.140
Dee	0.728	0.574	0.154	0.549	0.383	0.166
Jan	0.87	0.658	0.21	0.358	0.31	0.0485
Feb	0.62	0.378	0.228	0.306	0.240	0.0655
March	0.8	0.598	0.2	0.629	0.427	0.202
April	0.935	0.753	0.183	0.476	0.33	0.143
Yearly average	0.687	0.484	0.20	0.346	0.24	0.099
May 2000	1.132	0.817	0.339	0.397	0.294	0.103
June	0.976	0.66	0.320	0.380	0.198	0.192
July	0.49	0.305	0.169	0.216	0.172	0.044
Aug	0.41	0.320	0.150	0.187	0.117	0.07
Sept	0.32	0.113	0.26	0.219	0.17	0.049
Oct	0.395	0.202	0.093	0.364	0.305	0.058
Nov	0.471	0.358	0.112	0.288	0.223	0.065
Dee	0.74	0.495	0.245	0.563	0.415	0.148
Jan	0.795	0.470	0.327	0.396	0.252	0.144
Feb	0.716	0.547	0.169	0.348	0.196	0.152
March	0.547	0.377	0.169	0.511	0.409	0.103
April	0.829	0.509	0.282	0.601	0.353	0.247
Yearly Average	0.651	0.431	0.284	0.372	0.258	0.114
Two Years Average	0.669	0.457	0.242	0.357	0.252	0.107

In Talsande tank during first year of investigation G.P.P. varied from 0.329 g.c./m^3/h (September) to 0.935 g.c./m^3/h (April). The N.P.P. values ranged between 0.2.3 g.c./m^3/h (September) and 0.753 g.c./m^3/h (April). The community respiration values ranged from 0.0795 (September) to 0.28 g.c./m^3/h (May). The yearly average values of G.P.P., N.P.P. and community respiration were recorded as 0.687, 0.484 and 0.20 g.c./m^3/h respectively. The G.P.P. and N.P.P. showed bimodal peaks. The primary peak of G.P.P. was recorded in the month of April (0.935 g.c./m^3/h) and secondary peak was observed in the month of January (0.87 g.c./m^3/h). The primary peak of N.P.P. was recorded in April (0.753) and secondary peak was recorded in January (0.658 g.c./m^3/h).

During second year of study the productivity followed same seasonal pattern. The G.P.P. varied from 0.32 g.c./m^3/h (September) to 1.132 g.c./m^3/h (May). The N.P.P. values varied from 0.113 g.c./m^3/h (September) to.817g.c./m^3/h (May). The community respiration varied from 0.112 g.c./m^3/h (November) to 0.339 g.c./m^3/h (January). The yearly average values of G.P.P., N.P.P. and respiration were 0.651, 0.431 and 0.284 g.c./m^3/h respectively.

The two years average G.P.P., N.P.P. and respiration was 0.669, 0.457 and 0.242 g.c./m^3/h respectively.

The primary productivity in Talsande tank showed seasonal variation. During both the years of investigation the highest values of G.P.P., N.P.P. and respiration were recorded during summer seasons and lowest during monsoon season. The seasonal variations in primary productivity and various physicochemical and biological parameters in Talsande tank are shown in Figure 57.1.

The values of correlation coefficient (r) between gross productivity and some physico-chemical parameters, phytoplankton in Talsande and Vadgaon tank are shown in Table 57.2.

Table 57.2: Values of Correlation Coefficient Between Gross Primary Productivity and Some Physico-chemical and Biological Parameters of Talsande and Vadgaon Tank

Parameters	*G.P.P.*	
	Talsade	*Vadgaon*
Water temperature °C	0.36*	0.18
Secchidisc transparency (cm)	–0.65****	0.34*
Electrical conductivity μ mhos/cm	0.67*****	0.57****
pH	0.50***	0.69*****
Dissolved oxygen mg/l	0.63****	0.16
Alkalinity mg/l	0.62****	0.48**
Hardness mg/l	0.58****	0.52***
Chlorides mg/l	0.32	0.10
Total dissolved solids mg/l	0.62****	0.48**
Nitrate No3–N mg/l	–0.18	–0.26
Posphate P04–p mg/l	–0.12	–0.30
Phytoplankton No/l	0.76*****	0.61****

*****: $p < 0.001$; ****: $p < 0.01$; ***: $p < 0.02$; **: $p < 00.5$; *: $p < 0.10$.

The gross primary productivity showed significant direct correlation with water temperature r = 0.36, E.C. r = 0.67, pH r = 0.50, alkalinity r = 0.62, dissolved oxygen r = 0.63, total dissolved solids r =

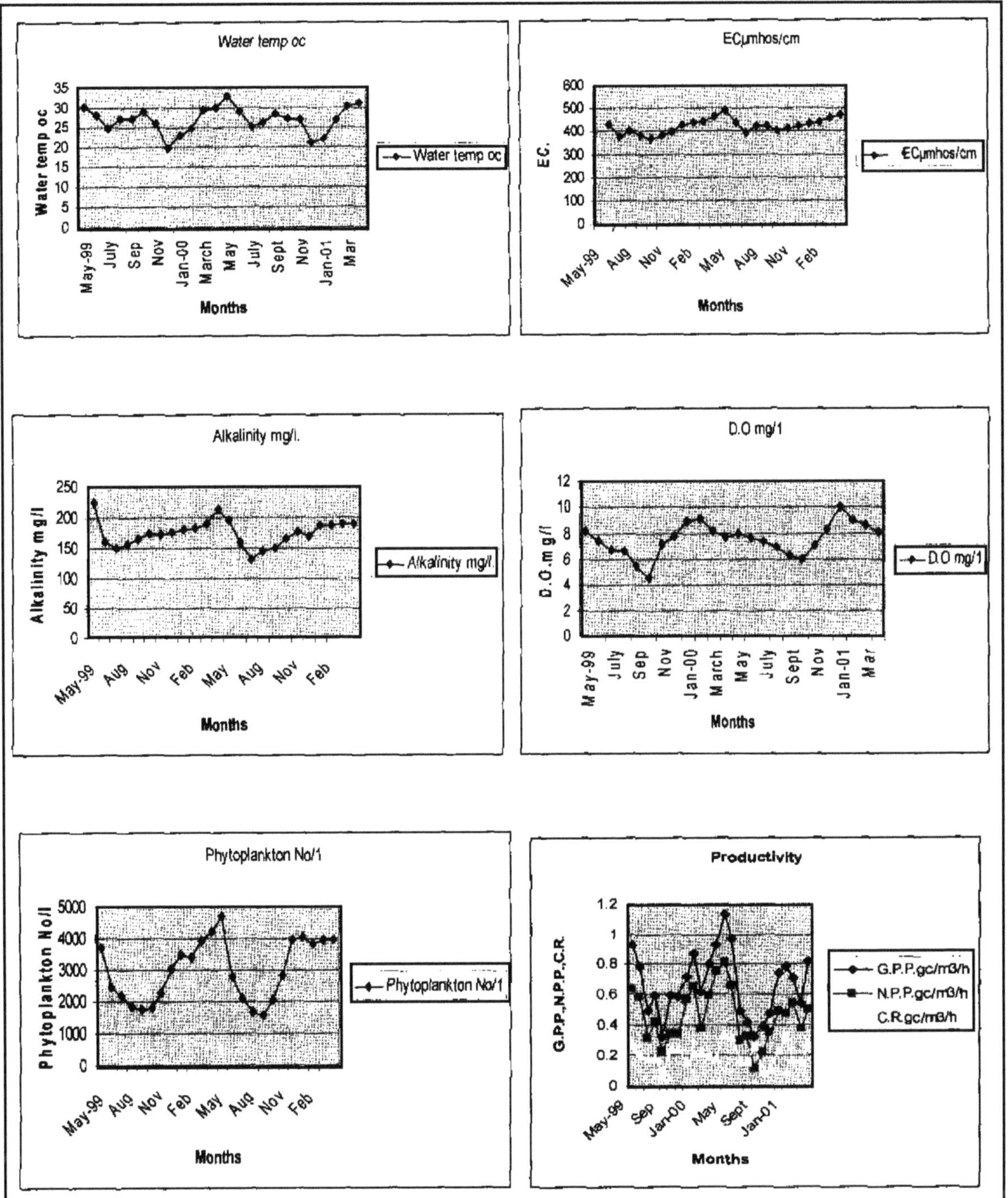

Figure 57.1: Seasonal Variation in Productivity, Some Physico-chemical and Biological Parameters in Talsande Tank

0.62, hardness r = 0.58 and phytoplantkton r = 0.76. Significant inverse relationship between G.P.P. and secchidisc transparency (r = –0.65) was observed.

In Vadgaon tank during first year of investigation the G.P.P. varied form 0.133 g.c./m^3/h (July) to 0.629 g.c./m^3/h (March). The N.P.P. ranged form 0.835 g.c./m^3/h (September) to 0.427 g.c./m^3/h (march). The community respiration ranged between 0.0385 g.c./m^3/h (July) and 0.202 g.c./m^3/h (March). The yearly average values of G.P.P., N.P.P. and C.R. for first year are recorded to be 0.346, 0.24 and 0.099 g.c./m^3/h respectively.

During second year the G.P.P. values in Vadgaon tank varied from 0.187 g.c./m^3/h (August) to 0.601 g.c./m^3/h (April). The N.P.P. varied from 0.117 g.c./m^3/h (August) to 0.415 g.c./m^3/h (December). The community respiration ranged between 0.044 g.c./m^3/h (July) to 6.247 g.c./m^3/h April. The yearly average values of G.P.P., N.P.P. and respiration are 0.372, 0.258 and 0.114 g.c./m^3/h.

The two years average values of G.P.P., N.P.P. and community respiration in Vadgaon tank are 0.357,0.252 and 0.107 g.c./m^3/h.respectively.

The productivity in Vadgaon tank showed seasonal pattern being highest in summer and lowest in monsoon. The seasonal variations in primary productivity and physico-chemical and biological factors of Vadgaon tank are depicted in Figure 57.2.

In the Vadgaon tank significant and direct correlation between gross productivity and transparency r = 0.34, pH r = 0.69, alkalinity r = 0.48, total dissolved solids r = 0.48 and phytoplankton r = 0.61 was observed (Table 57.2).

In the present investigation higher rate of primary productivity was recorded in Talsande tank than Vadgaon tank. This can be attributed to the comparable morphometric features of the two tanks. The Talsande tank is srpaller and shallower water body than Vadgaon tank. As the Talsande tank is located in dense human locality, intensive anthropogenic activities' are seen in and around Talsande tank. Sizable quantities of nutrients may enter through washing and bathing activities in Talsande tank. On the other hand the Vadgaon tank is located away from human locality, the anthropogenic activities are less and the only major source of nutrients may be the run-off. The Talsande tank shows vast littoral zone and some allocanthonous organic matter may also be introduced by aquatic vegetation growing in littoral zone. Ganapati and Sreenivasan (1970) observed higher productivity in smaller water bodies as compared to large lakes. This support our findings.

In the present study definite trend of seasonal. variation in primary productivity was observed in both tanks and bimodal pattern of increased productivity was noticed. The primary peak of production was observed during summer season, resulted in to highest gross production. The secondary peak was noticed in winter and the lowest rate of production was recorded during monsoon. The decline in the productivity during rainy seasons may be due to dilution of tank water and subsequent reduction in phytoplankton density. Moreover the cloudy weather in monsoon might have resulted in low production. Clear days generally in resulted in relatively greater photosynthesis than the cloudy days. Prasad and Nair (1963), Singh *et al.* (1996) also reported low productivity in rainy season.

In the present study direct and significant correlation between gross productivity and phytoplankton population was observed. The rate of production in both the tanks coincided with peak of phytoplankton population. The fall in phytoplankton density in monsoon reflected in decrease in primary production. Khan (1980) observed that population of phytoplankton is sufficiently related with the rate of production. A close relationship between temperature and primary productivity was reported by Singh (1990) and Vijay Kumar (1994). But in the present study the relatively smaller peak

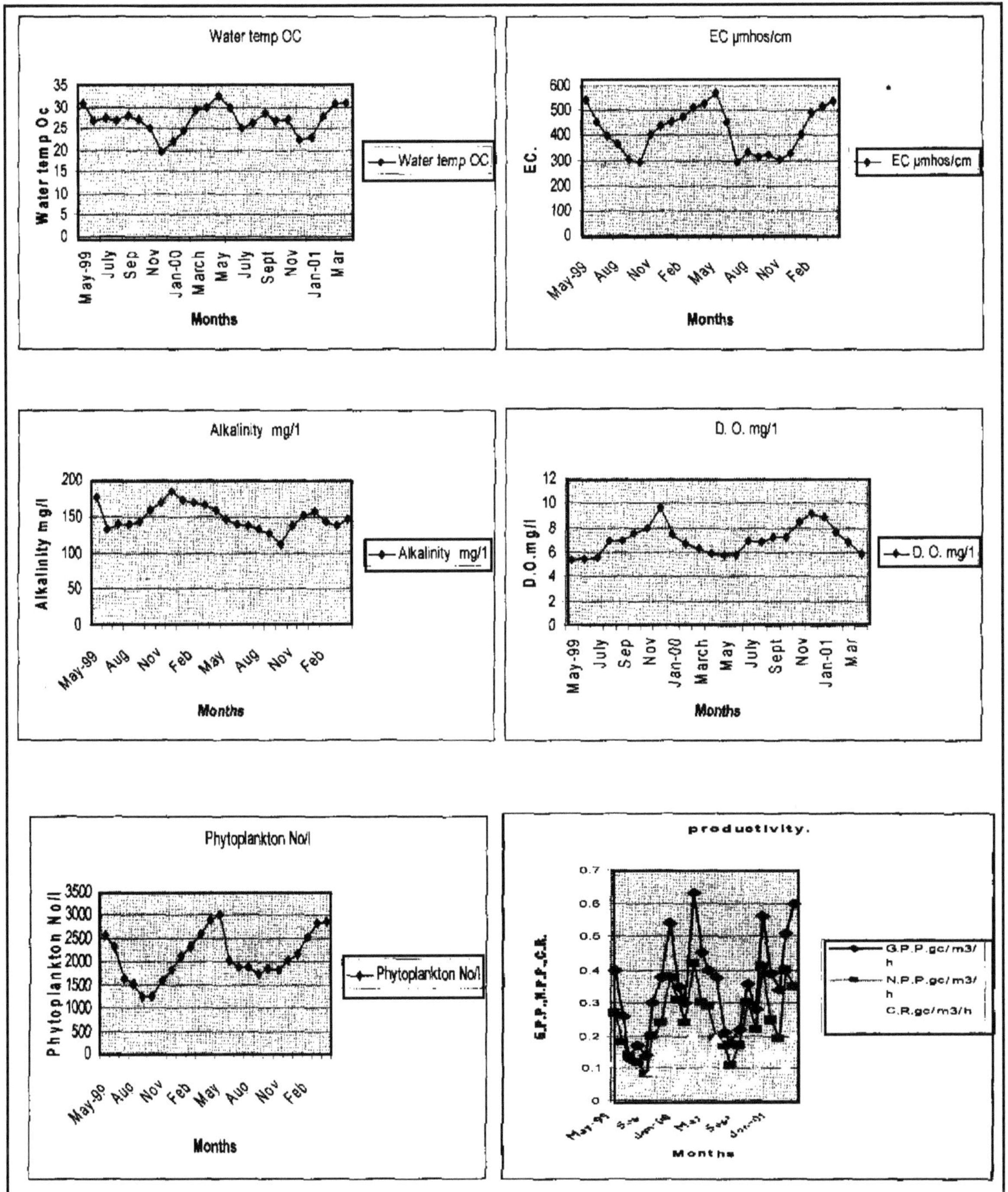

Figure 57.2: Seasonal Variation in Productivity, Some Physico-chemical and Biological Parameters in Vadgaon Tank

of production was observed in winter season, when water temperature was recorded minimum. Hence although direct relationship between G.P.P. and temperature was observed in Talsande tank, no such relationship was noticed in Vadgaon tank. The bimodal peak of productivity was also reported by Khan and Siddiqui (1971). Rao *et al.* (1981) reported winter peak of productivity in Wagholi pond and observed no direct relationship between temperature and productivity. According to Qasim *et al.* (1969) in tropical waters production remains moderate through out the year with little oscillations. This is clearly evident from the present study and temperature is not seems to be limiting factor under tropical conditions.

The alkalinity can be considered as a significant index for production.

According to Jackson (1961) alkalinity below 50 mg/l indicates low rate of photosynthetic rate. In Talsande and Vadgaon tank the alkalinity was much higher and both the tanks can be classified as alkaline tanks. However the alkalinity showed considerable decline during monsoon, remained moderate during winter and highest during summer and as such showed significant direct relationship with productivity. Similar relationship was also reported by Sreenivasan (1964) and Sumitra (1971).

Inverse relationship between gross production and transparency as observed in Talsande tank can be attributed to the increase in transparency during monsoon. As the Talsande tank sustains permanent bloom of *Microcystis,* the dilution of tank water during monsoon resulted into considerable reduction in *Microcystis* population and subsequently decline in gross productivity during monsoon. The direct relationship of gross productivity and transparency as noticed in Vadgaon tank may be due to decrease in transparency during monsoon resulting into fall in phytoplankton density which resulted in to low gross production in monsoon. A similar relationship was also reported by Khan (1980) and Singh (1990).

The range of electrical conductivity in both the tanks indicates their high productivity potential. The electrical conductivity is a measure of total dissolved solids. The E.C. and T.O.S. values of both tanks showed seasonal variation being highest in summer, when the rate of primary productivity was maximum. Thus in the present investigation gross productivity was found to be directly and significantly correlated with E.C. and T.O.S.

The pH of both the tanks remained alkaline throughout the study period. The annual fluctuations were small. The higher pH is normally associated with high photosynthetic activity in water (Goel *et al.,* 1986). In present study it was observed that pH gradually decreased in monsoon. The seasonal variations in pH coincided with gross productivity. Thus, direct and significant correlation between pH and gross productivity was observed.

The direct and significant relationship between gross productivity and dissolved oxygen in Talsande tank can be attributed to increase in dissolved oxygen during summer and winter due to higher photosynthetic activity of phytoplankton.

The nitrate (NO_3-N) and phosphate (PO_4-P) concentration was very poor in both tanks. The nutrients were not found to be limiting the productivity, because lower values of both the nutrients were recorded during summer and winter when the productivity was higher. This can be attributed to the faster recycling of nutrients. A similar observation has been made by Trivedy *et al.* (1993) who recorded high production rates under nutrient poor conditions.

The higher values of community respiration observed during summer month may be due to higher phytoplankton and zooplankton density. Goel and Chavan (1991) also reported higher community respiration in summer. Inspite of clear seasonal trend, in the Talsande and Vadgaon tank, the production was fairly uniform throughout the year except during few monsoon months. The low productivity during monsoon was probably due to heavy rain which resulted in dilution of water. The similar effect has been reported by Ray *et al.* (1966).

As the temperature changes in tropical region are not marked, different patterns of seasonal variation in productivity have been, observed in different water bodies depending up on the local conditions. Therefore generalization seems to be difficult. However Ganapati and Sreenivasan (1970) have tried to generalize the seasonal variations in the productivity of some Indian water bodies and observed that the order of magnitude of production, season-wise was almost same in different water bodies, maximum during summer and minimum during winter season. This is in contrast to present observations where minimum primary production was recorded during monsoon season.

References

APHA, AWWA, WPCE, 1985. *Standard Methods for the Examination of Water and Wastewater*, 20th edn. American Public Health Association, Washington D.C.

Adoni, A.D., Joshi, G., Ghosh, K., Chourasia, S.K., Vaishya, A.K., Yadav, M. and Verma, 1985. *Workbook on Limnology*. Pratibha Publ., Sagar, M.P., pp. 216.

Birasal, N.R., 1996. The primary productivity of the Supa reservoir during its filling phase. *J. Environ. Biol.*, 17(2): 127–132.

Cox Eileen, J., 1996. *Identification of Freshwater Diatoms from Live Material*. Chapman and Hall, 26 Boundary Row, London.

Fritch, F.E., 1994. The present day classification of algae. *Bot. Rev.*, 10.

Gaarder, T. and Gran, H.H., 1927. *Rappet. Proc. Verb Cons Expl. Mer.*, 42: 1–48.

Ganpati, S.V. and Sreenivasan, A., 1970. Energy flow in natural aquatic ecosystem in India. *Arch. Hydrobiol.*, 66: 458–498.

Ganapati, S.V. and Pathak, C.H., 1969. Primary productivity of Sayaji Sarovar (A man made lake) at Baroda. In: *Proc. Seminar on the Ecology and Fisheries of Freshwater Reservoirs*. Cenral Fish Res. Inst. Barrackpore, pp. 37–45.

Goel, P.K., Kulkarni, A.Y. and Trivedy, R.K., 1986. Limnological studies of a few freshwater bodies in South Western Maharashtra with special reference to their chemistry and pollution. *Poll. Res.*, 5M(2): 79–84.

Goel, P.K. and Chavan, V.R., 1991. Studies on the limnology of a polluted fresh water tank. *Aquatic Sciences in India*, p. 51–64.

Gurusamy, K. and Ramadoss, V., 2000. Seasonal variation of chlorophyll and phytoplankton productivity in man made ponds. *J. Ecobiol.*, 12(4): 281–287.

Jackson, D.F., 1961. Comparative studies on phytoplankton Photosynthesis in relation to total alkalinity. *Verh. Int. Ver. Limnol.*, 14: 125–133.

Khan, A.A. and Siddiqui, A.Q., 1971. Primary production in a tropical fish pond at Aligarh India. *Ibid*, 37: 447–456.

Khan Rashid, A., 1980. Primary productivity and tropical status of two tropical water bodies of Calcutta, India.

Bulletin of the Zoological Survey of India, 2 (2 and 3): 129–138.

Lacky, J.B., 1938. The manipulation and counting of river plankton and change in some organisms due to formalin preservation. *U.S. Public Health Reports*, 53: 2080–2093.

Paulsamy, Rathinasmy R. and Manian, S., 1993. Studies on primary productivity in freshwater ecosystem of Bhavanisagar reservoir. *Indian Bot. Contactor*, 10: 29–31.

Prasad, R.R. and Nair, P.V.R., 1963. Studies on aquatic production. I. Gulf of mannar. *J. Mar. Biol. Assoc., India*, 5: 1–26.

Qasim, S.Z., Weliershhaus, S., Bhattathiri, P. and Abidi, S.A.N., 1969. Organic production in a tropical estuary. *Proc. Indian Acadm. Sci.*, p. 51.

Rao, M. Babu, Muley, E.V. and Mukhopadhyay, S.K., 1981. Seasonal fluctuations of primary production in two freshwater ponds at Wagholi, Poona, India. *Geobios*, 8: 270–210.

Ray, P. Singh and Sehgal, K.L., 1966. A study of some aspects of the ecology of river of Ganga and Jamuna at Allahabad (U.P.). *Proc. Nat. Acad. Sci., India*, 37: 235–272.

Singh, Ravindra, 1990. Correlation between certain physico-chemical parameters and primary production of phytoplankton at Jamalpur, Munger. *Geobios*, 17: 229–234.

Sreenivasan, A., 1964. Limnological feature and primary production in a polluted moat at Vellore, Madras State. *Environmental Health*, 6: 237–245.

Sreenivasan, A., 1965. Limnology of tropical impoundments. II. Limnology and productivity of Amaravathy reservoir, (Madras State). *Hydrobiologia*, 26: 501–516.

Sumitra, V., 1971. Seasonal variations in primary production in three tropical ponds. *Hydrobiologia*, 38: 395–408.

Singh, Anil K., Shukla, Arvind N., Saxena, Pankaj and Mendhe, Kavita, 1996. Some observation on primary productivity in river Narmada (Western Zone) M.P., India. *J. Envron. Poll.*, 3: 203–206.

Trivedy, R.K., Goel, B. and Goel, P.K., 1993. Comparative study of primary Production and chlorophyll concentration in a non-polluted and sewage receiving reservoir. *Inter. J. Ecol. Environ. Sci.*, 19: 103–120.

Trivedy, R.K., Goel, P.K. and Trishal, C.L., 1998. *Practical Methods in Ecology and Environmental Science*. Enviromedia Publications, Karad.

Vijaykumar, K., 1994. Seasonal variations in the primary productivity of a tropical pond. *J. Ecobiol.*, 6(3): 207–211.

Chapter 58

Antimicrobial Activity of Aquatic Macrophytes

S. Reshma and V.R. Prakasam

Department of Environmental Sciences, University of Kerala, Kariavattom Thiruvananthapuram – 695 581

ABSTRACT

It is known that certain terrestrial or aquatic plants exhibit unique properties in decontaminating water. Extracts of several herbs or macrophytes possess antimicrobial activity and this property has been made good in water purification process. For this, root, stem, leaf, flower, fruit and seed of diverse species of plants have been made use of. In the present study bactericidal effects of leaf extracts of aquatic plants such as *Pistia stratiotes, Marsilea quadrifolia, Monochoria vaginalis, Hydrilla verticillata, Salvinia cuculata, Ipomoea aquatica* and *Lymnanthimum cristatum* were evaluated. It revealed that leaf extracts of all these plants caused bactericidal effects; antibacterial activity corresponded with concentration of leaf extracts; population of bacteria linearly decreased with increase in concentration of leaf extract; leaf extracts of *Pistia stratiotes* and *Salvinia cuculata* were more bactericidal than others but none of them produced total mortality of bacteria.

Keywords: *Aquatic Macrophytes, Antimicrobial activity.*

Introduction

A large number of herbs have been reported to be useful in purification of water (Varier, 1994; Abou Zeid *et al.*, 1975; Ryan *et al.*, 2001; Sadgir *et al.*, 2002; Babu *et al.*, 2002; Maya *et al.*, 2005; Maya *et al.*, 2006). But it is opined that so far only a small fraction of the plant species which exhibit the properties to act as microbicidal agents has been investigated (Parvatheesam and Shukla, 2004). Most of the investigations on antibacterial effects of plant extracts have been made using land plants especially with medicinal properties, and investigations using aquatic plants are rare. Therefore, a preliminary investigation was undertaken making use of certain aquatic plants that are locally found in Mayyanad of Kollam District in order to find out their bactericidal effects, if any. The aim of study was to identify aquatic plant species that could be used against coliforms in water.

Materials and Methods

In order to find out the effect of aquatic plants on aquatic bacteria, various local species were selected; *Pistia stratiotes, Marsilea quadrifolia, Monochoria vaginalis, Hydrilla verti illata, Salvinia cuculata, Ipomoea aquatica* and *Lymnanthimum cristatum.*

Preparation of Leaf Extracts

The leaf extracts were prepared following the method of Joy *et al.* (2004). Fresh leaves of each plant were removed, weighed and homogenized in sterile de-ionized water at a constant ratio of 5 ml liquid to 1 gm of leaf tissue. The suspension was then filtered through a muslin cloth and centrifuged for 10 minutes. The supernatant served as test substance.

Antimicrobial Assay

Dilutions of each test substance were prepared by mixing leaf extract with molten Mackonckey agar in the ratio of 1 : 10, 2 : 10, 3 : 10, 4 : 10, and 5 : 10. Ten ml each of the medium was then autoclaved

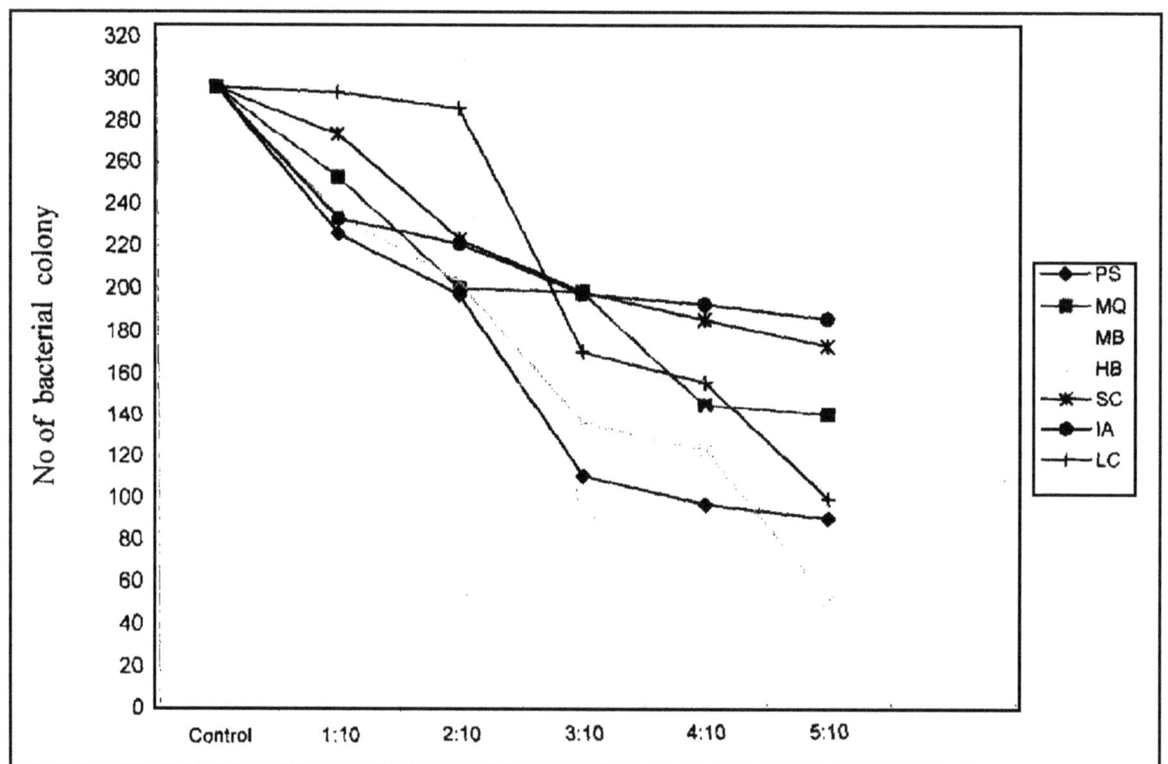

Figure 58.1: Antibacterial Activity of Leaf Extracts of Aquatic Plants in Drinking Water

P.S.: *Pistia stratiotes;* **M.Q.:** *Marsilea quadrifolia;* **M.V.:** *Monochoria vaginalis;* **H.V.:** *Hydrilla verticillata;* **S.C.:** *Salvinia cuculata;* **L.A.:** *Ipomoea aquatica* **and L.C.:** *Lymnanthimum cristatum.*

and subsequently poured into a sterile petri dish and allowed to set. One ml of fresh water sample was then pipetted into the plate and mixed thoroughly. Control plate consisted of 10 ml of agar and 1 ml of water sample without leaf extract. The plates were later incubated for 24 hrs at a temperature of 35°C and observed for bactedal growth. After incubation the number of colony was counted (Ryan *et al.*, 2001).

Results and Discussion

The results of effect of leaf extracts are given in Table 58.1 and Figure 58.1. It showed that (1) leaf extracts at all concentrations caused bactericidal action (2) antibacterial activity corresponded with concentration of leaf extracts (3) a linear decrease in number of bacteria with increase in concentration of leaf extract (4) leaf extracts of all the aquatic plants showed almost the same trend, however, leaf extracts of *S. cuculata* and *P. stratiotes* were more bactericidal than others (5) none of the extracts at the selected concentrations could produce total mortality of bacteria. However, it was concluded that many of the local fresh water macrophytes could be made use in the production of herbal disinfectants for water treatment.

Table 58.1: Effects of Various Leaf Extracts on Bacterial Count in Drinking Water Sample

Sl.No.	*Water Sample + Leaf Extract*	*Number of Bacterial Colony in Presence of Plant Extracts*						
		P.S.	*M.Q.*	*M.V.*	*H.V.*	*S.C.*	*L.A.*	*L.C.*
1.	Water sample (Control)	296	296	296	296	296	296	296
2.	Water sample + 1 ml of leaf extract	225	252	198	234	273	232	293
3.	Water sample + 2 ml of leaf extract	196	199	164	202	222	220	285
4.	Water sample + 3 ml of fleaf extract	109	1.97	143	135	197	196	169
5.	Water sample + 4 ml of leaf extract	95	143	139	121	184	191	154
6.	Water sample + 5 ml of leaf extract	88	138	120	49	171	184	97

P.S.: *Pistia stratiotes;* M.Q.: *Marsilea quadrifolia;* M.V.: *Monochoria vaginalis;* H.V.: *Hydrilla verticillata;* S.C.: *Salvinia cuculata;* L.A.: *Ipomoea aquatica* and L.C.: *Lymnanthimum cristatum.*

References

Abou-Zeid, A.Z., Hamsa, F.A. and Sheebeeny, 1975. Antimicrobial activities of *Hisbiscus sabdariffa, Pimpinella anisum, Cartina carawi, Coffea arabica, Thea sinunsis, Trigonella foenum graecum, Cinnamomum zeylanicum* and their therapeutic values. *Indian Journal of Microbiology*, 15(3): 114–117.

Babn, B., Jisha, V.K., Salitha, C.V., Mohan, S and Valsa, A.K., 2002. Antibacterial activity of different plant extracts. *Indian Journal of Microbiology*, 42(4): 361–363.

Joy, M., John, J., Smitha, K.P and Nair, R.V., 2004. Inhibitory effects of Cashew (*Anacardium occidentale* L) on poly pathogenic fungi. *Allelopathy Journal*, 13(1): 47–56.

Maya, S., Maragathan, T. and Iyer, C.S.P., 2005. Antimicrobial activity of three selected macrophytes of Southern India. *Indian Hydrobiology*, 8(2): 101–107.

Maya, S., Maragathan, T. and Iyer, C.S.P., 2006. Antimicrobial effect of some aquatic plant species of Kerala, India. *Indian Hydrobiology*, 8(2): 193–198.

Parvatesam, M. and Shukla, V., 2004. Antibacterial activities of extracts from leaves of 2 medicinally important plants *Citrus lemon* (L) and *Murraya koenigii* (L). *Journal of Ecology, Environment and Conservation*, 10(4): 537–539.

Ryan, W.J., Wilkinson, J.M. and Cavanagh, H.M.A., 2001. Antibacterial activity of Raspberry Cordial *in vitro*. *J. Research in Veterinary Science*, 71: 1–5.

Sadgir, P., Gargh, S.L., Patel, A.N and Jain, A., 2002. Use of different herbs as coagulant and coagulant aid for water purification. *Research Journal of Chemistry and Environment*, 6(2): 53–58.

Varier, P.S.V., 1994. *Indian Medicinal Plants: A Compendium of 500 Species*, Vol. 5. Orient Longman, Madras.

Chapter 59

Influence of Different Sources of Organic Manures and in Organic Fertilizer on Growth and Yield of Transplanted Rice (*Oryza sativa* L.) Variety–KMP 101

S. Shwetha[1], *J. Narayana*[1], *Narayana S. Mavarkar*[2] *and K.M. Mahadevan*[3]

[1]*Department of P.G. Studies and Research in Environmental Sciences, Kuvempu University, Jnana Sahyadri, Shimoga, Karnataka*
[2]*Department of Agronomy, Agricultural College, Navile, Shimoga – 577 204*
[3]*Department of Chemistry, Kuvempu University, Shankaraghatta – 577 451*

ABSTRACT

Field experiment was carried out under irrigated conditions during *Kharif* season of 2006–07 at Kuvempu University campus to evaluate the performance of organic manures (Vermicompost and FYM) and inorganic fertilizers on paddy crop and soil fertility. The paddy variety KMP 101 was used for the experiment. The paddy when supplied with 100 per cent of recommended dose of fertilizer (100 : 50 : 50 NPK kg ha^{-1}) along with vermicompost @10 t ha^{-1} produced significantly higher number of effective tillers m^{-2} and filled grains panicle^{-1}, grain yield (65.45 q ha^{-1}) and straw yield (70.73 q ha^{-1}) as compared with no application of fertilizer and 100 per cent Recommended Dose of Fertilizer (RDF) alone, but it was on par with the treatment that received vermicompost @15 t ha^{-1}. Despite higher soil reserve of available nitrogen (298 kg ha^{-1} available phosphorus (30 kg ha^{-1}) and available potassium (336 kg ha^{-1}) were found both in the application of vermicompost 15 t ha^{-1} and

100 per cent RDF along with vermicompost @10 t ha^{-1} than application of conventional RDF alone and no fertilizer application. Significantly higher number of grains (18.66 g plant $^{-1}$), Panicle length (25 cm), test weight (2.37 g $panicle^{-1}$) were also noticed in application of vermicompost combined with 100 per cent of recommended dose than rest of the treatments except in application of vermicompost @15 t ha^{-1}.. The result indicated that application of vermicompost @15t ha^{-1} and @10 t ha^{-1} + RDF increased the grain yield besides sustaining the fertility status of the soil as compared with 100 per cent of RDF alone.

Keywords: *Vermicompost, Chemical fertilizer, Farmyard manure, Paddy KMP 101.*

Introduction

Rice is the principal food crop of about 70 per cent of Asian population. Rice consumers world over are likely to increase more than 4 million in the next two decades. According to Rothschild (2006) the annual rice demand may rise, from the present 560 million tonnes to more than 900 million tonnes. World rice production in 2004 was 610 m t in an area of 147 m ha with the productivity of 3750 kg ha^{-1}. The rice productivity of India is lower than the world average, which is mainly due to existing low yielding varieties, susceptible to pests and diseases, lack of selection of suitable genotypes and imbalance nutrient management. Among these balance nutrient management Playa vital role improving the production of rice to a great extent. In recent years due to increased cost of fertilizers, detrimental effect on the soil health and water quality have been noticed. The reduction in the use of chemical fertilizer and supplementing the same through organic manures such as FYM, poultry manure, castor cake, press mud, vermicompost have become necessary to sustain productivity and profitability and to maintain the health of the soil.

Use of organic manures in integrated nutrient management system improve the soil properties and boost the existing yield levels. The supplementary and complementary use of organic manures along with chemical fertilizers besides, improving the soil physico-chemical properties they also enhance the efficiency of applied fertilizers. On the other hand, application of farm yard manure is practiced for many years, but it has become scarce due to scanty population of livestock. Therefore, vermicompost is one of the alternative source of organic manure.

Trials on application of vermicompost to rice fields showed enhanced beneficial microbial activities such as N_2 fixing bacteria and *mycorrhizae* as they release growthpromoting substance for betterment of crops. (Kale *et al.*, 1992).

Many agricultural and natural science scientists worked on vermicompost and its combination with chemical fertilizers on crop growth and yield of paddy. But, the information available on quantity of organic manure to be used to get maximum yield of rice is very meager. Hence keeping these points in view a study was initiated to determine the quantity of vermicompost required for maximization of rice yield.

Materials and Methods

Field trial was conducted at Kuvempu University, Shankaraghatta, Shimoga during kharif season of 2006–07 to study the effect of organic (vermicompost and farmyard manure) and chemical fertilizers (NPK) on rice crop. The rice variety used for the experimentation was KMP 101 (THANU).

The treatments comprised of no fertilizer application, Recommended Dose of Fertilizer (RDF), FYM and different doses of vermicompost. Totally there were 7 treatments were made and experiment was laid out in randomized complete block design with each treatment replicated thrice. Vermicompost and well-decomposed farmyard manure were applied in the rice field as per treatment imposed during final land preparation. While applying NPK fertilizer to the test crop 50 per cent of nitrogen and full dose of phosphorus and potassium were applied as basal before transplanting. The rest of the quantity of nitrogen was applied in two splits as top dress, *i.e.*, one at the active tillering stage and the other at the panicle initiation stage. Twenty-five days old seedlings were transplanted on 23rd July 2006 with a spacing of 20 cm × 10 cm and harvested on 22nd November 2006. The observations on growth attributes were recorded at 30, 60, 90, and 120 days after transplanting and yield attributes were recorded from the earmarked area.

Treatmental Details

T_1: Control

T_2: Vermicompost 5 t ha^{-1}

T_3: Vermicompost 10 t ha^{-1}

T_4: Vermicompost 15 t ha^{-1}

T_5: Farm yard manure 10 t ha^{-1}

T_6: Recommended dose of fertilizer @ 100 : 50 : 50 kg ha^{-1} (RDF)

T_7: Vermicompost 10 t ha^{-1} + RDF.

The data collected on different characters during the course of investigation were subjected to Fishers method of analysis of variance technique for interpretation of the data as given by Panse and Sukhatme (1967). The level of significance used in "f" and "t" test was $p = 0.05$. Critical Difference (CD) values were calculated for the $p = 0.05$ probability level wherever "f' test was found significant

Plants were dried and powdered finely for estimation of nutrients. The extract was prepared for the analyzing NPK contents. Plants analyses were made by using kjeldal method for nitrogen and for phosphorus by spectrophotometer and for potassium by flame photometer

Results and Discussion

Yield Parameter

Combination of vermicompost @ 10 t ha^{-1} + RDF recorded significantly higher grain yield (65.46 q ha^{-1}) over rest of the treatment except Vermicompost @ 15 t ha^{-1} (62.31 q ha^{-1}). The increase in yield might be due to significantly increase yield parameters *viz.*, panicle length, filled grains per panicle and number of panicles per plant. The results are in agreement with the findings of Bhattacharji *et al.*, 2001.

Growth Parameter

Different level of vermicompost and combined application of vermicompost and RDF affected the growth components significantly (Table 59.1). Combination of vermicompost 10 t ha^{-1} + RDF showed taller plants (107.46 cm) than rest of treatments except in 15 t ha^{-1} treatment (104.66 cm). Similar trend was also expressed in number of tillers, number of leaves per plant. The increased growth parameters might be due to the increased uptake of nutrient and more dry matter production. Further, application of vermicompost in conjunction with RDF promotes nutrient uptake for better growth. These results corroborates with the findings of Kale *et al.* (1992) and Zhoa Shi-Wei and Huong Fuzhen (1992).

Table 59.1: Effect of Organic Manure and Fertilizers on Plant Height (cm) of Rice

Treatment	*30 Days*	*60 Days*	*90 Days*	*120 Days*
T_1 Control	13.33	35.16	66.00	69.66
T_2 VC at 5 t/he	14.00	43.00	68.33	75.00
T_3 VC at 10 t/he	18.00	5.33	89.66	98.00
T_4 VC at 15 t/he	21.50	65.80	98.80	104.66
T_5 FYM at 10 t/he	17.00	50.66	74.66	85.00
T_6 RDF	20.00	63.76	86.33	99.90
T_7 VC at 10 t/he+RDF	24.83	70.16	100.33	107.46
SEM+	0.91	0.72	0.54	0.66
C.D at 5 per cent	2.81	2.21	1.67	2.04

Table 59.2: Number of Tillers per Plant in Transplanted Rice as Influenced by Different Organic Manure and Fertilizers

Treatment	*30 Days*	*60 Days*	*90 Days*	*120 Days*
T_1 Control	2.00	3.00	6.00	4.00
T_2 VC at 5 t/he	3.00	4.00	6.00	6.33
T_3 VC at 10 t/he	6.00	8.33	12.66	12.76
T_4 VC at 15 t/he	8.00	12.00	15.33	16.00
T_5 FYM at 10 t/he	5.00	7.00	11.00	11.50
T_6 RDF	7.00	9.00	12.00	13.33
T_7 VC at 10 t/h+RDF	9.00	14.00	16.23	16.86
SEM+	0.47	0.58	0.40	0.56
C.D at 5 per cent	1.45	1.79	1.21	1.73

Table 59.3: Influence of Organic Manure and Fertilizers on Number of Leaves per Plant of Rice

Treatment	*30 Days*	*60 Days*	*90 Days*	*120 Days*
T_1 Control	8.00	12.00	15.00	12.00
T_2 VC at 5 t/he	12.00	16.00	19.00	16.00
T_3 VC at 10 t/he	15.00	21.00	28.00	24.00
T_4 VC at 15 t/he	20.00	28.80	34.60	30.00
T_5 FYM at 10 t/he	14.00	19.33	29.00	25.00
T_6 RDF	18.00	25.00	30.00	20.00
T_7 VC at 10 t/he+RDF	22.00	29.80	36.00	20.00
SEM+	0.57	0.61	0.61	0.56
C.D at 5 per cent	1.77	1.90	1.81	1.73

Table 59.4: Yield and Yield Parameters of Rice as Influenced by Organic Manures and Recommended Dose of Fertilizers

Treatment	*Panicle Length (cm)*	*Filled Grain Panicle^{-1}*	*No. of Panicles Plant^{-1}*	*Test Weight (g)*	*Yield Plant^{-1} (g)*	*Grain Yield Plot^{-1} (g)*	*Grain Yield (q ha^{-1})*	*Straw Yield (q ha^{-1})*
T_1 Control	10.00	20.33	3.00	2.10	3.50	177.89	6.84	7.52
T_2 VC at 5 t/he	12.33	55.33	5.33	2.23	6.58	429.64	16.52	18.17
T_3 VC at 10 t/he	17.66	73.00	9.66	2.25	10.60	963.33	37.05	40.01
T_4 VC at 15 t/he	24.33	81.00	13.66	2.34	17.33	1620.13	62.31	69.18
T_5 FYM at 10 t/he	18.86	65.66	7.33	2.24	17.33	1620.13	33.99	37.45
T_6 RDF	19.06	70.00	9.33	2.29	12.66	1002.33	38.55	42.47
T_7 VC at 10 t/h+RDF	25.00	85.00	15.06	2.37	18.66	1702.08	65.46	70.73
SEM+	0.64	0.70	0.76	0.26	1.02	50.43	1.63	1.93
C.D at 5 per cent	1.96	2.10	2.21	0.081	3.07	151. 04	4.83	5.02

Nutrient uptake

In general nutrient uptake was greatly influenced with increase in the doses of vermicompost (Figure 59.1). Maximum nitrogen uptake (142.23 kg ha^{-1}) was observed with vermicompost @ 10t ha^{-1} + RDF than rest of treatments except in the treatment that received vermicompost @ 15 t ha^{-1} alone (136.88 kg ha-l). However, the phosphorus and potash were found significantly higher in treatment that received vermicompost @15 t ha^{-1} compare to the rest of the treatments. This might be due to the release of organic acids improved the availability phosphorus and potash.. It is interesting that other factors, such as the presence of beneficial micro organisms or biologically active plant growth

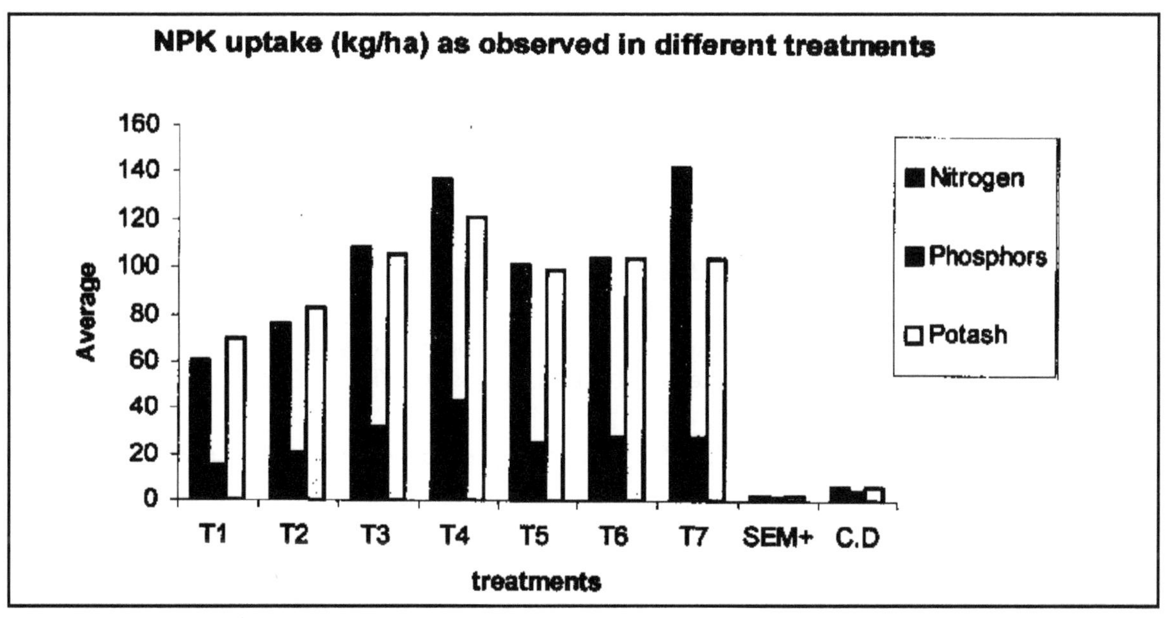

Figure 59.1: Influence of Organic Manure and Fertilizers on Uptake of Nutrients (Kg ha^{-1}) in Rice.

influencing substances/phytohormones released by beneficial micro organisms in the vermicompost might be involved in more uptake of nutrients (Krishnamoorthy and Vajranabhaiah, 1986., Tomati and Galli, 1995; Subler *et al.*, 1998 and Edward, 1998). Kale *et al.*, 1992 reported that increase in the rates of uptake of nutrients with increase in the symbiotic microbial associations in cereal and ornamental plants by using vermicompost. Similarly uptake of nutrients in the same concentrations in the mulberry plants grown on vermicompost or with chemical fertilizers was also reported by Kale (1998).

Conclusion

The scarcity of manures and associated high transport charges compel farmers to depend more on chemical fertilizers. However, in view of increasing reports on the adverse effect of long-term use of chemical fertilizers on soil health, the immediate need in the present situation of environmental degradation is to make proper use of available organic sources as plant nutrients. These organic wastes should be converted into compost within a minimum time span with the help of microbes and earthworms. Application of vermicompost not only reduces the dosage of NPK and cost of fertilizer but, also improves the soil health and increases the nutrient uptake. Thus, Vermicompost along with judicious use of chemical fertilizers bring down the cost of cultivation and present unique opportunities for sustainable agriculture.

References

Bhattacharjee, Gautam and Chaudhury, P.S., 2001. *Asian J. of Microbiol. Biotech. and Env. Sci.*, 3(3): 191–196.

Dominguez, J., Edwards, C.A. and Subler, S., 1997. A comparison of vermicomposting and composting. *Biocycle*, 38: 57–59 .

Edward's, A., 1998. Use of earthworms in breakdown and management of organic wastes. In: *Earthworm Ecology*, (Ed.) C.A. Edward. CRC Press LLC, Boca Raton, Florida, pp. 327–354.

Kale, R.D., 1998. Earthworms: Natures's gift for utilization of organic wastes. In: *Earthworm Ecology*, (Ed.) C.A. Edward. CRC Press LLC, Boca Raton, Florida, pp. 355–376

Kale, R.D., Bano, K., Sreenivasa, M.N. and Bagyarja, D.J., 1987. Influence of worm cast (Vee comp, E. UAS 83) on the growth and mycorrhizal colonization of two ornamental plants. *South Indian Horiticulture*, 35: 433–437.

Kale, R.D., Mahesh, B.C., Basno, K. and Bagyaraja, D.J., 1992. Influence of vermicompost application on the available macronutrients and selected microbial population in a paddy field. *Soil Biology and Biochemistry*, 24: 1317–1320.

Krishnamoorthy, R.V. and Vajranabhaih, S.N., 1986. Biological activity of earthworm casts: An assessment of plant growth promoter level in the cast. In: *Proc. Ind. Acd. Sci. (Anim. Sci.)*, 95: 341–351.

Subler, S., Edwards, C.A. and Metzer, J., 1998. Comparing vermicompost and composts. *Biocycle*, 39: 63–66.

Tomati, U. and Galli, 1995, Earthworms, soil fertility and plant productivity. *Acta Zoologica Fennica*, 196: 11–94

Zha Shi-Wei and Huong Fu-Zhem, 1992. The nitrogen uptake efficiency from N^{15} labelled chemcal fertilizer in the presence of earthworm manure (Cast.). In: *Advances in Management and Conservation of Soil Fauna*, (Eds.) G.K. Veeresh, D. Rajagopal, C.A. Viraktamath. Oxford and IBH, New Delhi, pp. 539–542.

Chapter 60

Effect of Metal Poisoning on Total Body Carbohydrate in *Sphaerodema rusticum* (Belostomatidae: Hemiptera)

***S. Mumtazuddin*[1] *and S. Ehteshamuddin*[2]**

[1]*University Department of Chemistry, B.R.A. Bihar University, Muzaffarpur – 842 001, India*
[2]*Pro Vice-Chancellor, Patna University, India*

ABSTRACT

Although many metals are required for normal physiological functions of animals in very low concentrations, altered physiological functions result when one or more of these reach sufficiently high concentrations in body cells. In the present investigation the effect of three metal compounds, *viz.*, copper sulphate, cobalt nitrate and Mohr's salt on the total body carbohydrate of an aquatic insect *Sphaerodema rusticum* has been studied. It was observed that both the male and the female insects showed decline in the total body carbohydrate concentrations on the treatment with all these three metal compounds. The decline became increasingly significant with the increase in the concentrations of the compounds, being most dramatic for total body carbohydrate in the case of copper sulphate. It is assumed that metal ions in high concentrations act as inhibitors to different enzyme systems responsible for carbohydrate syntheses.

Keywords: *Sphaerodema rusticum, Total body carbohydrate, Metal poisoning.*

Introduction

Many metals are required for normal physiological function of mammals, but only in very low concentrations. These include Copper, Iron, Zinc, Manganese, Cobalt, Selenium and Chromium (Harper *et al.*, 1977). It is assumed that most of these also probably have similar functions in fish and other aquatic animals. However, altered physiological functions result when one or more of these reach sufficiently high concentrations in body cells both in aquatic animals and mammals. Metals enter in the aquatic environment by a variety of industrial effluents and old mines. Acid precipitation also causes leaching of metals from surrounding soils (Norton, 1982).

In fact, aquatic environment is the ultimate sink for the pollutant and any compound either produced on the industrial scale or present in terrestrial environment is likely to reach there, sooner or later.

The study of physiological changes as means for understanding the responses of aquatic animals, like insects, fish, etc. to the stress of various types of pollutants is of great help in studying the nature and the extent of pollution. Eisler *et al.* (1979) prepared bibliographies on the biological effects of metals in aquatic environment.

The present investigation is an attempt to study the effects of three metal compounds, *viz.*, copper sulphate, cobalt nitrate and Mohr's salt on the total body carbohydrate of an aquatic insect *Sphaerodema rusticum.*

Investigation of body carbohydrate has been undertaken as it, alongwith protein and lipid, forms the principal class of organic compounds that are found in insect body. In general, carbohydrate metabolism in insects is similar to that found in vertebrates. However, insects, unlike vertebrates, possess an exoskeleton, chitin, which is rich in aminopolysaccharide. In addition, they also contain a disaccharide, trehalose, which acts as the storage form of glucose. Specific feature of insect carbohydrate metabolism, therefore, centres round the synthesis and hydrolysis of these compounds. Many workers have extensively studied carbohydrate metabolism in various insects (Chang *et al.*, 1964; Islam and Roy, 1981; Pant, 1985; Sharma and Ehteshamuddin, 1992).

Materials and Methods

The insects (*S. rusticum*) were treated separately in different batches by the following metal compounds in different concentrations.

Copper Sulphate ($CuSO_4 . 5H_2O$)

The LC_{50} of the compound for the insects was worked out to be one molar concentration (1 M) and as such ten different concentrations of the compound were taken from 0.1 M to 1 M. In each concentration of the solution, 30 to 50 insects were kept and biochemical assays were performed after 24 hours of each treatment.

Cobalt Nitrate [$CO(NO_3)_2$]

The LC_{50} for the insects in this case was found to be 0.055 M and 30 to 50 insects were placed in nine different concentrations ranging from 0.047 M to 0.055 M and the biochemical assays were performed in each case after 24 hours of treatment.

Mohr's Salt [$FeSO_4 . (NH_4)_2SO_4 . 6H_2O$]

The LC_{50} of Mohr's salt for the insects was found to be 0.1 M. Hence, ten different concentrations of the compound ranging from 0.01 M to 0.1 M were taken and in each 30 to 50 insects were placed for 24 hours for biochemical assays.

Quantitative estimation of total body carbohydrate of treated and control (untreated) insects was done by Anthrone reagent method (Carroll *et al.*, 1956) using Erma colorimeter. Mean and S.D. for each set of reading was calculated. Students-t-test, where necessary, was applied in order to find level of significance between two sets of readings.

Results and Discussion

Total Body Carbohydrate Concentration of 'Control' Insects

In the untreated male the total body carbohydrate concentration was found to be 173±5 S.D. mg/gm of the body weight and was a little higher, *i.e.*, 191±7 S.D. mg/gm of the body weight in the female insect. The adult female insect had greater body carbohydrate concentration than that of the male insect owing to greater metabolic demand for egg maturation.

Total Body Carbohydrate Concentration of 'Treated' Insects

Treatment with Copper Sulphate

On exposure to 0.1 M concentration of $CuSO_4$ itself the body carbohydrate concentration of both male and female insects showed a significant ($P < 0.01$) decline, maintaining steadily the same on treatment with 0.2 and 0.3 M concentrations. In 0.4 M concentration both male and female insects showed a very significant ($P < 0.001$) decline in concentration maintaining a gradual decline in every higher concentration of the treatment, reaching the lowest level, 15±1 S.D. mg/gm of body weight in male and 10±1 S.D. mg/gm of body weight in female insect at 1 M concentration of $CuSO_4$. The results are shown in Table 60.1.

Table 60.1: Body Carbohydrate Concentration in Insects on Treatment with Copper Sulphate

Concentration of Copper Sulphate (molar)	*Body Carbohydrate Concentration (mg/gm body weight*±S.D.)*	
	Male	*Female*
0.1	140±4	145±5
0.2	120±5	130±7
0.3	100±9	105±8
0.4	45±5	55±4
0.5	35±2	52±3
0.6	30±4	40±4
0.7	23±2	29±4
0.8	20±1	20±2
0.9	16±1	18±1
1.0**	15±1	10±1

*Mean wt. of insect:

Male insect: 250 mg±9 S.D. Female insect: 265 mg± 10 S.D.

**LC_{50}

Body carbohydrate in 'control':

Male insect: 173±5 Female insect: 191±7

Treatment with Cobalt Nitrate

In 0.047 M concentration of cobalt nitrate both male and female insects showed significant ($P < 0.05$) decline in body carbohydrate concentration maintaining a steady decline in concentration in every higher concentration of cobalt nitrate, reaching the lowest level (90±2 S.D. mg/gm of body weight in male and 88±3 S.D. mg/gm of body weight in female) in 0.055 M concentration of the compound. The results are shown in Table 60.2.

Table 60.2: Body Carbohydrate Concentration in Insects on Treatment with Cobalt Nitrate

Concentration of Cobalt Nitrate (molar)	Body Carbohydrate Concentration (mg/gm body weight*±S.D.)	
	Male	Female
0.047	155±6	158±8
0.048	145±5	148±6
0.049	130±4	136±5
0.050	125±5	125±6
0.051	120±6	115±4
0.052	110±5	115±5
0.053	105±4	102±6
0.054	100±5	95±4
0.055**	90±2	88±3

*Mean wt. of insect:

Male insect: 250 mg±9 S.D. Female insect: 265 mg± 10 S.D.

**LC_{50}

Body carbohydrate in 'control':

Male insect: 173±5 Female insect: 191±7

Treatment with Mohr's Salt

When exposed to 0.1 M concentration of Mohr's salt both male and female insects showed a significant ($P < 0.05$) decline in body carbohydrate concentration maintaining a steady decline upto 0.04 M concentration. However, in 0.05 M concentration both male and female insects showed a more significant ($P < 0.01$) decline in body carbohydrate concentration maintaining a steady declining trend upto 0.09 M concentration. In 0.1 M concentration of Mohr's salt the decline in body carbohydrate concentration of both male and female insects was highly significant ($P < 0.001$). The results are shown in Table 60.3.

The results clearly indicate that all the three compounds adversely affect the carbohydrate synthesising machinery of the insect, the effect increasing gradually from lower to higher concentrations in the case of cobalt nitrate and Mohr's salt but the effect becoming sharp from 0.4 M concentration onwards in the case of copper sulphate reaching a very low level in 1 M concentration of the compound. This is in accordance with the findings of the earlier workers on other insects (Singh, 1986; Islam and Roy, 1989; Shunmugavelu, 1993). Recently, some work has been done on the effect of organophosphorus insecticides on carbohydrate metabolism in some insects (Rekha *et al.*, 2000 a&b).

Some of the metal ions are known to act as inhibitors to different enzyme systems. In the present investigation also, it is assumed that copper, cobalt and iron in more than required concentrations act as inhibitors to various enzyme systems involved in the synthesis of carbohydrate.

Table 60.3: Body Carbohydrate Concentration in Insects on Treatment with Mohr's Salt

Concentration of Mohr's Salt (molar)	*Body Carbohydrate Concentration (mg/gm body weight*±S.D.)*	
	Male	*Female*
0.01	165±3	170±5
0.02	162±4	165±7
0.03	155±5	160±6
0.04	140±4	140±6
0.05	130±5	132±4
0.06	125±6	130±5
0.07	120±5	122±5
0.08	120±7	120±4
0.09	110±5	110±6
0.10**	92±6	85±5

*Mean wt. of insect:

Male insect: 250 mg±9 S.D. Female insect: 265 mg± 10 S.D.

**LC_{50}

Body carbohydrate in 'control':

Male insect: 173±5 Female insect: 191±7

Since the treatment with copper sulphate showed a greater decline in the concentration of carbohydrate as compared to those with cobalt nitrate and Mohr's salt, it is inferred that copper has a more potent role in enzyme system than the other two metals.

It may thus, be concluded that all the three metal compounds pollute the aquatic environment and even in low concentrations affect the physiology of the insect by inhibiting the synthesis of essential biochemical components like protein (Mumtazuddin and Ehteshamuddin, 2000), ascorbic acid (Mumtazuddin and Ehteshamuddin, 2005) and carbohydrate, thus affecting the production and growth of the insect. As the insect is an important constituent of the aquatic ecosystem in maintaining a normal functional life of pond, lake, etc. the decline in the growth and production of the insect would ultimately affect the normal life of a pond. Moreover, similar adverse effect on the physiology of fishes and other aquatic animals is also expected under the influence of these chemicals.

References

Carroll, N.V., Longley, R.W. and Roe, J.H., 1956. The determination of glycogen in liver and muscles by use of anthrone reagent. *J. Biol. Chem.*, 220: 583–593.

Chang, C.K., Liu, F. and Fang, H., 1964. Studies on the metabolism of Eri silk worm (*Samia ricini*) during metamorphosis (II). The properties of trehalose and its effect on metabolism. *Acta Ent. Sin.*, 13: 494–502.

Eisler, R., Rossoll, R.M. and Gaboury, G.A., 1979. Fourth annotated bibliography on biological effects of metals in aquatic environments. U.S. Environmental Protection Agency. Res. Rep. EPA-600/3-79084, Washington, D.C.

Harper, H.A., Rodwell, V. and Mayer, P.A., 1977. *Review of Physiological Chemistry*, 16th edition. Lange, Los Altos, Calif.

Islam, A. and Roy, S., 1981. Quantitative changes of carbohydrates, lipids and protein during the post-embryonic development and metamorphosis of *Tribolium castaneum* and *T. comfusum* (Coleoptera: Tenebrionidae: Insecta), *Proc. Indian Nat. Sci. Acad.*, B47: 313–320.

Islam, A. and Roy, S., 1983. Effect of cadmium chloride on the quantitative variation of carbohydrate, protein, amino acids and cholesterol in *Chrysocoris stolli. Curr. Sci.*, 52: 25.

Mumtazuddin, S. and Ehteshamuddin, S., 2000. Effect of metal poisoning on total body protein in *Sphaerodema rusticum* (Belostomatidae: Hemiptera). *Res. J. Chem. Environ.*, 4(4): 49–51.

Mumtazuddin, S. and Ehteshamuddin, S., 2005. Effect of metal poisoning on total body ascorbic acid in *Sphaerodema rusticum* (Belostomatidae: Hemiptera). *Indian J. Environ. and Ecoplan.*, 10(1): 83–86.

Norton, S.A., 1982. The effect of acidification on the chemistry of ground and surface water. In: *Acid Rain Fisheries*, (Ed.) R.E. Johnsonz. American Fisheries Society, Betbesda, Md, p. 93.

Pant, E.G., 1985. Dynamics of the total protein, lipid, carbohydrate and glycogen content in organs of the clorado beetle (*Leptinotarsa decelineata*) Stud. Univ. Bolyai. Biol., 30: 55–61.

Rekha, Misbahuddin, S. and Ehteshamuddin, S., 2000a. Effect of the systematic insecticide phosphamidon on LDH activity in the hemolymph of sap feeding insect pest *Aspongopus janus* and *Dysdercus cingulatus. Environ. Ecol.*, 18(2): 317–319.

Rekha, Shailesh, Sah, U.N. and Ehteshamuddin, S., 2000b. Effect of phosphamidon on the hemolymph glucose concentration in heteropteran insects *Dysdercus cingulatus* and *Aspongopus janus. Environ. Ecol.*, 18(4): 863–864.

Shunmugavelu, M., 1993. Effect of heavy metals on the biochemistry of *Mylabris pustulata, Environ. Ecol.*, 11: 386–390.

Singh, G.J., 1986. Haemolymph carbohydrate and lipid metabolization in *Locusta migratoria* in relation to the progress of poisoning following bicresmethrin treatment. *Pestic Biochem. Physiol.*, 25(2): 264–269.

Sharma, S. and Ehteshamuddin, S., 1992. Carbohydrate concentration during development and maturation in *Sphaerodema rusticum* (Belostomatidae, Hemiptera). *Comp. Physiol. Ecol.*, 17(3): 112–116.

Chapter 61

Investigation on Sub-Surface Water Quality of Tarikere Taluk with Special Reference to Physico-Chemical Characteristics

K. Harish Babu and E.T. Puttaiah

Department of P.G. Studies and Research in Environmental Science, Kuvempu University, Shankaraghatta –577 451, Shimoga Distt., Karnataka

ABSTRACT

The research study was carried out in Tarikere Taluk, Karnataka (India). This article mainly address the physico-chemical concentration of 56 groundwater sample during September 2004. The results of all the findings are discussed in details which reflect the present status of the groundwater quality of the study area.

Keywords: *Groundwater quality analysis, Tarikere taluk.*

Introduction

Rural India relies a mainly on groundwater for drinking and agriculture practices. Villages once relied on sources like wells, lakes, ponds and streams for their water needs. Contamination of most surface water sources has rendered them unfit for consumption. Also increase in water demand by an increasing population has necessitated resource to tapping groundwater. Unsustainable withdrawal of groundwater has led to the spectra of depleting the problem of water scarcity.

Every human society, be it rural or urban, industrially or technologically advanced, disposal of waste exceeds the limit of natural scavenging or removal process, they are bound to effect the normal functioning of the ecosystems and consequently they bear an adverse effect on the water resources (Miller, 1984).

Supplying inhabitants with safe and clean drinking water is one of the most common problems in developing countries like India, especially in arid and semi-arid regions. Hazardous pathogenic germs and anthropogenic substances do not only contaminate the available water quality but also geogenic substances is adversely affect the water supply of many regions.

The groundwater of Tarikere taluk had many threats such as anthropogenic activities, quality deterioration by agricultural activities and over exploitation. Persistence of continuous drought condition during 2001–2004 had declined the amount of groundwater in the region. With this background an effort has been made to know the quality of groundwater in this region.

Methodology

Study area

Tarikere taluk is located between 75°45′00″–75°52′30″ E and 13°30′00″–13°47′30″ N and it has a geographical area of 1,222 sq. kms. It comprises 45 Grama panchayats and 245 villages with a total population of 2,54,093 as per 2001 census. As a whole, the region is a flat plain land with an average elevation of about 590.58m msl (Gazetteer of India, 1981).

Meteorology

The climate of Tarikere taluk is semi arid and enjoys all the three seasons *viz.*, pre-monsoon (February to May), monsoon (June to September) and post-monsoon (October to January). The monthly mean temperature ranges from 8.9 to 38°C. The relative humidity varies between 50 to 84 per cent and it is highest during the month of August and September and lowest humidity observed during the months of January and February.

Precipitation

The precipitation and distribution of rainfall in Tarikere taluk is highly erratic. The annual average rainfall is 980 mm received over 55 rainy days. It varies from as low as 621 mm in the east and as high as 920 mm in west. About $2/3^{rd}$ of the geological area of Tarikere taluk receives less than 750 mm of annual rainfall. The study area had been receiving rainfall mainly from southwest monsoon and slightly from northeast monsoon.

Geology

Stratigraphically the study area comes under Bababudan belt of Dharwar super group. Bababudan belt is well known for its iron ore (horse shoe shape). Peninsular gneiss, which forms the basement to the younger green stone belts like Shimoga, Chitradurga and Tarikere (Swaminath and Ramakrishnan, 1981). An important gneissic exposure is near Tarikere that transect the structural trend lines of the schist belts while the gneissic foliation is conformable to the schist belt boundary. The tonalitic gneisses occur in Tarikere taluk (Naqvi, 1983).

Pillow breccias and conglomerates are also seen. Metagabbro with titaniferous magnatite is inter-sheeted with in the metabasites at north of the Tarikere dome of basement gneisses. Serpentinites and talcose schist form a minor component of the Meta volcanic association in the Bababudan group (Fareeddudin *et al.*, 1988). Geological status of Tarikere taluk were represented in Figure 61.1.

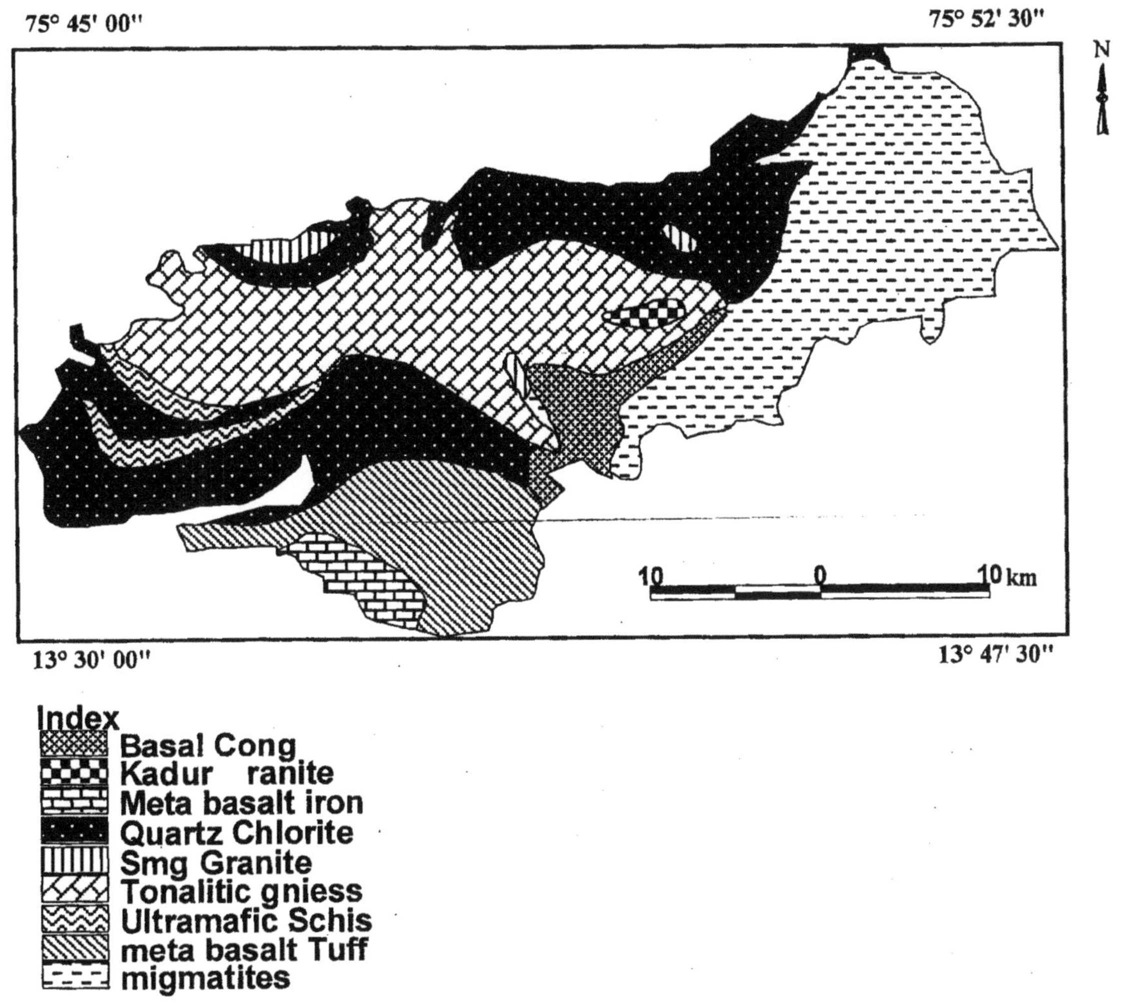

Figure 61.1: The Geological Status of Tarikere Taluk

Analysis of the Samples

The groundwater samples from the 56 sampling sites were collected and analyzed during September 2004 and Sampling locations have been shown in Figure 61.2.

Water samples from the sampling sites were collected from the bore wells. Initially the water was allowed to run for 15 minutes in order to flush out stationary water. Further, the sample bottles were also flushed with water before the samples were collected. The parameters of water such as dissolved oxygen; total dissolved solids, electrical conductivity and pH were analyzed on the spot with the help of water analysis kit (Elico). The remaining parameters were analyzed in the laboratory. Hence, the water was carried to the laboratory in suitable inert bottles. The samples were analyzed using analytical method of APHA (1995) and all analysis was done in triplicate.

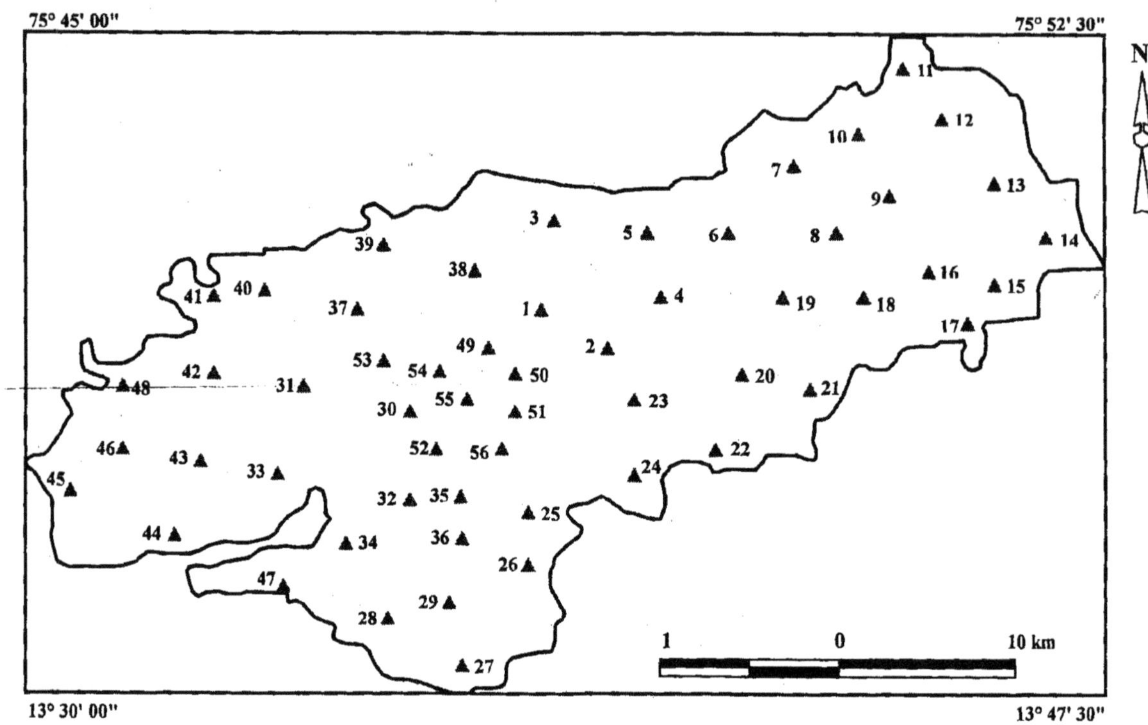

Figure 61.2: Map Showing the Groundwater Locations of the Tarikere Taluk

Results and Discussion

Physico-chemical Parameters

Water analysis was carried out, by taking 12 parameters, which are very essential to know the water qualities for drinking purpose. The findings of the present investigation are summarized in Table 61.1, and it has been compared with BIS 1998 drinking water standards (Table 61.2) which provides the comprehensive picture of the physico-chemical characteristics of groundwater in the study area.

Turbidity (TUR)

In the present study the turbidity values ranged between 0.15 and 86.1 NTU with a mean value of 9.03. The BIS (1998) acceptable limit for turbidity is 25 NTU. It is proved from the present study, 12.5 per cent of total number of samples cross their permissible limit with reference to the BIS (1998) standards. It is observed from the results that these parameters which crossed their permissible limit are unfit for drinking purposes. It causes health problems like gastro-intestinal disorders, headache and also associated with respiratory diseases (Maiti, 1982; Peter, 1974).

pH

In the present investigation, pH values found to be 6.7 to 8.7 with a mean value of 7.54. The recommended value of pH for drinking purposes is between 6.5 to 8.5 (BIS, 1998). In the present study all the water samples analyzed are all well with in the safer limits except in few sampling stations. However, higher values of pH hasten the scale formation in water heaters and reduce the germicidal potential of chlorine (Mohapathra and Purohit, 2000).

Table 61.1: Water Quality Data of Physico-chemical Parameters of the Study Sites During September 2004

Sam. No.	TUR	pH	EC	TDS	DO	TH	Ca^{2+}	Mg^{2+}	Cl^-	Alk	SO_4^{2-}	F^-
S1	11.5	8.35	1558	975	3.8	1325	327.5	218	678	815.5	170.5	1.8
S2	3.3	8.2	2338	1440	5.7	1775	364	228.5	700	295	135	1.55
S3	6.75	7.95	897.5	560	5.15	560	177.5	34.5	265	162	27.1	1.5
S4	0.15	7.85	1150	750	4.4	970.5	332.5	171	409	190	90.5	1.1
S5	7.6	7.2	1751	1051	5.4	1225	395	205.5	670.5	188.5	120.5	1.4
S6	1.2	6.85	1289	782	4.1	845	371	186.5	350	14.7	101.3	1.6
S7	0.15	7.15	449	276	4.5	350.5	119.5	102.1	75.35	146.5	104	1.25
S8	2.7	7.15	1125	718	5.3	790	130.5	108	281.5	231	125.5	1.4
S9	76.15	7.9	1600	970	5	947.5	164.5	173	515	156.5	170.6	1.55
S10	1.25	8.45	715	458	4	470	130	23.6	650	188.5	78.75	1.15
S11	1.7	7.4	544.5	352	4.7	508	186	27.6	616	120	21.8	1
S12	1.15	8.15	1136	771	4.15	970	191	170.5	451	537.5	126	1
S13	0.75	7.9	1699	1056	6.05	972.5	210.5	104.6	750	532.5	133	1
S14	0.2	7.3	1100	670	4.15	759.5	165	125.5	445	172.5	171	1.1
S15	36.5	8.15	1315	815	4.05	735	234.9	160	425	260	118.5	1.15
S16	41	8.3	633.5	421	4.55	395	165.5	110	330	184.5	143	0.9
S17	69.5	6.7	595	370	4.15	505	68	39.95	565	194	22.7	0.9
S18	2.15	8.7	1125	705	5	800	275	9.45	426.5	615	215.2	0.6
S19	1.3	8.4	955	606	4	695	206	21	163.5	119.5	176	0.7
S20	1.6	7.65	1469	950	3.8	1175	417.5	25.35	650	913	117	0.7
S21	1.4	6.7	101.8	68	4.7	85	26.5	3.5	25.65	116	174.5	0.3
S22	1.85	7.25	210	121	6.04	177	24.5	16.3	33.1	130	115.5	0.55
S23	0.3	7.6	1047	636	4.4	594	124	63	241.5	110	98.5	0.45
S24	0.2	6.8	227	138	5.6	270	45	37.5	110.1	46	98.95	0.6
S25	1.3	7.3	1278	758	4.6	970	220.5	108	553	130	165.1	0.725
S26	10.9	8.05	1373	830	4.8	986	180.5	115.5	367	96	136.3	0.625
S27	0.6	7.3	242	147	5.15	385	103	57.05	346	87	128	0.7
S28	0.25	7.15	805	482	5.35	625	135.5	102	467	110	171.5	0.85
S29	0.65	8.1	1608	995	3.65	960	137	131	547	132	100	0.975
S30	0.15	7.1	603	362	5.2	456	122.5	66.2	183.1	144	140	1
S31	0.9	7.25	466.5	280	3.9	400.5	105.5	32.85	285.5	110	246.5	1.05
S32	2.4	6.75	486.5	286	4.8	460	132	92.35	69.35	168	149.1	0.95
S33	6	6.7	778	467	4.12	623	305.5	128.7	70.2	86	104.1	1.05
S34	1.15	7.75	923	554	4.3	644	47.72	110.5	251	68	231	0.85
S35	3.15	7.25	1037	622	4.4	610	127.5	100.5	206	91.8	130.1	0.95
S36	0.2	7.45	987.5	593	4.63	570	177	91.55	266.5	180	145.1	1.05

Contd...

Table 61.1–Contd...

Sam. No.	TUR	pH	EC	TDS	DO	TH	Ca^{2+}	Mg^{2+}	Cl^-	Alk	SO_4^{2-}	F^-
S37	0.2	8.4	840	520	3.75	695	139	84.45	177	82.5	249	1.1
S38	3.45	7.85	545	315	4.05	1700	395.5	229.5	670	145.5	231.5	0.8
S39	1.6	7.95	360	260	3.95	595	83.9	63.35	200.5	79	132	1.55
S40	0.25	8.25	252.5	150	5.1	1160	189	170	735	138.5	215	1.075
S41	0.15	7.7	324	195	4.5	370	76.55	23.05	124.5	44.95	112.1	0.975
S42	10.55	8.35	322.5	178	3.85	1065	181.5	142	535	765.5	171	0.85
S43	1.55	7	452	263	4.6	1125	299.5	119.5	406	74	151.5	0.6
S44	0.5	8.15	784.5	478	4.15	580	279	11.1	285	66.5	141.5	0.95
S45	0.5	7.1	475.5	294	5.1	283	54.45	35.5	156.5	172	161.2	0.8
S46	11.65	7.05	1193	705	4.9	672	234.5	101.5	93	81	139.5	0.9
S47	0.35	6.85	693	425	4	397	108.1	56.7	118	92.1	265.5	1
S48	10.7	7.5	1550	925	3.55	882	212.5	186.2	86.95	75	262	1.05
S49	0.2	7	957	577	5.6	625	78.85	92.5	142	149	152.8	1
S50	1.65	7.05	789	477	4.4	440	91.35	43	175	110	162	1.1
S51	0.6	6.95	900	553	4.4	800	237	45.1	453	126	210.5	1
S52	0.15	7	520.5	315	5.15	640	156.5	56.05	538.5	101	222.5	1.15
S53	86.1	7.05	784.5	478	3.8	1505	520	46.1	652.5	86.7	172	1.1
S54	69.2	7.55	705	416	4.35	219	36.5	25.5	120.5	89	156.4	1.05
S55	1.25	7.9	815	585	4.1	695	215	9.7	685	255	170.3	1.4
S56	7.05	7.8	1200	749	4	695	125	116.5	470	126	57.3	1.2
Mean	9.029	7.542	894.3	551.7	4.552	727.4	185	92.12	361.8	191.1	150.7	1.012
Min	0.15	6.7	101.8	68	3.55	85	24.5	3.5	25.65	14.7	21.8	0.3
Max	86.1	8.7	2338	1440	6.05	1775	520	229.5	750	913	249	1.8

Electrical Conductivity (EC)

The values of electrical conductivity ranged between 101.8 μmhos/cm to 2338 μmhos/cm. Mean values of EC showed 885.31 μmhos/cm. Higher the concentration of acid, base and salts in water, higher will be the EC. The variability of EC could be explained to the natural concentration of ionised substances present in water (Kataria and Jain, 1995). However the higher values of electrical conductivity (>2000 μmhos/cm), may be due to long residence time and lithology. Ballukraya and Ravi (1999) had proved the variation of the conductivity of the water due to the residential times and the geographical features of the sites.

Total Dissolved Solids (TDS)

It is justified from the analytical report TDS values ranges from 68 mg/L to 1440 mg/L with a mean value of 551.53 mg/L. However groundwater chemistry changes as the water flows through the subsurface and the increase in geological environment and dissolved solids and major ions. Chebotarev (1985), Ramababu and Somashekara Rao, (1986) and Joseph (2001) expressed the dissolution of soil particles containing minerals under slightly alkaline condition; favour the TDS concentration in groundwater. However TDS concentration above the permissible limit (1500 ppm) causes gastrointestinal irritation (Shankar and Muttukrishnan, 1994).

Table 61.2: Comparison of Groundwater Quality Data with Drinking Water Standard (BIS, 1998)

Sl.No.	*Parameters*	*BIS (1998)*		*Observed Values*	
		P	*E*	*Minimum*	*Maximum*
1.	Turbidity	5	25	0.15	86.1
2.	pH	6.5	9.2	6.7	8.7
3.	EC	–	–	101.8	2338
4.	TDS	500	1000	68	1440
5	Dissolved oxygen	–	–	3.55	6.05
6.	Total hardness	300	600	85	775
7.	Calcium	75	200	24.5	520
8.	Magnesium	30	100	3.5	229.5
9.	Chloride	250	1000	25.65	750
10.	Alkalinity	200	600	14.7	913
11.	Sulfate	200	400	21.8	249
12.	Fluoride	0.6–1.2	1.5	0.3	1.8

Notes: P: Permissible limit; E: Excessive limit.

All parameters are expressed in mg/L except pH, Turbidity (NTU) and conductivity (μmohs/cm).

Dissolved Oxygen (DO)

The present research findings revealed for the values of DO varies from 3.55 mg/L to 6.05 mg/L, and observed mean value is 4.55 mg/L respectively.

In the present study, all the samples showed for DO values were well with in the permissible limits for drinking and domestic purposes. However, the presence of oxygen demanding pollutants (like organic wastes) causes rapid depletion of dissolved oxygen from water. Oxidizable inorganic substances like hydrogen, sulphide, ammonia, nitrites, iron etc. decrease the dissolved oxygen concentration (Sawant *et al.*, 2000). And also due to physical, chemical and biological activities in water the level of DO may vary (Jameel, 2002)

Total Hardness (TH)

Total hardness levels varied from 85 mg/L to 775 mg/L, associated with a mean value of 727.36 mg/L. The BIS (1998) acceptable limit for total hardness is 600 mg/L. In the present study, revealed that 60.7 per cent of total number of water samples crosses the permissible limits of BIS (1998) drinking water standards. Owing to fact that higher amount of hardness in the study area comes mainly from the leaching of igneous rock and carbonate rocks (dolomite, calcite and limestone). Water containing the soluble salts of calcium and magnesium such as chlorides, sulphates and bicarbonates are also governs the quality of water (Ramaswamy and Rajaguru, 1991). The adverse effects of total hardness are formation of kidney stone and the heart diseases (Sastry and Rathee, 1998). Nevertheless, groundwater chemistry is controlled by the composition of its recharge components as well as by geological and hydrological variations (Narayana and Suresh, 1989).

Calcium (Ca^{2+})

Present investigation reports stated for calcium values ranged from 24.5 mg/L to 520 mg/L, with a mean value of 184.99 mg/L. BIS (1998) acceptable limit for calcium is 200 mg/L. However, in the present study 30.2 per cent of water samples crosses the permissible limit. Presence of higher amount of calcium in the study area may be due to groundwater receives the calcium minerals leached from the rocks and other deposits like limestone, dolomites, calcite, gypsum, amphiboles, feldspar, and clay minerals leaching or weathering of igneous rocks. Sewage and domestic waste are also important sources of calcium (Mishra and Saxena, 1989).

Magnesium (Mg^{2+})

Investigation report reveals for the magnesium values ranged from 3.5mg/L to a 229.5 mg/L with a mean value of 92.11 mg/L.

BIS (1998) acceptable limit for magnesium is 100 mg/L and in the present study 48.2 per cent of the water samples crossed the permissible range. Magnesium arises principally from the weathering of rocks contain ferro-magnesium minerals and some carbonate rocks. High concentration of magnesium proves to be diuretic and laxative (Schroeder 1960).

Chloride (Cl^-)

Chloride is also one of the important parameter to know the quality of water. Anthropogenic sources of chlorides include fertilizer, road salt, human and animal waste. Concentration of chlorides is considered to be an indicator of organic pollution of animal origin (Kumara, 2002). Here Chloride values ranged from 25.65 mg/L to 750 mg/L, however the mean value observed in the present study is 361.82 mg/L.

The BIS (1998) acceptable limit for chloride is 1000 mg/L. In the present investigation, the values of chloride for all the sampling sites are with in the permissible range as prescribed by BIS (1998) drinking water standards. However, dissolving of the soil constituents had contributed the chloride into the groundwater and also the soil characteristics play an important role in contributing the chloride content in the groundwater (Shivasankaran, 1997).

Total Alkalinity

In the present study total alkalinity values ranged from a 14.7 mg/L to 913 mg/L and a mean value of 189.31 mg/L. The BIS (1998) acceptable limit for total alkalinity is 600 mg/L. In the present study, the data revealed that 10.7 per cent of water samples in the study area crossed the permissible limit. When alkalinity of water exceeds the permissible limits, it is likely to produce incrustation sediment deposits, difficulties in chlorination, certain physiological effects on human systems etc. (Raviprakash and Rao, 1989). The constituents of alkalinity result from dissolution of mineral substances in the soil and atmosphere contributes to alkalinity in groundwater (Mittal and Verma, 1997).

Sulphate (SO_4^-)

It is very interesting to note that, sulphate values ranged from 21.8 mg/L to 249 mg/L with a mean value of 150.2 mg/L. The BIS (1998) pemlissible range for sulphate is 400 mg/L. In the present investigation, the sulphate values for all the samples are within the prescribed limit of BIS (1998) drinking water standards. Sulphate in groundwater takes place the break down of organic substances in the soil. However, geological, hydrological and geomorphologic characteristics showed a remarkable variation in sulphur content (Alexander, 1961).

Fluoride (F^-)

In the present investigation, fluoride values varied from 0.3 mg/L to 1.8 mg/L with a mean value of 1.01 mg/L. The BIS (1998) acceptable limit for fluoride is 1.5 mg/L. In present study 7.14 per cent of total number of water samples in present investigations have crossed the permissible range as prescribed by BIS (1998) drinking water standards.

Degree of weathering and leachable fluoride in terrain is of great significance for the fluoride present in groundwater than the mere presence of fluoride bearing minerals in rocks (Kumar *et al.*, 2000).

Intake of excess fluoride causes dental, skeletal and non-skeletal fluorosis. The non-skeletal fluorosis can be observed such as gastrointestinal complaints, intermittent diarrhoea and flatulence in expectant and lactating mothers hardworking young adults and children. Therefore, fluorosis has been considered as one of the incurable diseases. Hence, prevention is the only solution for the disease (Hem, 1985).

Conclusion

Currently carried research investigation should give more precise answer on influence of Geomorphological condition than anthropogenic activities in the examined groundwater samples of the study area. Local geological settings may supports the increasing concentration of physico-chemical characteristics in groundwater. The factors like slow circulation, longer period of contact between aquifer and water, dissolving of minerals at the time of weathering, residential time, drainage pattern and surface water link. Porosity of the soil and rock also alters the characteristics of the groundwater.

The high level contents of the parameters observed may be minimized if the groundwater is recharged with the available water in the rainy season. This not only dilutes the constituents of the groundwater but also raises the groundwater level that depletes due to large-scale exploitation.

Groundwater is extremely important to the future economy and growth of rural India. If the resource is to remain available as high quality water for future generation it is important to protect from possible contamination. Hence it is recommended that suitable water quality management is essential to avoid any further contamination.

References

Alexander., 1961. *Introduction to Soil Microbiology*. Wiley, New York, London, p. 171–172.

APHA., 1995. *Standard Methods for the Examination of Water and Wastewater*; 18th edition, AWWA, WPCF, New York, pp. 1120.

Ballukraya, P.N. and R. Ravi., 1999. Characterization of Groundwater the unconfined aquifers of Chennai city. *Indian Journal of Geological Society of India*, 54: 1–11.

BIS., 1998. *Specifications for Drinking Water*, New Delhi, p. 171–178.

Chebotarev., 1985. Metamorphism of natural waters in the crust of weathering. *Geochem. Cosmochim. Acta*, 8: 22–28.

Fareeddudin, A.S. Janardhan and B. Basavalingu., 1988. Sedimentology, minerology and geochemistry of the Kalasapura conglomerate. *Geol. Soc. India. Mem.*, 9: 65– 82.

Gazetteer of India, Karnataka State, Chikmagalure district., 1981. p. 8–15 and 630–633.

Hem, J.D., 1985. *Study and Interpretation of the Chemical Characteristics of Natural Water*, 3rd ed. U.S. Geological Survey, Water Supply Paper, 225: 263.

Jameel, A., 2002. Physico-chemical studies in Vyakondan channel water of river Cauvery. *Pollution Research*, 17(2): 111–114.

Joseph, W., 2001. Groundwater chemistry in the valley De Yabucoa alluvial aquifer, South-eastern Pnerto Rico. *AWRA 3rd International Symposium on Tropical Hydrology*, San Juan U.S.A.

Kataria, H.C. and O.P. Jain., 1995. Physico-chemical analysis of river Ajhar. *Indian Journal of Environmental Protection*, 5: 569–571.

Kumar, P. Kumari and L.K.R. Singh., 2000. *Environmental Biology*. P.G. Dept. of Zoology, S.K. University, Dumka, Jharkand, p. 180–182.

Kumar., 2002. *Ecology of Polluted Waters*. 1: 144–180.

Maiti, T.C., 1982. *Science Reporter*, 19: 360–361.

Miller, D.R., 1984. Chemicals in the environment. In: *Effects of Pollutant at the Ecosystem Level*. John Wiley and Sons, Chinchester, pp. 7.

Mishra, S.R. and Saxena., 1989. Industrial effluent pollution at Birla Nagar, Gwalior. *Pollution Research*, 8(2): 77–86.

Mittal, S.K and N. Verma., 1997. Critical analysis of groundwater quality parameters. *Indian Journal of Environmental Protection*, 17: 426–429.

Mohapatra, T.K. and K.M. Purohit., 2000. Qualitative aspects of surface and groundwater for drinking purpose in Paradeep area. *Ecology of Polluted Waters*, 1: 144.

Naqvi, S.M., 1983. Geochemistry of gneisses from Hassan district and adjoining areas, Karnataka, India. In: *Precambrian of South India*, (Eds.) Naqvi, S.M. and J.J.W. Rogers. *Geological Society of India Mem.*, 4: 401–416.

Narayana, A.C. and G.C. Suresh., 1989. Chemical quality of groundwater of Mangalore City, Karnataka. *Indian Journal of Environmental Health*, 31: 228–236.

Ramaswamy, V. and P. Rajaguru., 1991. *Indian Journal of Environmental Health*, 33(2): 187–191.

Rambabu, C. and K. Somashekara Rao., 1986. Studies on the quality of bore well water by Nuzuid. *Indian Journal of Environmental Protection*, 16(7).

Raviprakash, S. and K.G. Rao., 1989. The chemistry of groundwater, Parvada area with regard to their suitability for domestic and irrigation purposes. *Indian Journal of Geochem.*, 4: 39–54.

Sastry, K.V. and P. Rathee., 1988. Physico-chemical and microbiological characteristics of water of village Kanneli, (Distt. Rohtak) Haryana. *Proc. Academic, Environmental Biology*, 7(1): 103–108.

Sawant, C.P., Saxena, G.C. and Shrivastava, V.S., 2000. Trace Metals in and around the Industrial belt. *Ecology Environment and Conservation*, 6(6): 135–137.

Schroeder, H.A., 1960. Relation between hardness of water and death rates from certain chronic and degeneration diseases in the US. *J. Chron. Dis.*, 12: 568–573.

Shankar and Muttukrishan., 1994. *In situ* bioremediation of contaminated groundwater. *Proceedings of National Seminar on EPCR–04*, UBDT Engineering College, Davangere, pp. 32.

Shivashankaran, M.A., 1997. Hydrogeochemical assessment and current status of pollutants in groundwater of Pondichery region, South India. *Ph.D. Thesis*, Anna University, Chennai, p. 80–87.

Swaminath, J. and M. Ramkrishnan., 1981. In: *Early Precambrian Supracrustals of South India*, (Eds.) Swaminath, J. and M. Ramakrishnan. Geol. Surv. Ind. Mem., 3: 23–38.

Chapter 62

Vehicular Pollution Abatement and Control Measures Implemented in Delhi

D.S. Kharat*, Paritosh Kumar and P.M. Ansari

Central Pollution Control Board, Ministry of Environment and Forests, East Arjun Nagar, Shahdara, Delhi – 110 032, India

ABSTRACT

Vehicular emissions are the main source of air pollution in Delhi. In an effort to address the air pollution problems caused by vehicles, series of steps have been taken up in recent past at the instance of the Hon'ble Supreme Court of India. During implementation of various pollution control measures, main emphasis was given to introduction of natural gas as an automotive fuel on large scale and necessary infrastructure facilities were strengthened. On account of various measures for abatement and control of pollution, the air quality in Delhi has indicated significant improvements. The present paper gives the details of the activities undertaken for implementation of the abatement and control measures between 1998 and 2005. The aspects covered include approach, action plan, different control measures, constrains, effect on ambient air quality and future plans.

Keywords: *CNG, Control measures, Delhi, Vehicular emission.*

Introduction

Besides being capital, Delhi is a major centre of commerce, industry and education. Rise in population and growth in economic activity have led to increase in air pollution in the city. Main sources of air pollution in Delhi are emissions from vehicles, coal based thermal power plants, industrial units and domestic activities. As per the published data, vehicular pollution contributes about 67 per

* *Corresponding Author.* E-mail: kharatds@hotmail.com.

cent of total air pollution of Delhi. With their contribution of 13 per cent, power plants are the second most polluting source in the city. The industries and domestic activities generate 12 per cent and 8 per cent of pollution. The air quality in the city is being monitored in terms of major pollutants since 1978. In the late 1980s, Delhi was reported to be one of the most polluted cities in the world. This assessment was mainly based on suspended particulates in the ambient air.

The data published by Delhi Transport Department indicate that vehicular population in 1991 was 1.9 million, which has increased to 4.1 million as on 1-4-2004. The average daily addition of new vehicle is 1679 (328 commercial and 1351 private). It is to be noted that Delhi's vehicular population is more than the number of vehicles of Mumbai, Kolkata and Chennai put together. Considering the severity of air pollution in Delhi, which is predominantly contributed by vehicular emissions, series of measurers have been taken up since 1998 at the instance of the Hon'ble Supreme Court of India. On account of various measures for abatement and control of pollution, the air quality in Delhi has indicated significant improvements.

Approaches and Action Plan

While framing the action plan for pollution control, more focused attention was given to tackle the emissions of commercial vehicles, as these vehicles are poorly maintained and heavily used as compared to private vehicles. Natural gas, which is inherently less polluting than diesel and petrol, was recommended for use as automotive fuel on large scale. While buses were directed to convert to Compressed Natural Gas (CNG) mode, use of CNG or clean fuel was recommended for three-wheeler autorikshaws and taxis. It was also decided to phase out the old commercial vehicles in a phased manner. For the non-commercial private vehicles, improved engine technology was made mandatory. In case of two-wheelers, no decision was taken. However, it was felt that use of LPG could be a long term solution to two-wheelers of the city. The approach for talking vehicular pollution is illustrated in Figure 62.1.

The abetment and control measures implemented in Delhi include improvement in vehicle technology and fuel quality, and instruction of CNG vehicles, which directly helps to cut down vehicular emissions. In addition, indirect control measures such as inspection and maintenance centers, pre-mixed 2T oil dispensers, interstate bus terminals at periphery of city etc. were also considered. The action plan with priority measures is given in Table 62.1.

While hearing the matter of writ petition no. 13029 of 1985; *M.C. Mehta* versus *Union of India and others* on July 28, 1998, the Hon'ble Supreme Court agreed to these priority measures. Most of the works were not completed as per the schedule. During the course of their implementation, the Hon'ble Supreme Court extended target dates on appeals from the concerned implementing agencies. From time to time many more directions were issued by the Hon'ble Court while hearing the matter and related interlocutory applications filed by concerned agencies. Other issues witch were examined by the Hon'ble Supreme Court for issuance of the direction include engine technology, fuel quality, non-destined transit trucks and express bye-pass roads.

Implementing Agencies

The concerted efforts for abatement and control of vehicular pollution were initiated after the Ministry of Environment and Forests, Government of India published the White Paper on Pollution With Action Plan in 1987. Subsequently, a committee called Environment Pollution (Prevention and Control) Authority for the National Capital Region (EPCA) was constituted on January 29, 1998 to monitor the implementation of the action plan at the instance of the Ministry of Environment and

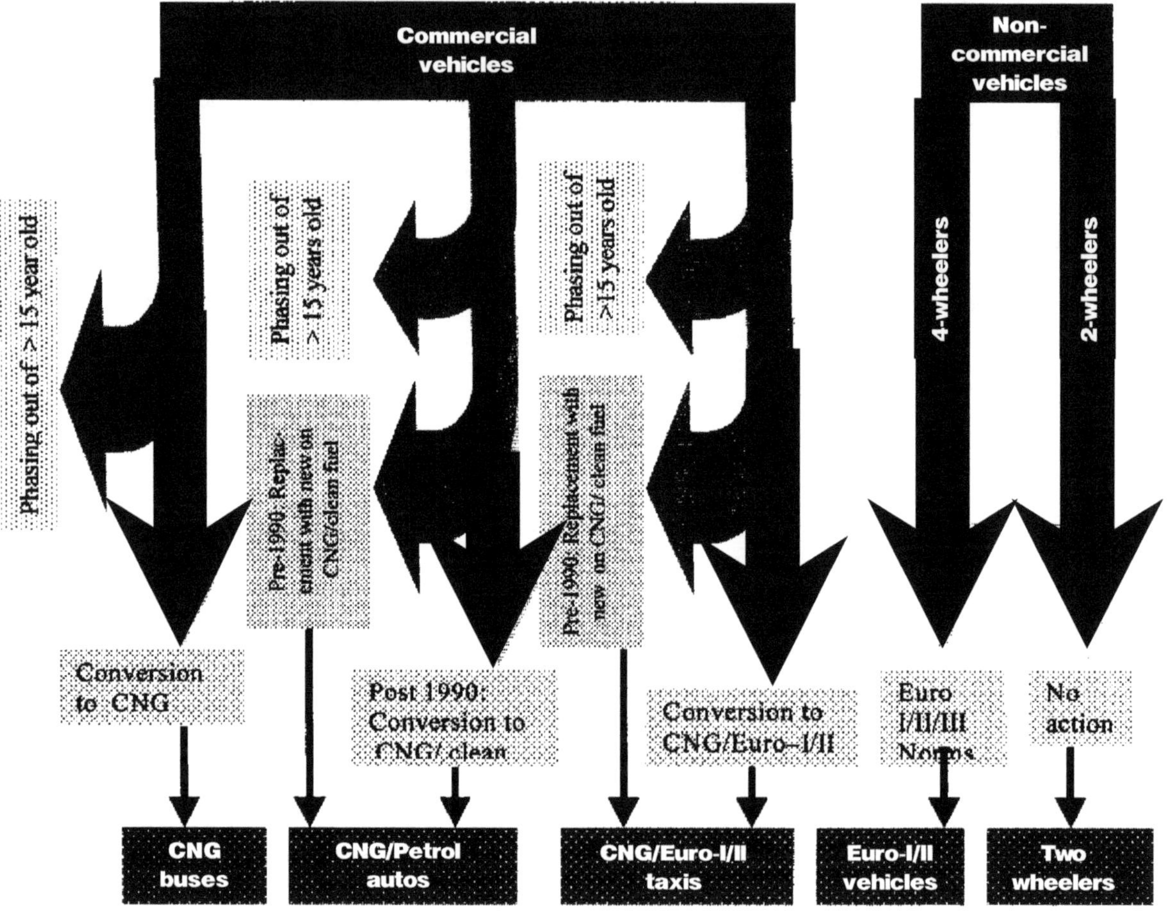

Figure 62.1: Approach for Containment of Vehicular Emissions

Forests and the Hon'ble Supreme Court of India. Central Pollution Control Board (CPCB) provided necessary technical and secretarial support to the EPCA. Main stakeholders in implementing the action plan include Indraprashtha Gas Ltd., Delhi Transport Department, oil companies and the automobile manufacturers.

Abatement and Control Measures

CNG Supply Network

Initially, there were 9 CNG refilling stations in Delhi in 1998. These stations were mostly located in South Delhi. With the decision to convert city buses, autorikashaws and taxies on CNG mode, it was required to increase the number of CNG stations and also to spread the same in the entire Delhi. Accordingly, in 1998, it was planned to increase the number of CNG stations to 80 based on a survey conducted by the Central Road Research Institute, New Delhi. An independent agency called Indraprashtha Gas Limited (IGL) was incorporated on December 23, 1998 to carry out work for installation of additional CNG refilling stations and other necessary infrastructure. The IGL is a joint venture of Gas Authority of India Limited, Bharat Petroleum Corporation and Government of National Capital Territory (NCT) of Delhi. In the initial stage, IGL succeeded in installation and commissioning of 44 CNG refilling stations by March 31, 2000. The target of 80 CNG station was achieved by IGL by the end of year 2001.

Table 62.1: Action Plan for Abatement and Control of Vehicular Pollution

Sl.No.	*Abatement and Control Measures*	*Target*
1.	Augmentation of public transport (stage carriage) to 10,000 buses from existing 6,600 buses	01-04-2001
2.	Elimination of leaded petrol from Delhi	01-09-1998
3.	Installation of pre-mix dispensers for supply of only pre-mix petrol in all petrol stations to two stroke engines	31-12-1998
4.	Replacement of all pre-1990 autos and taxis with new vehicles using clean fuel	31-03-1998
5.	Replacement with financial incentives of post 1990 autos and taxis with new vehicles on clean fuel	31-03-2001
6.	Ban on plying of buses more than 8 years old except on clean fuel	01-04-2000
7.	Entire city bus fleet (DTC and Private) to be steadily converted to single fuel mode on CNG	31-03-2001
8.	New ISBT to be built at North and South-West borders of NCT Delhi to avoid pollution due to entry of inter-state buses	31-03-2001
9.	Gas Authority of India Ltd., to ensure availability of CNG by increase of CNG supply outlets in the cities from 9 to 80	31-03-2001
10	Tow autonomous fuel testing laboratories to be established for monitoring fuel quality specifications and adulteration	01-06-1999
11.	Automated inspection and maintenance facilities to be set up for commercial vehicles in coordination with private sector, for which proposals received, *e.g.* Society of India Automobile Manufactures (SIAM)	Immediate

Till March 31, 2002, a total of 87 stations were installed and made operation. At this stage, it was realized that 87 CNG refilling stations were not adequate to meet the demands of CNG vehicles. Long queues of vehicles waiting for their turn could be seen at all the stations during busy hours primarily due to lack of CNG stations in all parts of the city. As such, the work of installation of new stations had to be continued.

In 2002, the gas transmission pipeline was available in South, North and East Delhi only. There was no facility available in western region. The entire bus fleet of western region was required to come to South or North Delhi for CNG filling leading to overloading of CNG stations in these areas. This also caused loss of mileage in traversing for the CNG filling. The demand of CNG in western Delhi was met by laying of gas pipe line connecting Dhaula Kuan to G.T. Karnal by-pass through West Delhi. By March 31, 2003, 106 numbers of CNG stations were installed which increased to 121 in March 2004. These stations are able to meet the CNG requirements of vehicles in the city. Figure 62.2 shows growth of CNG refilling station over the last six years.

The IGL has planned and installed the above CNG stations under three categories namely mother station, online station and daughter station. Mother stations and online stations have compression facilities, where natural gas is compressed to desired pressure of 200 to 250 bars. This pressure is

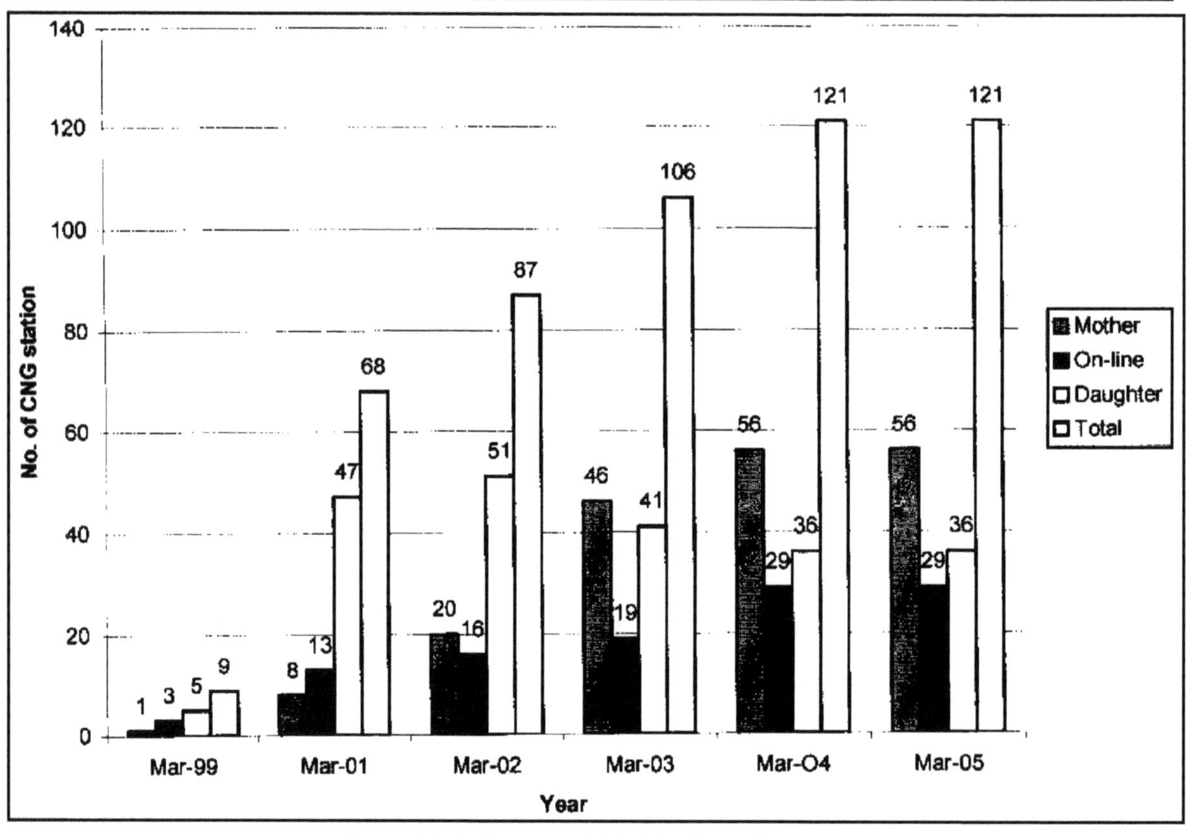

Figure 62.2: Growth of CNG Refilling Stations

sufficient for refilling CNG in the vehicles. The daughter stations do not have their own compression facilities but get the CNG cascades though trucks from the mother or online stations. These cascades are replaced once the CNG is exhausted. The mother stations and online stations can supply CNG to both heavy vehicles and light vehicles. The daughter stations cater to the needs of only light vehicles including taxis, cars and three-wheeler autorikshaws.

While providing the facility of CNG refilling stations, proper planning is needed so that the facility covers all the region and parts of Delhi and hassle-free supply of CNG to all CNG vehicles can be assured. As the city buses are the largest consumer of CNG, attempts were made to locate adequate number of CNG refilling station as close as possible to the bus-depots and wherever possible stations were installed at the bus-depot itself. Out of the existing 121 CNG stations, 56 are mother stations (19 for DTC at its depots), 29 are online stations and 36 are daughter stations. Among the 56 mother stations, 19 are installed in the bus-depots.

The combined CNG dispensing capacity of 121 CNG refilling stations works out to be 16.30 lakhs kg. per day against the initial target of 16.11 lakh kg per day. With increase in number of CNG refilling stations, sales of the CNG has proportionately increased. Actual sale of CNG in March 2004 was 7.96 lakh kg per day. The installed capacity and sale of CNG are given in Figure 62.3.

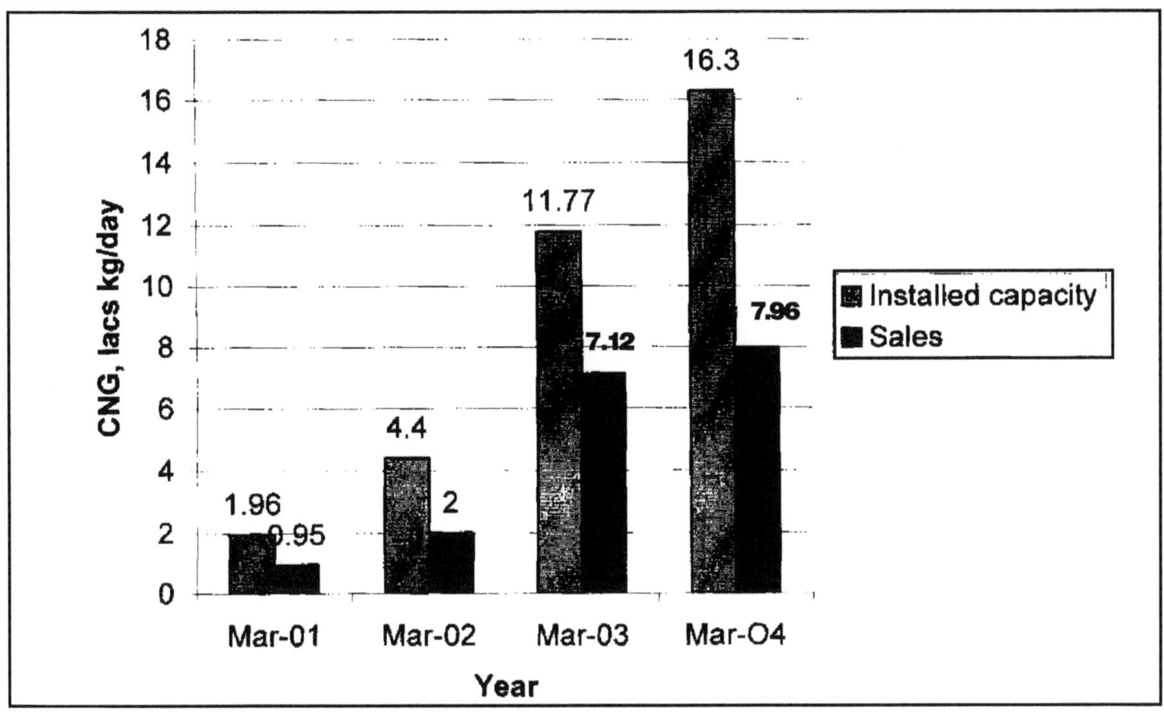

Figure 62.3: CNG Dispensing Capacity and Sales

Conversion of Vehicle Fleet to CNG Mode

Under the conversion programme, three approaches were adapted: (*a*) purchase of new dedicated CNG vehicles; (*b*) conversion of existing engine to operate on CNG mode by incorporating necessary modification; and (*c*) by retrofitting of new CNG engine in existing vehicles. In case of three-wheeler aoutorikshaws, which use two-stroke engine, conversion of the old petrol engine to CNG was not feasible. Introduction of CNG operated autorikshaws were possible when leading manufactures like Bajaj Autos and Scooter India started production of duel fuel autorikshaws, which operate on petrol as well as on CNG mode. Use of clean fuel, *i.e.*, lead-free petrol with low sulphur and benzene contents was allowed for autorikshaws.

Large commercial vehicle fleet was converted to CNG by using all the three approaches. Two principle bus suppliers TELCO and Ashok-Leyland strengthened their infrastructure to manufacture dedicated CNG buses. They also undertaken work of retrofitment of new CNG engines in their old buses. Few other leading workshops also modified old diesel engines to operate on CNG mode. Conversion of old buses to CNG was a cheaper proposition as compared to purchasing of new CNG buses. CNG being cheaper as compared to petrol and diesel, can partly off set cost involved in purchase or conversion operation. As in March 1999, only one bus and 1451 cars were playing on CNG mode in Delhi. There was no autorikshaw at this point of time. The growth of different CNG vehicles in Delhi is presented in Figure 62.4.

In conversion of vehicle fleet to CNG, there were some teething problems initially due to lack of technical know-how and defined regulatory mechanism to deal with CNG vehicle. While the conversion of a petrol engine to CNG required only minor modification and no elaborate procedure to certify its

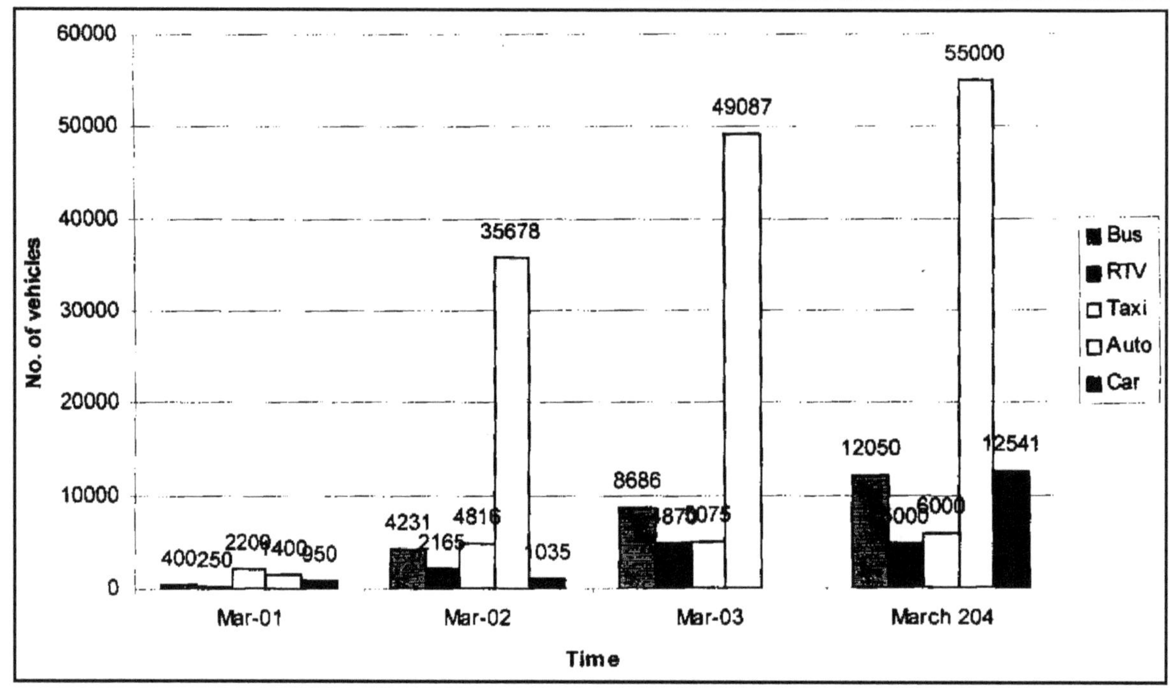

Figure 62.4: Population of Different CNG Vehicles

safety and road worthiness were required, a diesel engine had to undergo serious modifications and therefore had to be placed on a different footing for which knowledge on diesel conversion and kits were limited. The Ministry of Road Transport and Highways in association with the statutory recognized vehicle testing agencies, *i.e.*, Indian Institute of Petroleum, Dehradun and Automotive Research Association of India, Pune evolved and notified procedure for the type approval, safety and emission norms for CNG vehicles which helped concerned agencies to fulfil the task of conversion of vehicle fleet on CNG mode.

In beginning of transition, some CNG vehicles, particularly buses were reported to have caused problems like leakages of gas, miss-firing, seizure of engines, starting problem, inability to negotiate flyovers, heating up of engine etc. There have been few cases of fire due gas leakages. The defects were rectified by the vehicle manufacturers to ensure trouble-free operation of the vehicles.

Despite the enormous efforts by the concerned implementing agencies, the target set for the conversion of the entire bus fleet to CNG mode by the March 2001 was not met initially due to shortage of CNG buses and conversion facilities or delay in procurement or placement of orders for CNG buses. As such, diesel buses were allowed to ply after March 2001, if their owners filed affidavits that they had placed a firm order for either for same number of new CNG buses or for the conversion/retrofitment of their old buses. These diesel buses were directed to carry a sticker from the Delhi Transport Authority stating that this bus has been allowed to ply against a firm order for CNG bus that will replace it from September 30, 2001. The diesel buses which were left after September 30, 2001 were further allowed to ply subject payment of fine on daily basis. These relaxation were considered in the interest of commuters and to ensure that transition takes place as smoothly as possible without creating unnecessary disruption in to the public transport.

Phasing Out of Old Vehicles

For phasing out of old commercial vehicles, the schedule as indicated in Table 62.2 was directed. The commercial vehicles were phased out in stages. In the first stage, vehicles of age more than 20 years were phased out by October 2, 1998. The vehicles of age between 17 to 19 years were phased out till November 15, 1998. All the vehicles of 15 to 18 years were phased out by December 12, 1998. For vehicles below 15 years of age, conditional phasing out was implemented. Diesel buses of more than 8 years age were directed to convert to CNG by April 1, 2000. The date was subsequently extended. In case of taxis and autos different actions were taken for phasing out depending on their date of manufacture. Pre-1990 autos and taxis were directed for replacement with new vehicles using CNG or clean fuel. In case of post-1990 vehicles, replacement with financial incentives fro m the Government was effected by March 31, 2001.

Table 62.2: Schedule for Phasing Out of Old Vehicle

Sl.No.	*Age of vehicle*	*Target*
1.	More than 20 years old	2nd October 1998
2.	17–19 years	15th November 1998
3.	15–18 years	31st December 1998

The retrofitment of CNG kits in the post-1990 autos and taxis was allowed. However, retrofitment of CNG kits in pre-1990 autos and taxis, was not allowed because these vehicles were manufactured before the imposition of the emission standards of 1991, and were gross emitters.

The Delhi Government has to initiate stringent action including cancellation of permits and impounding of vehicles. Till 31st December 1998, 7696 numbers of three-wheeler autorikshaws were deposited with the Delhi Government. Out of these, 4403 were scrapped. The impounded vehicles were released only after the vehicle owners had given an undertaking to sale the vehicle outside Delhi or to scrape. Delhi Government also launched a scheme for purchase of new vehicles for the owners of phased out vehicles for post-1990 categories under the scheme, exemption of sale tax @ 8 per cent and subsidy of 4 per cent on interest rates on loan were given.

The transition of buses to CNG has not been smooth. The Action Committee of private schools and other institutions filed an interlocutory application to the Hon'ble Supreme Court for allowing use of their school buses more than 8 year old for transporting students. They submitted to the Hon'ble Supreme Court that these buses cater to the need of students only within a certain distance of the location of school and ply only for a shorter distance and therefore, do not strictly speaking, deserve to be treated as commercial buses. After considering the an interlocutory application, the Hon'ble Court vide its order dated April 24, 2000, directed that the buses be steadily converted to single fuel mode on CNG on or before March 31, 2001. The Hon'ble Supreme Court further directed that educational institution buses shall be got checked for pollution control to avail the benefit of the order. In the event it is found that any bus does not have pollution under control certificate, that bus shall not ply under the force of the order.

Improvement in Vehicle Technology

With increasing concern for controlling air pollution caused by automobiles, the emission norms for various categories of automobiles were progressively tightened. It was made mandatory the registration of Bharat Stage I (Euro I) compliant light vehicles from June 1, 1999. Subsequently, registration of only Bharat Stage II (Euro II) compliant light vehicles was started from April 1, 2000. In case of heavy vehicles, Bharat Stage II norms were enforced from October 23, 2001. Before the intervention of the Hon'ble Supreme Court, Bharat Stage I and Bharat Stage II norms were planed to be implemented from April 1, 2000 and April 1, 2005 respectively. In case of two-wheelers and three-wheelers, Barat Stage II norms are implemented from April 1, 2005. For the other vehicles, Bharat Stage III (Euro III) norms have been enforced. At present only Bharat Stage III compliant vehicles (except two-wheelers and three-wheelers which are Bharat Stage II) are being registered in Delhi. Further improvements in emission norms are expected in days to come. For compliance of the emission norms, the automobile manufacturers need to improve the engine technologies.

Improvement in Fuel Quality

Fuel quality is one of the predominant factors affecting the emissions and the ambient air quality. From time to time, the quality specifications have been laid down by the Bureau of Indian Standards and the Ministry of Road Transport and Highways. For the first time, a major initiative towards fuel quality improvement was taken when the emission related parameters were notified under the Environment Protection Act in 1996. Based on the local ambient quality and health effect of auto exhaust pollutants, critical parameters including lead, sulphur and benzene is considered for fuel quality enhancement, which was implemented as per the schedule given below.

In 1997, the sulphur content in diesel was 0.25 per cent, which was brought down to 0.05 per cent with effect from April 1, 2000 and made available at selected outlets in NCT of Delhi. Supply of this diesel was extended to the entire NCT of Delhi from March 31, 2001.

Petrol quality was also improved in respect of benzene, sulphur and lead content. Unleaded petrol was introduced in April 1995 for new four-wheeled vehicles in Delhi. Introduction of unleaded petrol enabled use the catalytic converters in the vehicles. Leaded petrol was completely phased out from National Capital Region (NCR) from September 1, 1998 and throughout country from January 1, 2000. Petrol supply with 0.05 per cent sulphur content from April 1, 2000 and 1 per cent benzene content from November 1, 2000 was made available in NCT of Delhi. In 1997, these parameters were 0.1 per cent and 5.0 per cent respectively.

Prevention of Fuel Adulteration

Adulteration of fuels is one of major causes of excess emission of pollutants. The Ministry of Petroleum and Natural Gas (MoP&NG) was asked to take effective measures to prevent fuel adulteration. The MoP&NG has taken preventive measures, which include issuance of the Solvent, Raffinate and Slop (Acquisition, Sale, Storage and Prevention of Use in Automobiles) Order, 2000; and the Naphtha (Acquisition, Sale, Storage and Prevention of Use in Automobiles) Order, 2000. Naptha and Solvent orders are in various stages of implementation. An Anti-adulteration Wing has also been set up under the aegis of the MoP&NG and Department of Food and Civil Supply, Delhi Government.

Besides the above regulatory mechanism, one independent fuel-testing laboratory has been commissioned at NOIDA in 2000 to check adulteration of fuels. The laboratory is maintained by Society for Petroleum Laboratory with members from Indian Oil, Oil Coordination Committee, Central Pollution Control Board, Indian Institute of Petroleum, Delhi Government, Bharat Peroleum, Hindustan Petroleum, IBP, SIAM and Tata Energy Research Institute. The capital cost of Rs. 13 crores for the laboratory was borne by the MoP&NG. The laboratory was made operational in 2000, but the issue of mobilizing funds for the recurring cost is yet to be resolved completely. The operation and maintenance cost of the laboratory is being presently shared by SIAM, Ministry of Road Transport and Highways,

Ministry of Heavy Industries, Ministry of Petroleum and Natural Gas, Government of NCT of Delhi and EPCA (Ministry of Environment and Forests). Since its commissioning in 2000, the laboratory has tested samples which have been mostly collected by the Anti-adulteration Wing. In the year 2003, an independent agency "Centre for Science and Environment, New Delhi" was also given the task to collect samples of fuels and perform testing to know the status of adulteration. It was found that 8.6 per cent of samples failed to confirm to the prescribed standards.

Premixed 2T Oil Dispensers

Before the pollution control programme was initiated, the two-stroke engined vehicles such as two-wheelers and three-wheeler autorikshaws were using manually mixed 2T engine oil with petrol while fuelling their vehicles. Manual mixing of 2T oil with the petrol does not ensure the required proportioning of fuel and the oil. For proper lubrication of two-stroke engine, about 2 per cent of 2T oil is sufficient. It was observed that the owners of two-stroke engined vehicles mostly use excessive quantity of 2T oil amounting to more than 5 per cent. This is one of causes for the higher emissions of pollutants. To check excessive use of 2T oil, it was decided to install pre-mixed oil dispensers at the petrol pump. It was also decided to ban the sale of loose 2T oil at petrol pumps, service garages etc. These decisions were implemented by the Ministry of Petroleum and Natural Gas and Oil Companies within a span of tow years. By the end of year 2000, 646 numbers of premixed 2T oil dispenses were installed at all the 294 petrol pumps located in Delhi.

Emission Warranty

The issue of emission warranty was taken up with Society of Indian Automobile Association. The SIAM agreed to give emission warranty for all the categories of vehicles including passenger cars, Multi-utility Vehicles (MUV), commercial vehicles and two-wheelers and three-wheelers. The warranty is valid for all passenger cars, multi-utility vehicles and two-wheelers and three-wheelers soled after July 1, 2001, and for commercial vehicles from the date of implementation of Bharat Stage II emission norms. To begin with, the warranty was started in Delhi, Mumbai, Kolkata and Chennai where Bharat Stage II emission norms were implemented. The warranty period of each vehicle category and criteria are given in Table 62.3. To avail the benefits of the emission warranty, the vehicle owner is required to sign an agreement with the manufacturers.

Table 62.3: Criteria for Emission Warranty

Sl.No.	*Vehicle Category*	*Criteria*
1.	2 wheelers	30,000 km or 3 years whichever occurs earlier
2.	3 wheelers	30,000 km or 1 year whichever occurs earlier
3.	Passenger cars	80,000 km or 3 years whichever occurs earlier
4.	Multi-utility Vehicles	80,000 km or 3 years whichever occurs earlier
5.	Commercial Vehicles	80,000 km or 1 year whichever occurs earlier

Provision of New ISBTs

In order to ensure that the interstate polluting diesel buses do not enter the city, it was proposed to have two Interstate Bus Terminals (ISBTs) at the periphery of the city. At present, there are two ISBTs in addition bus terminal at Kashmere Gate. These tow new bus terminals are at Anand Vihar and Sarai Kale Khan. The Anand Vihar Bus Terminal caters to the buses coming from the eastern side of the city The buses arriving towards the south Delhi side is terminated at Sarai Kale Khan Bus Terminal. For the buses arriving from North and South-West side of Delhi, two interstate bus terminals have been planned at Narela and Dwarka.

For the Dwarks ISBT payment of land charges amounting to Rs. 8.0 crores has been made by Transport Department to the Delhi Development Authority. In case of ISBT at North border, at Narela the land allotment is yet to be settled the by DDA and Transport Department.

Inspection and Maintenance Centre

Setting up of Inspection and Maintenance Centre for certification of fitness and emission is necessary to reduce emissions from inuse vehicles. Looking into the number of vehicles many Inspection and Maintenance Centres are required, which may not be possible for the government to create, operate and maintain the same. As such, it was desired to give permission to private independent bodies. One such new centre is planned to be set up at Burari, for which Delhi Government has tied up with ARAI, Pune for consultation.

Other Measures

There is ban on playing of good vehicles in day time, which helps to reduce traffic congestions as well as air pollution problems in pick hours. This direction is being implemented by Delhi Traffic Police. It is also to mention that over loading of goods vehicles has been strictly banned.

The Pollution Under Certificate (PUC) programme is also being enforced by the Delhi Government, for which computerized emission checking facilities have been set up at petrol pups and workshops in deferent parts of the city. The centres issue PUC certificate to the vehicles meeting emission standards. In case the vehicle is found polluting beyond prescribed emission norms, necessary repairs or tuning in the vehicle is required. At present, there are 280 centres for petrol and diesel vehicles and 73 centres for the diesel vehicles in Delhi.

Constraints

The work for setting up of 80 CNG stations was not accomplished within target date. Cause for the delay was mainly attributed to procedural delays in allotment and acquisition of land, and getting clearances for setting up of CNG stations and pipelines etc. In case of conversion of vehicle fleet on CNG also, the target date had to be extended. Initially, two independent fuel testing laboratory were planned. Out of these, one has been commissioned at NOIDA. The decision to set up other fuel testing laboratory has been kept under abeyance till the issue of mobilizing funds to meet the operating cost of existing laboratory at NOIDA is completely resolved. The action on setting up of inspection and maintenance centres has been delayed. Delhi Government has expressed its inability to initiate action due to stay order granted by the Calcutta High Court in the matter of All India Federation of Motor Vehicles Department Technical Executives Officer's Association and others versus Union of India and others [case no. 19138 (W) of 1992] against Section 56(2) of the Motor Vehicles Act regarding authorization of private testing station. The stay order has been vacated now.

In an effort to promote cleaner fuels other than natural gas, a project 'Introduction of Propane as an Automotive for Three-wheeler Autorikshaws' was commissioned by M/s G&T Resources Worldwide in 1998. Under the project, G&T Resources Worldwide was allowed to carry out experimentation on 25 new and 25 old autorikshaws using propane as fuel. The project did not reached to the conclusion.

Effect of Abetment and Control Measures on Ambient Air

The CPCB has been regularly monitoring ambient quality in Delhi for parameters including sulphur dioxide (SO_2), nitrogen dioxide (NO_2), Suspended Particulate Matters (SPM), carbon monoxide (CO_2) and lead (Pb). The annual average values of these parameters during and after implementation of control measures *i.e.*, 1998 to 2005 are presented in Figures 62.5(a) to (e). It can be seen that the concentrations of SO_2, CO_2 and Pb in the ambient air have shown decreasing trend since 1998. As regards the NO_2 and SPM are concerned, their concentrations in the ambient air have slightly increased during the period.

The reduction in SO_2 level in ambient air was observed to be 64 per cent on account of various measures for abatement and control of pollution. Similarly, CO level in ambient air has decreased by 53 per cent. As much as 79 per cent reduction in Pb level was also observed. The reduction of sulphur and lead content in the ambient air is primarily due to improvement in fuel quality as discussed earlier. Improved vehicle technology and three way catalytic converters contributed to reduction of CO emissions. As per the monitoring results, an increase of 31 per cent in NO_2 level and 20 per cent increase in SPM level in air were observed after implementation of control measures. Higher ignition point of CNG may be the reason for more NO_2 emissions in auto exhaust. Marginal increase in SPM level in ambient air indicates presence of other sources of SPM emissions. It needs to be noted that the improvement in air quality in terms of SO_2, CO_2 and Pb as observed during monitoring is despite the addition of new vehicles in Delhi, which is average 5,03,963 vehicles per year.

Future Plans of Implementing Agencies

Introduction of CNG as automotive fuel has been one of major step for containment of vehicular pollution in Delhi. Under this programme, emphasis has been given to conversion of commercial vehicles such as buses, three-wheeler autorikshaws and taxis to CNG mode. In near future emphasis will be given to conversion of light goods vehicles playing in Delhi for which time schedule has already been worked out. At present, CNG facility is not available in neighbouring towns of Delhi. For the supply of CNG in these areas, feasibility study is underway in towns of Noida, Greater NOIDA, Faridabad and Gurgaon in the first phase. IGL has proposed to cover towns of Ghaziabad and Bahadurgarh in the second phase.

In the future, high capacity bus system is to be introduced at selected routes. One such bus is presently being operated on experimental basis at Mulchand Hospital–Dr. Ambedkar Nagar route. There is also plan to operates electric trolley buses at least on two routes. One such route is Badarpur–Pragati Maidan. Mass rapid transport system is being implemented by Delhi Government with a view to place non-polluting and efficient rail based system, which will be integrated with other modes of transport.

At present, there is one inspection and maintenance system for commercial vehicles, which is inadequate. Additional inspection and maintenance centre has been planned at Burari. In order to avoid entry of non-destined vehicles, a plan for the construction of express by-pass road around Delhi has been approved. The bye-pass road will be 240 km long.

Conclusion

Most of the control measures discussed in the chapter have been tried and implemented for the first time in the country. Supply of lead free petrol was started in Delhi with few outlets and spread to the entire city and then in the entire Nation Capital Region. Similarly, the fuel with low sulhpur and

Figure 62.5(a–e): Ambient Air Quality

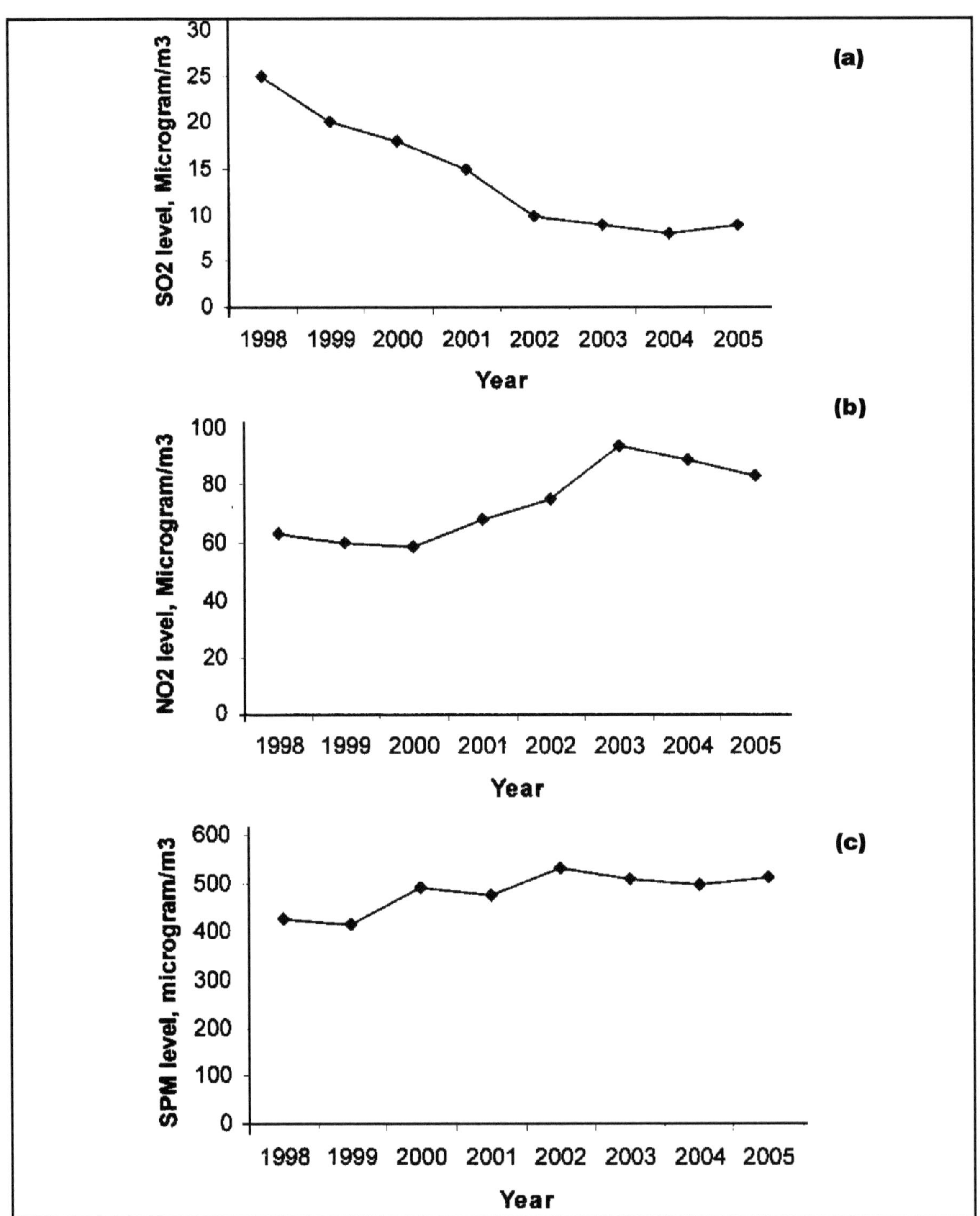

Contd...

Figure 62.5(a–e)–Contd...

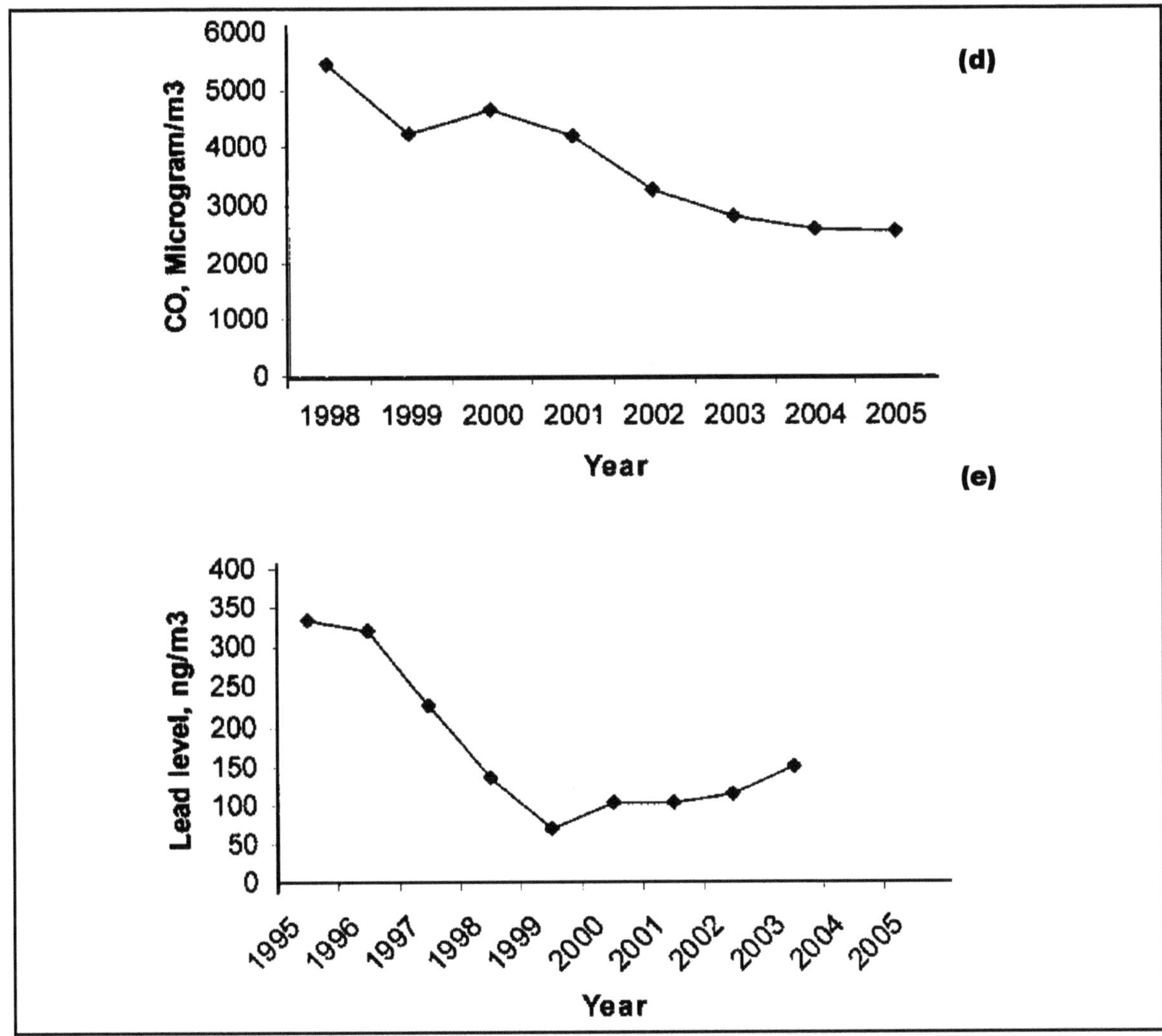

benzene content was introduced initially in Delhi. Registration of only Euro I (Bharat Stage I) and Euro II (Bharat Stage II) compliant vehicles have been started in Delhi. In case of CNG vehicles, experiment with few buses with modified engines to operate on CNG was initiated in Mumbai. However, introduction of CNG as an automotive fuel on mass scale has been taken place in Delhi. With more than 90,000 vehicles playing on CNG, Delhi is now one of the cities in the world to have large number of vehicles operating on CNG.

The GAIL India Limited is in the process of expanding CNG supply network in 22 different cities. Experience gained though the implementation of vehicular pollution control measures in Delhi will be useful in future. The practice followed in Delhi for the containment of vehicular pollution so far can serve as role model for the development of pollution control programme of other cities in India.

Acknowledgements

We wish to extend our appreciation to Shri Narayan Singh Mehra and Shri Lokesh Kumar for the assistance in preparation of this paper.

References

DTD Plan, 2002. *Tackling Urban Transport Pollution: Operating Plan for Delhi*. Delhi Transport Department, Delhi.

EPCA Report, 1998. Monitoring and priority measures for pollution control–March to June 1998. Environmental Pollution (Prevention and Control) Authority for the National Capital Region, Delhi, pp. 2–3.

EPCA Report, 1998. Third progress report–October to December 1998. Environmental Pollution (Prevention and Control) Authority for the National Capital Region, Delhi, pp. 2–3.

EPCA Report, 2000. Eighth progress report–March 2000 to June 2000. Environmental Pollution (Prevention and Control) Authority for the National Capital Region, Delhi, pp. 2–5.

MoEF Report, 1997. White paper on pollution with an action plan. Ministry of Environment and Forests, Government of India, New Delhi, pp. 15–17.

MoEF Notification, 1998. Notification No. S.O. 93 (E), dated January 29, 1998. Ministry of Environment and Forest, Government of India, New Delhi.

SC Order, 1998. Order dated July 28, 1998 in Writ Petition (Civil) No. 13029/1985, issued by the Hon'ble Supreme Court of India, New Delhi.

http://www. transport.delhigovt.nic/transport (Statistical information on vehicles).

Chapter 63

In vitro Plant Regeneration from Different Explants of *Centella asiatica* (L.)

Surjeet S. Bisht, Snehlata Bhandari and N.S. Bisht

Department of Botany, H.N.B. Garhwal University, Campus Pauri, Pauri (Garhwal) – 246 001, Uttarakhand

ABSTRACT

In vitro organogenesis study of *C. asiatica* (L.) member of family Apiaceae was tried considering its medicinal importance. MS medium containing Kn (1.0–3.0 mg/l^{-1}) and 2,4-D (1.5–4.0 mg/l^{-1}) is best for callusing in leaf and petiole explants. 1.5–2.0 mg/l^{-1} concentration of Kn and 2,4-D or IBA proved best for callus induction in nodal explant. BA (1.5 mg/l^{-1}) and IBA (1.0 mg/l^{-1}) combination in MS medium induced maximum number of shoots, while 1.0 mg/l^{-1} of auxin in MS full was suitable for higher root development. Controlled conditions and regular irrigation with nutrient media at initial stages are desirable for higher survival of plantlets.

Keywords: *Apiaceae, Callusing, Organogenesis, Rooting, Shooting.*

Introduction

Centella asiatica L. (Apiaceae), a weakly scented medicinal plant is occurring in part of India, Sri Lanka, Indonesia, Malaysia and Southern and Central Africa. Plant is used in several Ayurvedic preparations for the treatment of leprosy, varicose veins, ulcers, lupus and certain eczemas and of mental retardation since ancient times (Kartnig, 1988). The drug is also reported to possess anti-epileptic activity (Moharana and Moharana 1994). Clinical trials have shown that extracts of *C. asiatica* heal wounds, bums and ulcerous abnormalities of the skin, cure stomach and duodenal ulcers. Plant possesses anti-leprotic, anti-filarial, adaptogenic, antiviral and antibacterial and insecticidal properties. The requirement of *C. asiatica* is met from natural vegetation, leading to its gradual depletion. Tissue culture technique can play an important role in the rapid multiplication as well as germplasm conservation of the plant. Standardisation of an *in vitro* multiplication is a crucial pre-requisite for this. Shoot regeneration and somatic embryogenesis from different explants of *Brahmi* (*Bacopa monniera* L.) have been reported by Tiwari *et al.* (1998). Patra *et al.* (1998) had successfully regenerated *C. asiatica* from callus cultures. *In vitro* multiplication of *C. asiatica* from leaf explant has been studied by Banerjee *et al.* (1999) and Tiwari *et al.* (2000).

Material and Methods

Centella asiatica explants (leaf, petiole, node) collected from nature (Oak pine forest of Garhwal) are initially washed with tap water followed by a wash with 1 per cent (v/v) Labolene detergent for 15 minutes and then in running tap water for 30 minutes. The explants are then surface sterilized with an aqueous solution of 0.1 per cent $HgCl_2$ (w/v) for 2–3 min, followed by a final 5–6 rinses with sterilized double distilled water. The cut surfaces exhibiting mercuric chloride damage were asceptically trimmed with sharp, sterile surgical blade.

Murashige and Skoog (1962) basal medium (MS) supplemented with 3 per cent sucrose and 8 per cent agar is used in the present investigation. Auxins (Indole 3-acetic acid, Indole 3-butyric acid, Napthalene acetic acid or 2,4-Di chlorodiphenoxy acetic acid) and cytokinins (6-Benzyl Adenine or Kinetin) are incorporated into the basal medium in varying concentrations and combinations as indicated in the result. The pH of all the media combinations is adjusted to 5.8±0.1 using 0.1 N NaOH or 0.1 N HCl. Autoclaving is done at 1.06 kg cm^{-2} at 121° for 25 to 30 minutes. Cultures are incubated at 24°±2°C temperature and 60 per cent relative humidity with 16 : 8 hrs, light: dark photoperiod. All treatments having 4 to 8 replicates. Each culture flask contained 40 ml of culture medium and 2–4 explants. The experiments are repeated thrice.

Rooted plantlets removed from the culture media were thoroughly and gently washed under running tap water to remove agar from the roots, and planted in thermocol cups (5 cm. dia) having three different types of soil *viz.*, (*i*) sterilized mixture of garden soil and organic manure in the ratio of 1:1 (v/v), and (*ii*) sterilized mixture of sand and soil in the ratio of 1:1 (v/v), (*iii*) sterilized forest soil. Planted thermocol cups were kept in growth chamber for two to four weeks. Potted plantlets were irrigated either with (*i*) water only or (*ii*) 1/4 strength of growth regulator free MS medium subsequently with water or (*iii*) 1/2 strength of growth regulator free MS medium and subsequently with water during the plants growth in Growth Chamber. Plantlets were covered with 250 ml transparent glass beakers or polythene bags to provide high humidity. Temperature of growth chamber was maintained at 20°C±1°C with 16hr photoperiod.

Plantlets were transferred to earthenware pots (25 cm. dia) containing different types of soil mixture. Planted earthenware pots were kept in shade for two weeks and watered regularly. All the potted plants were then placed outdoor with the gradual increase in the periodicity under full sun. The plants transferred to full sun were irrigated regularly for a week and number of plants surviving till the end were counted and survival percentage was computed.

Results and Discussion

Callusing Response

MS medium containing either Kinetin or 2,4-D (0.5 mg/l^{-1}) did not show any callusing response. Poor callusing is observed when MS medium is fortified with auxin and 0.5 mg/l^{-1} Kn. Higher concentration of Kn (1.0–3.0 mg/l^{-1}) combined with higher concentration of 2,4-D (1.5–4.0 mg/l^{-1}) exhibited best callusing in leaf and petiole explants. Kn (1.0–1.5 mg/l^{-1}) and IBA (1.5 mg/l^{-1}) combination in MS proved most appropriate in the production of best callus mass in leaf explant.

Higher concentration of another cytokinin BA (2.0–2.5 mg/l^{-1}) produced best calli when combined with 0.5 to 1.5 mg/l^{-1} of NAA in leaf explants while 1.5–2.0 mg/l^{-1} concentration of both BA : NAA and Kn : 2,4-D or IBA are found to be most suitable of the development of best calli in petiole and nodal explants respectively.

Shooting Response

Development of petiole from the stolon part is considered as shooting in this plant. Periodic variations in the number of petioles were observed from 15th day after inoculation and at the interval of 15 days onwards up to 120 days. BA was combined with IBA in MS medium for shooting purpose. BA used was in the concentration range of 1.0 to 3.5 mg/l^{-1} and IBA in the range of 0.5 to 2.0 mg/l^{-1} and both were combined separately in different concentrations. Number of petiole ranged from 2.0±0.82 to 3.55±1.28 initially at 15 day stage. The rate of appearance of new petioles in different concentrations varied during study period. At 120-day stage maximum number of petioles developed is 1.5 : 1.0 mg/l^{-1} followed by 2.0 : 1.0 mg/l^{-1} combination (31.1±3.3). Higher concentrations of BA and IBA in MS medium resulted in the slowest growth rate where the petiole number was 10.0±1.63.

Lowest concentration of these chemicals used in the present study also exhibited lower number of petioles (11.6±2.28). 2.5 : 1.5 mg/l^{-1} combination was found to be better than the combinations of either lowest or highest concentrations of cytokinin and auxin. Thus most favourable combination for best growth was 1.5 : 1.0 mg/l^{-1} of BA and IBA (Table 63.1).

Table 63.1: Periodic Shooting Variation in *C. asiatica*

MS Basal + Growth Regulators (mg/l^{-1})	*Number of Days*							
Cytokinin : Auxin	*15*	*30*	*45*	*60*	*75*	*90*	*105*	*120*
1.0 (BA) : 0.5 (IBA)	3.33±1.24	5.0±1.63	5.66±1.24	7.0±1.63	7.12±1.63	8.0±1.63	10.0±1.63	11.6±2.28
1.5 (BA) : 1.0 (1BA)	3.0±0.82	8.33±1.25	13.33±2.28	19.0±1.63	21.3±2.05	26.6±2.15	32.6±1.71	37.0±1.63
2.0 (BA) : 1.0 (IBA)	3.0±0.81	7.33±1.24	12.33±2.05	16.6±2.54	21.0±1.63	24.6±2.05	26.3±2.05	31.3±3.32
2.5 (BA) : 1.5 (IBA)	2.0±0.82	4.66±1.69	7.0±l.63	9.66±1.69	12.3±1.88	15.3±2.05	17.3±3.09	20.6±2.62
3.0 (BA) : 1.5 (IBA)	3.12±1.20	5.4±1.76	6.0±1.61	6.29±1.32	7.0±1.63	7.43±1.62	8.0±1.63	10.0±1.63
3.5 (BA) : 2.0 (IBA)	3.55±1.28	5.7±1.82	5.79±1.32	7.12±1.61	7.33±1.63	8.12±1.69	8.52±1.75	10.0±1.63

±: SD.

Rooting Response

MS medium was fortified with 0.5 to 1.5 mg/l^{-1} of auxin (NAA or IBA) in separate sets of experiment to induce rooting in the small fragments having well developed shoots that were transferred in the fresh medium for rooting purpose. Root number was periodically recorded at the interval 15 days upto 120 days.

Rooting behaviour was studied in MS medium fortified with 0.5 or 1.0 mg/l^{-1} of NAA or IBA as well as in MS half strength combined with 1.5 mg/l^{-1} of auxins. Requirement of 1.0 mg/l^{-1} auxin in MS full for the development of greater number of roots is well evident. However, good results were noticed when half strength of MS medium was used with 1.5 mg/l^{-1} of auxin. Lower number of roots were observed in the nutrient medium containing lower concentration of auxins (Table 63.2).

Table 63.2: Periodic Variation in Number of Roots of *C. asiatica*

MS Auxins (mg/l^{-1})	*Number of Days*							
	15	*30*	*45*	*60*	*75*	*90*	*105*	*120*
MS full + 0.5 NAA	3.33±1.24	7.33±2.05	11.0±3.74	16.0±3.74	18.0±2.94	20.0±0.82	34.3±3.3	38.6±2.87
MS full + 1.0 NAA	6.0±0.82	10.3± 1.0 1	14.6±2.05	22.0±1.63	27.6±2.05	32.0±2.16	43.3±3.39	50.6±3.09
MS half + 1.5 NAA	2.66± 1.69	8.33±1.24	16.0±2.94	21.0±2.94	26.3±5.74	35.0±4.08	41.0±2.94	45.6±4.94
MS full + 0.5 IBA	3.0±0.8	6.66± 1.69	11.0±2.44	15.0±4.32	17.0±2.94	22.0±4.96	33.0±3.26	37.6±2.89
MS full + 1.0 IBA	3.66± 1.24	9.33±1.2	13.3±1.07	21.3± 1.24	26.3±2.05	35.3±2.62	42.3±2.86	49.0±2.16
MS half + 1.5 IBA	4.66± 1.69	9.33±2.05	15.3±2.05	24.0±2.94	32.0±2.49	35.3±5.31	41.3±4.19	43.3±4.78

±: SD.

Field Transfer

Observation indicates that higher percentage of survival can be achieved by keeping the plantlets under controlled conditions (Growth Chamber) for a longer period with the supply of nutrie:1ts (half or quarter MS). Irrigation with water only increased the mortality rate thus, reducing survival percentage.

It was important to maintain low temperature of 18°C to 20°C because at 25°C±2°C most of the plantlets died. Similar findings have been reported for *Valeriana wallichii* (Mathur *et al.*, 1988) and *Swertia chirata* (Wawrosch *et al.*, 1999), the Himalayan plants which grow at an altitude of 1500–3000 m.

Thus it indicates the requirement of sufficient nutrients during initial stages is important for enhancing the survival rate. Results indicate that hardening of the plants weaned away from culture vessels at least for a minimum period of 2 weeks was a pre-requisite for successful establishment, because all plantlets transferred directly to earthen pots could not survive while those hardened for 2–4 weeks in the Growth Chamber with high humidity showed higher percentage of survival.

Irrigation soon after planting and intermittent misting at 4 hour intervals was the most suitable to achieve high percentage establishment of the plants (Sreekumar *et al.*, 2000). These workers have reported the highly desirable condition to rear the hardened plants at least for two weeks in the net house with 50 per cent shade and regular irrigation before field planting.

80 to 85 per cent survivality of *C. asiatica* was reported by Banerjee *et al.* (1999) during transfer from growth chamber to glass house. Field transfer trials made in the present study indicate that garden soil with organic compost potted plants kept in controlled environment for a indicate that garden soil with organic compost potted plants kept in controlled environment for a longer duration, irrigation with half strength MS solution and water are the desirable conditions to achieve high survival percentage for *C. asiatica*.

No callusing has been seen without cytokinins and auxins in the culture media, irrespective of the explant used. Kukreja (1996) had reported that there was no morphogenetic response on MS medium and similarly the leaf explant of *Azadirachta indica* had no callusing in MS medium alone (Eeswara *et al.*, 1998).

Saikia *et al.* (2001) have reported the best concentration and combination for callus induction was 1.0–1.0 mg/l^{-1} of 2,4-D and BAP with increase in the concentration of both the phytohormones resulted in poor growth of friable callus in *Hydrocotyle rotundifolia.* Leaf explant of *C. asiatica* in the present study also reflected best growth of callus in 1.0 : 1.0 mg/l^{-1} of Kn and IBA but similar response was noticed with 1.5 : 1.5 mg/l^{-1} combination as well. Higher concentration of Kn (1.5–2.5 mg/l^{-1}) in combination with higher concentration of 2,4-D (1.5–3.5 mg/l^{-1}) also responded towards best callusing in the present case. The difference in the response in our study may be due to the age of explant as they were collected from the plants growing in the field. Reports on callus induction in *Psoraiea corylifolia* using mature leaves and stem in MS medium supplemented with combination of cytokinins (BA and Kn) and auxins (NAA and 2,4-D) all at 0.0–4.0 mg/l^{-1} by Saxena *et al.* (1997) are substantiating our findings.

Lower concentration of BA (0.5 mg/l^{-1}) in combination with 0.5–2.5 mg/l^{-1} of NAA showed poor callusing while higher concentration of BA (2.0–2.5 mg/l^{-1}) with lower concentration of NAA (0.5–1.5 mg/l^{-1}) reflected fast callus proliferation resulting into best callus mass during growth period. Similar finding were earlier reported by Matsubara *et al.* (1996) while studying *in vitro* callus formation of *Ocimum basilicum* cultivars and *Perilla* spp.

Generally a cytokinin or a combination of cytokinin and auxin is required for *in vitro* shoot proliferation (Thrope and Patel, 1984). Fracaro and Echeverrigaray (2001) also observed that the presence of cytokinin on the culture media positively influence the micropropagation of *Cunila galioides,* and that BA gave significantly more shoot per explants than the other cytokinins tested (Tiwari *et al.*, 1998; Banerjee *et al.*, 1999; Saikia *et al.*, 2001). The combination of cytokinins and auxins stimulated the *in vitro* multiplication and the growth of shoots of several plant species (George, 1993).

Islam *et al.* (1996) had observed that number of shoot/seedling increased with increase in BA concentration and 2.0 mg/l^{-1} BA showed the maximum response. Maximum multiple shoots were noticed in 1.5 mg/l^{-1} BA concentration with IBA (1.0 mg/l^{-1}) in the present study.

Inhibitory effect of increased auxin concentration on shoot bud formation was reported by Nandawani and Ramawat (1991). Increase in the concentration of auxin suppresses the bud differentiation. This may be explained by assuming that the amount of auxin synthesized endogenously was more than sufficient to interact with added cytokinin to achieve the balance necessary for bud induction. The present findings have the supports of observation made by Mandal and Gadgil (1979).

Banerjee *et al.* (1999) have reported that the initial sprouting required the presence of BAP at 2.0 mg/l^{-1} and IBA at 0.1 mg/l^{-1} concentration. Tiwari *et al.* (2000) also observed that the maximum frequency and optimum number of shoots was attained on MS medium containing high BA and low NAA concentrations. High shoot numbers obtained in the present investigation in high BA (1.5–2.5 mg/l^{-1}) concentrations find support from above findings.

Excised shoots were transferred to rooting medium. Earlier workers have reported the use of full or half or quarter strength of MS medium supplemented with auxin (IAA, IBA, NAA etc.). The use of half or full strength of MS media for root induction can be attributed to the favourable effect of low concentrations of macro and micro nutrients on rooting is probably due to the decreased requirement of nitrogen for rhizogenesis in some plants, while it may not be the case with other plant species. Low level of sucrose with auxin was reported optimum in *C. asiatica* by Patra *et al.* (1998) and in *Punica granatum* by Nataraja and Neelambika (1996).

Nandwani and Ramawat (1991) have observed the increased rooting with increased concentration of IBA up to 3.0 mg/l^{-1} and NAA up to 5.0 mg/l^{-1} incorporated singly in half strength of MS. In the present study also maximum number of roots were induced on MS half with higher concentration of auxin.

Banerjee *et al.* (1999) while studying *in vitro* micropropagation using leaf explant of *Centella asiatica* have reported root initiation on full or half strength MS medium with or without an auxin. The average number of roots was good on half strength MS + 1.5 mg/l^{-1} auxin. In the present study, however, maximum number of roots were observed on MS full + 1.0 mg/l^{-1} NAA or IBA (Table 63.2).

References

Banerjee, S., Mehra, Z. and Kumar, S., 1999. *In vitro* multiplication of *Centella asiatica*: a medicinal herb from leaf explant. *Current Science*, 76(2): 147–148.

Eeswara, J.P., Stuchbury, T., Allan, E.J. and Mordue, A.J. (Luntz), 1998. A standard procedure for the micropropagation of the neem tree (*Azadirachta indica* A. Juss). *Plant Cell Reports*, 17: 215–219.

Fracaro, F. and Echeverrigaray, S., 2001. Micropropagation of *Cunila galioides*: A popular medicinal plant of South Brazil. *Plant Cell Tissue and Organ Culture*, 64: 1–4.

George, E.F., 1993. *Plant Propagation by Tissue Culture, Part 1: The Technology.* Exegetics Ltd., Edington.

Islam, R., Hossain, M., Reza, M.A., Mamun, A.N.K. and Jorder, O.I., 1996. Adventitious shoot regeneration from root tips of intact seedlings of *Aegle marmelos*. *J. Hort. Sci.*, 11(6): 995–1000.

Kartnig, T., 1988. Clinical application of *Centella asiatica* (L.) Urb. In: *Herb Spices and Medicinal Plants*, (Eds.) Craker, L.E. and Simson, J.E. Oxyx Press, Phoenix, p. 145–173.

Kukreja, A.K., 1996. Micropropagation and shoot regeneration from leaf and nodal explant of peppermint (*Mentha pipperita* L.). *Journal of Spices and Aromatic Crops*, 5(20): 111–119.

Matsubara, S., Ino, M., Murakami, K., Kamada, M. and Ishihara, I., 1996. Callus formation and plant regeneration of herbs in *Perilla* family. *Scientific Report of the Faculty of Agriculture*, Okayama University, 85: 23–30.

Mandal, S. and Gadgil, V.N., 1979. Differentiation of shoot buds and roots in fruit callus culture of *Solanum nigrum* L. *Indian Journal of Experimental Biology*, 17: 876–878.

Mathur, J., Ahuja, P.S., Mathur, A., Kukreja, A.K. and Shah, N.C., 1988. *In vitro* propagation of *Valeriana wallishii*. *Planta Medica*, 39: 187, 277, 1–3.

Moharana, D. and Moharana, S., 1994. A clinical trial of mental in patients with various types of epilepsy. *Probe* 33: 160–162.

Murashige, T. and Skoog, F., 1962. A revised medium for rapid growth and bioassys with tobacco tissue cultures. *Physiol. Plant*, 15: 473–497.

Nandwani, D. and Ramawat, K.G., 1991. Callus culture and plantlets formation from nodal explants of *Prosopis juliflora* (Swartz) DC. *Indian Journal of Experimental Biology*, 29(5): 523–527.

Nataraja, K. and Neelambika, G.K., 1996. Somatic embrogenesis and plantlet from petal cultures of pomegranate, *Punica granatum* L. *Indian Journal of Experimental Biology*, 34(7): 719–721.

Patra, A., Rai, E., Rout, G.R. and Das, P., 1998. Successful plant regeneration from callus culture of *C. asiatica* (Linn.) Urban. *Plant Growth Regul.*, 24: 13–16.

Saikia, F.R., Baruah, C.C., Geka, A.C. and Kalita, M.C., 2001. Micropropagation of *Hydrocotyle rotundifolia*: An indigenous medicinal plant of north east. *Indian Journal of Medicinal and Aromatic Plant Sciences*, 22(4A), 23(1A): 291–293.

Saxena, C., Palai, S.K., Samantaray, S., Rout, G.R., and Das, P., 1997. Plant regeneration from callus cultures of *Psoralea corylifolia* Linn. *Plant Growth Regulation,* 22(1): 13–17.

Sreekumar, S., Seeni, S. and Pushpangadan, P., 2000. Micropropagation of *Hemidesmus indicus* for cultivation and production of 2-hydroxy, 4-methoxy benzaldehyde. *Plant Cell Tissue and Organ Culture,* 62: 211–218.

Thorpe, T.A. and Patel, K.R., 1984. Clonal propagation: Adventitious buds. In: *Cell Culture and Somatic Cell Genetics of Plant Cells, Vol. 1: Laboratory Procedures and Their Application,* (Ed.) IX. Vasil. Academic Press, Orlando, Florida, p. 49–60.

Tiwari, K.N., Sharma, N.C., Tiwari, V. and Singh, B.D., 2000. Micopropagation of *C. asiatica* (L.): A valuable medicinal herb. *Plant Cell Tissue and Organ Culture,* 63: 179 –185.

Tiwari, V., Singh, B.D. and Tiwari, K.N., 1998. Shoot regeneration and somatic embryogenesis from different explants of Brahmi [*Bacopa monniera* (L.) Wettsty]. *Plant Cell Rep.,* 17: 538–543.

Wawrosch, C. Maskay, N. and Kopp, B., 1999. Micropropagation of the threatened Nepalese medicinal plant *Swertia chirata* Buch.-Ham. Ex Wall. *Plant Cell Report,* 18: 997–1001.

Chapter 64

Evaluation of the Efficacy of Garlic Bulb Extract and Selected Animal Excrements on the Mycelial Growth, Sporulation and Conidial Germination of *Helminthosporium oryzae* (Breda de Haan) Subram and Jain (*in vitro*)

M. Thamarai Selvi, P. Balabaskar and V. Kurucheve

Department of Plant Pathology, Faculty of Agriculture, Annamalai University, Annamalainagar – 608 002, Chidambaram, Tamil Nadu

ABSTRACT

Efficacy of certain natural products like garlic bulb extract, sheep urine, buffalo urine, goat urine, cow urine, pig dung, goat dung, cow dung, buffalo dung and hen litter were tested against *Helminthosporium oryzae* under *in vitro* conditions. The results revealed that sheep urine (5 per cent), buffalo urine (20 per cent), goat urine (20 per cent), cow urine (40 per cent), hen litter (80 per cent), goat dung (100 per cent) and garlic bulb extract (20 per cent) were found to completely inhibit the mycelial growth of *Helminthosporium oryzae*. The natural products at their respective MIC's showed complete inhibitory effect on the sporulation and conidial germination of the test fungus.

Keywords: *Helminthosporium oryzae, Percent inhibition, Animal excreta, Inverted Petri plate method, Reinoculation, Minimum Inhibitory Concentration (MIC's).*

Introduction

Rice brown spot disease caused by *Helminthosporium oryzae* Cav. is generally considered as the principal disease of rice because of its wide occurrence and destructiveness under favourable conditions. Eventhough the chemical fungicides contributed greatly for the management of the disease, (Fawcett, and Spencer, 1970) the possible pollutions and residual effects due to chemicals necessitated the search for non-chemical methods of managing the disease. Among the non-chemical methods, use of plant products and animal excrements has gained importance in the recent years. With this background the present experiment was conducted to test the efficacy of garlic bulb extract, sheep urine, buffalo urine, hen litter and goat dung extracts against *Helminthosporium oryzae* under *in vitro* conditions. The products were tested for the presence of non-volatile and volatile antifungal principles using poisoned food technique and inverted Petri plate techniques and the reduction of sporulation and conidial germination by cavity slide method.

Materials and Methods

Preparation of Garlic Bulb Extract

For the preparation of garlic bulb extract, the method suggested by Gerard Ezhilan *et al.* (1994) was followed. Fresh bulbs were collected, washed with tap water followed by sterile distilled water, processed @ 1 ml per g of tissue (1 : 1 v/w) with pestle and mortar and filtered through a double layered cheese cloth. This formed the standard bulb extract solution (100 per cent).

Preparation of Animal Dung Extracts

For the preparation of animal dung extract, the method of Sundarraj *et al.* (1996) was followed. Animal dung were collected fresh, shade dried for one week and made into powder. The powdered animal dung was soaked in sterile distilled water @ 5 ml g^{-1} (5: 1 v/w) and kept ever night; blended and filtered through cheese cloth. This formed the standard animal excreta solution (100 per cent). After extraction, they were subjected to low speed centrifugation (5000 rpm for 20 minutes) to avoid any contamination (Jagarinathan and Narasimhan, 1998).

Urine

Freshly collected urine was used as such forming the standard extract (100 per cent) (Raja and Kurucheve, 1998).

Poisoned Food Technique

PDA medium was prepared in 250 mo Erlenmeyer flasks and sterilized. Aqueous extracts of 2.5 and 10 ml and were added to 47.5 and 40 ml of broth respectively in the. flask so as to get the final concentration of 5 and 20 per cent of the extracts in the medium. For getting 100 per cent concentration of extract, the standard extract as such was used. The fungicide Minosan 50 per cent EC at 0.2 per cent concentration was used for comparison. PDA medium without any extract served as control. Each Petriplate was poured with 15 ml of natural product amended medium. The culture disc (9 mm) of *Helminthosporium oryzae* obtained from the periphery of the actively growing seven days old culture was inoculated in the centre of the Petri plates aseptically and incubated for seven days. Five replications were maintained for each treatment. The diameter of the colony was measured when the mycelium fully covered the Petri plates on anyone of the treatments. The percent inhibition of growth was calculated as per Vincent (1947) for each treatment and expressed in mm.

$$\text{Percent inhibition} = \left[\frac{C-T}{C}\right] \times 100$$

where,

C: Diameter of growth in control

T: Diameter of growth in treatment.

Inoculum discs showing no growth due to treatment with garlic bulb extract and animal excrements were tested again for their viability by re inoculating them in to fresh medium.

Inverted Petri Plate Method

Fifteen ml of PDA medium was seeded with three ml of spore suspension (1 × 10″ spores/ml) and allowed to solidify in Petri plates. These plates were then inverted upside down. Extracts of garlic bulb and animal excrements (20 ml) at different concentrations *viz.*, 10, 20, 30, 40, 60, 80 and 100 were poured into the lower lid of the inverted Petri plates. The Petri plates were then incubated at room temperature (25±3°C) for seven days. Sterile distilled water was poured in control plates. Fungicide Hinosan 50 per cent EC (0.2 per cent) was used for comparison. After the incubation period, mycelial growth was recorded. Inoculum discs showing no growth due to treatment with garlic bulb extract and animal excrements were tested again for their viability by re inoculating them in to fresh medium.

Sporulation and Conidial Germination of *H. oryzae* (Cavity Slide Method)

Sporulation

The effect of garlic bulb extract and animal excrements on the sporulation of *H. oryzae* was determined by the method of Bera and Saba (1983) with slight modification.

Medium containing natural products at their respective MIC were poured into Petri plates aseptically at 15 ml/plate. Fungal disc of nine mm size was inoculated in the centre of Petri and plate and incubated at 25±3°C for seven days. Five replications were maintained for each treatment and a suitable control was also maintained. A single mycelial disc from the periphery of Petri plate was removeq with the aid of a sterile cork borer after incubation and transferred into one ml of sterile distilled water and shaken vigorously. A drop of spore suspension was placed on a glass slide and the spore produced was estimated using a haemocytometer.

Spore Germination

Double the required strength aguous solution of the test fungicides were prepared, one drop of spore suspension (2.5 × 10^4 Conidia ml^{-1}) was placed in the well of each cavity slide together with a drop of individual fungicidal solutions and incubated at 22°C in moist Petri dishes for 12 h. Hinosan 50 per cent EC at 02 per cent concentration was used for comparison. Five replications were kept for each treatment. Observations were taken from ten random fields under microscope and per cent inhibition of conidial germination was calculated and the inhibition of germ tube length was expressed as per cent reduction over control (Mustaq Ahmad, 1992).

Results and Discussion

Poisoned Food Technique

The results on the effect of water extracts of natural products at various concentrations on the mycelial growth of *H. oryzae* are presented in Table 64.1. Among the various natural products tested,

Table 64.1: Effect of Some Natural Products Against Mycelial Growth of *P. oryzae*

Sl.No.	*Sources*	*Diameter Mycelial Growth (mm)*									*% Decrease (–) Over control*							
		2.5%	*5%*	*10%*	*20%*	*40%*	*60%*	*80%*	*100%*	*Mean*	*2.5%*	*5%*	*10%*	*20%*	*40%*	*60%*	*80%*	*100%*
1.	Garlic bulb extract	81.67	71.31	20.00	0.00	0.00	0.00	0.00	0.00	21.62	9.25	20.77	77.78	100	100	100	100	100
2.	Sheep urine	20.00	0.00	0.00	0.00	0.00	0.00	0.00	0.00	3.90	65.43	100	100	100	100	100	100	100
3.	Goat urine	88.00	35.56	28.00	0.00	0.00	0.00	0.00	0.00	18.95	2.22	60.48	68.89	100	100	100	100	100
4.	Cow urine	86.00	45.00	33.00	15.00	0.00	0.00	0.00	0.00	22.38	4.44	50.00	63.33	83.33	100	100	100	100
5.	Buffalo urine	82.00	40.00	24.00	0.00	0.00	0.00	0.00	0.00	18.25	8.89	55.56	73.33	100	100	100	100	100
6.	Hen letter	89.00	60.00	46.00	36.00	33.00	30.00	0.00	0.00	39.75	1.11	33.33	48.89	60.00	63.00	66.67	100	100I
7.	Pig dung	90.00	61.00	45.00	37.00	31.00	27.00	24.00	13.00	41.00	0.00	32.22	50.00	58.89	65.56	70.00	73.33	100
8.	Goat dung	90.00	60.00	30.00	28.00	26.00	18.00	13.00	0.00	33.13	0.00	33.33	66.67	68.89	71.11	80.00	85.53	100
9.	Cow dung	85.53	62.00	39.00	37.00	30.00	26.00	21.00	11.00	38.94	4.97	31.11	56.67	58.89	66.67	71.11	76.67	87.78
10.	Buffalo dung	87.00	60.00	41.00	39.00	38.00	36.00	15.00	13.00	41.13	3.33	33.33	54.44	56.67	57.78	60.00	83.33	85.56 I
11.	Hinosan 50% EC (0.2%)	0.00	–	–	–	–	–	–	–	0.00	–	–	–	–	–	–	–	–
12.	Control	90.00	–	–	–	–	–	–	–	90.00	–	–	–	–	–	–	–	–
	Mean	75.02	41.24	25.50	16.00	13.171	11.42	8.08	3.08									

Factors	S.E	CD (p=0.05)
Main treatment	0.2480	0.7260
Sub-treatment	0.1864	0.5166
ST × MT	0:6456	1.7895
MT × ST	0.6528	1.8242

* Mean of five replications

Garlic bulb extract at 20 per cent concentration completely inhibited the mycelial growth of *H. oryzae* and it was on par with Hinosan (0.2 per cent). Among the animal excrements tested, sheep urine at 5 per cent, buffalo and goat urine at 20 per cent concentrations completely inhibited the mycelial growth of *H. oryzae* and all these were on par with Hinosan (0.2 per cent). In case of dung, hen litter and goat dung extracts gave 100 per cent inhibition of mycelial growth at '100 per cent concentration. The fungicide Hinosan at 0.2 per cent recorded complete inhibition of the growth of the test fungus and the untreated control recorded the maximum mycelial growth (90 mm). All the inhibited discs were inoculated in fresh medium and observed for mycelial growth for 7 days as given Table 64.2. No growth was observed indicating the fungitoxic nature of all the tested natural products.

Table 64.2: Evaluation of Fungitoxicity of Selected Natural Products by Reinoculating the Inhibited Discs of *H. oryzae* into Fresh Medium

Sl.No.	*Sources*	*Conc. (per cent)*	*Mycelial Growth**				*Nature of Toxicity*
			1	*2*	*3*	*4*	
1.	Garlic bulb	20	0	0	0	0	Fungicidal
2.	Sheep urine	5	0	0	0	0	Fungicidal
3.	Buffalo urine	20	0	0	0	0	Fungicidal
4.	Hen litter	100	0	0	0	0	Fungicidal
5.	Goat dung	100	0	0	0	0	Fungicidal
6.	Hinosan (0.2%)	0.2	0	0	0	0	Fungicidal
7.	Control	–	90				

*: Mean of five replications.

Raja and Kurucheve (1997) found that buffalo urine even at the lowest concentration tested (2.5 per cent) recorded 82 per cent inhibition, of *M. phaseolina.* The effect of cow urine on the in *vitro* inhibition of *F. subglutinans* was reported by Alonoso *et al.* (1994). Hen litter extract (10 per cent) recorded the maximum inhibition of the mycelial growth of *R. solani* (Sundarraj *et al.*, 1996). Buffalo urine (10 per cent) was found to have fungicidal activity against *M. phasolina* (Raja and Kurucheve, 1997). *F. oxysporum* f. sp. Lycopersici (Raja and Kurucheve, 1998). Pig dung extract (40 per cent) exhibited inhibitory action against *P. aphanidernafum* (Sivaprakash and Kurucheve, 1999), inhibited disc of *S. oryzae* replanted into fresh medium there was no growth indicating the fungicidal nature of cattle urine (Rajendra Prasad and Kurucheve, 2002). The difference in the inhibitory effect of natural products at various concentrations may be due to the quantitative differences in the antifungal principles present in them.

Inverted Petri Plate Method

All the treatments *viz.*, sheep urine and buffalo urine (60 per cent), hen litter and goat dung (100 per cent) and garlic (10 per cent) were found to inhibit the growth of *H. oryzae* when compared to control. Garlic bulb extract completely inhibited the growth of *H. oryzae* at 10 per cent concentration while all the cattle urines needed 40 per cent concentration for complete inhibition of mycelial growth of the test fungus. In case of goat dung and hen litter, the mycelial growth was completely inhibited at 80 and 100 per cent concentrations. Animal urines needed higher concentrations than their MICs for complete inhibition except in garlic bulb extract, which needed only 100 per cent concentration for complete inhibition of pathogen on the upper lid of the Petri plate (Table 64.3). The results of the present study the efficacy of natural products to inhibit the pathogen through volatile compounds similarly Wani (2001) reported that cow urine, buffalo urine, sheep urine and garlic bulb extract completely inhibited the mycelial growth of *A. niger* and *A. flavus* by inverted petri plate method. This might be due to the emission of volatile substances from these natural products.

Table 64.3: Effect of Garlic Bulb Extract and Animal Excrements on the Growth of *H. oryzae* (Inverted Petri plate method)

Sl.No.	*Sources**	*Mycelial Growth**					
		Conc. 10%	*20%*	*40%*	*60%*	*80%*	*100%*
1.	Garlic	–	–	–	–	–	–
2	Sheep urine	++	+	+	–	–	–
3.	Buffalo urine	++	++	+	–	–	–
4.	Hen litter	+++	+++	++	++	+	–
5.	Goat dung	+++	– +++	++	++	+	–
6.	Hinosan (0.2%)	–					
7.	Control		+++				

+++: Profuse growth; ++: Moderate growth; +: Scare growth; –: No growth.

*: Mean of five replications.

Effect of Garlic Bulb Extract and Animal Excrements on Sporulation and Conidial Germination of *H. oryzae* (Cavity Slide Method)

Garlic bulb extract and buffalo urine at 20 per cent concentration, sheep urine at 5 per cent concentration and hen litter and goat dung at 100 per cent concentration were found to completely inhibit the sporulation of *H. oryzae* and they were on par with Hinosan 0.2 per cent. Among the natural products garlic bulb extract at 10 per cent, sheep urine at 5 per cent, buffalo urine at 20 per cent, hen litter 60 per cent and goat dung at 80 per cent concentrations recorded complete inhibition of the conidial germination of *H. oryzae*. Several earlier workers have reported about the efficacy of natural products to inhibit spore germination and sporulation (Table 64.4).

Santhosh Kumar (2000) reported that bulb extract of garlic leaf extract of lawsonia and buffalo urine completely inhibited the conidial germination and sporulation of *C. falcatum*. Resmy Jayaraj (2002) reported that garlic bulb extract, cow urine, buffalo urine, goat urine and hen litter completely inhibited the conidial germination. sporulation of *H. oryzae*. The same products inhibited the conidial germination and sporulation of *C. capsid* (Krishna Kumar, 2002). These reports are in line with our findings. Reduction in spore germination and growth rate might be due to the presence of some inhibitory substances in aqueous extracts, which affected the physiological metabolism of the fungus as reported by Bera and Saba (1983).

Table 64.4: Effect of Natural Products on the Conidial Germination and Sporulation of *P. oryzae*

Sl.No.	Sources Concentration (%)	Conidial Germination (per cent) *								*Sporulation at MICs (after 7 days)
		2.5%	5%	10%	20%	40%	60%	80%	100%	
1.	Garlic	24.54 (29.69)	22.14 (28.06)	0	0	0	0	0	0	–
2.	Sheep urine	22.52 (28.32)	0	0	0	0	0	0	0	–
3.	Buffalo urine	67.72 (55.37)	51.58 (45.92)	34.61 (36.03)	0	0	0	0	0	–
4.	Hen litter	98.34 (82.58)	82.26 (65.09)	75.51 (60.36)	62.82 (54.42)	38.68 (38.47)	0	0	0	–
5.	Goat dung (82.08)	98.12 (70.73)	89.12 (62.65)	78.92 (56.04)	68.82 (45.71)	51.24 (34.94)	32.78	0	0	–
6.	Hinosan (0.2%)	0.00								–
7.	Control	98.34 (82.51)								+++

Factors	S.E.	CD (p=0.05)
Main treatment	0.0102	0.0309
Sub treatment	0.0095	0.0265
ST × MT interaction	0.0267	0.0749
MT × ST interaction	0.0268	0.0759

Figures in parenthesis are transformed values.

*: Mean of five replications; –: No sporulation; +++– Heavy inoculum

MIC's Garlic (bulb extract)	20%
Sheep urine	5%
Buffalo urine	20%
Hen litter	80%
Goat dung	100%

References

Alonso, S.K. de, Baretto, M., Costa, A.F., Silva Acuna, R., Zambolim, L., Ventura, J.A. and De Alonoso, S.K., 1994. Effect of cow urine on the growth of *Fusarium subglutinans in vitro. Fitopatologia brasileria,* 19: 235–238.

Bera, A.K. and Saha, A., 1983. Antifungal activity of an aqeous extract of *Catharanthus roseus. J. Pl. Dis. Prot.,* 90: 363–365.

Fawcett, C.H. and Spencer, O.M., 1970. Plant chemotherapy with natural products. *Ann. Rev. Phytopathol.,* 8: 403–416.

Gerard Ezhilan, J., Chandrasekhar, V. and Kurucheve, V., 1994. Effect of six selected plant products and oil cakes on the sclerotial production and germination of *Rhizoctonia solani. Indian Phytopath.,* 47: 183–185.

Jagannathan, R. and Narasimban, V., 1998. Effect of plant extracts/products on two fungal pathogens of finger millet. *Indian J. Mycol. Pl. Pathol.,* 18: 250–254.

Krishnakumar, B., 2002. *In vitro* evaluation of efficacy of some natural products against. *Colletotrichum capsici* (Syd.) Batter and Bisby, the incitant of fruit rot disease of chilli. *M.Sc. (Ag.) Thesis,* Annamalai University, India.

Ahmad, Mustaq, 1992. Control of rice blast by foliar fungicidal sprays. *Plant. Dis. Res.,* 7(1): 24–32.

Raja, J. and Kuruchevc, V., 1997. Antifungal properties some animal products against *Macrophomina phaseolina* causing dry not of cotton. *Plant Dis. Res.,* 12: 11–14.

Raja, J. and Kurucheve, V., 1998. Influence of plant extracts and buffalo urine on the growth and sclerotial germination of *Maqrophomina phaseolina. Indian Phytopathology,* 51: 102–103.

Rajendraprasad, K. and Kurucheve, L., 2002. Effect of human and animal excreta on sheath rot of rice. In: *54th Annual Meeting and National Symposium on Crop Protection and WTO: An Indian Perspective,* January 22–25. Organized by Central Plantation Crops Research Institute, Kasaragod, Kerala, India.

Resmy, J., 2002. *In vitro* evaluation of antifungal activity of some natural products against *Helminthosporium oryzae. M.Sc. (Ag.) Thesis,* Annamalai University, Annamalai Nagar, India.

Santhoshkumar, K., 2000. *Mycotoxic* activity of natural products against *Collelotrichum falcatum* (went): The incitant of sugarcane red rot. *M.Sc. (Ag.) Thesis,* Annamalai University, Annamalai Nagar, India.

Sivaprakash, K.E. and Kurucheve, V., 1999. There any possibility to manage sheath blight of rice without synthetic fungicides? In: *Agro Annual Meeting China 99, Plant Protection and Plant Nutrition,* April 13–19. Organized by Beijing International Convention Centre, Beijing, China.

Sundarraj, T., Kurucheve, V. and Jayaraj, J., 1996. Screening of higher plants and animals faeces for the fungi toxicity against *Rhizoctonia solani. Indian Phytopath.,* 49: 398–403.

Vincent, J.M., 1947. Distortion of fungal hyphae in the presence of certain inhibitors. *Nature,* 159: 850.

Wani, M.A., 2001. *In vitro* evaluation of some natural products against storage fungi of paddy *Aspergillus niger* Van and Teigh and *Aspergillus flavus* Link. *M.Sc. (Ag.) Thesis,* Annamalai University, Annamalainagar, India.

Chapter 65

Manifestation of Physiological and Biochemical Variations in Rice Cultivars Under Effluent Irrigation During Early Emergence

G. Panduranga Murthy[1], *G. Chidananda Murthy*[2] *and Shivalingaiah*[3]

[1]*P.G. Department of Studies and Research in Biotechnology, J.S.S. College, (Autonomous) (Affiliated to University of Mysore), Mysore – 570 025, Karnataka, India*

[2]*Lecturer, Department of Botany, Siddaganga College for Women, Karnataka, India*

[3]*Lecturer, P.FS. College, Mandya, Karnataka, India*

ABSTRACT

A study was conducted to assess the variations in different concentrations of (0 per cent, 5 per cent, 10 per cent, 15 per cent, 25 per cent, 50 per cent, 75 per cent and 100 per cent) paper mill effluent on physiological parameters, relative water turgidity, pigment contents and biochemical constituents like total contents of proteins, amino acids, soluble sugars, phenol and enzyme, amylase activity in the seedlings of two paddy cultivars, Jaya and IR64. All the parameters were declined with an increase in effluent concentrations. In both cultivars, at the highest concentrations (50 per cent, 75 per cent and 100 per cent), the deleterious effects on relative water turgidity and pigment (Chlorophyll 'a', 'b' and total and carotenoid) contents in leaves, and total content of protein, amino acid, soluble sugar, phenol and on the amylase activity were noticed. All these parameters were drastically decreased along with increased concentrations. The effluent in the lower concentrations enhanced all the above parameters of rice seedlings and are significantly superior over other treatments. So, it can be concluded that, the paper mill effluent can be used for irrigating the fields only after the proper dilution *i.e.,* upto 15 per cent, and finally reduce the problem of scarcity of water during cultivation of the crop plants.

Keywords: *Effluent, Paddy cultivars, Germination, Physiology, Biochemistry.*

Introduction

Among major industries, in India; pulp and paper is one of those that contribute to water pollution to a significant level. The paper industry is one among the 20 highly polluting industries in our country as notified by the department of Environment and forest within the purview of Industrial licensing. The effluent mixed water not only contains Nutrients that enhance the growth of Various Crop Plants, but due to presence of heavy metals and other toxic substances which disturb the Physiological process thus, ultimately affects the biochemical constituents of Crop Plants irrigated with effluents (Bhatnagar, 1986; Agarwal, 1991; Hari *et al.,* 1994; Dutta, 1999; Krishnan, 2001 and Panduranga Murthy *et al.,* 2004). Besides, the undiluted paper mill effluent not only effects the growth of crop plants but can also cause severe problems to environment as well as public health (Ramana *et al.,* 2002 and Asamudo *et al.,* 2005). Hence, keeping the above fact in mind, the study was initiated to investigate the effects of paper mill effluents on some physiological and biochemical constituents of two rice cultivars during early growth respectively.

Materials and Methods

Two cultivars of Paddy Crop Plant was selected for the present study *i.e.,* Jaya and IR64, from local farmers, Nanjangud, Mysore, Karnataka.

The effluent was subjected to various physico-chemical analyses according to standard procedures. (APHA, 1995)

The experiments were conducted in laboratory condition, nearly 100 seeds of paddy cultivars were arranged at equal distance in sterilised petri plates lined with filter paper (Top of the paper method). A known and equal volume of various concentrations of paper mill effluent [5 per cent (T1), 10 per cent (T2), 15 per cent (T3), 25 per cent (T4), 50 per cent (T5), 75 per cent (T6) and 100 per cent (17)] were poured into respective petriplates. The seedling raised in the tap water treated as control (T0). For effluent irrigation all the sets were subjected to germination in germinator at 28±2°C.

On termination day (ISTA, 1999), the physiological parameters like relative water turgidity in leaves, chlorophyll and carotenoids and stability index of chlorophyll were estimated.

The biochemical parameters like total contents of protein, amino acids, soluble sugars, phenol coment were estimated according to the standard procedures. Besides, the enzyme, amylase activity. was also recorded.

Statistical Analysis of Data

All the experiments were repeated four times at different time of intervals and the data were collected from all the experiments and were subjected to two way analysis of variance (*Anova*).

Results and Discussion

Hence in the present investigation, it was aimed to find out the effects of paper mill effluent on physiological, biochemical and enzymatic changes induced by the afore said effluent treatments during early seedling growth of Paddy (*Oryza sativa* L.) seed varieties under different concentrations.

Physico-chemical Analysis

The results of various physico-chemical parameters of paper mill effluents was represented in the Table 65.1.

Table 65.1: Physico-chemical Characterstics of the Paper Mill Effluent Used in the Present Investigation

Sl.No.	Parameters	Effluent
	I. Physical Characters	
1.	Color	Dark Brown
2.	pH	8.2
3.	Temperature (°C)	36
4.	Suspended solids	
	Total (mg/l)	940.00
	Fixed (mg/l)	501.00
	Volatile (mg/l)	419.00
5.	Dissolved solids	
	Total (mg/l)	1665.00
	Fixed (mg/l)	601.00
	Volatile (mg/l)	89.00
6.	Total volatile solids (mg/l)	96,88
7.	Dissolved Oxygen (DO)	0.81
8.	BOD 5 pays 24°C (mg/l)	191.56
9.	COD (mg/l)	540.23
10.	Total alkalinity (mg/l as Na_2O)	626
11.	Oil and grease (mg/l)	09
	II. Chemical Characters	
1.	Ammonical Nitrogen (mg/l as N)	26.01
2.	Nitrates (mg/l as N)	Nil
3.	Chlorides (mg/l as Cl)	1471
4.	Phosphates (mg/l as P)	Nil
5.	Phenolic Compounds (mg/l as Phenol)	0.07
6.	Cyanides (mg/l as Cn)	Nil
7.	Sulphides (mg/l as S)	32
8.	Sodium (mg/l as Sd)	119
8.	Aourides (mg/l as F)	1.03
9.	Percent Sodium	4.6
	III. Heavy Metals	
1.	Copper (mg/l as Cu)	0.7
2.	Zinc (mg/l as Zn)	1.06
3.	Other heavy metals	Nil

Physiological Study

Relative Water Turgidity in Leaves

A noticeable reduction was recorded at higher dosages (T6 and T7) of effluent. The reduction of 40.65 per cent in Jaya and 39.60 per cent in IR64 were observed over the lower dosage (T3) of 86.68 per cent and 83.14 per cent respectively (Table 65.2).

Table 65.2: Relative Water Turgidity in Leaves of Paddy Cultivars Exposed to Different Concentrations of Paper Mill Effluent

Treatments		*Paddy Cultivars*		*Mean*
		Jaya	*IR-64*	
Control	T0	61.00	62.55	61.78[c]
5 per cent	T1	65.54	63.00	64.27[c]
10 per cent	T2	68.51	70.15	69.33[b]
15 per cent	T3	86.68	83.14	84.91[a]
25 per cent	T4	66.00	71.08	68.54[b]
50 per cent	T5	55.00	58.60	56.80[d]
75 per cent	T6	52.46	52.64	52.55[e]
100 per cent	T7	47.80	43.20	45.50[f]
	Mean	62.07[b]	63.045[a]	62.55

F Value: *: Varieties; **: Treatments; **: Interaction.

*: Significant at 5 per cent level; **: Significant at 1 per cent level.

The observed reduction in relative water turgidity in leaves may be due to loss of membrane integrity in response to higher concentrations (T4–T7) of effluent, causing partial or complete desiccation in the plant system leading to complete or partial death of the plant. Similar observations were also made by Eayers and Weatherley (1968), Patil (1990) and Murthy *et al.* (2004).

Chlorophyll Content and Stability Index of Chlorophyll (CSI)

When compared to the other parameters, there was an anomalous contrariety in the pigment content. A non-significant decrease in chlorophyll 'a' chlorophyll 'h' and total chlorophyll was noticed at increased effluent dosages (T4–T7). There was an influence of Chlorophyll 'b' in both the varieties was noticed respectively (Tables 65.3–65.5).

In both the cultivars of Paddy, the chlorophyll stability index was decreased marginally at increased concentrations of paper mill effluent. The reduction of 48.21 per cent in Jaya and 47.10 per cent was noticed in IR 64. The lower dosages (T1–T3) are significantly higher over other treatments (Table 65.6).

Increasing concentrations of the effluent correspondingly decreased the chlorophyll pigments in the treated seedlings of both varieties of paddy. Increasing salinity levels are known to gradually reduce the pigment content and chloroplast activity. The other possible reasons for the decrease in chlorophyll content in response to the higher concentrations may be due to the formation of *'chlorophyllase'* enzyme that is responsible for chlorophyll degradation. Similar facts were reported by Murthy and Majumdar (1962), Sabater and Rodrique (1978), Rodriquez *et al.* (1987), Majumdar *et al.* (1991), Gadallah (1986), Kumar and Singh (1996) and Murthy *et al.* (2004).

Table 65.3: Chlorophyll 'a' Content (mg.g^{-1}) of Paddy Cultivars Exposed to Different Concentrations of Paper Mill Effluent

Treatments		*Paddy Cultivars*		*Mean*
		Jaya	*IR-64*	
Control	T0	0.0656	0.0541	0.0599^{c}
5 per cent	T1	0.0711	0.0613	0.0662^{b}
10 per cent	T2	0.0726	0.0629	0.0677^{b}
15 per cent	T3	0.0841	0.0791	0.0816^{a}
25 per cent	T4	0.0691	0.0541	0.0616^{c}
50 per cent	T5	0.0512	0.0490	0.0501^{d}
75 per cent	T6	0.0421	0.0444	0.0432^{e}
100 per cent	T7	0.0337	0.0365	0.0351^{f}
	Mean	0.0611^{a}	0.0551^{b}	0.0581

F Value: *: Varieties; **: Treatments; **: Interaction.

*: Significant at 5 per cent level; **: Significant at 1 per cent level.

Table 65.4: Chlorophyll 'b' Content (mg.g^{-1}) of Paddy Cultivars Exposed to Different Concentrations of Paper Mill Effluent

Treatments		*Paddy Cultivars*		*Mean*
		Jaya	*IR-64*	
Control	T0	0.0751	0.0639	0.0697^{e}
5 per cent	T1	0.0806	0.0824	0.0815^{c}
10 per cent	T2	0.0819	0.0910	0.0865^{b}
15 per cent	T3	0.0966	0.1041	0.1003^{a}
25 per cent	T4	0.0714	0.0806	0.0761^{d}
50 per cent	T5	0.0656	0.0529	0.0592^{f}
75 per cent	T6	0.0490	0.0501	0.0496^{g}
100 per cent	T7	0.0424	0.0479	0.0451^{g}
	Mean	0.0703^{b}	0.0716^{a}	0.0709

F Value: *: Varieties; **: Treatments; **: Interaction.

*: Significant at 5 per cent level; **: Significant at 1 per cent level.

Carotenoid

Both the cultivars of Paddy were displayed highest carotenoid content at increased (T4–T7) dosages of the effluent and with the least at lower dosages (T1–T3) (Table 65.7).

The increased carotenoid content was due to influence of increased nitrogen content in response to various effluent treatments (Nirmala Devi and Janardhanan, 1988 and Murthy *et al.*, 2004).

Table 65.5: Total Chlorophyll ($mg.g^{-1}$) of Paddy Cultivars Exposed to Different Concentrations of Paper Mill Effluent

Treatments		*Paddy Cultivars*		*Mean*
		Jaya	*IR-64*	
Control	T0	0.1411	0.1180	0.1295[d]
5 per cent	T1	0.1517	0.1433	0.1475[c]
10 per cent	T2	0.1545	0.1541	0.1543[b]
15 per cent	T3	0.1803	0.1832	0.1817[a]
25 per cent	T4	0.1407	0.1348	0.1377[d]
50 per cent	T5	0.1163	0.1019	0.1091[c]
75 per cent	T6	0.0904	0.951	0.0927[f]
100 per cent	T7	0.0764	0.0845	0.0804[f]
	Mean	0.131[a]	0.1261[b]	0.128

F Value: *: Varieties; **: Treatments; **: Interaction.

*: Significant at 5 per cent level; **: Significant at 1 per cent level.

Table 65.6: Chlorophyll Stability Index of Paddy Cultivars Exposed to Different Concentrations of Paper Mill Effluent

Treatments		*Paddy Cultivars*		*Mean*
		Jaya	*IR-64*	
Control	T0	81.81	84.64	83.22[b]
5 per cent	T1	79.65	80.00	79.83[c]
10 per cent	T2	82.90	83.33	83.11[b]
15 per cent	T3	92.00	90.73	91.37[a]
25 per cent	T4	73.11	77.70	75.40[c]
50 per cent	T5	66.01	61.54	63.77[d]
75 per cent	T6	56.44	60.00	58.22[c]
100 per cent	T7	51.66	48.64	50.15[f]
	Mean	72.94[b]	73.32[a]	73.13

F Value: *: Varieties; **: Treatments; **: Interaction.

*: Significant at 5 per cent level; **: Significant at 1 per cent level.

Biochemical Study

The biochemical analyses were done on 8th day old seedlings of both Jaya and IR64 at all the effluent treatments *viz.*, 5 per cent, 10 per cent, 15 per cent, 25 per cent, 50 per cent, 75 per cent and 100 per cent respectively.

Table 65.7: Carotenoides (mg.g⁻¹) of Paddy Cultivars Exposed to Different Concentrations of Paper Mill Effluent

Treatments		Paddy Cultivars		Mean
		Jaya	*IR-64*	
Control	T0	0.865	0.765	0.815[a]
5 per cent	T1	0.749	0.619	0.684[b]
10 per cent	T2	0.811	0.520	0.666[b]
15 per cent	T3	0.605	0.465	0.535[c]
25 per cent	T4	0.629	0.545	0.587[c]
50 per cent	T5	0.515	0.702	0.608[d]
75 per cent	T6	0.544	0.660	0.602[d]
100 per cent	T7	0.660	0.606	0.633[c]
	Mean	0.672[a]	0.610[b]	0.641

F Value: *: Varieties; **: Treatments; **: Interaction.

*: Significant at 5 per cent level; **: Significant at 1 per cent level.

Total Content of Protein and Free Amino Acid

The data of total protein content (mg/gm of fresh weight) as affected by different concentrations of effluent on 8th day after treatment are furnished in the Table 65.8. It was observed that, the total protein content of seedlings of both the cultivars decreased with an increased concentrations (T4–T7) of effluents and a linear increase in IR64 than Jaya at lower dosage (T3) was observed.

Table 65.8: Total Protein Content of Paddy Cultivars Exposed to Different Concentrations of Paper Mill Effluent (mg/gm of fresh weight)

Treatments		Paddy Cultivars		Mean
		Jaya	*IR-64*	
Control	T0	6.5160	7.638	7.077[c]
51ft	T1	7.041	8.091	7.566[c]
10 per cent	T2	7.161	8.556	7.858[b]
15 per cent	T3	8.55'4	9.164	8.859[a]
25 per cent	T4	6.721	6.564	6.642[d]
50 per cent	T5	4.109	4.654	4.381[e]
75 per cent	T6	3.296	3.344	3.321[f]
100 per cent	T7	2.996	3.014	3.005[f]
	Mean	5.799[b]	6.378[a]	6.088

F Value: *: Varieties; **: Treatments; **: Interaction.

*: Significant at 5 per cent level; **: Significant at 1 per cent level.

The highest protein content of 8.554 mg in Jaya and 9.164 mg in IR64 per gram fresh weight of seedlings at 15 per cent effluent treatment (T3) was recorded than any other concentrations (T1, T2 and T4–T6) and untreated set (T0) on 8th day of treatment respectively (Table 65.8).

The total free amino acid was marginally decreased along with the increase in concentrations (T4– T1) in the seedlings of Paddy cultivars. In control (T0), there was a slight increase in free amino acid content on 8th day after treatment (1.481 mg in Jaya and 1.503 mg in IR64 per gram fresh weight). The highest content of amino acid at lower dosage (T3) of effluent was noticed *i.e.*, 1.685 mg in Jaya and 1.702 mg in IR64 respectively (Table 65.9).

Table 65.9: Total Free Amino Acid of Paddy Cultivars Exposed to Different Concentrations of Paper Mill Effluent (mg/gm of fresh weight)

Treatments		*Paddy Cultivars*		*Mean*
		Jaya	*IR-64*	
Control	T0	1.481	1.503	1.492[b]
5 per cent	T1	1.304	1.561	1.432[b]
10 per cent	T2	1.369	1.499	1.434[b]
15 per cent	T3	1.685	1.702	1.693[a]
25 per cent	T4	1.299	1.412	1.355[c]
50 per cent	T5	0.910	1.014	0.962[d]
75 per cent	T6	0.721	0.888	0.804[e]
100 per cent	T7	0.289	0.679	0.484[f]
	Mean	1.132[b]	1.282[a]	1.207

F Value: *: Varieties; **: Treatments; **: Interaction.

*: Significant at 5 per cent level; **: Significant at 1 per cent level.

The reduction in total protein content may be due to destruction in the protease activity at increased concentrations of effluent during germination. The stored food materials in the cotyledons are degraded due to imbibition of the toxic effluent water thus inhibits the protein synthesis. Further, the inhibition of proteolytic enzyme leads to destruct the translocation of reserve food materials to the growing embryonic axis also causing reduction in the total free amino acid needed for normal protein synthesis and thus growth was reduced. Similar results are reported by Somashekar *et al.* (1984), Behera and Misra (1996), Vasantha and Perumal Samy (1998) and Panduranga Murthy *et al.* (2004).

Total Soluble Sugar

The data on total soluble sugar content of seedlings indicated that the effluent with different concentrations has differential response in both the cultivars of *Paddy i.e.*, Jaya and IR64.

The total soluble sugar content was increased at lower dosage (T3) and was decreased at increased dosages (T4–T7) on 8th day after effluent treatment. But the lowest content was recorded in Jaya (9.6/ mg) and IR64 (12.03 mg) at 100 per cent (17) effluent treatment (Table 65.10).

The reduction of total soluble sugar may be perhaps because of both phytotoxic and phototoxic effects on many metabolic and physiological process like reduced Chlorophyll content. In addition, the decrease in total soluble sugar are mainly related to decreased starch degradation. The toxic elements present in the effluents inhibits the hydrolytic enzymes, which are responsible for the break-down of reserve food materials. These results are in close conformity with the findings of Patil (1990), Behera and Misra (1996) and Vasamha and Perumalsami (1998).

Table 65.10: Total Soluble Sugars of Paddy Cultivars Exposed to Different Concentrations of Paper Mill Effluent (mg/gm of fresh weight)

Treatments		*Paddy Cultivars*		*Mean*
		Jaya	*IR-64*	
Control	T0	27.65	33.84	30.74[c]
5 per cent	T1	30.97	37.51	34.24 [c]
10 per cent	T2	32.82	40.09	72.91[a]
15 per cent	T3	42.69	49.50	46.24[b]
25 per cent	T4	24.13	32.04	28.08[d]
50 per cent	T5	19.60	26.68	23.14[e]
75 per cent	T6	14.34	19.11	16.72[f]
100 per cent	T7	09.61	12.03	10.82[g]
	Mean	25.22[b]	31.38[a]	28.30

F Value: *: Varieties; **: Treatments; **: Interaction.

*: Significant at 5 per cent level; **: Significant at 1 per cent level.

Total Phenol Content

The data indicated that, the total phenol content in seedlings is differed significantly. Among the treatments. It was observed that the content was increased at lower dosages. (T1–T3) and untreated set (T0) on 8th day after treatment. The decreased trend was noticed at higher dosages of the effluent. (T4–T7) *i.e.,* 0.540 mg in Jaya and 0.477 in IR64 as compare to control (T0) (Table 65.11).

Table 65.11: Total Phenol Content of Paddy Cultivars Exposed to Different Concentrations of Paper Mill Effluent (mg/gm of fresh weight)

Treatments		*Paddy Cultivars*		*Mean*
		Jaya	*IR-64*	
Control	T0	1.246	1.971	1.308[c]
5 per cent	T1	1.304	1.416	1.360[b]
10 per cent	T2	1.369	1.384	1.376[b]
15 per cent	T3	1.426	1.614	1.520[a]
25 per cent	T4	1.299	1.306	1.302[c]
50 per cent	T5	0.981	1.016	0.998[d]
75 per cent	T6	0.871	0.801	0.836[e]
100 per cent	T7	0.540	0.477	0.509[f]
	Mean	1. 129[b]	1.173[a]	1.151

F Value: *: Varieties; **: Treatments; **: Interaction.

*: Significant at 5 per cent level; **: Significant at 1 per cent level.

Phenols are aromatic compounds with an hydroxyl group substituent. They are common constituents of many plants and are known to participate in a number of physiological process which are essential for the growth and development. Phenols play an important role in determining the resistance or susceptibility of plants to any type of toxicity. The resistant varieties have been found to contain more phenols than susceptible ones. The injurious action of toxic materials present in the effluent may inhibited the poly-phenoloxidase, which is responsible for the conversion of phenols into quinones. Thus, this reduces the total amount of phenol in the plant system. Similar findings are reported by Vanja *et al.* (2000) and Panduranga Murthy *et al.* (2004).

Enzyme Study

Amylase Activity

Amylase is of prime importance in the initial stage of starch degradation in germinating seeds. Any deviation like effluent toxicity may disturb the activity of this enzyme, this intern affects the rate of germination. The data on amylase activity in seedlings of paddy cultivars showed non-significant reduction at increased dosages (T4–T7) of effluent. After that, there was a increase in the activity at lower dosages (T1–T3) was observed. The reduction of 47.69 mg in Jaya and 36.51 mg of glucose realesed in IR64 per gram fresh weight of seedlings on 8th day after treatment at higher dosage (T7) of effluent. Further marginal reduction was observed along with increased (T4–T7) concentrations of the effluents (Table 65.12).

Table 65.12: Amylase Activity in Two Paddy Cultivars Exposed to Different Concentrations of Paper Mill Effluent (mg of glucose/30 min/gram fresh weight)

Treatments		*Paddy Cultivars*		*Mean*
		Jaya	*IR-64*	
Control	T0	63.996	74.210	69.100^{f}
5 per cent	T1	81.614	90.601	86.107^{d}
10 per cent	T2	93.065	109.59	101.62^{b}
15 per cent	T3	114.15	127.50	120.83^{a}
25 per cent	T4	89.06	91.54	90.31^{c}
50 per cent	T5	71.53	84.65	78.og^{e}
75 per cent	T6	55.82	49.71	52.76^{g}
100 per cent	T7	47.69	36.51	42.10^{h}
	Mean	77.11^{b}	83.16^{a}	80.13

F Value: *: Varieties; **: Treatments; **: Interaction.

*: Significant at 5 per cent level; **: Significant at 1 per cent level.

The major enzyme involved in the starch degradation is amylase, this may be the predominant enzyme involved during germination. Mainly, the inhibition of protein synthesis may result in the decreased amylase production in response to increased doses of the effluent (T4–T7). Behera and Misra (1996).

Conclusion

From the discussion, it is observed that the paper mill effluent had an evident toxic effect on physiological and biochemical parameters in both varieties of paddy. Among both cultivars, the relative susceptibility of cultivar Jaya than IR64 at higher dosages (T4–T7) of effluent with respect to all the parameters have been noticed.

The study reveals that, the effluent contains various toxic chemicals, suspended solids, dissolved solids, heavy metals like copper, zinc etc. and other unwanted materials. Besides, an excess of various forms of *cations and anions* which are injurious to germination and subsequent growth of the plant. The concentrations of these constituents are not in favour on both physiological and biochemical parameters. It should be reduced to beneficial level by diluting the effluents upto 15 per cent which can be used for irrigation purposes as substitutes for chemical fertilizers and can minimise the application of chemical pesticides also.

Acknowledgement

The Authors thank the Research supervisor for his valuable guidance, encouragement and support in completing the research paper.

References

Anon., 1999. *International Rules for Seed Testing, Seed Science and Technology*, 27: 421–463.

APHA, AWWA and WPCF, 1995. *Standard Methods for the Analysis of Water and Wastewater*. APHA Inc., New York.

Arnon, D.I., 1949. Copper enzymes in isolated chloroplast, poly-phenoloxidase in *Beta vulgaris. Plant Physiol.*, 24: 1–15.

Asamudo N.V., Daba, A.S. and Ezeronye, O.U., 2005. Bioremediation of textile effluent using *Phanerochaete chrysosporium., African Journal of Biotechnology*, 4(13): 1548–1553.

Barrs, H.D. and Weatherley, P.E., 1962. A re-examination of the relative turgidity technique for estimating water deficits in leaves. *Austr. J. Biol. Sci.*, 15(3): 413–428.

Behera, B.K. and Misra, B.N., 1985. The effect of sugar mill effluent on enzyme activities of rice seedlings. *Industrial Research*, 37: 390–398.

Clegg, K.M., 1956. The application of the anthrone reagent to the estimation of starch in cereals. *J. Sci. Food Agric.*, 7: 40–44.

Gadallah, M.A.A., 1996. Phytotoxic effects of industrial and sewage wastewater on growth, chlorophyll content, transpiration rate and relative water content of potted sunflower plants. *Water, Air and Soil Pollution*, 89: 33–47.

ISI, 1974. Tolerance limits for industrial effluents discharged into Inland, surface waters. *Indian Standard Institutions*, New Delhi, I.S.–2490.

Kannabiran, B. and Prakasam, A., 1993. Effect of distillery effluent on seed germination, seedling growth and pigment content of *Vigna mungo* (L.) Hepper (CV19). *Geobios*, 20: 108–112.

Kauskik, A., Bala, R. Kadyan and Kaushik, C.P., 1994. Sugar mill effluent effects on growth, photosynthetic pigments and nutrient uptake in wheat seedlings in aqueous Vs soil medium. *Water, Air and Soil Pollution*, 87(1–4): 39–46.

Kirk, J.T.O. and Allen, R.I., 1965. Dependence of chloroplast pigment synthesis on protein synthesis, effect of acitidione. *Biochemical and Biophysical Research Communications*, 21(6): 523–530.

Lowry, O.H., Rosebrough, N.J., Farr, A. and Rendall, R.J., 1951. Protein measurement with the folin phenol reagent. *J. Biol. Chem.*, 193: 265–275.

Madappan, K., 1993. Impact of tannery effluent on seed germination, morphological characters and pigment concentrations of *Phaseolus mungo* L. and *Phaseolus aureus* L. *Poll. Res.*, 12(3): 159–163.

Majumdar, S., Ghosh, S., Glick, B.R. and Dumbroff, E.B., 1991. Activities of chlorophyllase, phosphoenol, pyruvate, carboxylase and ribulase, 1,5 bi-phosphatecarboxylase in the primary leaves of soybean during scenesence and drought. *Physiol. Plant*, 81: 473–480.

Malick, C.P. and Singh, M.B., 1980. In: *Plant Enzymology and Histo-Enzymology*. Kalyani Publishers, New Delhi, pp. 286.

Miller, G., 1972. *Annal. Chem.*, 31: 456.

Moore and Stein, W.H., 1948. In: *Methods in Enzymology*, (Eds.) S.P. Olowkk and N.D. Kalpan. Academic Press, New York, 3: 468.

Murthy, K.S. and Majumder, S.K., 1962. Modification of the technique for determination of chlorophyll stability index in relation to studies of drought resistance in rice. *Curr. Sci.*, 13(11): 470–471.

Neelam, S. and Sahai, R., 1988. Effect of fertilizer factory effluent on seed germination, seedling growth, pigment content and biomass of *Sesamum indicum* Linn. *J. Environ. Biol.*, 9: 45–50.

Nirmala Devi, J. and Janardhanan, K., 1988. Effect of South India viscose factory effluent on seed germination, seedling growth and chloroplast pigments content in five varieties of Maize (*Zea mays* L). *Madras Agric. J.*, 75(1–2): 41– 47.

Pandit, B.R., Prasanna Kumar, P.G. and Mahesh Kumar, R., 1996. Effect of diary effluent on seed germination, seedling growth and pigment of *Pennisetum typhoides* Barm. and *Sorghum bicolor* L. *Poll. Res.*, 15(2): 103–106.

Panduranga Murthy, G., Sudarshana, M.S., Prakasha and Manjunatha, R.A., 2004. Studies on the physico-chemical properties of sugar mill effluent and their effects on germination, seedling morphology and biochemical constituents of some commercial crops. *Biosciences, Biotech. Research Asia*, 2(1): 59–63.

Rodriquez, M.T., Gozatez, M.P. and Linares, J.M., 1987. Degradation of chlorophyll and cholorophyllase in senescing barley leaves. *J. Plant Physiol.*, 127: 369–374.

Sahai, R., Jabeen, S. and Saxena, P.K., 1983. Effect of distillery waste on seed germination, seedling growth and pigment contents of rice. *Indian J. Ecol.*, 10: 7–10.

Thamizhiniyan, P., Sundar, Moorthy P. and Lakshamanachary, A.S., 2000. Effect of Sugar mill effluent on germination, growth and pigment contents of ground nut (*Arachis hypogea* L.) and paddy (*Oryza sativa* L.). *Indian J. Applied and Pure Biol.*, 15(2): 151–155.

Vasanthy, M. and Lakshamana Perumal Samy, 1998. Effect of untreated and carbon treated dyeing industry effluent on seed germination and biochemical parameters of Maize. *Poll. Res.*, 17(2): 173–176.

Velo, V., Arumugam, A. and Arunachalam, G., 1999. Impact of paper mill effluent irrigation on the yield, nutrient content and their uptake in rice. *Madras Agric. J.*, 86(1–3): 118–124.

Wuncheng Wang and Keturi, Paul H., 1990. Comparative seed germination tests using ten plant species for toxicity assessment of a metal engraving effluent sample. *Water, Air and Soil Pollution*, 52: 369–376.

Chapter 66

Assessment of Pollution Soils Near Aluminium Industry (Industrial Estate) Warangal, Andhra Pradesh

B. Lalitha Kumari and M.A. Singara Charya

Department of Botany, Kakatiya University, Warangal – 506 009, Andhra Pradesh

ABSTRACT

An investigation of polluted soils amended with aluminium industry was under taken during 1995–96. Various physical, chemical and biological characteristics were assessed and discussed to understand the pollution in the soils. High pH values were recorded in January, and the mean value of bulk density was 1.24 gm/cm^3. The maximum alkalinity was recorded in March with a range of variation in between 0.01 to 0.62 mg/g. No significant quantitative changes were seen in nitrate, nitrates and ammonia contents. The variation for fungi was $0.7 \times 1.10 \times 10^5$ in the polluted or 0.5–2.3 $\times 10^5$ in control soils. The soil enzymes varied differently with different seasonal variations.

Introduction

Aluminium industry is located in industrial estate Warangal, Andhra Pradesh. It was established in the year 1980 and concentrating on various activities on manufacturing, aluminium materials which includes, coasting, rolling and spinning and washing materials. During the process the heavy amounts of costic soda and HNO_3 are used and they appear equally in high amounts in the effluents too. The effluents discharged from the industry without any treatment were applied on soils causing a major threat to the environment. The soil quality is being degraded and not used for any cultivation purpose.

Materials and Methods

The soils from the sewage irrigated soils adjacent to sewage canals near aluminium industry were selected and surface soils were collected for analysis. Soils of Kakatiya University Campus which were unpolluted and physically similar in nature, served as control. The soils was air dried and crushed to parts in 2 mm screen. The physical, chemical and biological characteristics were analysed as methods suggested by Trivedi *et al.* (1987). The data is presented in Tables 66.1–66.4.

Results and Discussion

The physical properties of the soils from July 1995 to June 1996 was presented in Table 66.1. From the table it was evident that the maximum soil pH was recorded in January and May (8.0) and minimum (7.0) in polluted sites and in unpolluted site (Kakatiya University Campus) it was usually 7.0. Nannipieri *et al.* (1978), Kay and Lilly (1970) measured the stability and kinetic properties of accumulated soil enzymes and their correlation with that of pH. The variation of bulk density was in between 1.13 to 1.32 gm/cm^3 with the mean value of 1.24 gm/cm^3 and Carter (1990), Wallace and Wallace (1990) a1so drawn some conclusions on the characterization of the soils based on the property of bulk density and micro-porosity with that of some enzyme kinetics. The average value of pore space is 53.8 per cent in aluminium industry, and 50.4 per cent in soils at Kakatiya University campus. The fluctuations in water holding capacity at polluted aluminium industry site was 20.6–25.6 per cent whereas this variation was between 19.7–24.9 in the control soils. The mean value of conductivity was 1.19 μmho in the soils at aluminium industry. The minimum conductivity level was recorded in the

Table 66.1: Physical Characteristics of Polluted Soils of Aluminium Industry in Warangal During 1995–96

	1995						1996						Mean Value	Standard Error
	July	Aug.	Sept.	Oct.	Nov.	Dec.	Jan.	Feb.	March	April	May	June		
pH														
A:	7.5	7.0	7.0	7.5	7.5	7.0	8.0	7.5	7.0	7.5	8.0	7.0	7.4	±0.27
B:	7.0	7.0	7.0	7.0	7.0	7.0	7.0	7.0	7.0	7.0	7.0	7.0	7.0	±0.00
Bulk Density (gm/cm^3)														
A:	1.16	1.24	1.31	1.17	1.26	1.31	1.26	1.13	1.32	1.23	1.16	1.28	1.24	±0.64
B:	1.19	1.23	1.26	1.39	1.26	1.36	1.42	1.41	1.23	1.32	1.38	1.28	1.32	±0.52
Pore Space (%)														
A:	56.2	53.2	50.6	55.9	55.1	50.6	75.1	57.4	50.2	53.6	56.2	51.7	53.8	±0.76
B:	55.1	53.6	52.5	47.6	52.5	46.7	46.4	46.7	53.6	50.2	47.9	51.7	50.4	±0.95
Water Holding Capacity (%)														
A:	20.6	21.4	23.2	22.8	23.2	21.2	20.8	22.4	24.6	24.7	25.1	25.4	22.9	±0.52
B:	24.2	23.4	22.8	20.3	20.3	21.9	19.7	20.2	21.8	20.6	24.8	24.9	22.4	±1.73
Conductivity (□ m Mho)														
A:	0.95	0.65	0.54	0.22	0.40	0.32	0.68	1.28	4.23	0.92	1.42	2.68	1.19	±0.44
B:	0.62	0.46	0.17	0.37	0.19	0.20	0.68	0.72	1.02	0.98	0.44	1.28	0.49	±0.33
Permanent Wilting Co-efficient														
A:	7.13	7.38	8.00	7.89	8.00	7.31	7.27	7.72	8.48	8.52	8.68	8.93	7.94	±0.34
B:	8.35	8.07	7.85	7.00	7.55	6.29	6.96	7.52	7.79	8.38	8.89	8.93	7.84	±0.28

A: Polluted; B: Unpolluted.

month of June in the soils collected at control site. Lavado and Tabaoda (1987) established soil salinity and sodicity which have a detrimental effect on both chemical and physical aspects of soil quality. The wilting coefficient values were recorded in maximum in the month of June (8.93). This range of variation in control soils was 6.29 to 8.93. Meghraj *et al.* (1986) established and concluded that the highest wilting coefficient was in the polluted soils to unpolluted soils.

The chemical properties of the soils from 1995 July to 1996 June was presented in Table 66.2. The alkalinity of the soils varied among the different soil samples (polluted and unpolluted) during 1995–96. Generally, it was observed that, the maximum alkalinity was recorded in March. The range of variation was in between 0. 01 to 0. 62 mg/g and 0.01–0.20 mg/g in control soils. The chloride range of variation was in between 0.85 to 6.12 mg/g in polluted soil at aluminium industry and 0.92–2.13 mg/g in control soils. Slobe and De (1986) in their study on effects of plant use, agriculture and forestry compared chloride concentrations with that of degradation of environment and soil pollution. The calcium range in the soils varied from 1.46 to 10.5 mg/g in industrial amended soils effluent and were recorded 1.46–9.40 mg/g range in control soils. Ross (1973) studied the glycoside hydrolase activities and its significant relation with calcium. The magnesium varied in between 1.9 to 4.09 mg/g in aluminium industry soils. The calcium carbonate range was in between 24.0–60.5 per cent in polluted and 20.0–42.5 per cent in control soils. The aluminium content was high in September month, and the range was between 0.02 to 0.46 mg/g in aluminium industry effected soils and 0.01 to 0.11 mg/g in control soils. The range of variation in the content of silicate were high in the month of July in polluted soils with the range of 0.29–0.74 mg/g. The sodium content was high (22.4 mg/g) in polluted site near aluminium industry. The potassium content was high in the month of June with a range between 1.4 to 24.3 mg/g in polluted sites. The iron; content was high in the month of July 1.2 mg/g in polluted soils and 0.9 mg/g in September in control soils. The sulphates were high in the month of April and the range of variation was 20 to 186 mg/g in polluted soils and 17 to 95 mg/g in unpolluted soils. Foth (1990) in his extensive study on fundamentals of soil science, demonstrated sulphates as basic nutrients in changing the productivity levels and which were constantly and rapidly decreasing through industrial effluents. The maximum phosphorus content was recorded in the month of February and may months and the minimum amounts were recorded during July month in polluted soils. The range was between 9.0–28.0 mg/g in polluted and 9.0–14.0 mg/g in control soils. Sanchez and Uehera (1980) suggested the management practices to avoid soils with high phosphorus fixation. Raghava *et al.* (1980), Bhattachalya and Deshpandey (1993) successfully demonstrated the relations between polluted and unpolluted soils in various industrial belts on development and yield of crops studied by them. The highest nitrate was recorded in the month of February in polluted soils and March for unpolluted soils and the variation was in between 0.32 to 0.98 mg/g in polluted soils. Somasekhar *et al.* (1984), observed that the industrial waste water treatment increases the pH, organic carbon, nitrogen of the soil and decreased the water holdings capacity. No significant quantitative changes were seen in nitrites in between polluted and unpolluted, soils. Mc Clung *et al.* (1983) explained the nitrification inhibition and decrease nitrite content in soils amended with sewage sludge and certain industrial effluents. The ammonia contents was high in control compared to unpolluted soils. In polluted soils the range was in between 0.20 to 0.50 mg/g. Maximum ammonia was recorded in the month June in control soils. Parr (1972) and Brady (1995) reviewed the status of ammonia in chemical and biochemical considerations for minimising the efficiency of fertiliser nitrogen. The highest amounts of organic matter recorded in the soils flooded with effluents from aluminium industry, where the range was 10.44 to 89.9 per cent. The heavy metals such as mercury, cadmium, cobalt, chromium, lead were analysed and found to be very meagre in their concentrations.

Table 66.2: Chemical Characteristics of Polluted Soils of Aluminium Industry in Warangal During 1995–96

	1995						*1996*						*Mean Value*	*Standard Error*
	July	*Aug.*	*Sept.*	*Oct.*	*Nov.*	*Dec.*	*Jan.*	*Feb.*	*March*	*April*	*May*	*June*		
Alkalinity														
A:	0.08	0.14	0.12	0.06	0.08	0.12	0.46	0.46	0.62	0.08	0.01	0.38	0.21	± 0.05
B:	0.06	0.10	0.08	0.04	0.06	0.14	0.08	0.20	0.16	0.06	0.01	0.20	0.09	± 0.01
Chlorides														
A:	1.98	1.91	1.42	2.27	1.42	2.55	3.97	2.55	0.85	1.27	1.49	6.12	2.32	± 0.41
B:	1.77	2.05	1.42	1.27	1.56	1.42	1.98	1.98	2.13	1.21	0.92	2.13	1.66	± 0.11
Calcium														
A:	1.60	2.00	1.92	1.46	1.76	4.32	10.5	4.80	9.45	7.40	5.31	7.47	2.32	± 0.41
B:	4.84	2.48	5.29	1.44	1.76	1.92	6.41	9.45	7.85	4.64	7.05	9.29	1.66	±0.17
Magnesium														
A:	0.29	0.87	0.97	0.19	0.87	0.68	1.16	0.87	0.97	1.20	1.77	4.09	1.16	±0.29
B:	3.89	1.26	3.99	4.71	0.48	1.55	1.36	5.83	0.49	1.12	1.36	0.48	2.21	±0.53
Calcium Carbonate														
A:	51.0	60.5	24.0	31.0	46.5	37.5	34.5	46.0	51.5	49.0	43.5	35.0	42.5	±2.98
B:	37.5	28.0	27.5	22.5	36.0	42.5	30.0	36.0	31.0	29.0	40.5	20.0	31.7	±2.01
Aluminium														
A:	0.06	0.17	0.46	0.42	0.17	1.16	0.02	0.22	0.02	0.29	0.11	0.09	0.18	±0.04
B:	0.05	0.04	0.04	0.11	0.10	0.06	0.11	0.01	0.01	0.07	0.11	0.08	0.07	±0.01
Silica														
A:	0.74	0.35	0.42	0.38	0.29	0.68	0.44	0.50	0.60	0.43	0.74	0.35	4.93	±0.45
B:	0.45	0.25	0.45	0.35	0.2i	0.37	0.71	0.43	0.54	0.74	0.72	0.56	4.81	±0.51
Sodium														
A:	6.4	7.9	10.6	10.6	6.8	4.3	5.3	10.1	12.2	16.3	7.8	22.4	1.92	±0.47
B:	0.9	2.3	1.2	0.6	0.9	0.6	4.3	0.9	2.3	1.5	2.0	0.6	1.50	±0.31
Potassium														
A:	0.28	0.30	0.32	0.22	0.35	0.28	0.32	0.31	0.68	0.29	0.14	2.03	0.46	±0.14
B:	0.39	0.38	0.44	0.42	0.43	0.44	0.48	0.52	1.08	0.69	0.74	0.54	0.54	±0.05
Iron														
A:	0.12	0.06	0,02	0,09	0.02	0,05	0.02	0.09	0.03	0,09	0.04	0.08	0.04	±0.01
B:	0.01	0.07	0.09	0.08	0.01	0.01	0.01	0.01	0.01	0.02	0.01	0.04	0.02	±0.01
Sulphates														
A:	4.4	1.6	4.7	3.2	6.5	2.6	5.8	4.1	2.0	18.6	15.0	17.0	7.95	±0.17
B:	7.5	9.5	6.5.	4.3	5.5	1.7	3.5	3.3	6.1	5.5	6.8	5.6	5.48	±0.06
Phosphates														
A:	9.0	27.0	54.0	36.0	46.0	68.0	68.0	96.0	280.0	190.0	41.0	234.0	168.0	±2.60
B:	9.0	22.0	78.0	25.0	48.0	58.0	58.0	140.0	128.0	86.0	46.0	20.0	72.0	±1.11

Contd...

Table 66.2–Contd...

	1995						1996						Mean Value	Standard Error
	July	Aug.	Sept.	Oct.	Nov.	Dec.	Jan.	Feb.	March	April	May	June		
Nitrates														
A:	0.52	058	0.55	0.36	0.42	0.48	0.62	0.98	0.32	0.62	0.52	0.49	0.54	±0.05
B:	0.34	0.28	0.41	0.45	0.56	0.52	0.74	0.72	0.96	0.48	0.62	0.46	0.54	±0.01
Nitrites														
A:	0.24	0.28	0.18	0.16	0.40	0.20	0.60	0.19	0.24	0.16	0.16	0.18	0.25	±0.04
B:	0.15	0.22	0.14	0.19	0.25	0.21	0.38	0.09	0.16	0.22	0.17	0.22	0.19	±0.02
Organic Matter (%)														
A:	70.4	51.7	89.9	36.2	49.7	24.8	10.4	49.6	38.2	48.7	31.0	36.6	44.8	±6.00
B:	52.8	47.6	54.8	53.8	24.8	27.9	53.8	58.9	71.4	41.7	36.6	25.9	45.8	±4.22

A: Polluted; B: Unpolluted.

The soil microorganisms which determines the form and arrangement of the soils were enumerated and presented in Table 66.3. The variation for fungi was 0.7 to 1.10 × 10^5 in the soils effected by aluminium effluents and the minimum and maximum number of colonies fluctuated in between 0.5 to 2.3 × 10^5 in control soils. Anders *et al.* (1986) in their studies established the relationship between pollution of the soils and soil microflora. The range of variation of bacterial density was in between 3.4 to 25.0 × 10^6 g among polluted soils whereas it was 2.8 to 12.3 × 10^6 g in control soils.

Table 66.3: Biological Characteristics of Polluted Soils of Aluminium Industry in Warangal During 1995–96

	1995						1996						Mean Value	Standard Error
	July	Aug.	Sept.	Oct.	Nov.	Dec.	Jan.	Feb.	March	April	May	June		
Fungi (1 ५ 10⁶/g)														
A:	1.8	0.7	1.1	1.2	1.8	1.8	1.3	0.9	1.1	0.9	2.1	1.7	1.36	±0.15
B:	1.1	0.7	0.5	2.3	1.2	1.5	0.6	0.9	0.7	0.6	1.1	1.3	1.04	±0.16
Bacteria (1 ५ 10⁶/g)														
A:	10.4	12.4	11.2	3.4	6.9	8.8	7.9	6.3	5.9	8.2	21.2	23.0	10.46	±1:39
B:	42	6.1	4.6	2.8	3.4	6.2	6.9	4.1	3.4	12.3	4.6	9.4	5.66	±0.83
Actinomycetes (1 ५ 10⁵/g)														
A:	2.1	6.3	0.9	1.3	0.9	0.7	0.8	0.6	1.3	1.6	1.9	2.4	1.72	±0.46
B:	0.9	1.4	0.6	0.9	1.1	0.7	1.6	1.3	1.1	0.9	1.3	1.7	1.12	±0.09

A: Polluted; B: Unpolluted.

Talashilkar (1989) working on recycling of urban waste and relation to agriculture reviewed the application of sewage effluent, industrial waste water and sludge compost on agricultural land. The minimum and maximum population of actinomycetes were 0.6–6.3 × 10^5 gill in polluted soils and their range was 0.6–1.7 × 10^5/g m in control soils.

The soil enzyme which determines the physiological status of the soils were estimated (Table 66.4). The cellulase activity (Relative enzyme activity) was recorded in between 136.4 to 476.2 in polluted soils and the minimum and maximum range was in between 140.9 to 307.7 for control. Preece (1992) and Baize (1993) suggested that cellulase is to contribute the breakdown of some hazardous, and toxic insecticide. The catalase enzyme showed its maximum accumulation during the month of December. The maximum and minimum values of enzymes activity were 0.7 to 5.9 units in the polluted soils and this variation was 0.8 to 5.7 units in unpolluted soils. Broadman (1985) measured five soil types for catalase activity incubated with and without adding organic matter and Bentomite. The proteolytic activity was maximum in the month of March with the range of 0.23 to 3.56 in polluted soils and 0.12 to 3.59 in unpolluted soils. Gray and Williams (1971) in their studies on microbial, productivity in soil ascertained the affinity of pyrophosphate for extraction of proteases from three different soil varied differently during the study period. The amylase activity was in between 0.05 to 1.0 units in the soils treated with aluminium industry effluents. The range of activity in the control soils was in between 0.08 to 3.42 units.

Table 66.4: Enzyme Characteristics of Polluted Soils of Aluminium Industry in Warangal During 1995–96

	1995						*1996*						*Mean Value*	*Standard Error*
	July	*Aug.*	*Sept.*	*Oct.*	*Nov.*	*Dec.*	*Jan.*	*Feb.*	*March*	*April*	*May*	*June*		
Cellulase–Relative Enzyme Activity (REA)														
A:	311.5	242.1	306.8	191.6	190.5	187.9	476.2	142.9	136.4	220.8	294.9	306.8	250.7	±27.5
B:	248.8	284.1	307.7	294.1	232.6	229.9	142.9	140.9	141.8	250.0	278.6	306.8	238.16	±18.4
Catalase Units														
A:	3.1	2.2	2.5	1.4	1.9	5.9	0.8	0.7	0.8	2.4	2.7	2.1	2.04	±0.4.1
B:	1.8	2.3	1.8	2.2	2.1	5.9	0.9	0.8	1.8	1.9	1.4	0.8	1.95	±0.39
Protease (mg/g)														
A:	0.23	0.24	1.0	0.31	1.32	1.19	1.41	1.51	3.56	0.84	0.72	0.48	1.07	±0.26
B:	0.62	0.12	0.34	0.56	1.00	1.11	1.92	2.66	3.59	0.56	0.41	0.28	1.05	±0.25
α-amylase units														
A:	0.98	0.65	0.13	0.17	0.05	0.12	0.90	1.00	3.70	0.28	0.31	0.72	0.71	±0.29
B:	0.36	0.17	0.43	0.13	0.96	0.50	0.48	0.80	1.20	0.52	0.68	0.74	0.58	±0.09

A: Polluted; B: Unpolluted.

The Nanniperi *et al.* (1980) signified the activities of amylase with extracted C and N suggested that a major proportion of the humic compounds were removed from soils with the amylases. The urease activity ranged between 0.29 to 1.24 mg/g in polluted soils. Kaspar and Bland (1992), Maun Taj *et al.* (1992) studied the relative effect of sewage study on urease activity, which was highly sensitive to small quantities of trace ions and because its susbstrate urea is added to soil as a synthetic fertilizer and in animal excreta.

Acknowledgement

Authors are thankful to Prof V. Thirupathaiah, Head, Department of Botany, Kakatiya University for provided laboratory facilities.

References

Barman, S.S. and Lal, M.M., 1994. Accumulation of heavy metals (Zn, Cu, Cd and, Pb) in soil and cultivated vegetables and weeds grown in judiestrainally polluted fields. *J. Env. Biol.*, 15: 107–115.

Bhattacharya, B. and Bhattacharya, A.K., 1993. Ecological study of plant species diversity in Delhi region. *J. Ecobiol.*, 5: 207–211.

Brady, N.C., 1995. *The Nature and Properties of Soils*, 10th Edition. Organisms of the Soil, Prentice Hall, India Pvt. Ltd., New Delhi, p. 253–277.

Chadha, J. and Pandey, S.N., 1990. *Industrial Pollution and Plants.* Ashish Publishers, New Delhi.

David, I., Gregory, S., Tylka, L. and Lewis, C., 1996. Effects of fertilizers on virulence of *Steinernene Carpocap. Applied Soil Ecology,* 3: 27–34.

Gupta, A., 1991. A characteristic soil actinomycetes of Agra region. *Acta Botanica Indica,* 19: 81–83.

Guru Prasada Rao, M. and Nanda Kumar, N.V., 1981. Analysis of irrigation reservoir contaminated by tannery effluents. *Ind. J. Env. Hlth.,* 23: 239–241.

Hattingh, A.M., 1990. Role of clay minerology in the choice of an empluric equation for predicting field capacity of soils. *Appl. Plant Sci.,* 4: 21–24.

Kaspar, T.C. and Bland, W.L., 1992. Soil temperature and root growth. *Soil Sci.,* 154: 290–299.

Kay, G. and Lilly, M.D., 1970. The chemical attachment of chimolypsin to water insomable polymers using 2-Amino-4, 6-Dichloro-5-Triazine. *Acta,* 198: 276–289.

Lal, K., H.K. Parwana and Verma, P., 1993. Effect of waste water irrigation on soil properties. *Indian Env. Prot.,* 13: 274–378.

Mauntay, S.D. and Mishra, R.P., 1992. Dehydrogenase and urease activities of earthworm caste and the surrounding soil. *J. Indian Bot. Soc.,* 71: 303–304.

Messing, I. and Jarwis, N.T. Agropotentiality of paper mills waste water. *Soil Poll.,* 7: 97–120.

Mishra, B. and Srivastava, L.L., 1990. Physico-chemical characteristics of humic substances of major soil associations of Bihar. *Plant Soil.* 122: 185–192.

Omar, S.A., Abdul Sattar, M.A., Khathul, A.M. and Alle, M.A., 1994. Growth and enzyme activities of fungi and bacteria in soil salinized with sodium chloride. *Soil Micro Biologa,* 39: 23–28.

Page, A.L., 1974. Fate and effects of trace elements in sewage sludge when applied to agricultural land: A Literature Review. *Env. Prot. Agency* (EPA) Cincinnati, Ohio, USA.

Patel, J.D. and Shakunthala Devi., 1986. Variations in chloroplasts of bafmesophyll cells of *Zuzyobium cuminil* and *Tamarindus indica* L. growing under air pollution stress of a fertilizer complex. *Indian J. Ecol.,* 13: 124–130.

Patil, H.S., 1985. Studies on the physico-chemical and biological characteristics of sewage stabilisation pond. *J. Env. Biol.,* 6: 93–102.

Rainer, G.J., 1966. The fumigation extraction method to estimate soil microbial biomass, calibration of KEC values. *Soil Biol. Biochem.,* 28: 25–31.

Rajeswari, S., 1993. Effect of pesticides on soil organisms. *Indian J. Env. Hlth.,* 35: 227–231.

Ray, M. and Benerjee, S., 1986. Cytological studies of the water contaminated with industrial effluents effects of Tanla Nallah water on *Allium salvum*. *Genetics*, 5: 475–483.

Ross, D.J., Burean, S., Hutt, L. and Roberts, H.S., 1973. Biochemical activities in a soil profile under hard beech forest I. Invertase and amylase activities and relationships with other properties. *N.Z. J. Sci.*, 16: 209–224.

Selvarangan, R., 1981. Prevention of pollution from tannery. *Leather Sci.*, 28: 192–199.

Signa, G.C., Isensec, A.R. and Sadsshi, A.D., 1993. Influence of rainfall intensity and Crop residense leaching of atrazing through intact no till soil cores. *Soil Sci.*, 166: 225–232.

Trivedi, R.K., Goel, P.K. and Trisal, C.L., 1987. *Practical Methods in Ecology and Environmental Science*. Environ Media Publications, Karad, India.

Yadav, K., Jha, K.K., Prasad, C.R. and Sinha, L., 1989. Kinetics of carbon mineralisation from poultry manural sewage sludge in two soils at field capacity and submergence moisture. *J. Indian Soc. Soil. Sci.*, 37: 240–243.

Chapter 67

Microbial Contamination in Drinking Water: Cause, Detection and Remedy

M.K. Bhutra and Ambica Soni*

Department of Chemistry, J.N.V. University, Jodhpur (Rajasthan)

ABSTRACT

Most microorganisms in water derive from air, soil, living and decaying plants or animals, and fecal excrements of humen and other warm blooded animals. Many of these organisms come from unknown sources and have no sanitary significance because they are widely distributed in the natural environment and have no suspected association with human or animal wastes. This discussion will consider only those disease producing bacteria, which have been associated with recognized outbreaks in India and abroad. This chapter also discusses merits and demerits of existing microbiological analytical techniques and highlights importance of rapid analytical techniques in effective implementation of microbiological quality assurance. Some remedial measures for control of bacterial population in water have also been discussed.

Keywords: *Microorganisms, Coliforms, Quality assurance, Waterborne disease outbreak.*

Introduction

Water is in fact a vehicle for the transfer of a wide range of diseases of microbial origin. The widespread prevalence of waterborne diseases continues to pose a very serious public health issue. Microorganisms get into water from air, soil, sewage, organic wastes, dead plants and animals etc. The disease producers most frequently found are:

----

* *Corresponding Author.*

1. *Salmonella*, which causes typhoid fever and other gastrointestinal disorders.
2. *Shigella*, which are causative agents for diarrhoeal diseases.
3. *Entamoeba histolytica*, which produces dysentery, and
4. *Enteric viruses*, like infectious hepatitis, polioviruses, coxsackie, ECHO and adenoviruses which are associated with enteric and other complex diseases.

Waterborne Disease Outbreaks

Frequent outbreaks of waterborne diseases are being reported from different parts of the world due to absence of proper sanitation and rapid monitoring facilities. Some of them, which drew the attention of health workers, are cited below.

The outbreak of cholera, attributed to the Broad street pump in London in 1854 and was probably the first time that water was implicated by epidemiological methods, in transmitting a specific disease. Cholera, an illness caused by ingestion of the bacterium *Vibrio cholerae*, is characterized by intense diarrhoea, which results rapidly to massive fluid depletion and death in a very large percentage of untreated patients. Thousands of Indians continue to die each year even at present due to this disease. The epidemic of typhoid fever that occurs in Lausanne, Switzerland in 1872 was the first case to attract general attention. Since then more than 275 outbreaks have been observed throughout the world. In America the major well-documented epidemic of typhoid was attributed contaminated municipal water supply of Keene in December 1959. The Riverside outbreak of salmonellosis in May 1965 was attributed to the community water supply. The basis of this conclusion was the epidemiological investigation that eliminated all other likely possibilities except water. An outbreak of bacillary dysentery occurred in Newton, Kan in 1942, when sewage entered the water distribution system through frost-proof hydrants and water closet valves. An outbreak of water borne gastroenteritis also occurred in California, USA in August 1965. The city obtained its water from deep wells. There was considerable evidence that the outbreak was a result of sewage contamination of the city's water supply through one of the wells.

Apparently the transmission of amoebiasis has been traced directly to drinking water on only few occasions. In the first epidemic in 1933 and 1934 at two Chicago hotels three sources of direct and fecal contamination of drinking water were found, *viz.*, back siphonage form toilets, draining from a defective sewer over an open water cooler and direct cross connections between sever lines and water supply. Another outbreak occurred under similar circumstances in the Mantetsu apartment building in Tokyo. All these epidemics followed fecal contamination of drinking water in the distribution systems.

Infectious hepatitis is currently the only viral disease for which a waterborne route of infection is generally accepted, although person-to-person contact is considered the most frequent type of transmission. In 1971 Indo-Pak war more casualties at western border of Rajasthan (India) were due to spread of this infectious, disease for which root cause was considered contaminated water.

The status of water supply in India, till today, is far from satisfactory. Frequent outbreaks of waterborne diseases are being reported from different parts of country in absence of proper sanitation and rapid monitoring facilities. It is reported by WHO that about 15 million children are said to die every year in the world because of waterborne disease and 50 per cent of world hospital beds in underdeveloped and developing countries are filled with people suffering from waterborne disease.

Some Facts about Water-Related Diseases

1. Each year more than ten million people die from water-related disease.
2. The leading cause of child death in the world is diarrhoea.
3. UNICEF estimates that in 1993 alone, 3.8 million children under the age of five died from diarrhoea resulting from ingesting waterborne pathogens.
4. Of the 37 major diseases in developing countries, 21 are water and sanitation related.
5. The World Health Organization (WHO) estimates that 80 per cent of all sickness in the world is attributed to unsafe water and sanitation.
6. WHO also estimates that each year, throughout the world, children under five suffer 1.5 billion episodes of diarrhoea, approximately 4 million of which are fatal.
7. If no action is taken to address unmet basic human need for water, as many as 135 million people will die from water related diseases by 2020.
8. Even if the explicit Millennium Goals announced by the United Nations in 2000 are achieved, between 34 and 76 million people will perish from water-related diseases by 2020.
9. No single type of intervention has greater overall impact upon the national development and public health than does the provision of safe drinking water and the proper disposal of human excreta.
10. Human health improvements are influenced not only by the use of clean water, but also by personal hygiene habits and the use of sanitation facilities.
11. Water, sanitation and hygiene interventions have been shown to reduce sickness from diarrhoea on average between a quarter and a third.
12. Six to nine million people are estimated to be blind from trachoma. The population at risk for this disease is 500 million. This disease can be reduced 25 per cent by provision of adequate quantities of water.
13. In reducing the toll of sickness and death from diarrhoea, the supply of adequate quantities of water is usually more important than improving its quality. This is because the organisms that cause diarrhoea can be spread through many routes besides drinking water, and increased quantities of water can improve household and personal hygiene to prevent these.
14. About one-third of the population of the developing world are infected with intestinal worms that can be controlled through better water, hygiene, and sanitation. These parasites can lead to malnutrition, anaemia, and retarded growth.
15. 200 million people in the world are infected with schistosomiasis, of which 20 million suffer severe consequences. This disease can be reduced 77 per cent from well-designed water and sanitation interventions.
16. At any given time, half the people in developing countries are suffering from water-related diseases.
17. More people die each year from unsafe water from all forms of violence, including war.

Microbiological Analytical Techniques

The objective of the bacteriological examination of water is usually to estimate the hazards due to fecal pollution, and therefore, the probability of disease producing organisms being present. Because tests for the detection of pathogens are expensive and time consuming, so normally occurring bacteria in the intestines of warm blooded animals have been used as indicators of fecal pollution; and, indirectly of disease producing potential. The coliform group and fecal coliforms have been used at various times-in making these estimations.

Coliforms

The coliform group (total coliforms) as described in USPHS Drinking Water Standards includes organisms that differ in biochemical and serological characteristics and in natural sources and habitats. This group of bacteria is used world wide as an indicator of the microbiological quality of the water (WHO, 1993). The coliforms include all aerobic and facultative anaerobic, gram negative, non-spore forming, rod shaped bacilli, which ferment lactose within 48 hours at 35±2°C and which are native to the intestinal tract of mammals. They are members of the family Enterobacteriaceae and like the other members can be cultured from mammalian feces. The group mainly consists of several genera of bacteria belonging to this family, mostly *Escherichia, Klebsiella, Enterobacter, Citrobacter*, and *Serratia*. Among these *Escherichia* (*E. coli*) is of fecal origin while others are frequently found on vegetation. The detection of coliforms in tap water is supposed to indicate to utilities and health authorities that a breach has occurred in treatment or in the distribution system, allowing microbial (fecal) contamination of water to occur and thereby rendering it inappropriate for drinking.

Fecal Coliforms (Thermo Tolerant Coliforms)

Studies reveal that fecal coliform analysis is a more definitive test for recent fecal pollution than the total coliforms. The word fecal coliform primarily defines *E. coli* and occasionally *Klebssiella*. It is relatively harmless bacterial specie and is almost always present in water that contains enteric pathogens. Thus, because they are relatively easy to isolate and because they normally survive longer than the disease producing organisms, fecal coliforms are a useful indicator of possible presence of enteric pathogens and viruses. In most cases water that is free from fecal coliforms is considered free of disease producing bacteria. So nowadays estimation of fecal coliforms is preferred over total coliforms.

Fecal coliforms are distinguished from total coliforms by having the capability of fermenting lactose even at elevated temperature like 44.5±0.2°C. At this temperature due to heat shock the non-fecal coliforms do not grow. In sewage, the fecal coliforms may constitute 30–40 per cent of total coliforms.

Bacteriological Standards

For Water Entering the Distribution System in Piped Supply after Chlorination or Disinfection

Fecal coliform–zero per 100 ml of the sample.

Total coliforms–zero per 100 ml of the sample.

A sample of water that does not conform to this standard, calls for immediate investigation into the efficiency of the purification process and the method of sampling.

For Untreated Water Maximum Allowable Limit is:

Fecal coliforms–zero per 100 ml of the sample.

Total coliforms–8–10 per 100 ml of the sample.

If the number exceeds, the supply should be disinfected.

Conventional Method for Detection of Fecal Coliforms

Multiple Tube Fermentation Technique (MTFT Technique)

It is a 3-stage techniques, which can be used for estimation of total coliforms as well as fecal coliforms. In this procedure 3 sets of tubes (each set comprising of 5 test tubes) were taken. Water (under test) in 10 ml, 1 ml and 0.1 ml amount is inoculated into lactose broth tubes. The tubes are incubated at 37°C for maximum 48 hours and coliform organisms may be identified by their production of gas from the lactose broth. By referring to a MPN table (Table 67.1) a statistical range of the number of coliforms may be determined by observing how many broth tubes showed gas formation and turbidity.

Table 67.1: MPN Index and 95 per cent Confidence Limit for Various Combinations of Positive Results when 5 Tubes are Used per Dilution (1.0 ml, 1.0 ml and 0.1 ml)

Combination of Positives	*MPN Index/ 100 ml*	*95% Confidence Limits*		*Combination of Positives*	*MPN Index/ 100 ml*	*95% Confidence Limits*	
		Lower	*Upper*			*Lower*	*Upper*
0–0–0	< 2	–	–	4–2–0	22	9.0	56
0–0–1	2	1.0	10	4–2–1	26	12	65
0–1–1	2	1.0	10	4–3–0	27	12	67
0–2–0	4	1.0	13	4–3–1	33	15	77
				4–4–0	34	16	80
1–0–0	2	1.0	11	5–0–0	23	9.0	86
1–0–1	4	1.0	16	5–0–1	30	1.0	110
1–1–0	4	1.0	15	5–0–2	40	20	140
1–1–1	6	2.0	18	5–1–0	30	10	120
1–2–0	6	2.0	18	5–1–1	50	20	150
				5–1–2	60	30	180
2–0–0	4	1.0	17	5–2–0	50	20	170
2–0–1	7	2.0	20	5–2–1	70	30	210
2–1–0	7	2.0	21	5–2–2	90	40	250
2–1–1	9	3.0	24	5–3–0	80	30	250
2–2–0	9	3.0	25	5–3–1	110	40	300
2–3–0	12	5.0	29	5–3–2	140	60	360
3–0–0	8	3.0	24	5–3–3	170	80	410
3–0–1	11	4.0	29	5–4–0	130	50	390
3–1–0	11	4.0	29	5–4–1	170	70	480
3–1–1	14	6.0	35	5–4–2	220	100	580
3–2–0	14	6.0	35	5–4–3	280	120	690
3–2–1	17	7.0	40	5–4–4	350	160	820
4–0–0	13	5.0	38	5–5–0	240	1.00	940
4–0–1	17	7.0	45	5–5–1	300	100	1300
4–1–0	17	7.0	46	5–5–2,	500	200	2000
4–1–1	21	9.0	55	5–5–3	900	300	2900
4–1–2	26	12	63	5–5–4	1600	600	5300
				5–5–5	1600		

This technique is applicable to the analysis of even saline of brackish water but method utilizes five days till completion test, so there is a risk that the water will be used by many consumers before analysis is completed. Moreover different culture media are required during different stages, which are of perishable nature.

Results of the examination are reported in terms of Most Probable Number (MPN) and for unusual combinations (not cited in the table) following formula, which is suggested by Thomas, can be used:

$$\text{MPN}/100\text{ml} = \frac{\text{No. of Positive Tubes} \times 100}{\sqrt{\text{ml of Sample in negative tubes} \times \text{ml sample in all tubes}}}$$

Membrane Filter Technique (MF Technique)

The MF technique is highly reproducible and yield numerical results more rapidly than the MTFT technique. The method employs a filtration unit. About 100 ml water sample is filtered through a membrane filter and the filter pad is then transferred to a plate of bacteriological medium. Bacteria trapped in the filter will form colonies, which may be counted. For total coliform and fecal coliform different culture media are used. Total number of bacteria can be calculated by following statistical formula:

$$\text{MPN}/100\text{ml} = \frac{\text{Colonies counted} \times 100}{\text{ml of sample filtered}}$$

The test may be carried out in field also. But the test is quite expensive because both filtration unit as well as media are imported and the media used are unstable too.

Standard Plate Count (SPC)

Water sample is diluted in sterile buffer. Measured amounts are pipetted into petri dishes. Agar medium is added and plates incubated. Colony counts are made and multiplied by the reciprocal dilution factor to have total bacteria per ml water. This method is similar to that used for milk. This method requires skilled operators and the media used are unstable too.

Need for Rapid Technique

Enumeration and isolation of organisms using conventional method involves steps like enrichment, isolation on selective agar medium, biotyping and serotyping. Although these techniques are serving well over the years but there are some inherent limitations with them. One of the major limitation is the long time period for the completion of test and another is that media used for growth of bacteria are of perishable nature. So there is a strong need to develop methods for rapid detection of indicator organisms as well as pathogens (using chemically defined culture media which are generally non-perishable and cheap) in order to assess quality of water without compromising on the sensitivity and accuracy of test methods.

Treatment for Removal of Organisms

Water and sewage harbor a variety of organisms responsible for diseases in man. It is thus, necessary to remove or rather interrupt the disease cycle in nature through treatment of water and sewage. Water purification involves 3 basic steps.

Sedimentation

The objective is to remove bulky objects such as leaves, sand and, gravel particles and the materials that have run off form soil. It is done in large reservoirs or in restricted area of settling tank. Aluminium Sulphate (alums) or iron sulphate may be added to hasten sedimentation process. After mixing, chemicals form jelly like masses of coagulated materials called flocs. They fall through the water and cling to organic particles and microorganisms, dragging a major portion to the bottom sediment in a process called flocculation.

Filtration

The process removes the remaining microorganisms from the water. Although different types of filter materials are available, but mostly a layer of sand and gravel is utilized to trap the microorganisms.

Disinfection

It is the last step to ensure 100 per cent killing of microorganisms. Mostly bleaching powders are used. Studies reveal that enough chlorine is added which should, not only satisfy chlorine demand of water but should leave some residual free chlorine to ensure complete disinfection.

Now-a-days chloramination, ozonation, UV irradiation and use of hydrogen peroxide is preferred over chlorination, because their bactericidal power is high and long lasting. The efficacy of silver as a bactericidal against *E. coli* has also been established.

Conclusion

Waterborne diseases are major health problem in many countries. Fecal contamination of drinking water is one of the main cause of these disease outbreaks as many pathogens are carried through human feces. At present, the quality of drinking water is ensured by testing the water sample for total coliforms and thermo tolerant fecal coliforms or *E. coli*. However when applied to remote areas the standard methods have various limitations. The distance to laboratory facilities and the high cost of analysis restrict water quality assessment in remote areas and in developing countries. Under such circumstances as on-site bacteriological test is considered to be highly essential for routine water quality testing.

The combination of a more effective rapid detection method and an analytical tool integrating a variety of parameters will greatly help water utilities in identifying the causes of water quality problems and consequently, allow them to apply corrective measures to eliminate coliform occurrence from their distribution system.

References

Ahmed, F. and M.E. Sikder., 1997. Performance of pond sand filter in saline problem areas. *J. IPHE*, p. 10–16.

Anand, S.K., 2002. Role of rapid analytical techniques in effective implementation of microbiological quality assurance. *Dairy Microbiology*, p. 10–15.

Battling Waterborne ills in a Sea of 950 Million, February 17, 1997. *The Washington Post*.

Brion, G.M. *et al.*, 2000. New approach to use of total coliform test for watershed management. *Water Sci. & Technol.*, 42: 1–65.

Chalmers, R.M. *et al.*, 2000. Waterborne *Escherichia coli* 0157. *Jour. Applied Microbiol.*, 88: 124S.

Chan, M.S., 1997. The global burden of intestinal nematode infections: Fifty years on. *Parasitology Today*, 13(11): 438–443.

Chatterjee, D., 1986. Some rapid but sensitive methods for routine determination of drinking water quality. *J. IPHE*, 1: 29–35.

Craun, G.F. and Calderon, R.L., 1999. *Waterborne Disease Outbreaks: Their Causes, Problems and Challenges to the Treatment Barriers*, 15th edn. Waterborne Pathogens, AWWA Manual M48, AWW A, Denver.

Craun, G.F. and Calderon, R.L., 2001. Waterborne disease outbreaks caused by distribution system deficiencies. *J. AWWA*, 93: 9–64.

Craun, G.F. *et al.*, 1997. Coliform bacteria and waterborne disease outbreaks. *J. AWWA*, 89: 3–96.

Easterbrook, G. September 11, 1994. Forget PCBs. Radon. Alar. The World's Greatest Environmental Dangers Are Dung Smoke and Dirty Water, *The New York Times Magazine.*

Edberg, S.C. *et al.*, 2000. *Escherichia coli*: The best biological drinking water indicator for public health protection. *J. Applied Microbiol.*, 88: 106S.

Esrey, S.A., J.B. Potash, Roberts, L. and Shiff. C., 1991. Effects of improved water supply and sanitation on ascariasis, diarrhoea, dracunculiasis, hookworm infection, schistosomiasis, and trachoma, WHO, 69(5): 609–621.

Furtado, C. *et al.*, 1998. Outbreaks of waterborne diseases in England 1992–95. *Epidemiol. Infect.*, 121: 1–109.

Gale, P., 1996. Coliforms in the drinking water supply: What information do the 0/100 ml Samples Provide? *J. Water SRT-Aqua.*, 45: 4–155.

Gatel, D. *et al.*, 1995. Forecasting the Number of Coliform–positive Sample. Proc., AWW A WQTC, New Orleans.

Geldreich, E.E., 1996, Pathogenic agents in freshwater resources. *Hydrological Processes*, 10: 2–315.

Gleick, Peter H., August 15,2002. "Dirty Water: Estimated Deaths from Water-Related Diseases 2000–2020".

Greenberg, A.E. and Ongerth, H.J., 1966. Salmonellosis in riverside California. *J. AWWA*, 58: 145.

Healy, W.A and Grossman, R.P., 1961. Waterborne typhoid fever epidemic at Keene, New Hampshire. *J. NEWWA*, 75: 38.

Henry, F. *et al.*, 1990. Environmental sanitation, food and water contamination and diarrhoea in rural Bangladesh. *Epidemiol. Infect.*, 104: 253–259.

Kinnman, C.H. and Beelman, F.C., 1944. An epidemic of 3000 cases of bacillary dysentery involving a war Industry and members of armed forces. *Am. J. Public Health*, 34: 948.

Lenore, S.C., Arnold, E.D. and Andrew, D.E., 1998. *Standard Methods for Examination of Water and Waste Water*, 20th edition.

Michael, A.B., Lance Adling and Michael, P.L., 2003. The efficacy of silver as a bactericidal agent: Advantages, limitations and considerations for future use. *Aqua*, 522: 6.

Momba, M.N.B., Cloeta, T.E. and Kaflf, R., 2000. Influence of disinfection processes on the microbial quality of potable ground water in laboratory-scale system model. *Aqua*, 49: 1.

Mosley, J.W., 1959. Waterborne infectious hepatitis. *New Eng. J. Med.*, 261: 703.

Parry-Jones, S. and P. Kolsky, WELL Technical Brief. Some global statistics for water and sanitation related diseases.

Power and Daginwala., 1999. *General Microbiology*, Vol. 2. Himalaya Publishing House.

Rodier, J., 1975. *Analysis of Water*. John Wiley and Sons, New York.

Rodriquez, M.J., Milot, J. and Serodes, J.B., 2003. Predicting trihalomethane formation in chlorinated waters using multivariate regression and neutral networks. *Aqua*, 52: 3.

Ross, A.I. and Gillespie, E.H., 1952. An outbreak of waterborne gastroenteritis and some dysentery. *Gr. Brit. Ministry of Health, Bull.*, 11: 36.

Sharma, P.D., 1994. *Microbiology*, 1st Ed. Vivek Rastogi and Company Educational Publishers, Meerut.

Water and Sanitation for Health project, sponsored by the US Agency for International Development., 1993. Lessons Learned in Water, Sanitation and Health.

World Health Organization and Harvard University., 1996. *Global Burden of Disease*.

World Health Organization Fact Sheet, November 1996. Water and Sanitation, Water-related diseases include not only those contracted by ingesting unsafe water but also those that result from improper sanitation.

World Health Organization., 1992. Our Planet. Our Health: Report of the WHO Commission on Health and Environment.

Index

www.ingramcontent.com/pod-product-compliance
Ingram Content Group UK Ltd.
Pitfield, Milton Keynes, MK11 3LW, UK
UKHW051504310726
14059UKWH00009BB/107

9 789351 240822